In addition to making the illustrations more colorful and visually pleasing, the use of four-color artwork makes the illustrations easier to understand. For example, forces and moments can be readily distinguished from all other information on the illustration because they are always designated by a big red arrow. Similarly, position vectors are always designated by a medium blue arrow; velocity and acceleration vectors, by a standard green arrow; and unit vectors by a small black arrow. Coordinate axes and dimension lines (linear and angular) are drawn with thin black lines. This still leaves many colors for depicting the various parts of objects.

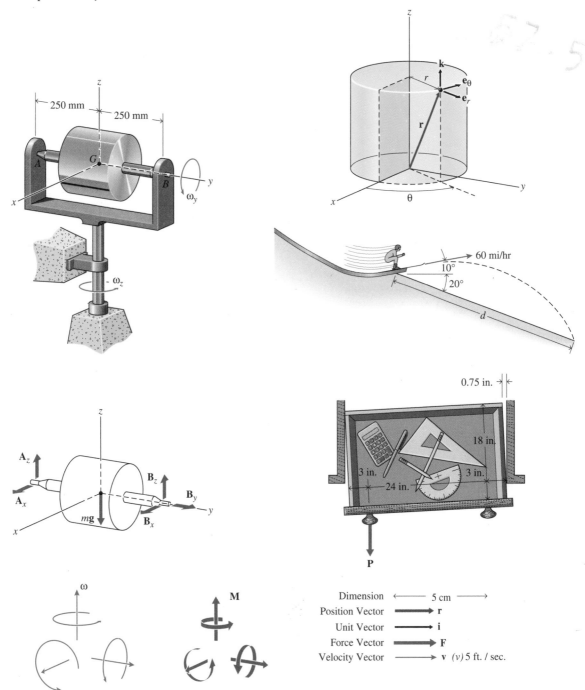

ENGINEERING MECHANICS
STATICS

ENGINEERING MECHANICS
STATICS

WILLIAM F. RILEY
Professor Emeritus
Iowa State University

LEROY D. STURGES
Iowa State University

JOHN WILEY & SONS, INC.
New York · Chichester · Brisbane · Toronto · Singapore

COVER: Designed by Laura Ierardi
Photograph by Alan Weitz

ACQUISITIONS EDITOR Charity Robey
DEVELOPMENTAL EDITOR Christine Peckaitis
MARKETING MANAGER Debra Riegert
PRODUCTION SUPERVISOR Charlotte Hyland
DESIGN SUPERVISOR Ann Marie Renzi
MANUFACTURING MANAGER Andrea Price
COPY EDITING SUPERVISOR Marjorie Shustak
PHOTO RESEARCHER Hilary Newman
ILLUSTRATION COORDINATOR Sigmund Malinowski
ILLUSTRATION DEVELOPMENT Boris Starosta
ELECTRONIC ILLUSTRATIONS Precision Graphics

This book was set in Palatino by York Graphic Services and printed and bound by
Von Hoffmann Press. The cover was printed by Phoenix Color Corp.

Library of Congress Cataloging in Publication Data:
Riley, William F. (William Franklin), 1925–
    Engineering mechanics : statics / William F. Riley, Leroy D. Sturges.
      p. cm.
    Includes index.
      ISBN 0-471-51241-9 (alk. paper)
      1. Statics. I. Sturges, Leroy D. II. Title.
    TA351.R55 1993
      620.1'03—dc20   92-30352
                                                                                  CIP

Printed in the United States of America

10 9 8 7 6 5 4 3 2 1

PREFACE

Our purpose in writing this statics book, together with the companion dynamics book, was to present a fresh look at the subject and to provide a more logical order of presentation of the subject material. We believe our order of presentation will give students a greater understanding of the material and will better prepare students for future courses and later professional life.

INTRODUCTION

This text has been designed for use in undergraduate engineering programs. Students are given a clear, practical, comprehensible, and thorough coverage of the theory normally presented in introductory mechanics courses. Application of the principles of statics to the solution of practical engineering problems is demonstrated. This text can also be used as a reference book by practicing aerospace, automotive, civil, mechanical, mining, and petroleum engineers.

Extensive use is made in this text of prerequisite course materials in mathematics and physics. Students entering a statics course that uses this book should have a working knowledge of algebra, geometry, and trigonometry, and should have taken an introductory course in calculus and vector algebra.

Vector methods do not always simplify solutions of two-dimensional problems in statics; therefore, they are used only in instances where they provide an efficient solution to a problem. For three-dimensional problems, however, vector algebra provides a systematic procedure that often eliminates errors that might occur with a less systematic approach. Students are encouraged to develop the ability to select the mathematical tools most appropriate for the particular problem that they are attempting to solve.

ORGANIZATION

This volume on statics is divided into 11 chapters. The first six chapters are used to develop fundamental concepts and the principle of equilibrium. The principle of equilibrium is then applied to a wide variety of problems in Chapters 7, 8, and 9. Second moments of area and moments of inertia are developed in Chapter 10. The method of virtual work and the principle of potential energy are developed in Chapter 11. Since the book is divided into an extensive number of subdivisions, the material can be presented in a different order, at the discretion of the instructor, with little or no loss in continuity.

An introduction to mechanics and a discussion of units is presented in Chapter 1. Included is a discussion of computational accuracy and the significance of results.

Concurrent force systems are introduced in Chapter 2. While the forces may be expressed in terms of vectors, resultants are normally calculated in terms of components of forces. Vector dot (scalar) products are introduced as a means of determining rectangular components of a force. Chapter 2 also shows that a coordinate system is not an intrinsic part of the problem; it is an aid, used by the problem solver, to facilitate solution of the problem. Forces and resultants, together with free-body diagrams, are then used in Chapter 3 to solve problems involving equilibrium of particles.

The concepts of moment of a force about a point and moment of a force about a line are introduced in Chapter 4. Vector cross products and triple scalar products are introduced as means of determining moments about a point and moments about a line, respectively. Chapter 4 also contains a discussion of equivalent force systems that focuses on properties common to all force systems without emphasizing the numerous special cases.

Chapter 5 contains a general discussion of distributed forces and their resultants together with the related topics of centroids and center of mass. The discussion of distributed forces follows naturally from the discussion of equivalent force systems. Introduction of the discussion of distributed forces at this location is also desirable since it allows use of distributed loads in the equilibrium problems in the chapters that follow.

Rigid-body equilibrium and a further development of free-body diagrams is presented in Chapter 6. Statically indeterminate reactions and partial constraints are also discussed in this chapter. In Chapter 7, the principle of equilibrium is applied to problems involving internal joint forces in pin-connected structures. Specific applications considered are trusses, frames, and simple machines. Internal force distributions in bars, shafts, beams, and flexible cables are discussed in Chapter 8. The discussion includes axial force and torque diagrams as well as shear force and bending moment diagrams.

Frictional forces and their effects are introduced in Chapter 9. The discussions include sliding friction, belt friction, rolling resistance, and friction in journal and thrust bearings.

Second moments of area and mass moments of inertia are discussed in Chapter 10. Although this material is closely related to the material on centroids discussed in Chapter 5, it is not used further in statics. It is included for those who wish to cover this material in a statics course for later use in Dynamics and Mechanics of Materials.

Finally, the method of virtual work and the principle of potential energy are developed and applied to the solution of equilibrium problems in Chapter 11.

FEATURES

Engineering Emphasis

Throughout this book, strong emphasis has been placed on the engineering significance of the subject area in addition to the mathematical methods of analysis. Many illustrative example problems have been integrated into the main body of the text at points where the presentation of a method can be best reinforced by the immediate illustration of the method. Students are usually more enthusiastic about a subject if they can see and appreciate its value as they proceed into the subject.

We believe that students can progress in a mechanics course only by understanding the physical and mathematical principles jointly, not by mere memorization of formulas and substitution of data to obtain answers to simple problems. Furthermore, we think that it is better to teach a few fundamental principles for solving problems than to teach a large number of special cases and trick procedures. Therefore the text aims to develop in the student the ability to analyze a given problem in a simple and logical manner and to apply a few fundamental, well-understood principles to its solution.

A conscientious effort has been made to present the material in a simple and direct manner, with the student's point of view constantly in mind.

Free-body Diagrams

Most engineers consider the free-body diagram to be the single most important tool for the solution of mechanics problems. Mastering the concept of the free-body diagram is fundamental to success in this course. Students frequently have difficulty with the concept, and coverage in this book has been carefully designed to ensure student understanding. A step-by-step procedure walks the student through the process of developing a complete and correct free-body diagram. Whenever an equation of equilibrium is written, we recommend that it be accompanied by a complete, proper free-body diagram.

Problem-solving Procedures

Success in engineering mechanics courses depends, to a surprisingly large degree, on a well-disciplined method of problem solving and on the solution of a large number of problems. The student is urged to develop the ability to reduce problems to a series of simpler component problems that can be easily analyzed and combined to give the solution of the initial problem. Along with an effective methodology for problem decomposition and solution, the ability to present results in a clear, logical, and neat manner is emphasized throughout the text. A first course in mechanics is an excellent place to begin development of this disciplined approach that is so necessary in most engineering work.

Worked-out Examples

Worked-out example problems are invaluable to students. Example problems were carefully chosen to illustrate the concepts being discussed. When a concept is presented in this book, a worked-out example problem follows to illustrate the concept. We have included approximately 150 worked-out examples in this book.

Homework Problems

This book contains a large selection of problems that illustrate the wide application of the principles of statics to the various fields of engineering. The problems in each set represent a considerable range of difficulty. We believe that a student gains mastery of a subject through application of basic theory to the solution of problems that appear somewhat difficult. Mastery, in general, is not achieved by solving a large number of simple but similar problems. The problems in this text require an understanding of the principles of statics without demanding excessive time for computational work.

Significant Figures

Results should always be reported as accurately as possible. However, results should not be reported to 10 significant figures merely because the calculator displays that many digits. One of the tasks in all engineering work is to determine the accuracy of the given data and the expected accuracy of the final answer. Results should reflect the accuracy of the given data.

In a textbook, however, it is not possible for students to examine or question the accuracy of the given data. It is also impractical for the authors to place error bounds on every number. An accuracy greater than 0.2 percent is seldom possible in engineering work, since physical data is seldom known with any greater degree of accuracy. A practical rule for "rounding off" numbers, that provides approximately this degree of accuracy, is to retain four significant figures for numbers beginning with the figure 1 and three significant figures for numbers beginning with any figure from 2 through 9. In this book, all given data, regardless of the number of figures shown, are assumed to be sufficiently accurate to permit application of this practical rule. Therefore, answers are given to three significant figures, unless the number lies between 1 and 2 or any decimal multiple thereof, in which case four significant figures are reported.

Computer Problems

Many students come to school with computers as well as programmable calculators. In recognition of this fact, we include problems at the ends of most chapters that can be best solved using these tools. These problems are more than just an exercise in crunching numbers; each has been chosen to illustrate how the solution to the problem depends on some specific parameter of the problem. Computer problems appear at the end of most chapters, and are marked with a C before the problem number.

Review Problems

A set of review problems is provided at the end of each chapter. These problems are designed to test students on all the concepts covered in the chapter. Since the problems are not directly associated with any particular section, they often integrate topics covered in the chapter and thus can deal with more realistic applications than can a problem designed to illustrate a single concept.

SI vs. US Units

Most large engineering companies deal in an international marketplace. In addition, the use of the International System of Units (SI) is gaining acceptance in the United States. As a result, most engineers must be proficient in both the SI system and the U.S. Customary System (USCS) of units. In response to this need, both U.S. Customary units and SI units are used in approximately equal proportions in the text for both illustrative examples and homework problems. As an aid to the instructor in problem selection, all odd-numbered problems are given in USCS units and even-numbered problems in SI units.

Chapter Summaries

As an aid to students we have written a summary that appears at the end of each chapter. These sections provide a synopsis of the major concepts that are explained in the chapter and can be used by students as a review or study aid.

Answers Provided

Answers to about half of the problems are included in the back of the book. We believe that the first assignment on a given topic should include some problems for which the answers are given. Since the simpler problems are usually reserved for this first assignment, answers are provided for the first few problems of each article and thereafter are given for approximately half of the remaining problems. The problems whose answers are provided are indicated by an asterisk after the problem number.

DESIGN

Use of Color

One of the first things you'll notice when you open this book is that we have used a variety of colors. We believe that color will help students learn mechanics more effectively for two reasons: First, today's visually oriented students are more motivated by texts that depict the real world more accurately. Second, the careful color coding makes it easier for students to understand the figures and text.

Following are samples of figures found in the book. As you can see, force and moment vectors are depicted as red arrows; velocity and acceleration vectors are depicted as green arrows. Position vectors appear in blue; unit vectors in bold black; and dimensions as a thin black line. This pedagogical use of color is standard throughout this book and its companion dynamics book.

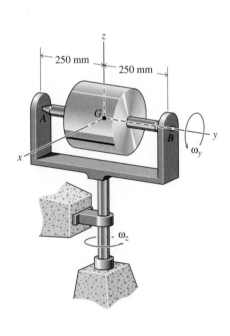

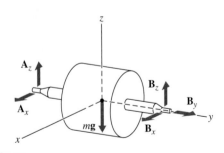

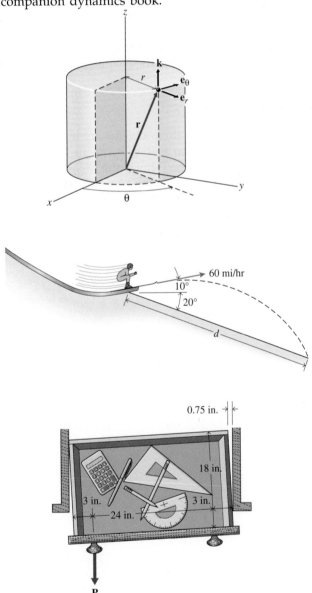

We have also used color to help students identify the most important study elements. For instance, example problems are always outlined in red and important equations appear in a green box.

Illustrations

One of the most difficult things for students to do is to visualize engineering problems. Over the years, students have struggled with the lack of realism in mechanics books. We think that mechanics illustrations should be as colorful and three-dimensional as life is. To hold students' attention, we developed the text illustrations with this point in mind.

We started with a basic sketch. Then a specialist in technical illustration added detail. Then the art studio created the figures using *Adobe Illustrator*©. All of these steps enabled us to provide you with the most realistic and accurate illustrations on the market.

Accuracy

After many years of teaching, we appreciate the importance of an accurate text. We have made an extraordinary effort to provide you with an error-free book. Every problem in the text has been worked out at least twice independently; many of the problems have been worked out a third time independently.

Development Process

This book is the most extensively developed text ever published for the engineering market. The development process involved several steps.

1. **Market Research**   A Wiley marketing specialist team of six senior sales representatives was formed to gather information to help focus and develop the text. An extensive market research survey was also sent to over 3,000 professors teaching Statics and Dynamics to home in on key market issues. Two focus groups consisting of professors teaching Statics and Dynamics were conducted to gain a clearer understanding of classroom needs as the texts took shape.
2. **Reviews**   Professors from the United States and Canada carefully reviewed each draft of this manuscript. Their suggestions were carefully considered and incorporated whenever possible. Six additional reviewers were commissioned to evaluate one of the key components of the text—the problem sets.
3. **Manuscript and Illustration Development**   A developmental editor worked with the authors to hone both the manuscript and the art sketches to their highest potential. A special art developer worked with the authors and the art studio to enhance the illustrations.

TECHNICAL PACKAGE FOR THE INSTRUCTOR

Solution Manual

After years of teaching, we realize the importance of an accurate solution manual that matches the quality of the text. For that reason, we have prepared the manual ourselves. The manual includes a complete solution for every problem in the book, and especially challenging problems are marked with an asterisk. Each solution appears with the original problem statement and, where appropriate, the problem figure. We do this for the convenience of the instructor, who no longer will have to refer to both book and solution manual in preparing for class. The manual also contains transparency masters for use in preparing overhead transparencies.

FOR THE STUDENT

Software

Our reviewers told us that they are generally dissatisfied with publisher-provided software. They also told us that students need software that is easy to use, provides reinforcement of basic concepts, and is highly interactive. With this in mind, we have worked with Intellipro, an engineering software developer, to produce a package that satisfies all these demands. The software consists of 30 problems, 10 from *Statics* and 20 from *Dynamics*. The software reinforces the importance of free-body diagrams by giving students practice in drawing them. The dynamics problems are animated to aid student visualization.

Study Guide

Mechanics can be a tough course, and sometimes students need extra help. Our study guide is written as a tool for developing student understanding and problem-solving skills. This study guide provides reinforcement of the major concepts in the text.

ACKNOWLEDGMENTS

Many people participated directly and indirectly in the preparation of this book. In particular we wish to thank Rebecca Sidler for her careful review of the manuscript and for solving many problems in the two books. In addition to the authors, many present and former colleagues and students contributed ideas concerning methods of presentation, example problems, and homework problems. Final judgments concerning organization of material and emphasis of topics, however, were made by the authors. We will be pleased to receive comments from readers and will attempt to acknowledge all such communications.

We'd like to thank the following people for their suggestions and encouragement throughout the reviewing process.

H. J. Sneck	Rensselaer Polytechnic Institute
Thomas Lardiner	University of Massachusetts
K. L. DeVries	University of Utah
John Easley	University of Kansas
Brian Harper	Ohio State University
Kenneth Oster	University of Missouri–Rolla
D. W. Yannitell	Louisiana State University

James Andrews	University of Iowa
D. A. DaDeppo	University of Arizona
Ed Hornsey	University of Missouri–Rolla
William Bingham	North Carolina State University
Robert Rankin	Arizona State University
David Taggart	University of Rhode Island
Allan Malvick	University of Arizona
Gaby Neunzert	Colorado School of Mines
Tim Hogue	Oklahoma State University
Bill Farrow	Marquette University
Matthew Ciesla	New Jersey Institute of Technology
William Lee	US Naval Academy
J. K. Al-Abdulla	University of Wisconsin
Erik G. Thompson	Colorado State University
Dr. Kumar	University of Pennsylvania
William Walston	University of Maryland
John Dunn	Northeastern University
Ron Anderson	Queen's University (Canada)
Duane Storti	University of Washington
Jerry Fine	Rose-Hulman Institute of Technology
Ravinder Chona	Texas A & M
Bahram Ravani	University of California–Davis
Paul C. Chan	New Jersey Institute of Technology
Wally Venable	West Virginia University
Eugene B. Loverich	North Arizona University
Kurt Keydel	Montgomery College
Francis Thomas	University of Kansas
Colonel Tezak	U.S. Military Academy

William F. Riley
Leroy D. Sturges

CONTENTS

LIST OF SYMBOLS

Unit Vectors

$\mathbf{i}, \mathbf{j}, \mathbf{k}$ Unit vectors in the x, y, z directions (rectangular coordinates)

$\mathbf{e}_n, \mathbf{e}_t$ Unit vectors in the n, t directions (normal and tangential coordinates)

$\mathbf{e}_r, \mathbf{e}_\theta$ Unit vectors in the r, θ directions (polar coordinates)

Miscellaneous Physical Constants

m	Mass of a particle or rigid body
W	Weight of a particle or rigid body
k	Spring constant
μ_s	Coefficient of static friction
μ_k	Coefficient of dynamic friction
$I_x, I_y, I_{xy}, \ldots$	Moments and products of inertia
k	Radius of gyration
G	Universal gravitational constant
M_e	Mass of the Earth
R_e	Radius of the Earth

GENERAL PRINCIPLES

The builders of ancient monuments such as Stonehenge probably understood and used most of the basic principles of statics.

1-1 INTRODUCTION TO MECHANICS

Mechanics is the branch of the physical sciences that deals with the response of bodies to the action of forces. The subject matter of this field constitutes a large part of our knowledge of the laws governing the behavior of gases and liquids as well as the laws governing the behavior of solid bodies. The laws of mechanics find application in astronomy and physics as well as in the study of the machines and structures involved in engineering practice. For convenience, the study of mechanics is divided into three parts: namely, the mechanics of rigid bodies, the mechanics of deformable bodies, and the mechanics of fluids.

A study of the mechanics of rigid bodies can be further subdivided into three main divisions: statics, kinematics, and kinetics. Statics is concerned with bodies that are acted on by balanced forces and hence are at rest or have uniform motion. Such bodies are said to be in equilibrium. Statics is an important part of the study of mechanics because it provides methods for the determination of support reactions and relationships between internal force distributions and external loads for stationary structures. Many practical engineering problems involving the loads carried by structural components can be solved using the relationships developed in statics. The relationships between internal force distributions and external loads that are developed in statics play an important role in the subsequent development of deformable body mechanics.

Kinematics is concerned with the motion of bodies without considering the manner in which the motion is produced. Kinematics is sometimes referred to as the geometry of motion. Kinematics forms an important part of the study of mechanics, not only because of its application to problems in which forces are involved, but also because of its application to problems that involve only motions of parts of a machine. For many motion problems, the principles of kinematics, alone, are sufficient for the solution of the problem. Such problems are discussed in "Kinematics of Machinery" books, where the motion of machine elements such as cam shafts, gears, connecting rods, and quick-return mechanisms are considered.

Kinetics is concerned with bodies that are acted on by unbalanced forces; hence, they have nonuniform or accelerated motions. A study of kinetics is an important part of the study of mechanics because it provides relationships between the motion of a body and the forces and moments acting on the body. Kinetic relationships may be obtained by direct application of Newton's laws of motion or by using the integrated forms of the equations of motion that result in the principles of work–energy or impulse–momentum. Frequently the term *dynamics* is used in the technical literature to denote the subdivisions of mechanics with which the idea of motion is most closely associated, namely, kinematics and kinetics.

The branch of mechanics that deals with internal force distributions and the deformations developed in actual engineering structures and machine components when they are subjected to systems of forces is known as mechanics of deformable bodies. Books covering this part of mechanics commonly have titles like "Mechanics of Materials" or "Mechanics of Deformable Bodies."

The branch of mechanics that deals with liquids and gases at rest or in motion is known as fluid mechanics. Fluids can be classified as compressible or incompressible. A fluid is said to be compressible if the density of the fluid varies with temperature and pressure. If the volume of a fluid remains constant during a change in pressure, the fluid is said to be incompressible. Liquids are considered incompressible for most engineering applications. A subdivision of fluid mechanics that deals with incompressible liquids is commonly known as hydraulics.

In this book on statics and in the companion volume on dynamics only rigid-body mechanics will be considered. The two books will provide the foundation required for follow-on courses in many fields of engineering.

1-2   HISTORICAL BACKGROUND

The portion of mechanics known as statics developed early in recorded history because many of the principles are needed in building construction. Ancient Egyptian and Assyrian monuments contain pictorial representations of many kinds of mechanical implements. The builders of the pyramids of Egypt probably understood and used such devices as the lever, the sled, and the inclined plane. An early history of mechanics was published by Dr. Ernst Mach of the University of Vienna in 1893.[1] The milestone contributions to mechanics presented in this brief review were obtained from this source.

Archytas of Tarentum (circa 400 B.C.) founded the theory of pulleys. The writings of Archimedes (287–212 B.C.) show that he understood the conditions required for equilibrium of a lever and the principle of buoyancy. Leonardo da Vinci (1452–1519) added to Archimedes' work on levers and formulated the concept of moments as they apply to equilibrium of rigid bodies. Copernicus (1473–1543) proposed that the Earth and the other planets of the solar system revolve about the sun. From the time of Ptolemy in the second century A.D., it had been assumed that the Earth was the center of the universe. Stevinus (1548–1620) first described the behavior of a body on a smooth inclined plane and employed the parallelogram law of addition for forces. Varignon (1654–1722) was the first to establish the equality between the moment of a force and the moment of its components. Both Stevinus and Galileo (1564–1642) appear to have understood the principle of virtual displacements (virtual work), but the universal applicability of the principle to all cases of equilibrium was first perceived by John Bernoulli (1667–1748), who communicated his discovery to Varignon in a letter written in 1717.

The portion of mechanics known as dynamics developed much later since velocity and acceleration determinations require accurate time measurements. Galileo experimented with blocks on inclined planes, pendulums, and falling bodies; however, he was handicapped

---

[1] Dr. Ernst Mach, "Die Mechanik in ihrer Entwickelung historisch-kritisch dargestellt," Professor an der Universitat zu Wien. Mit 257 Abbildungen. Leipzig, 1893. First translated from the German by Thomas J. McCormack in 1902. *The Science of Mechanics*, 9th ed. The Open Court Publishing Company, LaSalle, Ill., 1942.

by his inability to accurately measure the small time intervals involved in the experiments. Huygens (1629–1695) continued Galileo's work with pendulums and invented the pendulum clock. He also investigated the motion of a conical pendulum and made an accurate determination of the acceleration of gravity. Sir Isaac Newton (1642–1727) is generally credited with laying the true foundation for mechanics with his discovery of the law of universal gravitation and his statement of the laws of motion. Newton's work on particles, based on geometry, was extended to rigid-body systems by Euler (1707–1793). Euler was also the first to use the term *moment of inertia* and developed the parallel-axis theorem for moment of inertia. More recent contributions to mechanics include Max Planck's (1858–1947) formulation of quantum mechanics and Albert Einstein's (1879–1955) formulation of the theory of relativity (1905). These new theories do not repudiate Newtonian mechanics; they are simply more general. Newtonian mechanics is applicable to the prediction of the motion of bodies where the speeds are small compared to the speed of light.

1-3 FUNDAMENTAL QUANTITIES OF MECHANICS

The fundamental quantities of mechanics are space, time, mass, and force. Three of the quantities—space, time, and mass—are absolute quantities. This means that they are independent of each other and cannot be expressed in terms of the other quantities or in simpler terms. The quantity known as a force is not independent of the other three quantities but is related to the mass of the body and to the manner in which the velocity of the body varies with time. A brief description of these and other important concepts follows.

Space is the geometric region in which the physical events of interest in mechanics occur. The region extends without limit in all directions. The measure used to describe the size of a physical system is known as a length. The position of a point in space can be determined relative to some reference point by using linear and angular measurements with respect to a coordinate system whose origin is at the reference point. The basic reference system used as an aid in solving mechanics problems is one that is considered fixed in space. Measurements relative to this system are called absolute.

Time can be defined as the interval between two events. Measurements of this interval are made by making comparisons with some reproducible event such as the time required for the earth to orbit the sun or the time required for the earth to rotate on its axis. Solar time is earth rotation time measured with respect to the sun and is used for navigation on earth and for daily living purposes.

Any device that is used to indicate passage of time is referred to as a clock. Reproducible events commonly used as sensing mechanisms for clocks include the swing of a pendulum, oscillation of a spiral spring and balance wheel, and oscillation of a piezoelectric crystal. The time required for one of these devices to complete one cycle of motion is known as the period. The frequency of the motion is the number of cycles occurring in a given unit of time.

Matter is any substance that occupies space. A body is matter bounded by a closed surface. The property of a body that causes it to resist any change in motion is known as inertia. Mass is a quantitative measure of inertia. The resistance a body offers to a change in translational motion is independent of the size and shape of the body. It depends only on the mass of the body. The resistance a body offers to a change in rotational motion depends on the distribution of the mass of the body. Mass is also a factor in the gravitational attraction between two bodies.

A force can be defined as the action of one body on another body. Our concept of force comes mainly from personal experiences in which we are one of the bodies and tension or compression of our muscles results when we try to "pull" or "push" the second body. This is an example of force resulting from direct contact between bodies. A force can also be exerted between bodies that are physically separated. Gravitational forces exerted by the earth on the moon and on artificial satellites to keep them in earth orbit are examples. Since a body cannot exert a force on a second body unless the second body offers a resistance, a force never exists alone. Forces always occur in pairs, and the two forces have equal magnitude and opposite sense. Although a single force never exists, it is convenient in the study of motions of a body to think only of the actions of other bodies on the body in question without taking into account the reactions of the body in question. The external effect of a force on a body is either acceleration of the body or development of resisting forces (reactions) on the body.

A particle has mass but no size or shape. When a body (large or small) in a mechanics problem can be treated as a particle, the analysis is greatly simplified since the mass can be assumed to be concentrated at a point and the concept of rotation is not involved in the solution of the problem.

A rigid body can be represented as a collection of particles. The size and shape of the body remain constant at all times and under all conditions of loading. The rigid-body concept represents an idealization of the true situation since all real bodies will change shape to a certain extent when subjected to a system of forces. Such changes are small for most structural elements and machine parts encountered in engineering practice; therefore, they have only a negligible effect on the acceleration produced by the force system or on the reactions required to maintain equilibrium of the body. The bodies dealt with in this book, with the exception of deformable springs, will be considered to be rigid bodies.

1-3-1 Newton's Laws

The foundations for studies in engineering mechanics are the laws formulated and published by Sir Isaac Newton in 1687. In a treatise called *The Principia*, Newton stated the basic laws governing the motion of a particle as follows:[2]

---

[2] As stated in Dr. Ernst Mach, *The Science of Mechanics,* 9th ed. The Open Court Publishing Company, LaSalle, Ill., 1942.

> **Newton's Laws of Motion**
>
> *Law 1:*   Every body perseveres in its state of rest or of uniform motion in a straight line, except in so far as it is compelled to change that state by impressed forces.
>
> *Law 2:*   Change of motion is proportional to the moving force impressed, and takes place in the direction of the straight line in which such force is impressed.
>
> *Law 3:*   Reaction is always equal and opposite to action; that is to say, the actions of two bodies upon each other are always equal and directly opposite.

These laws, which have come to be known as "Newton's Laws of Motion," are commonly expressed today as follows:

Law 1.   In the absence of external forces, a particle originally at rest or moving with a constant velocity will remain at rest or continue to move with a constant velocity along a straight line.

Law 2:   If an external force acts on a particle, the particle will be accelerated in the direction of the force and the magnitude of the acceleration will be directly proportional to the force and inversely proportional to the mass of the particle.

Law 3:   For every action there is an equal and opposite reaction. The forces of action and reaction between contacting bodies are equal in magnitude, opposite in direction, and collinear.

Newton's three laws were developed from a study of planetary motion (the motion of particles); therefore, they apply only to the motion of particles. During the eighteenth century, Leonhard Euler (1707–1783) extended Newton's work on particles to rigid-body systems.

The first law of motion is a special case of the second law and covers the case where the particle is in equilibrium. Thus, the first law provides the foundation for the study of statics. The second law of motion provides the foundation for the study of dynamics. The mathematical statement of the second law that is widely used in dynamics is

$$\mathbf{F} = m\mathbf{a} \tag{1-1}$$

where

   $\mathbf{F}$ is the external force acting on the particle,
   m is the mass of the particle, and
   $\mathbf{a}$ is the acceleration of the particle in the direction of the force.

The third law of motion provides the foundation for an understanding of the concept of a force since in practical engineering applications the word "action" is taken to mean force. Thus, if one body exerts a force on a second body, the second body exerts an equal and opposite force on the first.

The law that governs the mutual attraction between two isolated bodies was also formulated by Newton and is known as the "Law of Gravitation." This law can be expressed mathematically as

$$F = G\frac{m_1 m_2}{r^2} \qquad (1\text{-}2)$$

where

F is the magnitude of the mutual force of attraction between the two bodies.
G is the universal gravitational constant,
m_1 is the mass of one of the bodies,
m_2 is the mass of the second body, and
r is the distance between the centers of mass of the two bodies.

Approximate values for the universal gravitational constant that are suitable for most engineering computations are

$G = 3.439(10^{-8})$ ft^3/(slug $\cdot$ s^2)   in the U. S. Customary system of units
$G = 6.673(10^{-11})$ m^3/(kg $\cdot$ s^2)   in the SI system of units

The mutual forces of attraction between the two bodies represent the action of one body on the other; therefore, they obey Newton's third law, which requires that they be equal in magnitude, opposite in direction, and collinear (lie along the line joining the centers of mass of the two bodies). The law of gravitation is very important in all studies involving the motion of planets or artificial satellites.

Some of the quantities and constants that may be of interest in applying the law of universal gravitation are listed in Table 1-1.

TABLE 1-1   SOLAR SYSTEM MASSES AND DISTANCES

Mass:		
of the Earth	$= 4.095(10^{23})$ slug	$= 5.976(10^{24})$ kg
of the moon	$= 5.037(10^{21})$ slug	$= 7.350(10^{22})$ kg
of the sun	$= 1.364(10^{29})$ slug	$= 1.990(10^{30})$ kg
Mean or average radius:		
of the Earth[a]	$= 2.090(10^{7})$ ft	$= 6.371(10^{6})$ m
of the moon	$= 5.702(10^{6})$ ft	$= 1.738(10^{6})$ m
of the sun	$= 2.284(10^{9})$ ft	$= 6.960(10^{8})$ m
Mean or average distance from the earth:		
to the moon	$= 1.261(10^{9})$ ft	$= 3.844(10^{8})$ m
to the sun	$= 4.908(10^{11})$ ft	$= 1.496(10^{11})$ m

[a]Radius of a sphere of equal volume.

Polar radius	$= 2.0856(10^{7})$ ft	$= 6.357(10^{6})$ m
Equatorial radius	$= 2.0925(10^{7})$ ft	$= 6.378(10^{6})$ m

1-3-2   Mass and Weight

The mass m of a body is an absolute quantity that is independent of the position of the body and independent of the surroundings in which the body is placed. The weight W of a body is the gravitational attraction exerted on the body by the planet Earth or by any other massive body such as the moon. Therefore, the weight of the body depends on the position of the body relative to some other body. Thus for Eq. 1-2, at the surface of the earth:

$$W = G\frac{m_e m}{r_e^2} = mg \tag{1-3}$$

where

m_e is the mass of the earth,
r_e is the mean radius of the earth, and
$g = Gm_e/r_e^2$ is the gravitational acceleration.

Approximate values for the gravitational acceleration that are suitable for most engineering computations are

$$g = 32.17 \text{ ft/s}^2 \qquad = 9.807 \text{ m/s}^2$$

A source of some confusion arises because the pound is sometimes used as a unit of mass and the kilogram is sometimes used as a unit of force. In grocery stores in Europe, weights of packages are marked in kilograms. In the United States, weights of packages are often marked in both pounds and kilograms. Similarly, a unit of mass called the pound or the pound mass, which is the mass whose weight is one pound under standard gravitational conditions, is sometimes used.

Throughout this book on statics and the companion book on dynamics, without exception, the pound (lb) will be used as the unit of force and the slug will be used as the unit of mass for problems and examples when the U. S. Customary System of units is used. Similarly, the newton (N) will be used as the unit of force and the kilogram (kg) will be used as the unit of mass for problems and examples when the SI System of units is used.

A body weighs 250 lb at the earth's surface. Determine

a. The mass of the body.
b. The weight of the body 500 mi above the earth's surface.
c. The weight of the body on the moon's surface.

SOLUTION

a. The weight of a body at the earth's surface is given by Eq. 1-3 as

$$W = mg$$

Thus,

$$m = \frac{W}{g} = \frac{250}{32.17} = 7.77 \frac{\text{lb} \cdot \text{s}^2}{\text{ft}} = 7.77 \text{ slug} \qquad \text{Ans.}$$

b. The force of attraction between two bodies is given by Eq. 1-2 as

$$W = F = G\frac{m_1 m_2}{r^2}$$

or

$$Wr^2 = Gm_1 m_2 = \text{constant}$$

The mean radius of the earth (see Table 1-1) is $r_e = 2.090(10^7)\text{ft} = 3958$ mi. Thus, for the two positions of the body

$$Wr_e^2 = W_{500}(r_e + 500)^2 = Gm_1 m_2 = \text{constant}$$

$$W_{500} = \frac{Wr_e^2}{(r_e + 500)^2}$$

$$= \frac{250(3958)^2}{(3958 + 500)^2} = 197.1 \text{ lb} \qquad \text{Ans.}$$

c. On the moon's surface, the weight of the body is given by Eq. 1-2 as

$$W = G\frac{mm_m}{r_m^2}$$

The mean radius and mass of the moon (see Table 1-1) are $r_m = 5.702(10^6)$ ft and $m_m = 5.037(10^{21})$ slug. Also, $G = 3.439(10^{-8})$ ft³/(slug · s²). Thus,

$$W = G\frac{mm_m}{r_m^2}$$

$$= 3.439(10^{-8})\frac{7.77(5.037)(10^{21})}{[5.702(10^6)]^2} = 41.4 \text{ lb} \qquad \text{Ans.}$$

PROBLEMS

1-1* Calculate the mass m of a body that weighs 500 lb at the surface of the Earth.

1-2* Calculate the weight W of a body at the surface of the Earth if it has a mass m of 575 kg.

1-3* If a man weighs 180 lb at sea level, determine the weight W of the man

a. At the top of Mt. McKinley (20,320 ft above sea level).
b. At the top of Mt. Everest (29,028 ft above sea level).

1-4* Calculate the weight W of a navigation satellite at a distance of 20,200 km above the Earth's surface if the satellite weighs 9750 N at the earth's surface.

1-5 Compute the gravitational force acting between two spheres that are touching each other if each sphere weighs 1125 lb and has a diameter of 20 in.

1-6 Two spherical bodies have masses of 60 kg and 80 kg, respectively. Determine the force of gravity acting between them if the distance from center to center of the bodies is 500 mm.

1-7 At what distance from the surface of the Earth, in miles, is the weight of a body equal to one-half of its weight on the Earth's surface?

1-8 Calculate the gravitational constant g, in SI units, for a location on the surface of the moon.

1-9* If a woman weighs 125 lb when standing on the surface of the Earth, how much would she weigh when standing on the surface of the moon?

1-10* The gravitational acceleration at the surface of Mars is 3.73 m/s^2 and the mass of Mars is 6.39(10^{23}) kg. Determine the radius of Mars.

1-11* The planet Venus has a diameter of 7700 mi and a mass of 3.34(10^{23}) slug. Determine the gravitational acceleration at the surface of the planet.

1-12* Calculate the gravitational force, in kilonewtons, exerted by the Earth on the moon.

1-13 At what distance, in miles, from the surface of the Earth on a line from center to center would the gravitational force of the Earth on a body be exactly balanced by the gravitational force of the moon on the body?

1-14 At what distance, in kilometers, from the surface of the Earth on a line from center to center would the gravitational force of the Earth on a body be three times the gravitational force of the moon on the body?

1-4 UNITS OF MEASUREMENT

The building blocks of mechanics are the physical quantities used to express the laws of mechanics. Some of these quantities are mass, length, force, time, velocity, and acceleration. Physical quantities are often divided into fundamental quantities and derived quantities. Fundamental quantities cannot be defined in terms of other physical quantities. The number of quantities regarded as fundamental is the minimum number needed to give a consistent and complete description of all the physical quantities ordinarily encountered in the subject area. Examples of quantities viewed as fundamental in the field of mechanics are length and time. Derived quantities are those whose defining operations are based on measurements of other physical quantities. Examples of derived quantities in mechanics are area, volume, velocity, and acceleration. Some quantities may be viewed as either fundamental or derived. Mass and force are examples of such quantities. In the SI system of units, mass is regarded as a fundamental quantity and force as a derived quantity. In the U. S. Customary System of units, force is regarded as a fundamental quantity and mass as a derived quantity.

The magnitude of each of the fundamental quantities is defined by an arbitrarily chosen unit or "standard." The familiar yard, foot, and inch, for example, come from the practice of using the human arm, foot, and thumb as length standards. However, for any sort of precise calculations, such units of length are unsatisfactory. The first truly international standard of length was a bar of platinum-iridium alloy, called the standard meter,[3] which was kept at the International Bureau of Weights and Measures in Sèvres, France. The distance between two fine lines engraved on gold plugs near the ends of the bar is defined to be one meter. Historically, the meter was intended to be one ten-millionth of the distance from the pole to the equator along the meridian line through Paris. Accurate measurements made after the standard meter bar was constructed show that it differs from its intended value by approximately 0.023 percent.

In 1961 an atomic standard of length was adopted by international agreement. The wavelength in vacuum of the orange-red line from the spectrum of isotope krypton 86 was chosen. One meter (m) is now defined to be 1,650,763.73 wavelengths of this light. The choice of an atomic standard offers advantages other than increased precision in length measurements. Krypton 86 is available everywhere, the material can be obtained relatively easily and cheaply, and all atoms of the material are identical and emit light of the same wavelength. The particular wavelength chosen is uniquely characteristic of krypton 86 and is very sharply defined. The definition of the yard, by international agreement, is 1 yard = 0.9144 m, exactly.[4] Thus, 1 in. = 25.4 mm, exactly; and 1 ft = 0.3048 m, exactly.

Similarly, time can be measured in a number of ways. Since the earliest times, the length of a day has been an accepted standard of time measurement. The internationally accepted standard unit of time, the second (s), has been defined in the past as 1/86,400 of a mean solar day or 1/31,557,700 of a mean solar year. Time defined in terms of the rotation of the earth must be determined by astronomical observations. Since these observations require at least several weeks, a good secondary terrestrial measure, calibrated by astronomical observations, is used. Quartz crystal clocks, based on the electrically sustained natural periodic vibrations of a quartz wafer, have been used as secondary time standards. The best of these quartz clocks have kept time for a year with a maximum error of 0.02 seconds.

To meet the need for a better time standard, an atomic clock has been developed that uses the periodic atomic vibrations of isotope cesium 133. The second based on this cesium clock was adopted as the time standard by the Thirteenth General Conference on Weights and Measures in 1967. The second is defined as the duration of 9,192,631,770 cycles of vibration of isotope cesium 133. The cesium clock provides an improvement over the accuracy associated with astronomical methods by a factor of approximately 200. Two cesium clocks will differ by no more than one second after running 3000 years.

The standard unit of mass, the kilogram (kg), is defined by a bar of

[3]The United States has accepted the meter as a standard of length since 1893.

[4]*Guide for the Use of the International System of Units*, National Institute of Standards and Technology (NIST) Special Publication 811, September 1991.

platinum-iridium alloy that is kept at the International Bureau of Weights and Measures in Sèvres, France.

1-4-1 The U. S. Customary System of Units

Until very recently, almost all engineers in the United States used the U. S. Customary System of Units (sometimes called the British gravitational system) in which the base units are foot (ft) for length, the pound (lb) for force, and the second (s) for time. In this system, the foot is defined as 0.3048 m, exactly. The pound is defined as the weight at sea level and at a latitude of 45 degrees of a platinum standard, which is kept at the Bureau of Standards in Washington, D. C. This platinum standard has a mass of 0.453,592,43 kg. The second is defined in the same manner as in the SI system.

In the U. S. Customary System, the unit of mass is derived and is called a slug. One slug is the mass that is accelerated one foot per second squared by a force of one pound, or 1 slug equals 1 lb · s²/ft. Since the weight of the platinum standard depends on the gravitational attraction of the earth, the U. S. Customary System is a gravitational system of units rather than an absolute system of units.

1-4-2 The International System of Units (SI)

The original metric system provided a set of units for the measurement of length, area, volume, capacity, and mass based on two fundamental units: the meter and the kilogram. With the addition of a unit of time, practical measurements began to be based on the meter–kilogram–second (MKS) system of units. In 1960 the Eleventh General Conference on Weights and Measures formally adopted the Système International d'Unites (International System of Units), for which the abbreviation is SI in all languages, as the international standard. Thirty-six countries, including the United States, participated in this conference.

The International System of Units adopted by the conference includes three classes of units: (1) base units, (2) supplementary units, and (3) derived units. The system is founded on the seven base units listed in Table 1-2.

Certain units of the international system have not been classified under either base units or derived units. These units, listed in Table 1-3, are called supplementary units and may be regarded as either base units or derived units.

Derived units are expressed algebraically in terms of base units and/or supplementary units. Their symbols are obtained by means of

TABLE 1-2 BASE UNITS AND THEIR SYMBOLS

Quantity	Name of Unit	Symbol
Length	meter	m
Mass	kilogram	kg
Time	second	s
Electric current	ampere	A
Thermodynamic temperature	kelvin	K
Amount of substance	mole	mol
Luminous intensity	candela	cd

TABLE 1-3 SUPPLEMENTARY UNITS AND THEIR SYMBOLS

Quantity	Name of Unit	Symbol
Plane angle	radian	rad
Solid angle	steradian	sr

TABLE 1-4 DERIVED UNITS AND THEIR SYMBOLS AND SPECIAL NAMES

Quantity	Derived SI Unit	Symbol	Special Name
Area	square meter	m^2	—
Volume	cubic meter	m^3	—
Linear velocity	meter per second	m/s	—
Angular velocity	radian per second	rad/s	—
Linear acceleration	meter per second squared	m/s^2	—
Frequency	(cycle) per second	Hz	hertz
Density	kilogram per cubic meter	kg/m^3	—
Force	kilogram · meter per second squared	N	newton
Moment of force	newton · meter	N · m	—
Pressure	newton per meter squared	Pa	pascal
Stress	newton per meter squared	Pa	pascal
Work	newton · meter	J	joule
Energy	newton · meter	J	joule
Power	joule per second	W	watt

the mathematical signs of multiplication and division. For example, the SI unit for velocity is meter per second (m/s) and the SI unit for angular velocity is radian per second (rad/s). In the SI system, the unit of force is derived and is called a newton. One newton is the force required to give one kilogram of mass an acceleration of one meter per second squared. Thus, $1 \text{ N} = 1 \text{ kg} \cdot m/s^2$. For some of the derived units, special names and symbols exist; those of interest in mechanics are listed in Table 1-4.

Prefixes are used to form names and symbols of multiples (decimal multiples and submultiples) of SI names. The choice of the appropriate multiple is governed by convenience and should usually be chosen so that the numerical values will be between 0.1 and 1000. Only one prefix should be used in forming a multiple of a compound unit, and prefixes in the denominator should be avoided. Approved prefixes with their names and symbols are listed in Table 1-5.

As the use of the SI system becomes more commonplace in the United States, engineers will be required to be familiar with both the SI system and the U. S. Customary System in common use today. As an aid to interpreting the physical significance of answers in SI units for those more accustomed to the U. S. Customary System, some conversion factors for the quantities normally encountered in mechanics are provided in Table 1-6.

TABLE 1-5   MULTIPLES OF SI UNITS

Factor by Which Unit Is Multiplied	Prefix Name	Prefix Symbol
10^{18}	exa	E
10^{15}	peta	P
10^{12}	tera	T
10^{9}	giga	G
10^{6}	mega	M
10^{3}	kilo	k
10^{2}	hecto[a]	h
10	deca[a]	da
10^{-1}	deci[a]	d
10^{-2}	centi[a]	c
10^{-3}	milli	m
10^{-6}	micro	μ
10^{-9}	nano	n
10^{-12}	pico	p
10^{-15}	femto	f
10^{-18}	atto	a

[a]To be avoided when possible.

TABLE 1-6   CONVERSION FACTORS BETWEEN THE SI AND U.S. CUSTOMARY SYSTEMS

Quantity	U. S. Customary to SI	SI to U. S. Customary
Length	1 in. = 25.40 mm	1 m = 39.37 in.
	1 ft = 0.3048 m	1 m = 3.281 ft
	1 mi = 1.609 km	1 km = 0.6214 mi
Area	1 in.^{2} = 645.2 mm^2	1 m^2 = 1550 in.^{2}
	1 ft^2 = 0.0929 m^2	1 m^2 = 10.76 ft^2
Volume	1 in.^{3} = 16.39(10^3) mm^3	1 mm^3 = 61.02(10^{-6}) in.^{3}
	1 ft^3 = 0.02832 m^3	1 m^3 = 35.31 ft^3
	1 gal = 3.785 L[a]	1 L = 0.2642 gal
Velocity	1 in./s = 0.0254 m/s	1 m/s = 39.37 in./s
	1 ft/s = 0.3048 m/s	1 m/s = 3.281 ft/s
	1 mi/h = 1.609 km/h	1 km/h = 0.6214 mi/h
Acceleration	1 in./s^2 = 0.0254 m/s^2	1 m/s^2 = 39.37 in./s^2
	1 ft/s^2 = 0.3048 m/s^2	1 m/s^2 = 3.281 ft/s^2
Mass	1 slug = 14.59 kg	1 kg = 0.06854 slug
Second moment of area	1 in.^{4} = 0.4162(10^6) mm^4	1 mm^4 = 2.402(10^{-6}) in.^{4}
Force	1 lb = 4.448 N	1 N = 0.2248 lb
Distributed load	1 lb/ft = 14.59 N/m	1 kN/m = 68.54 lb/ft
Pressure or stress	1 psi = 6.895 kPa	1 kPa = 0.1450 psi
	1 ksi = 6.895 MPa	1 MPa = 145.0 psi
Bending moment or torque	1 ft · lb = 1.356 N · m	1 N · m = 0.7376 ft · lb
Work or energy	1 ft · lb = 1.356 J	1 J = 0.7376 ft · lb
Power	1 ft · lb/s = 1.356 W	1 W = 0.7376 ft · lb/s
	1 hp = 745.7 W	1 kW = 1.341 hp

[a]Both L and 1 are accepted symbols for liter. Because "1" can be easily confused with the numeral "1", the symbol "L" is recommended for United States use by the National Institute of Standards and Technology (see NIST special publication 811, September 1991).

For the foreseeable future, engineers in the United States will be required to work with both the U. S. Customary and SI systems of units; therefore, we have used both sets of units in examples and problems in this book on statics and the companion volume on dynamics.

EXAMPLE PROBLEM 1-2

A foreign manufacturer lists the fuel consumption for a new automobile as 15 kilometers per liter (km/L). Determine the fuel consumption in miles per gallon.

SOLUTION

One accepted procedure for converting units is to write the associated units in abbreviated form with each of the numerical values used in the conversion. Like unit symbols can then be canceled in the same manner as algebraic symbols. The conversion factors (see Table 1-6) needed for this exercise are:

$$1 \text{ km} = 0.6214 \text{ mi}$$
$$1 \text{ gal} = 3.785 \text{ L}$$

Thus

$$15 \frac{\text{km}}{\text{L}} \times 0.6214 \frac{\text{mi}}{\text{km}} \times 3.785 \frac{\text{L}}{\text{gal}} = 35.3 \text{ mi/gal} \qquad \text{Ans.}$$

EXAMPLE PROBLEM 1-3

The value of G (universal gravitational constant) used for engineering computations in the U.S. system of units is $G = 3.439(10^{-8}) \text{ ft}^3/(\text{slug} \cdot \text{s}^2)$. Use the conversion factors listed in Table 1-6 to determine a value of G with units of $\text{m}^3/(\text{kg} \cdot \text{s}^2)$ suitable for computations in the SI system of units.

SOLUTION

The conversion factors (see Table 1-6) needed for this example are:

$$1 \text{ ft}^3 = 0.02832 \text{ m}^3$$
$$1 \text{ kg} = 0.06854 \text{ slug}$$

Thus

$$G = 3.439(10^{-8}) \frac{\text{ft}^3}{\text{slug} \cdot \text{s}^2} \times 0.02832 \frac{\text{m}^3}{\text{ft}^3} \times 0.06854 \frac{\text{slug}}{\text{kg}}$$
$$= 6.675(10^{-11}) \frac{\text{m}^3}{\text{kg} \cdot \text{s}^2} \qquad \text{Ans.}$$

PROBLEMS

1-15* Determine the weight W, in U. S. Customary units, of a 75-kg steel bar under standard conditions (sea level at a latitude of 45 degrees)

1-16* Determine the mass m, in SI units, for a 500-lb steel beam under standard conditions (sea level at a latitude of 45 degrees).

1-17 Verify the conversion factors listed in Table 1-6 for converting the following quantities from U. S. Customary units to SI units by using the values listed for length and force as defined values:

a. Volume
b. Acceleration
c. Mass
d. Distributed load

Use 1 gallon = 231 in.^{3} and 1 liter = 0.001 m^3.

1-18 Verify the conversion factors listed in Table 1-6 for converting the following quantities from SI units to U. S. Customary units by using the values listed for length and mass as defined values.

a. Area
b. Velocity
c. Second moment of area
d. Pressure or stress

1-19* Express the density, in SI units, of a specimen of material that has a specific weight of 0.025 lb/in.^{3}.

1-20* Express the specific weight, in U. S. Customary units, of a specimen of material that has a density of 8.86 Mg/m^3.

1-21 The velocity of light in space is approximately 186,000 mi/s. What is the velocity of light in SI units?

1-22 The viscosity of crude oil under conditions of standard temperature and pressure is $7.13(10^{-3})$ N · s/m^2. What is the viscosity of crude oil in U. S. Customary units?

1-23* An automobile has a 440-cubic inch engine displacement. What is the engine displacement in liters?

1-24* The fuel consumption of an automobile is 10 kilometers per liter. What is the fuel consumption in miles per gallon?

1-25 One acre equals 43,560 ft.^{2} One gallon equals 231 in.^{3}. Determine the number of liters of water in 2000 acre · ft of water.

1-26 How many barrels of oil are contained in 84 kiloliters of oil? One barrel (petroleum) equals 42.0 gal.

1-27* Express the following in appropriate SI units:

a. 80 statute miles
b. 20 nautical miles (1 nautical mile = 6076 ft)
c. 40 fathoms (1 fathom = 6 ft)

1-28* The specific heat of air under standard atmospheric pressure, in SI units, is 1003 N · m/kg · K. What is the specific heat of air under standard atmospheric pressure in U. S. Customary units (ft · lb/slug · °R)?

1-5 DIMENSIONAL CONSIDERATIONS

All the physical quantities encountered in mechanics can be expressed dimensionally in terms of the three fundamental quantities: mass, length, and time, denoted, respectively by M, L, and T. The dimensions of quantities other than the fundamental quantities follow from definitions or from physical laws. For example, the dimension of velocity L/T follows from the definition of velocity, rate of change of position with time. Similarly, acceleration is defined as the rate of change of velocity with time and has the dimension $L/T.^2$ From Newton's second law, force is defined as the product of mass and acceleration; therefore, force has the dimension ML/T^2. The dimensions of a number of other physical quantities commonly encountered in mechanics are given in Table 1-7.

1-5-1 Dimensional Homogeneity

An equation is said to be dimensionally homogeneous if the form of the equation does not depend on the units of measurement. For exam-

TABLE 1-7   DIMENSIONS OF THE PHYSICAL QUANTITIES OF MECHANICS

Physical Quantity	Dimension	Common Units SI System	Common Units U. S. Customary System
Length	L	m, mm	in., ft
Area	L^2	m^2, mm^2	in.^{2}, ft^2
Volume	L^3	m^3, mm^3	in.^{3}, ft^3
Angle	$1\ (L/L)$	rad, degree	rad, degree
Time	T	s	s
Linear velocity	L/T	m/s	ft/s
Linear acceleration	L/T^2	m/s^2	ft/s^2
Angular velocity	$1/T$	rad/s	rad/s
Angular acceleration	$1/T^2$	rad/s^2	rad/s^2
Mass	M	kg	slug
Force	ML/T^2	N	lb
Moment of a force	ML^2/T^2	N $\cdot$ m	ft $\cdot$ lb
Pressure	M/LT^2	Pa, kPa	psi, ksi
Stress	M/LT^2	Pa, MPa	psi, ksi
Energy	ML^2/T^2	J	ft $\cdot$ lb
Work	ML^2/T^2	J	ft $\cdot$ lb
Power	ML^2/T^3	W	hp
Linear impulse	ML/T	N $\cdot$ s	lb $\cdot$ s
Momentum	ML/T	N $\cdot$ s	lb $\cdot$ s
Specific weight	M/L^2T^2	N/m^3	lb/ft^3
Density	M/L^3	kg/m^3	slug/ft^3
Second moment of area	L^4	m^4, mm^4	in.^{4}, ft^4
Moment of inertia	ML^2	kg $\cdot$ m^2	slug $\cdot$ ft^2

ple, the equation describing the distance h a body released from rest has fallen is $h = gt^2/2$, where h is the distance traveled, t is the time since release, and g is the gravitational acceleration. This equation is valid whether length is measured in feet, meters, or inches and whether time is measured in hours, years, or seconds, provided g is measured in the same units of length and time as h and t. Therefore, by definition, the equation is dimensionally homogeneous.

If the value $g = 32.2$ ft/s^2 is substituted in the previous equation, the equation obtained is $h = 16.1t^2$ ft/s^2. This equation is not dimensionally homogeneous since the equation applies only if length is measured in feet and time is measured in seconds. Dimensionally homogeneous equations are usually preferred because of the potential confusion connected with the unknown units of constants appearing in dimensionally inhomogeneous equations. Dimensionally homogeneous equations also eliminate the explicit use of unit conversion factors.

All like dimensions in a given equation should be measured with the same unit. For example, if the length dimension of a beam is measured in feet and the cross-sectional dimensions are measured in inches, all measurements should be converted to either feet or inches before they are used in a given equation. If this is done, the terms of the equation can be combined after the numerical values for the variables are substituted.

EXAMPLE PROBLEM 1-4

Determine the dimensions of I, R, w, M, and C in the dimensionally homogeneous equation

$$EIy = Rx^3 - P(x - a)^3 - wx^4 + Mx^2 + C$$

in which x and y are lengths, P is a force, and E is a force per unit area.

SOLUTION

The equation can be written dimensionally as

$$\frac{F}{L^2}(I)(L) = R(L^3) - F(L - a)^3 - w(L^4) + M(L^2) + C$$

For this equation to be dimensionally homogeneous a must be a length; hence, all terms must have the dimensions FL^3. Thus,

$$(I)\frac{F}{L} = (R)L^3 = (w)L^4 = (M)L^2 = C = FL^3$$

The dimensions for each of the unknown quantities is obtained as follows:

$$I = \frac{L}{F}(FL^3) = L^4 \qquad \text{Ans.}$$

$$R = \frac{1}{L^3}(FL^3) = F \qquad \text{Ans.}$$

$$w = \frac{1}{L^4}(FL^3) = \frac{F}{L} \qquad \text{Ans.}$$

$$M = \frac{1}{L^2}(FL^3) = FL \qquad \text{Ans.}$$

$$C = FL^3 \qquad \text{Ans.}$$

PROBLEMS

1-29* The angle of twist for a circular shaft subjected to a twisting moment is given by the equation $\theta = TL/GJ$. What are the dimensions of J if θ is an angle in radians, T is the moment of a force, L is a length, and G is a force per unit area?

1-30* The elongation of a bar of uniform cross section subjected to an axial force is given by the equation $\delta = PL/AE$. What are the dimensions of E if δ and L are lengths, P is a force, and A is an area?

1-31 The period of oscillation of a simple pendulum is given by the equation $T = k(L/g)^{1/2}$, where T is in seconds, L is in feet, g is the acceleration due to gravity, and k is a constant. What are the dimensions of k for dimensional homogeneity?

1-32 The equation $x = Ae^{-t/b} \sin(at + \alpha)$ is dimensionally homogeneous. If A is a length and t is time, determine the dimensions of x, a, b, and α.

1-33* In the dimensionally homogeneous equation $w = x^3 + ax^2 + bx + a^2b/x$, if x is a length, what are the dimensions of a, b, and w?

1-34* In the dimensionally homogeneous equation $d^5 = Ad^4 + Bd^3 + Cd^2 + D/d^2$, if d is a length, what are the dimensions of A, B, C, and D?

1-6 METHOD OF PROBLEM SOLVING

The principles of mechanics are few and relatively simple; however, the applications are infinite in their number, variety, and complexity. Success in engineering mechanics depends to a large degree on a well-disciplined method of problem solving. Experience has shown that the development of good problem-solving methods and skills results from solving a large variety of problems. Professional problem solving consists of three phases: problem definition and identification, model development and simplification, and mathematical solution and result interpretation. The problem-solving method outlined in this section will prove useful for the engineering mechanics courses that follow and for most situations encountered later in engineering practice.

Problems in engineering mechanics (statics, dynamics, and mechanics of deformable bodies) are concerned with the external effects of a system of forces on a physical body. The approach usually followed in solving an engineering mechanics problem requires identification of all external forces acting on the "body of interest." A carefully prepared drawing that shows the "body of interest" separated from all other interacting bodies and with all external forces applied is known as a free-body diagram (FBD).

> Most engineers consider an appropriate free-body diagram to be the single most important tool for the solution of mechanics problems.

Since the relationships between the external forces applied to a body and the motions or deformations that they produce are stated in mathematical form, the true physical situation must be represented by a mathematical model in order to obtain the required solution. Often it is necessary to make assumptions or approximations in setting up this model in order to simplify the solution. The most common approximation is to treat most of the bodies in statics and dynamics problems as rigid bodies. No real body is absolutely rigid; however, the changes in shape of a real body usually have a negligible effect on the acceleration produced by a force system or on the reactions required to maintain equilibrium of the body. Considerations of changes in shape under these circumstances would be an unnecessary complication of the problem. Similarly, the weights of many members can be neglected since they are small with respect to the applied loads, and a distributed force, which acts over a small area, can be considered to be concentrated at a point.

Usually, an actual physical problem cannot be solved exactly or completely. However, even in complicated problems, a simplified model can provide good qualitative results. Appropriate interpretation of such results can lead to approximate predictions of physical behavior or be used to verify the "reasonableness" of more sophisticated analytical, numerical, or experimental results. The engineer must always be aware of the actual physical problem under consideration and of any limitations associated with the mathematical model used. Assumptions must be continually evaluated to ensure that the mathematical problem solved provides an adequate representation of the physical process or device of interest.

As stated previously, the most effective way to learn the material contained in engineering mechanics courses is to solve a variety of problems. In order to become an effective engineer, the student must develop the ability to reduce complicated problems to simple parts that can be easily analyzed and to present the results of the work in a clear, logical, and neat manner. This can be accomplished by following the sequence of steps listed below.

1. Read the problem carefully.
2. Identify the result requested.
3. Identify the principles to be used to obtain the result.
4. Prepare a scaled sketch and tabulate the information provided.
5. Draw the appropriate free-body diagrams.
6. Apply the appropriate principles and equations.
7. Report the answer with the appropriate number of significant figures and the appropriate units.
8. Study the answer and determine if it is reasonable.

The development of an ability to apply an orderly approach to problem solving constitutes a significant part of an engineering education. Also, the problem identification, model simplification, and result interpretation phases of engineering problem solving are often more important than the mathematical solution phase.

1-7   SIGNIFICANCE OF NUMERICAL RESULTS

The accuracy of solutions to real engineering problems depends on three factors:

1. The accuracy of the known physical data
2. The accuracy of the physical model
3. The accuracy of the computations performed

An accuracy greater than 0.2 percent is seldom possible for practical engineering problems since physical data are seldom known with any greater accuracy. A practical rule for "rounding off" the final numbers obtained in the computations involved in engineering analysis, which provides answers to approximately this degree of accuracy, is to retain four significant figures for numbers beginning with the figure "1" and to retain three significant figures for numbers beginning with any figure from "2" through "9."

Pocket electronic calculators are widely used to perform the numerical computations required for solution of engineering problems. The number of significant figures obtained when these calculators are used, however, should not be taken as an indication of the accuracy of the solution. As noted previously, engineering data are seldom known to an accuracy greater than 0.2 percent; therefore, calculated results should always be "rounded off" to the number of significant figures that will yield the same degree of accuracy as the data on which they are based. Three significant figures are used for most of the data provided in this book for example and homework problems.

For closed-form analytical predictions, the accuracy of the data and adequacy of the model determine the accuracy of the results. For numerical predictions, the computational accuracy of the algorithms used also influences the accuracy of the results.

An error can be defined as the difference between two quantities. The difference, for example, might be between an experimentally measured value and a computed theoretical value. An error may also result from the rounding off of numbers during a calculation. One method for describing an error is to state a percent difference (%D). Thus, for two numbers A and B, if it is desired to compare number A with number B, the percent difference between the two numbers is defined as

$$\%D = \frac{A - B}{B}(100)$$

In this equation, B is the reference value with which A is to be compared. The percent difference resulting from round-off error is illustrated in the following example.

EXAMPLE PROBLEM 1-5

Round off the number 12,345 to two, three, and four significant figures. Find the percent difference between the rounded-off numbers and the original number by using the original number as the reference.

SOLUTION

Rounding off the number 12,345 to two, three, and four significant figures yields 12,000, 12,300, and 12,350. The percent difference for each of these numbers is

$$\%D = \frac{A - B}{B}(100)$$

For 12,000:

$$\%D = \frac{12,000 - 12,345}{12,345}(100) = -2.79\% \qquad \text{Ans.}$$

For 12,300:

$$\%D = \frac{12,300 - 12,345}{12,345}(100) = -0.36\% \qquad \text{Ans.}$$

For 12,350:

$$\%D = \frac{12,350 - 12,345}{12,345}(100) = +0.041\% \qquad \text{Ans.}$$

The minus signs associated with the above percent differences indicate that the rounded-off numbers are smaller than the reference number. Similarly, a positive percent difference indicates that the rounded-off number is larger than the reference number.

PROBLEMS

Round off the numbers in the following problems to two significant figures. Find the percent difference between each rounded-off number and the original number by using the original number as the reference.

1-35*   (a)   0.015362   (b)   0.034739   (c)   0.056623

1-36*   (a)   0.837482   (b)   0.472916   (c)   0.664473

1-37   (a)   1.839462   (b)   3.462948   (c)   6.752389

Round off the numbers in the following problems to three significant figures. Find the percent difference between each rounded-off number and the original number by using the original number as the reference.

1-38*   (a)   26.39473   (b)   74.82917   (c)   55.33682

1-39   (a)   374.9371   (b)   826.4836   (c)   349.3378

1-40   (a)   6471.907   (b)   3628.729   (c)   7738.273

Round off the numbers in the following problems to four significant figures. Find the percent difference between each rounded-off number and the original number by using the original number as the reference.

1-41*   (a)   63,746.27   (b)   27,382.84   (c)   55,129.92

1-42   (a)   937,284.9   (b)   274,918.2   (c)   339,872.8

1-43   (a)   91,827,364   (b)   28,473,992   (c)   34,269,174

SUMMARY

The foundations for studies in mechanics are the laws formulated by Sir Isaac Newton in 1687. The first law deals with conditions for equilibrium of a particle; therefore, it provides the foundation for the study of statics. The second law, which establishes a relationship between the force acting on a particle and the motion of the particle, provides the foundation for the study of dynamics. The third law provides the foundation for understanding the concept of a force. In addition to the basic laws of motion, Newton also formulated the law of gravitation that governs the mutual attraction between two isolated bodies.

Physical quantities used to express the laws of mechanics can be divided into fundamental quantities and derived quantities. The magnitude of each fundamental quantity is defined by an arbitrarily chosen unit or "standard." The units used in the SI system are the meter (m) for length, the kilogram (kg) for mass, and the second (s) for time. The unit of force is a derived unit called a newton (N). In the U. S. Customary System of units, the units used are the foot (ft) for length, the pound (lb) for force, and the second (s) for time. The unit of mass is a derived unit called a slug. The U. S. Customary System is a gravitational rather than an absolute system of units.

The terms of an equation used to describe a physical process should not depend on the units of measurement (should be dimensionally homogeneous). If an equation is dimensionally homogeneous, the equation is valid for use with any system of units provided all quantities in the equation are measured in the same system. Use of dimensionally homogeneous equations eliminates the need for unit conversion factors.

Success in engineering depends to a large degree on a well-disciplined method of problem solving. Professional problem solving consists of three phases:

1. Problem definition and identification
2. Model development and simplification
3. Mathematical solution and result interpretation

Problems in mechanics are concerned primarily with the effects of a system of forces on a physical body. As a result, an extremely important part of the solution of any problem involves identification of the external forces acting on the body. This is accomplished efficiently and accurately by using a free-body diagram. In obtaining a solution to most problems, the true physical situation must be represented by a mathematical model. A common approximation made in setting up this model is to treat the body as a rigid body. Even though no real body is absolutely rigid, changes in shape usually have a negligible effect on the accelerations produced by a force system or on the reactions required to maintain equilibrium of the body; therefore, considerations of changes in shape usually result in an unnecessary complication of the problem. Anytime a mathematical model is used in solving a problem, care must be exercised to ensure that the model and the associated mathematical problem being solved provides an adequate representation of the physical process or device that it represents.

The accuracy of solutions to real engineering problems depends on three factors:

1. The accuracy of the known physical data
2. The accuracy of the physical model
3. The accuracy of the computations performed

An accuracy greater than 0.2 percent is seldom possible. Calculated results should always be "rounded off" to the number of significant figures that will yield the same degree of accuracy as the data on which they are based.

REVIEW PROBLEMS

1-44* Determine the force of attraction in SI units between (a) the Earth and the moon, (b) the Earth and the sun.

1-45* On the surface of the Earth the weight of a body is 150 lb. At what distance from the center of the Earth would the weight of the body be (a) 100 lb? (b) 50 lb?

1-46 At what distance from the center of the Earth would the force of attraction between two 1-m diameter spheres in contact equal the force of attraction of the Earth on one of the spheres? The mass of each sphere is 100 kg.

1-47 The weight of a satellite on the surface of the Earth prior to launch is 250 lb. When the satellite is in orbit 3500 miles from the surface of the Earth, determine the force of attraction between the Earth and the satellite.

1-48* A fluid has a dynamic viscosity of $1.2(10^{-3})$ N · s/m^2. Express its dynamic viscosity in U. S. Customary units.

1-49* Convert 640 acres (1 square mile) to hectares if 1 acre equals 4840 yd^2 and 1 hectare equals 10^4 m^2.

1-50 The stress equation for eccentric loading of a short column is

$$\sigma = -\frac{P}{A} - \frac{Pey}{I}$$

If P is a force, A is an area, and e and y are lengths, what are the dimensions of stress σ and second moment of area I?

1-51 When a body moves through a fluid it experiences a resistance to its motion that can be represented by the equation $F = \frac{1}{2}C_D\rho V^2 A$, where F is a force, ρ is the density of the fluid, V is the velocity of the body relative to the fluid, and A is the cross-sectional area of the body. Show that the drag coefficient C_D is dimensionless.

Computer Problems

C1-52 A common practice in rounding answers is to report numbers whose leading digit is 1 to an accuracy of 4 significant figures and all other numbers to an accuracy of 3 significant figures. Although this practice probably started with the accuracy with which slide rules could be read, it also reflects the fact that an accuracy of greater than 0.2 percent is seldom possible. This project will examine the error introduced by this and some other rounding schemes.

For each of the rounding schemes below,

1. Generate 20,000 random numbers between 1 and 10. For example, Number = 10 * RND+1
2. Round each number to the specified number of significant figures. For example, to round to n+1 significant figures:

$$RoundNumber = 10^{-n} * INT(Number * 10^n + .5)$$

3. Calculate the percent relative error for each number.

$$RelError = \left| \frac{Number - RoundNumber}{Number} \right| * 100$$

4. Plot RelError versus Number.
5. Comment on the maximum round-off error and the distribution of round-off error.

   a. Round all numbers to an accuracy of 3 significant figures.
   b. Round numbers less than 2 to an accuracy of 4 significant figures and numbers greater than 2 to an accuracy of 3 significant figures.
   c. Round numbers less than 3 to an accuracy of 4 significant figures and numbers greater than 3 to an accuracy of 3 significant figures.
   d. Round numbers less than 5 to an accuracy of 4 significant figures and numbers greater than 5 to an accuracy of 3 significant figures.

2

CONCURRENT FORCE SYSTEMS

The six cables supporting the unit and the helicopter lift cable form a
concurrent force system.

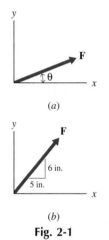

(a)

(b)

Fig. 2-1

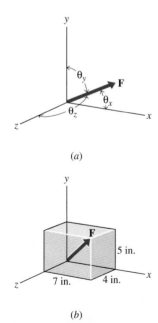

(a)

(b)

Fig. 2-2

2-1 INTRODUCTION

A force was defined in Section 1-3 as the action of one body on another. The action may be the result of direct physical contact between the bodies or it may be the result of gravitational, electrical, or magnetic effects for bodies that are separated.

A force exerted on a body has two effects on the body: (1) the external effect, which is the tendency to change the motion of the body or to develop resisting forces (reactions) on the body, and (2) the internal effect, which is the tendency to deform the body. In many problems, the external effect is significant but the internal effect is not of interest. This is the case in many statics and dynamics problems when the body is assumed to be rigid. In other problems, when the body cannot be assumed to be rigid, the internal effects are important. Problems of this type are discussed in textbooks on Mechanics of Materials or Mechanics of Deformable Bodies.

When a number of forces are treated as a group, they are referred to as a force system. If a force system acting on a body produces no external effect, the forces are said to be in balance and the body that experiences no change in motion is said to be in equilibrium. If a body is acted on by a force system that is not in balance, a change in motion of the body must occur. Such a force system is said to be unbalanced or to have a resultant.

Two force systems are said to be equivalent if they produce the same external effect when applied in turn to a given body. The resultant of a force system is the simplest equivalent system to which the original system will reduce. The process of reducing a force system to a simpler equivalent system is called composition. The process of expanding a force or a force system into a less simple equivalent system is called resolution. A component of a force is one of the two or more forces into which the given force may be resolved.

2-2 FORCES AND THEIR CHARACTERISTICS

The properties needed to describe a force are called the characteristics of the force. The characteristics of a force are as follows:

1. Its magnitude
2. Its direction (slope and sense)
3. Its point of application

The magnitude (positive numerical value) of a force is the amount or size of the force. In this book, the magnitude of a force will be expressed in newtons (N) or kilonewtons (kN) when the SI System of units is used and in pounds (lb) or kilopounds (kip) when the U. S. Customary System of units is used.

The direction of a force is the slope and sense of the line segment used to represent the force. In a two-dimensional problem, the slope can be specified by providing an angle as shown in Fig. 2-1a or by providing two dimensions as shown in Fig. 2-1b. In a three-dimensional problem, the slope can be specified by providing three angles as shown in Fig. 2-2a or by providing three dimensions as shown in Fig. 2-2b. The sense of the force can be specified by placing an arrowhead on the appropriate end of the line segment used to represent the force.

Alternatively, a plus or minus sign can be used with the magnitude of a force to indicate the sense of the force.

The point of application of a force is the point of contact between the two bodies. A straight line extending through the point of application in the direction of the force is called its line of action.

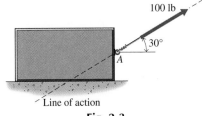

Fig. 2-3

The three characteristics of a force are illustrated on the sketch of a block shown in Fig. 2-3. In this case, the force applied to the block can be described as a 100-lb (magnitude) force acting 30° upward and to the right (direction: slope and sense) through point A (point of application). A discussion of the manner in which these characteristics influence the reactions developed in holding a body at rest forms an important part of the study of statics. In a similar manner, a discussion of the manner in which these characteristics influence the change in motion of a body forms an important part of the study of kinetics.

2-2-1 Scalar and Vector Quantities

Scalar quantities can be completely described with a magnitude (number). Examples of scalar quantities in mechanics are mass, density, length, area, volume, speed, energy, time, and temperature. In mathematical operations, scalars follow the rules of elementary algebra.

A vector quantity has both a magnitude and a direction (line of action and sense) and obeys the parallelogram law of addition. Examples of vector quantities in mechanics are force, moment, displacement, velocity, acceleration, impulse, and momentum. Vectors can be classified into three types: free, sliding, or fixed.

1. A free vector has a specific magnitude, slope, and sense but its line of action does not pass through a unique point in space.
2. A sliding vector has a specific magnitude, slope, and sense and its line of action passes through a unique point in space. The point of application of a sliding vector can be anywhere along its line of action.
3. A fixed vector has a specific magnitude, slope, and sense and its line of action passes through a unique point in space. The point of application of a fixed vector is confined to a fixed point on its line of action.

Vector quantities are indicated in typeset print by the use of boldface type (**A**). Since it is not feasible to produce boldface characters with pencil or chalk, vector quantities are frequently indicated in these instances by using an arrow over the vector quantity ($\vec{A}$). When text material is prepared with a typewriter, vector quantities are frequently indicated by underlining the vector quantity (A).

The use of scalars and vectors to represent physical quantities is a simple example of the modeling of physical quantities by mathematical methods. An engineer must be able to construct good mathematical models and to correctly interpret their physical meaning.

2-2-2 Principle of Transmissibility

In most statics and dynamics problems, the assumption is made that the body is rigid. As a result, only the external effects of any force applied to the body are of interest. When this is the case, the force can be applied at any point along its line of action without changing the

Fig. 2-4

external effects of the force. For example, a stalled automobile (see Fig. 2-4) can be moved by pushing on the back bumper or pulling on the front bumper. If the magnitude, direction, and line of action of the two forces are identical, the external effect will be the same. Clearly in this case, the point of application of the force has no effect on the external effect (motion of the automobile).

This fact is formally expressed by the principle of transmissibility as, "The external effect of a force on a rigid body is the same for all points of application of the force along its line of action." It should be noted that only the external effect remains unchanged. The internal effect of a force (stress and deformation) may be greatly influenced by a change in the point of application of the force along its line of action. For those cases where the principle of transmissibility applies (rigid-body mechanics), force can be treated as a sliding vector.

2-2-3   Classification of Forces

A force has been defined as the action of one physical body on another. Since the interaction can occur when there is contact between the bodies or when the bodies are physically separated, forces can be classified under two general headings: (1) contacting or surface forces, such as a push or a pull produced by mechanical means, and (2) noncontacting or body forces, such as the gravitational pull of the earth on all physical bodies.

Forces may also be classified with respect to the area over which they act. A force applied along a length or over an area is known as a distributed force. The distribution can be uniform or nonuniform. The weight of a concrete bridge floor of uniform thickness (see Fig. 2-5) is an example of a uniformly distributed load. Any force applied to a relatively small area compared with the size of the loaded member can be assumed to be a concentrated force. For example, the force applied by a car wheel to the longitudinal members of a bridge (see Fig. 2-6) can be considered to be a concentrated load.

Any number of forces treated as a group constitute a force system. Force systems may be one-, two-, or three-dimensional. A force system is said to be concurrent if the action lines of all forces intersect at a common point (see Fig. 2-7a) and coplanar when all the forces lie in the same plane (see Fig. 2-7b). A parallel force system is one in which the action lines of the forces are parallel (see Fig. 2-7c). In a parallel force system, the senses of the forces do not have to be the same. If the forces of a system have a common line of action, the system is said to be collinear (see Fig. 2-7d).

2-2-4   Free-Body Diagram

A concept that is fundamental to the solution of problems in mechanics is the free-body diagram. A free-body diagram (FBD) is a carefully prepared drawing that shows the "body of interest" separated from all

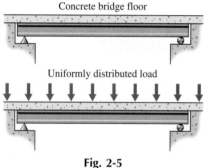

Concrete bridge floor

Uniformly distributed load

Fig. 2-5

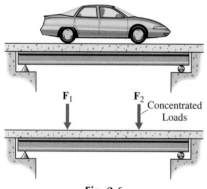

F_1 F_2
                    Concentrated
                    Loads

Fig. 2-6

other interacting bodies and with all external forces applied. Since a force is the action of one body on another, the number of forces on the free-body diagram is determined by noting the number of bodies that exert forces on the "body of interest." It is important to note that these forces may be either surface forces or body forces. The actual procedure for drawing a free-body diagram consists of two essential steps:

1. Make a decision regarding what body (or part of a body or group of bodies) is to be isolated and analyzed. Prepare a sketch of the external boundary of the body selected.
2. Represent all forces, known and unknown, that are applied by other bodies to the isolated body with vectors in their correct positions.

Each force on a complete free-body diagram should be labeled either with its known magnitude or with a symbol to identify the particular force when it is unknown. The slope or angle of inclination of all forces should be shown. The sense of an unknown force can be assumed if it is not known. After all calculations are completed, a plus sign with an answer will indicate that the force acts in the direction assumed, and a minus sign will indicate that the force acts in a direction opposite to that assumed.

Further discussion of free-body diagrams will be provided in Chapters 3 and 5 when equilibrium of a particle (Chapter 3) and equilibrium of a rigid body (Chapter 5) are discussed.

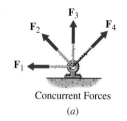

Concurrent Forces

(a)

Coplanar Forces

(b)

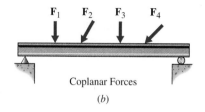

Parallel Forces

(c)

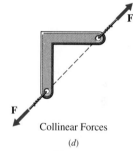

Collinear Forces

(d)

Fig. 2-7

2-3 RESULTANT OF TWO CONCURRENT FORCES

Any two concurrent forces $\mathbf{F}_1$ and $\mathbf{F}_2$ acting on a body can be replaced by a single force, called the resultant $\mathbf{R}$, which will produce the same effect on the body as the original two forces. The resultant of the two forces can be determined by adding the two forces vectorially using the parallelogram law. Mathematically, the sum of the two forces is given by the vector equation

$$\mathbf{F}_1 + \mathbf{F}_2 = \mathbf{R}$$

The process by which two forces are added graphically using the parallelogram law is illustrated in Fig. 2-8.

The resultant $\mathbf{R}$ of two forces can also be determined graphically by using one-half of a parallelogram. Since one-half of a parallelogram is a triangle, the method is called the triangle law for addition of vectors. When the triangle law is used to determine the resultant $\mathbf{R}$ of two

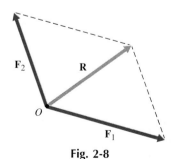

Fig. 2-8

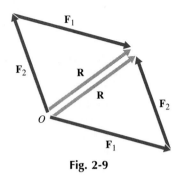

Fig. 2-9

forces $\mathbf{F}_1$ and $\mathbf{F}_2$, force $\mathbf{F}_1$ is drawn to scale first; then force $\mathbf{F}_2$ is drawn to scale from the tip of force $\mathbf{F}_1$ in a head-to-tail fashion. The closing side of the triangle, drawn from the origin O to the tip of force $\mathbf{F}_2$, is the resultant $\mathbf{R}$ of the two forces. The process by which two forces are added using the triangle law is illustrated in Fig. 2-9. The triangle constructed using this method is called a force triangle.

Alternatively, force $\mathbf{F}_2$ can be drawn first; then force $\mathbf{F}_1$ can be drawn from the tip of force $\mathbf{F}_2$ in a head-to-tail fashion. Again, the resultant $\mathbf{R}$ of the two forces is the closing side of the triangle. As indicated in Fig. 2-9,

$$\mathbf{F}_1 + \mathbf{F}_2 = \mathbf{F}_2 + \mathbf{F}_1 = \mathbf{R}$$

The results shown in Fig. 2-9 indicate that the resultant $\mathbf{R}$ does not depend on the order in which forces $\mathbf{F}_1$ and $\mathbf{F}_2$ are selected. Figure 2-9 is a graphical illustration of the commutative law of vector addition.

Graphical methods for determining the resultant of two forces require an accurate scaled drawing if accurate results are to be obtained. In practice, numerical results are obtained by using trigonometric methods based on the law of sines and the law of cosines in conjunction with sketches of the force system. For example, consider the triangle shown in Fig. 2-10, which is similar to the force triangles illustrated in Figs. 2-8 and 2-9. For this general triangle, the law of sines is

$$\frac{a}{\sin \alpha} = \frac{b}{\sin \beta} = \frac{c}{\sin \gamma}$$

and the law of cosines is

$$c^2 = a^2 + b^2 - 2ab \cos \gamma$$

The procedure for determining the resultant $\mathbf{R}$ of a force system by using the law of sines and the law of cosines is demonstrated in the following example.

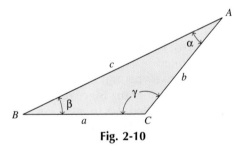

Fig. 2-10

Two forces are applied to an eye bracket as shown in Fig. 2-11a. Determine the magnitude of the resultant **R** of the two forces and the angle θ between the x-axis and the line of action of the resultant.

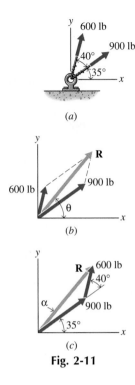

Fig. 2-11

SOLUTION

The two forces, the resultant **R**, and the angle θ are shown in Fig. 2-11b. The triangle law for the addition of the two forces can be applied as shown in Fig. 2-11c. Applying the law of cosines to the triangle yields,

$$R^2 = 900^2 + 600^2 - 2(900)(600) \cos (180° - 40°)$$

where R is the magnitude of the resultant. Thus,

$$R = |\mathbf{R}| = 1413.3 = 1413 \text{ lb} \qquad\qquad \text{Ans.}$$

Applying the law of sines to the triangle yields,

$$\sin \alpha = \frac{600}{1413.3} \sin (180° - 40°) = 0.2729$$

from which

$$\alpha = 15.84°$$

Thus,

$$\theta = 15.84 + 35 = 50.84 = 50.8° \qquad\qquad \text{Ans.}$$

PROBLEMS

Use the Law of Sines and the Law of Cosines, in conjunction with sketches of the force triangles, to solve the following problems. Determine the magnitude of the resultant **R** and the angle θ between the x-axis and the line of action of the resultant for the following:

2-1*   The two forces shown in Fig. P2-1.

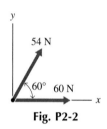

Fig. P2-1

2-2*   The two forces shown in Fig. P2-2.

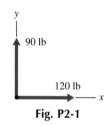

Fig. P2-2

2-3   The two forces shown in Fig. P2-3.

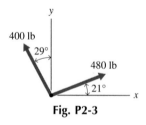

Fig. P2-3

2-4   The two forces shown in Fig. P2-4.

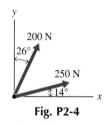

Fig. P2-4

2-5*   The two forces shown in Fig. P2-5.

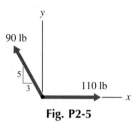

Fig. P2-5

2-6*   The two forces shown in Fig. P2-6.

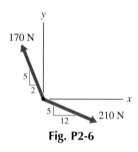

Fig. P2-6

2-7   The two forces shown in Fig. P2-7.

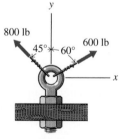

Fig. P2-7

2-8   The two forces shown in Fig. P2-8.

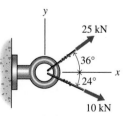

Fig. P2-8

2-4 RESULTANT OF THREE OR MORE CONCURRENT FORCES

In the previous section, use of the parallelogram and triangle laws to determine the resultant **R** of two concurrent forces $\mathbf{F}_1$ and $\mathbf{F}_2$ was discussed. The method can easily be extended to cover three or more forces. As an example, consider the case of three coplanar, concurrent forces acting on an eye bolt as shown in Fig. 2-12. Application of the parallelogram law to forces $\mathbf{F}_1$ and $\mathbf{F}_2$, as shown graphically in Fig. 2-13, yields resultant $\mathbf{R}_{12}$. Then, combining resultant $\mathbf{R}_{12}$ with force $\mathbf{F}_3$, through a second graphical application of the parallelogram law, yields resultant $\mathbf{R}_{123}$, which is the vector sum of the three forces.

In practice, numerical results for specific problems involving three or more forces are obtained algebraically by using the law of sines and the law of cosines in conjunction with sketches of the force system similar to those shown in Fig. 2-14. The sketches shown in Fig. 2-14 are known as force polygons. The order in which the forces are added can be arbitrary, as shown in Figs. 2-14a and 2-14b, where the forces are added in the order $\mathbf{F}_1$, $\mathbf{F}_2$, $\mathbf{F}_3$ in Fig. 2-14a and in the order $\mathbf{F}_3$, $\mathbf{F}_1$, $\mathbf{F}_2$ in Fig. 2-14b. Although the shape of the polygon changes, the resultant force remains the same. The fact that the sum of the three vectors is the same, regardless of the order in which they are added, illustrates the associative law of vector addition.

If there are more than three forces (as an example, Fig. 2-15 shows the eye bolt of Fig. 2-12 with four forces), the process of adding additional forces can be continued, as shown in Fig. 2-16, until all of the forces are joined in head-to-tail fashion. The closing side of the polygon, drawn from the tail of the first vector to the head of the last vector is the resultant of the force system.

Application of the parallelogram law to more than three forces requires extensive geometric and trigonometric calculation. Therefore, problems of this type are usually solved by using the rectangular-component method, which is developed in Section 2-7 of this text.

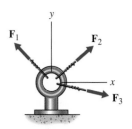

Fig. 2-12

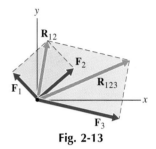

Fig. 2-13

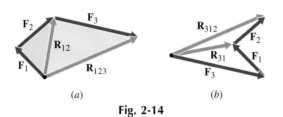

(a)   (b)

Fig. 2-14

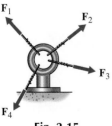

Fig. 2-15

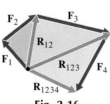

Fig. 2-16

PROBLEMS

Use the Law of Sines and the Law of Cosines, in conjunction with sketches of the force polygons, to solve the following problems. Determine the magnitude of the resultant **R** and the angle θ between the x-axis and the line of action of the resultant for the following:

2-9* The three forces shown in Fig. P2-9.

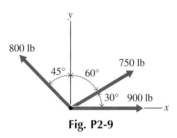

Fig. P2-9

2-10* The three forces shown in Fig. P2-10.

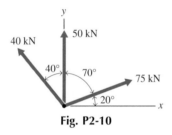

Fig. P2-10

2-11 The three forces shown in Fig. P2-11.

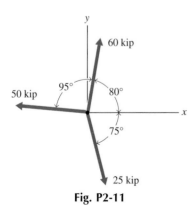

Fig. P2-11

2-12 The three forces shown in Fig. P2-12.

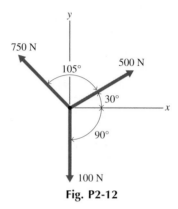

Fig. P2-12

2-13* The three forces shown in Fig. P2-13.

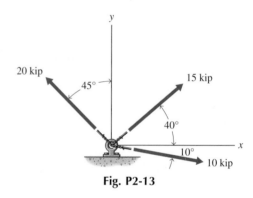

Fig. P2-13

2-14* The three forces shown in Fig. P2-14.

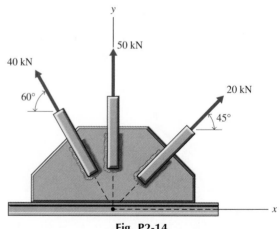

Fig. P2-14

2-15 The four forces shown in Fig. P2-15.

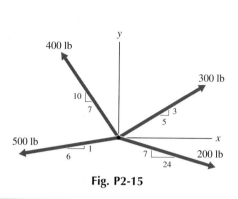

Fig. P2-15

2-16 The four forces shown in Fig. P2-16.

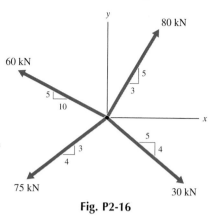

Fig. P2-16

2-5 RESOLUTION OF A FORCE INTO COMPONENTS

In the previous two sections of this chapter, use of the parallelogram and triangle laws to determine the resultant **R** of two concurrent forces $\mathbf{F}_1$ and $\mathbf{F}_2$ or three or more concurrent forces $\mathbf{F}_1$, $\mathbf{F}_2$, $\cdots$, $\mathbf{F}_n$ was discussed. In a similar manner, a single force **F** can be replaced by a system of two or more forces $\mathbf{F}_a$, $\mathbf{F}_b$, $\cdots$, $\mathbf{F}_n$. Forces $\mathbf{F}_a$, $\mathbf{F}_b$, $\cdots$, $\mathbf{F}_n$ are called components of the original force. In the most general case, the components of a force can be any system of forces that can be combined by the parallelogram law to produce the original force. Such components are not required to be concurrent or coplanar. Normally, however, the term *component* is used to specify either one of two coplanar concurrent forces or one of three noncoplanar concurrent forces that can be combined vectorially to produce the original force. The point of concurrency must be on the line of action of the original force. The process of replacing a force by two or more forces is called resolution.

The process of resolution does not produce a unique set of vector components. For example, consider the four coplanar sketches shown in Fig. 2-17. It is obvious from these sketches that

$$\begin{aligned} \mathbf{A} + \mathbf{B} &= \mathbf{R} & \mathbf{E} + \mathbf{F} &= \mathbf{R} \\ \mathbf{C} + \mathbf{D} &= \mathbf{R} & \mathbf{G} + \mathbf{H} + \mathbf{I} &= \mathbf{R} \end{aligned}$$

where **R** is the same vector in each expression. Thus, an infinite number of sets of components exist for any vector.

Use of the parallelogram and triangle laws to resolve a force into components along any two oblique lines of action is illustrated in the following examples.

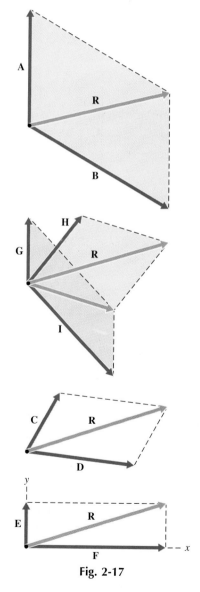

Fig. 2-17

37

Determine the magnitudes of the u- and v-components of the 900-N force shown in Fig. 2-18a.

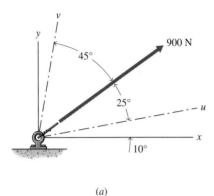

(a)

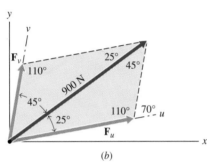

(b)

Fig. 2-18

SOLUTION

The magnitude and direction of the 900-N force are shown in Fig. 2-18b. The components $\mathbf{F}_u$ and $\mathbf{F}_v$ along the u- and v-axes can be determined by drawing lines parallel to the u- and v-axes through the head and tail of the vector used to represent the 900-N force. From the parallelogram thus produced, the law of sines can be applied to determine the forces $\mathbf{F}_u$ and $\mathbf{F}_v$ since all the angles for the two triangles that form the parallelogram are known. Thus

$$\frac{F_u}{\sin 45°} = \frac{F_v}{\sin 25°} = \frac{900}{\sin 110°}$$

from which

$$F_u = |\mathbf{F}_u| = \frac{900 \sin 45°}{\sin 110°} = 677 \text{ N} \qquad \text{Ans.}$$

$$F_v = |\mathbf{F}_v| = \frac{900 \sin 25°}{\sin 110°} = 405 \text{ N} \qquad \text{Ans.}$$

Two forces are applied to an eye bracket as shown in Fig. 2-19a. The resultant **R** of the two forces has a magnitude of 1000 lb and its line of action is directed along the x-axis. If the force $\mathbf{F}_1$ has a magnitude of 250 lb, determine:

a. The magnitude of force $\mathbf{F}_2$.
b. The angle α between the x-axis and the line of action of the force $\mathbf{F}_2$.

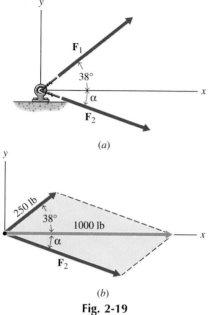

(a)

(b)
Fig. 2-19

SOLUTION

The two forces $\mathbf{F}_1$ and $\mathbf{F}_2$, the resultant **R**, and the angle α are shown in Fig. 2-19b. The force triangle was drawn by using $\mathbf{F}_1$, **R**, and the 38° angle. Completing the parallelogram identifies force $\mathbf{F}_2$ and the angle α.

a. Applying the law of cosines to the top triangle of Fig. 2-19b yields,

$$F_2^2 = 250^2 + 1000^2 - 2(250)(1000) \cos 38°$$

from which

$$F_2 = |\mathbf{F}_2| = 817.6 = 818 \text{ lb} \qquad \text{Ans.}$$

b. Applying the law of sines to the top triangle yields,

$$\frac{250}{\sin \alpha} = \frac{817.6}{\sin 38°}$$

Thus,

$$\sin \alpha = \frac{250}{817.6} \sin 38° = 0.18825$$

from which

$$\alpha = 10.85° \qquad \text{Ans.}$$

PROBLEMS

Use the Law of Sines and the Law of Cosines in conjunction with sketches of the force triangles to solve the following problems. Determine the magnitudes of the *u*- and *v*-components of

2-17* The 1000-lb force shown in Fig. P2-17.

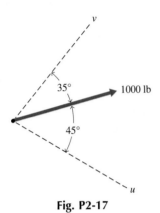

Fig. P2-17

2-18* The 750-N force shown in Fig. P2-18.

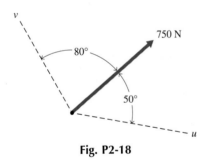

Fig. P2-18

2-19 The 650-lb force shown in Fig. P2-19.

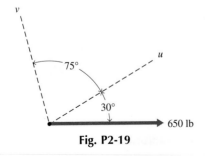

Fig. P2-19

2-20 The 25-kN force shown in Fig. P2-20.

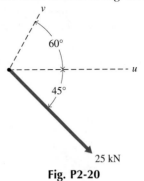

Fig. P2-20

2-21* Two cables are used to support a stop light as shown in Fig. P2-21. The resultant **R** of the cable forces $\mathbf{F}_u$ and $\mathbf{F}_v$ has a magnitude of 300 lb and its line of action is vertical. Determine the magnitudes of forces $\mathbf{F}_u$ and $\mathbf{F}_v$.

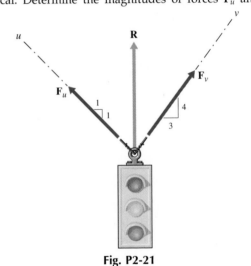

Fig. P2-21

2-22* Two ropes are used to tow a boat upstream as shown in Fig. P2-22. The resultant **R** of the rope forces $\mathbf{F}_u$

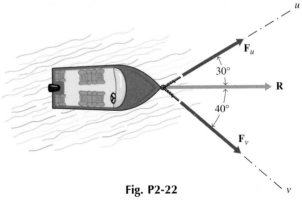

Fig. P2-22

and $\mathbf{F}_v$ has a magnitude of 1500 N and its line of action is directed along the axis of the boat. Determine the magnitude of forces $\mathbf{F}_u$ and $\mathbf{F}_v$.

2-23 A 3000-lb force is resisted by two struts as shown in Fig. P2-23. Determine the component $\mathbf{F}_u$ of the force along the axis of strut AB and the component $\mathbf{F}_v$ of the force along the axis of strut BC.

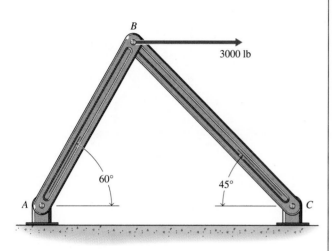

Fig. P2-23

2-25* A 25-kip force is resisted by an eye bar and a strut as shown in Fig. P2-25. Determine the component $\mathbf{F}_u$ of the force along the axis of eye bar AB and the component $\mathbf{F}_v$ of the force along the axis of strut BC.

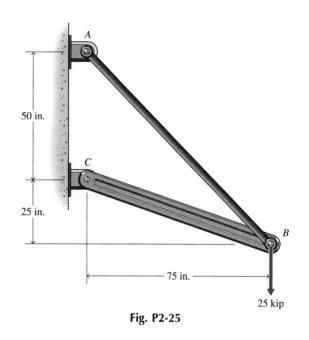

Fig. P2-25

2-24 A 15-kN force is resisted by an eye bar and a strut as shown in Fig. P2-24. Determine the component $\mathbf{F}_u$ of the force along the axis of eye bar AB and the component $\mathbf{F}_v$ of the force along the axis of strut BC.

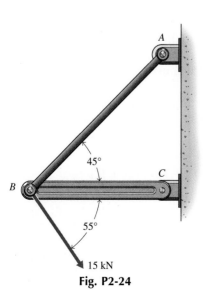

Fig. P2-24

2-26* A 100-kN force is resisted by an eye bar and a strut as shown in Fig. P2-26. Determine the component $\mathbf{F}_u$ of the force along the axis of eye bar AB and the component $\mathbf{F}_v$ of the force along the axis of strut AC.

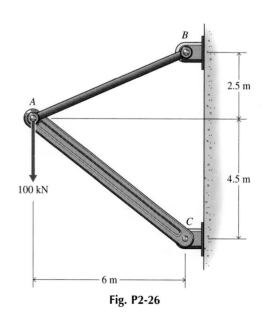

Fig. P2-26

2-27 Three forces are applied to an eye bracket as shown in Fig. P2-27. The magnitude of the resultant **R** of the three forces is 50 kips. If the force **F**$_1$ has a magnitude of 30 kips, determine the magnitudes of forces **F**$_2$ and **F**$_3$.

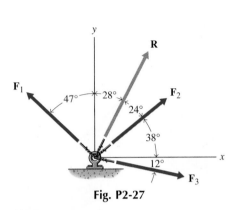

Fig. P2-27

2-28 A gusset plate is used to transfer forces from three bars to a beam as shown in Fig. P2-28. The magnitude of the resultant **R** of the three forces is 100 kN. If the force **F**$_1$ has a magnitude of 20 kN, determine the magnitudes of forces **F**$_2$ and **F**$_3$.

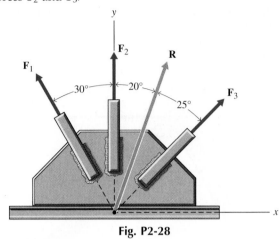

Fig. P2-28

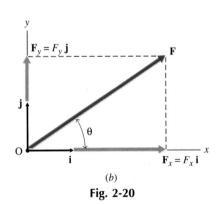

(a)

(b)

Fig. 2-20

2-6 RECTANGULAR COMPONENTS OF A FORCE

General oblique components of a force are not widely used for solving most practical engineering problems. Mutually perpendicular (rectangular) components, on the other hand, find wide usage. The process for obtaining rectangular components is greatly simplified since the parallelogram used to represent the force and its components reduces to a rectangle and the law of cosines used to obtain numerical values of the components reduces to the Pythagorean theorem.

A force **F** can be resolved into a rectangular component **F**$_x$ along the x-axis and a rectangular component **F**$_y$ along the y-axis as shown in Fig. 2-20a. The forces **F**$_x$ and **F**$_y$ are the vector components of the force **F**. The x- and y-axes are usually chosen horizontal and vertical as shown in Fig. 2-20a; however, they may be chosen in any two perpendicular directions. The choice is usually indicated by the geometry of the problem.[1]

The force **F** and its two-dimensional vector components **F**$_x$ and **F**$_y$ can be written in Cartesian vector form by using unit vectors **i** and **j** directed along the positive x- and y-coordinate axes as shown in Fig. 2-20b. Thus,

$$\mathbf{F} = \mathbf{F}_x + \mathbf{F}_y = F_x\mathbf{i} + F_y\mathbf{j} \qquad (2\text{-}1)$$

where the scalars F_x and F_y are the x and y scalar components of the force **F**. The scalar components F_x and F_y are related to the magnitude

[1]Coordinate systems and axes are tools that may be used to advantage by the analyst. Machine components and structural elements do not come inscribed with x-y-axes; therefore, the analyst is free to select directions that are convenient for his or her work.

$F = |\mathbf{F}|$ and the angle of inclination θ (direction) of the force $\mathbf{F}$ by the following expressions:

$$F_x = F \cos \theta \qquad F = \sqrt{F_x^2 + F_y^2}$$

(2-2)

$$F_y = F \sin \theta \qquad \theta = \tan^{-1} \frac{F_y}{F_x}$$

The scalar components F_x and F_y of the force $\mathbf{F}$ can be positive or negative depending on the sense of the vector components $\mathbf{F}_x$ and $\mathbf{F}_y$. A scalar component is positive when the vector component has the same sense as the unit vector with which it is associated and negative when the vector component has the opposite sense.

Similarly, for problems requiring analysis in three dimensions, a force $\mathbf{F}$ in space can be resolved into three mutually perpendicular rectangular components $\mathbf{F}_x$, $\mathbf{F}_y$, and $\mathbf{F}_z$ along the x-, y-, and z-coordinate axes as shown in Fig. 2-21. The force $\mathbf{F}$ and its three-dimensional vector components $\mathbf{F}_x$, $\mathbf{F}_y$, and $\mathbf{F}_z$ can also be written in Cartesian vector form by using unit vectors $\mathbf{i}$, $\mathbf{j}$, and $\mathbf{k}$ directed along the positive x-, y-, and z-coordinate axes as shown in Fig. 2-22. Thus

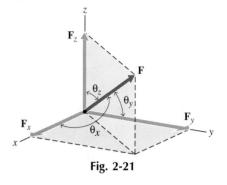

Fig. 2-21

$$\begin{aligned}
\mathbf{F} &= \mathbf{F}_x + \mathbf{F}_y + \mathbf{F}_z \\
&= F_x \mathbf{i} + F_y \mathbf{j} + F_z \mathbf{k} \\
&= F \cos \theta_x \mathbf{i} + F \cos \theta_y \mathbf{j} + F \cos \theta_z \mathbf{k}
\end{aligned}$$

(2-3)

Thus, the scalar components F_x, F_y, and F_z are related to the magnitude F and the direction of the force $\mathbf{F}$ by the following expressions:

$$F_x = F \cos \theta_x \qquad F_y = F \cos \theta_y \qquad F_z = F \cos \theta_z$$

$$\theta_x = \cos^{-1} \frac{F_x}{F} \qquad \theta_y = \cos^{-1} \frac{F_y}{F} \qquad \theta_z = \cos^{-1} \frac{F_z}{F} \quad \text{(2-4)}$$

$$F = \sqrt{F_x^2 + F_y^2 + F_z^2}$$

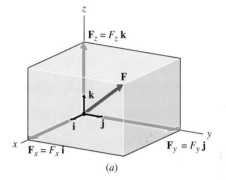

(a)

The angles θ_x, θ_y, and θ_z are the angles ($0 \le \theta \le 180°$ between the force $\mathbf{F}$ and the positive coordinate axes. The cosines of these angles, called direction cosines, must satisfy the equation

$$\cos^2 \theta_x + \cos^2 \theta_y + \cos^2 \theta_z = 1$$

If an angle is greater than 90°, the cosine is negative, indicating that the sense of the component is opposite to the positive direction of the coordinate axis. Thus, Eqs. 2-4 provide the sign as well as the magnitude of the scalar components of the force and hold for any value of the angle.

The rectangular component of a force $\mathbf{F}$ along an arbitrary direction n can be obtained by using the vector operation known as the dot or scalar product (see Appendix A). For example, the scalar component F_x of a force $\mathbf{F}$ is obtained as

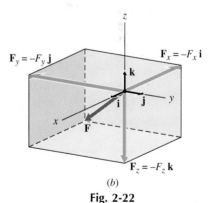

(b)

Fig. 2-22

$$\begin{aligned}
F_x &= \mathbf{F} \cdot \mathbf{i} = (F_x \mathbf{i} + F_y \mathbf{j} + F_z \mathbf{k}) \cdot \mathbf{i} \\
&= F_x (\mathbf{i} \cdot \mathbf{i}) + F_y (\mathbf{j} \cdot \mathbf{i}) + F_z (\mathbf{k} \cdot \mathbf{i}) = F_x
\end{aligned}$$

since

$$\mathbf{i} \cdot \mathbf{i} = \mathbf{j} \cdot \mathbf{j} = \mathbf{k} \cdot \mathbf{k} = 1$$

and

$$\mathbf{i} \cdot \mathbf{j} = \mathbf{j} \cdot \mathbf{i} = \mathbf{i} \cdot \mathbf{k} = \mathbf{k} \cdot \mathbf{i} = \mathbf{j} \cdot \mathbf{k} = \mathbf{k} \cdot \mathbf{j} = 0$$

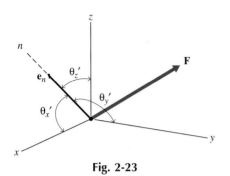

Fig. 2-23

In more general terms, if $\mathbf{e}_n$ is a unit vector in a specified direction n, then the rectangular component F_n of the force $\mathbf{F}$ is

$$F_n = \mathbf{F} \cdot \mathbf{e}_n = (F_x\mathbf{i} + F_y\mathbf{j} + F_z\mathbf{k}) \cdot \mathbf{e}_n$$

Since the angles between the direction n and the x-, y-, and z-axes are θ'_x, θ'_y, and θ'_z as shown in Fig. 2-23, the direction cosines for the unit vector $\mathbf{e}_n$ are $\cos \theta'_x$, $\cos \theta'_y$, and $\cos \theta'_z$; therefore, the unit vector $\mathbf{e}_n$ can be written in Cartesian vector form as

$$\mathbf{e}_n = \cos \theta'_x \mathbf{i} + \cos \theta'_y \mathbf{j} + \cos \theta'_z \mathbf{k}$$

Thus, the force F_n is

$$F_n = \mathbf{F} \cdot \mathbf{e}_n = (F_x\mathbf{i} + F_y\mathbf{j} + F_z\mathbf{k}) \cdot (\cos \theta'_x \mathbf{i} + \cos \theta'_y \mathbf{j} + \cos \theta'_z \mathbf{k})$$
$$= F_x \cos \theta'_x + F_y \cos \theta'_y + F_z \cos \theta'_z \qquad (2\text{-}5)$$

Substituting Eqs. 2-4 into Eq. 2-5 yields an expression for the scalar component F_n in terms of F and the direction cosines associated with $\mathbf{F}$ and n. Thus

$$F_n = \mathbf{F} \cdot \mathbf{e}_n = F (\cos \theta_x \cos \theta'_x + \cos \theta_y \cos \theta'_y + \cos \theta_z \cos \theta'_z) \qquad (2\text{-}6)$$

The rectangular component F_n of the force $\mathbf{F}$ can be expressed in Cartesian vector form by multiplying the scalar component F_n by the unit vector $\mathbf{e}_n$. Thus,

$$F_n = (\mathbf{F} \cdot \mathbf{e}_n) \mathbf{e}_n = F_n\mathbf{e}_n$$
$$= F_n(\cos \theta'_x \mathbf{i} + \cos \theta'_y \mathbf{j} + \cos \theta'_z \mathbf{k}) \qquad (2\text{-}7)$$

The angle α between the line of action of the force $\mathbf{F}$ and the direction n can be determined by using the dot-product relationship and the definition of a rectangular component of a force. Thus

$$F_n = F \cos \alpha = \mathbf{F} \cdot \mathbf{e}_n$$

and therefore,

$$\alpha = \cos^{-1} \frac{\mathbf{F} \cdot \mathbf{e}_n}{F} = \cos^{-1} \frac{F_n}{F} \qquad (2\text{-}8)$$

Equation 2-8 can be used to determine the angle α between any two vectors $\mathbf{A}$ and $\mathbf{B}$ or between any two lines by using the unit vectors $\mathbf{e}_1$ and $\mathbf{e}_2$ associated with the lines. Thus,

$$\alpha = \cos^{-1} \frac{\mathbf{A} \cdot \mathbf{B}}{AB} \qquad (2\text{-}9)$$

or

$$\alpha = \cos^{-1} (\mathbf{e}_1 \cdot \mathbf{e}_2) \qquad (2\text{-}10)$$

The dot-product relationships represented by Eqs. 2-8, 2-9, and 2-10 apply to nonintersecting vectors as well as to intersecting vectors.

EXAMPLE PROBLEM 2-4

A force **F** is applied at a point in a body as shown in Fig. 2-24.

a. Determine the x and y scalar components of the force.
b. Determine the x' and y' scalar components of the force.
c. Express the force **F** in Cartesian vector form for the xy and $x'y'$ axes.

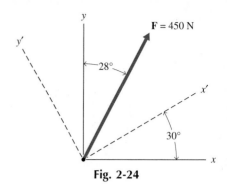

Fig. 2-24

SOLUTION

a. The magnitude F of the force is 450 N. The angle θ_x between the x-axis and the line of action of the force is

$$\theta_x = 90° - 28° = 62°$$

Thus

$$F_x = F \cos \theta_x = 450 \cos 62° = +211 \text{ N} \qquad \text{Ans.}$$
$$F_y = F \sin \theta_x = 450 \sin 62° = +397 \text{ N} \qquad \text{Ans.}$$

b. The magnitude F of the force is 450 N. The angle $\theta_{x'}$ between the x'-axis and the line of action of the force is

$$\theta_{x'} = \theta_x - 30° = 62° - 30° = 32°$$

Thus

$$F_{x'} = F \cos \theta_{x'} = 450 \cos 32° = +382 \text{ N} \qquad \text{Ans.}$$
$$F_{y'} = F \sin \theta_{x'} = 450 \sin 32° = +238 \text{ N} \qquad \text{Ans.}$$

As a check note that

$$F = \sqrt{F_x^2 + F_y^2} = \sqrt{F_{x'}^2 + F_{y'}^2} = \sqrt{211^2 + 397^2} = \sqrt{382^2 + 238^2} = 450 \text{ N}$$

c. The force **F** expressed in Cartesian vector form for the xy- and $x'y'$-axes are

$$\mathbf{F} = 211\mathbf{i} + 397\mathbf{j} \text{ N} \qquad \mathbf{F} = 382\mathbf{e}_{x'} + 238\mathbf{e}_{y'} \text{ N} \qquad \text{Ans.}$$

EXAMPLE PROBLEM 2-5

A force **F** is applied at a point in a body as shown in Fig. 2-25.

a. Determine the x, y, and z scalar components of the force.
b. Express the force in Cartesian vector form.

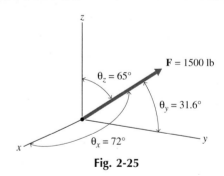

F = 1500 lb

Fig. 2-25

SOLUTION

a. The magnitude F of the force is 1500 lb. Thus

$$F_x = F \cos \theta_x = 1500 \cos 72.0° = +464 \text{ lb} \qquad \text{Ans.}$$
$$F_y = F \cos \theta_y = 1500 \cos 31.6° = +1278 \text{ lb} \qquad \text{Ans.}$$
$$F_z = F \cos \theta_z = 1500 \cos 65.0° = +634 \text{ lb} \qquad \text{Ans.}$$

As a check note that

$$F = \sqrt{F_x^2 + F_y^2 + F_z^2} = \sqrt{464^2 + 1278^2 + 634^2} = 1500 \text{ lb}$$

b. The force **F** expressed in Cartesian vector form is

$$\mathbf{F} = F_x\mathbf{i} + F_y\mathbf{j} + F_z\mathbf{k} = 464\mathbf{i} + 1278\mathbf{j} + 634\mathbf{k} \text{ lb} \qquad \text{Ans.}$$

A force $\mathbf{F}$ is applied at a point in a body as shown in Fig. 2-26. Determine

a. The angles θ_x, θ_y, and θ_z.
b. The x, y, and z scalar components of the force.
c. The rectangular component F_n of the force along line OA.

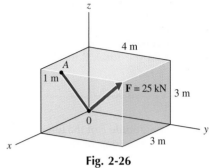

Fig. 2-26

SOLUTION

a. The angles θ_x, θ_y, and θ_z can be determined from the geometry of the
dotted box shown in Fig. 2-26. The length of a diagonal of the box is

$$d = \sqrt{x^2 + y^2 + z^2} = \sqrt{3^2 + 4^2 + 3^2} = 5.831 \text{ m}$$

Thus

$$\theta_x = \cos^{-1}\frac{x}{d} = \cos^{-1}\frac{3}{5.831} = 59.0° \qquad \text{Ans.}$$

$$\theta_y = \cos^{-1}\frac{y}{d} = \cos^{-1}\frac{4}{5.831} = 46.7° \qquad \text{Ans.}$$

$$\theta_z = \cos^{-1}\frac{z}{d} = \cos^{-1}\frac{3}{5.831} = 59.0° \qquad \text{Ans.}$$

b. The magnitude F of the force is 25 kN. Thus

$$F_x = F \cos \theta_x = 25\left(\frac{3}{5.831}\right) = +12.86 \text{ kN} \qquad \text{Ans.}$$

$$F_y = F \cos \theta_y = 25\left(\frac{4}{5.831}\right) = +17.15 \text{ kN} \qquad \text{Ans.}$$

$$F_z = F \cos \theta_z = 25\left(\frac{3}{5.831}\right) = +12.86 \text{ kN} \qquad \text{Ans.}$$

c. The angles θ'_x, θ'_y, and θ'_z between the n direction (along OA) and the x-, y-,
and z-axes can also be determined from the geometry of the dotted box
shown in Fig. 2-26. The length of the line from O to A is

$$d' = \sqrt{(x')^2 + (y')^2 + (z')^2} = \sqrt{3^2 + 1^2 + 3^2} = 4.359 \text{ m}$$

Thus

$$\theta'_x = \cos^{-1}\frac{x'}{d'} = \cos^{-1}\frac{3}{4.359} = 46.5°$$

$$\theta'_y = \cos^{-1}\frac{y'}{d'} = \cos^{-1}\frac{1}{4.359} = 76.7°$$

$$\theta'_z = \cos^{-1}\frac{z'}{d'} = \cos^{-1}\frac{3}{4.359} = 46.5°$$

The unit vector $\mathbf{e}_n$ along line OA is

$$\mathbf{e}_n = \cos \theta'_x \mathbf{i} + \cos \theta'_y \mathbf{j} + \cos \theta'_z \mathbf{k} = 0.6882\mathbf{i} + 0.2294\mathbf{j} + 0.6882\mathbf{k}$$

The force $\mathbf{F}$ expressed in Cartesian vector form is

$$\mathbf{F} = 12.86\mathbf{i} + 17.15\mathbf{j} + 12.86\mathbf{k}$$

Therefore

$$\begin{aligned}
F_n &= \mathbf{F} \cdot \mathbf{e}_n \\
&= (12.86\mathbf{i} + 17.15\mathbf{j} + 12.86\mathbf{k}) \cdot (0.6882\mathbf{i} + 0.2294\mathbf{j} + 0.6882\mathbf{k}) \\
&= 12.86(0.6882) + 17.15(0.2294) + 12.86(0.6882) \\
&= 21.64 = 21.6 \text{ kN} \qquad \text{Ans.}
\end{aligned}$$

PROBLEMS

Determine the x- and y-components of

2-29*   The force shown in Fig. P2-29.

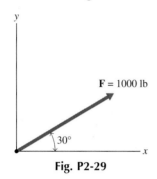

Fig. P2-29

/ **2-30***   The force shown in Fig. P2-30.

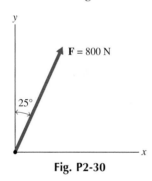

Fig. P2-30

2-31   The force shown in Fig. P2-31.

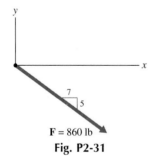

F = 860 lb
Fig. P2-31

2-32   The force shown in Fig. P2-32.

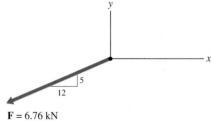

F = 6.76 kN
Fig. P2-32

Determine (a) The x- and y-components and (b) the x'- and y'-components

/ **2-33***   For each of the forces shown in Fig. P2-33.

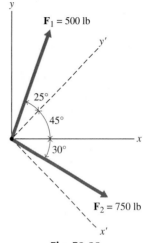

Fig. P2-33

2-34*   For each of the forces shown in Fig. P2-34.

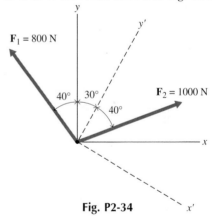

Fig. P2-34

2-35   Determine the x- and y-components for each of the forces shown in Fig. P2-35.

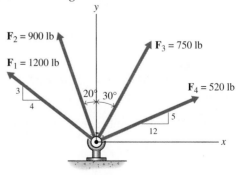

Fig. P2-35

47

2-36 Determine the *x*- and *y*-components for each of the forces shown in Fig. P2-36.

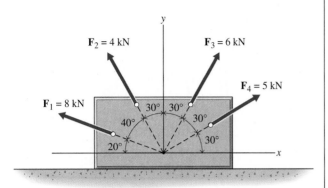

Fig. P2-36

2-37* A force is applied to an eye bolt as shown in Fig. P2-37. If $F = 10$ kips, $\theta_x = 60°$, $\theta_y = 70°$, and $\theta_z = 37.3°$

a. Determine the *x*-, *y*-, and *z*-components of the force.

b. Express the force in Cartesian vector form.

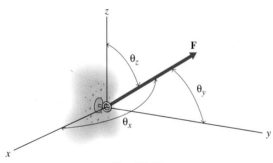

Fig. P2-37

2-38* Solve Problem 2-37 if $F = 15$ kN, $\theta_x = 75°$, $\theta_y = 130°$, and $\theta_z = 43.9°$.

2-39 Solve Problem 2-37 if $F = 28$ kip, $\theta_x = 120°$, $\theta_y = 130°$, and $\theta_z = 54.5°$.

2-40 Solve Problem 2-37 if $F = 36$ kN, $\theta_x = 70°$, $\theta_y = 110°$, and $\theta_z = 28.9°$.

2-41* A force of 800 lb is applied to an eye bolt as shown in Fig. P2-41.

a. Determine the angles θ_x, θ_y, and θ_z.

b. Determine the *x*-, *y*-, and *z*-components of the force.

c. Express the force in Cartesian vector form.

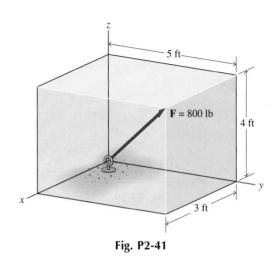

Fig. P2-41

2-42* A force of 50 kN is applied to an eye bolt as shown in Fig. P2-42.

a. Determine the angles θ_x, θ_y, and θ_z.

b. Determine the *x*-, *y*-, and *z*-components of the force.

c. Express the force in Cartesian vector form.

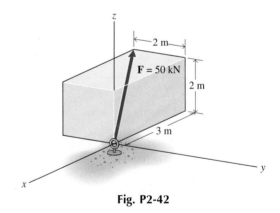

Fig. P2-42

2-43* Two forces are applied to an eye bolt as shown in Fig. P2-43.

a. Determine the *x*-, *y*-, and *z*-components of force F_1.

b. Express force F_1 in Cartesian vector form.

c. Determine the magnitude of the rectangular component of force F_1 along the line of action of force F_2.

d. Determine the angle α between forces F_1 and F_2.

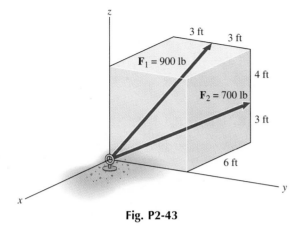

Fig. P2-43

2-44* Two forces are applied to an eye bolt as shown in Fig. P2-44.

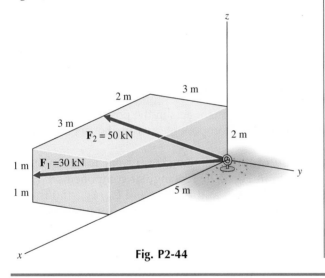

Fig. P2-44

a. Determine the x-, y-, and z-components of force F_1.

b. Express force F_1 in Cartesian vector form.

c. Determine the magnitude of the rectangular component of force F_1 along the line of action of force F_2.

d. Determine the angle α between forces F_1 and F_2.

2-45 Two forces are applied to an eye bolt as shown in Fig. P2-43.

a. Determine the x-, y-, and z-components of force F_2.

b. Express force F_2 in Cartesian vector form.

c. Determine the magnitude of the rectangular component of force F_2 along the line of action of force F_1.

2-46 Two forces are applied to an eye bolt as shown in Fig. P2-44.

a. Determine the x-, y-, and z-components of force F_2.

b. Express force F_2 in Cartesian vector form.

c. Determine the magnitude of the rectangular component of force F_2 along the line of action of force F_1.

2-7 RESULTANTS BY RECTANGULAR COMPONENTS

Previous sections of this chapter discussed the use of the parallelogram and triangle laws to determine the resultant R of two or more concurrent coplanar forces $F_1, F_2, F_3, \cdots, F_n$. Using the parallelogram law to add more than two forces is time consuming and tedious since the procedure requires extensive geometric and trigonometric calculation in order to determine the magnitude and locate the line of action of the resultant R. Problems of this type, however, are easily solved using the rectangular components of a force discussed in Section 2-6.

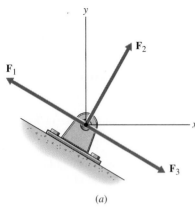

(a)

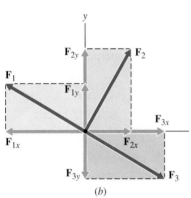

(b)

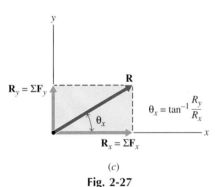

(c)

Fig. 2-27

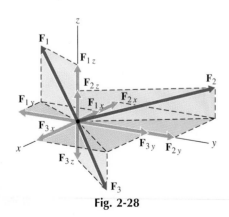

Fig. 2-28

For any system of coplanar concurrent forces, such as the three shown in Fig. 2-27a, rectangular components $\mathbf{F}_{1x}$ and $\mathbf{F}_{1y}$, $\mathbf{F}_{2x}$ and $\mathbf{F}_{2y}$, $\mathbf{F}_{3x}$ and $\mathbf{F}_{3y}$, $\cdots$, and $\mathbf{F}_{nx}$ and $\mathbf{F}_{ny}$ can be determined as shown in Fig. 2-27b. Adding the respective x and y components yields

$$\mathbf{R}_x = \Sigma \mathbf{F}_x = \mathbf{F}_{1x} + \mathbf{F}_{2x} + \mathbf{F}_{3x} + \cdots + \mathbf{F}_{nx}$$
$$= (F_{1x} + F_{2x} + F_{3x} + \cdots + F_{nx})\mathbf{i} = R_x\mathbf{i}$$
$$\mathbf{R}_y = \Sigma \mathbf{F}_y = \mathbf{F}_{1y} + \mathbf{F}_{2y} + \mathbf{F}_{3y} + \cdots + \mathbf{F}_{ny}$$
$$= (F_{1y} + F_{2y} + F_{3y} + \cdots + F_{ny})\mathbf{j} = R_y\mathbf{j}$$

By the parallelogram law

$$\mathbf{R} = \mathbf{R}_x + \mathbf{R}_y = R_x\mathbf{i} + R_y\mathbf{j}$$

The magnitude R of the resultant can be determined from the Pythagorean theorem:

$$R = \sqrt{R_x^2 + R_y^2} = \sqrt{(\Sigma F_x)^2 + (\Sigma F_y)^2}$$

The angle θ_x between the x-axis and the line of action of the resultant $\mathbf{R}$, as shown in Fig. 2-27c, is

$$\theta_x = \tan^{-1}\frac{R_y}{R_x} = \tan^{-1}\frac{\Sigma F_y}{\Sigma F_x}$$

The angle θ_x can also be determined, if it is more convenient, from the equations

$$\theta_x = \cos^{-1}\frac{\Sigma F_x}{R} \qquad \text{or} \qquad \theta_x = \sin^{-1}\frac{\Sigma F_y}{R}$$

The sense of each component must be designated in the summations by using a plus sign if the component acts in the positive x- or y-direction, and a minus sign if the component acts in the negative x- or y-direction.

In the general case of three or more concurrent forces in space, such as the three shown in Fig. 2-28, rectangular components $\mathbf{F}_{1x}$, $\mathbf{F}_{1y}$, and $\mathbf{F}_{1z}$; $\mathbf{F}_{2x}$, $\mathbf{F}_{2y}$, and $\mathbf{F}_{2z}$; $\mathbf{F}_{3x}$, $\mathbf{F}_{3y}$, and $\mathbf{F}_{3z}$; $\cdots$; and $\mathbf{F}_{nx}$, $\mathbf{F}_{ny}$, and $\mathbf{F}_{nz}$ can be determined. Adding the respective x, y, and z components yields

$$\mathbf{R}_x = \Sigma \mathbf{F}_x = \mathbf{F}_{1x} + \mathbf{F}_{2x} + \mathbf{F}_{3x} + \cdots + \mathbf{F}_{nx}$$
$$= (F_{1x} + F_{2x} + F_{3x} + \cdots + F_{nx})\mathbf{i} = R_x\mathbf{i}$$
$$\mathbf{R}_y = \Sigma \mathbf{F}_y = \mathbf{F}_{1y} + \mathbf{F}_{2y} + \mathbf{F}_{3y} + \cdots + \mathbf{F}_{ny}$$
$$= (F_{1y} + F_{2y} + F_{3y} + \cdots + F_{ny})\mathbf{j} = R_y\mathbf{j}$$
$$\mathbf{R}_z = \Sigma \mathbf{F}_z = \mathbf{F}_{1z} + \mathbf{F}_{2z} + \mathbf{F}_{3z} + \cdots + \mathbf{F}_{nz}$$
$$= (F_{1z} + F_{2z} + F_{3z} + \cdots + F_{nz})\mathbf{k} = R_z\mathbf{k}$$

The resultant $\mathbf{R}$ is then obtained from the expression

$$\mathbf{R} = \mathbf{R}_x + \mathbf{R}_y + \mathbf{R}_z = R_x\mathbf{i} + R_y\mathbf{j} + R_z\mathbf{k}$$

Once the scalar components R_x, R_y, and R_z are known, the magnitude R of the resultant and the angles θ_x, θ_y, and θ_z between the line of action of the resultant and the positive coordinate axes can be obtained from the expressions

$$R = \sqrt{R_x^2 + R_y^2 + R_z^2}$$

and

$$\theta_x = \cos^{-1}\frac{R_x}{R} \qquad \theta_y = \cos^{-1}\frac{R_y}{R} \qquad \theta_z = \cos^{-1}\frac{R_z}{R}$$

Determine the magnitude R of the resultant of the four forces shown in Fig. 2-29a and the angle θ_x between the x-axis and the line of action of the resultant.

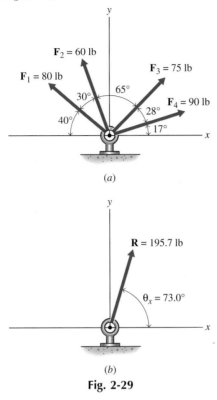

(a)

(b)

Fig. 2-29

SOLUTION

The magnitude R of the resultant will be determined by using the rectangular components F_x and F_y of each of the forces. Thus

$$
\begin{array}{ll}
F_{1x} = 80 \cos 140° = -61.28 \text{ lb} & F_{1y} = 80 \sin 140° = +51.42 \text{ lb} \\
F_{2x} = 60 \cos 110° = -20.52 \text{ lb} & F_{2y} = 60 \sin 110° = +56.38 \text{ lb} \\
F_{3x} = 75 \cos 45° = +53.03 \text{ lb} & F_{3y} = 75 \sin 45° = +53.03 \text{ lb} \\
F_{4x} = 90 \cos 17° = +86.07 \text{ lb} & F_{4y} = 90 \sin 17° = +26.31 \text{ lb}
\end{array}
$$

Once the rectangular components of the forces are known, the components R_x and R_y of the resultant are obtained from the expressions

$$
\begin{aligned}
R_x = \Sigma F_x &= F_{1x} + F_{2x} + F_{3x} + F_{4x} \\
&= -61.28 - 20.52 + 53.03 + 86.07 = +57.30 \text{ lb} \\
R_y = \Sigma F_y &= F_{1y} + F_{2y} + F_{3y} + F_{4y} \\
&= +51.42 + 56.38 + 53.03 + 26.31 = +187.14 \text{ lb}
\end{aligned}
$$

The magnitude R of the resultant is

$$
R = \sqrt{R_x^2 + R_y^2} = \sqrt{(57.30)^2 + (187.14)^2} = 195.7 \text{ lb} \qquad \text{Ans.}
$$

The angle θ_x is obtained from the expression

$$
\theta_x = \tan^{-1} \frac{R_y}{R_x} = \tan^{-1} \frac{+187.14}{+57.30} = 73.0° \qquad \text{Ans.}
$$

The resultant $\mathbf{R}$ of the four forces shown in Fig. 2-29a is shown in Fig. 2-29b.

Determine the magnitude R of the resultant of the three forces shown in Fig. 2-30 and the angles θ_x, θ_y, and θ_z between the line of action of the resultant and the positive x-, y-, and z-coordinate axes.

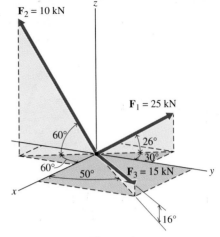

Fig. 2-30

SOLUTION

The magnitude R of the resultant will first be determined by using the rectangular components F_x, F_y, and F_z of each of the forces. Thus

$$F_{1x} = 25 \cos 26° \cos 120° = -11.235 \text{ kN}$$
$$F_{1y} = 25 \cos 26° \sin 120° = +19.459 \text{ kN}$$
$$F_{1z} = 25 \sin 26° = +10.959 \text{ kN}$$

$$F_{2x} = 10 \cos 60° \cos (-60°) = +2.500 \text{ kN}$$
$$F_{2y} = 10 \cos 60° \sin (-60°) = -4.330 \text{ kN}$$
$$F_{2z} = 10 \sin 60° = +8.660 \text{ kN}$$

$$F_{3x} = 15 \cos 16° \cos 50° = +9.268 \text{ kN}$$
$$F_{3y} = 15 \cos 16° \sin 50° = +11.046 \text{ kN}$$
$$F_{3z} = 15 \sin 16° = +4.135 \text{ kN}$$

Once the rectangular components of the forces are known, the components R_x, R_y, and R_z of the resultant are obtained from the expressions

$$R_x = \Sigma F_x = F_{1x} + F_{2x} + F_{3x} = -11.235 + 2.500 + 9.268 = +0.533 \text{ kN}$$
$$R_y = \Sigma F_y = F_{1y} + F_{2y} + F_{3y} = +19.459 - 4.330 + 11.046 = +26.175 \text{ kN}$$
$$R_z = \Sigma F_z = F_{1z} + F_{2z} + F_{3z} = +10.959 + 8.660 + 4.135 = +23.754 \text{ kN}$$

The magnitude R of the resultant is

$$R = \sqrt{R_x^2 + R_y^2 + R_z^2} = \sqrt{(0.533)^2 + (26.175)^2 + (23.754)^2} = 35.4 \text{ kN} \quad \text{Ans.}$$

The angles θ_x, θ_y, and θ_z are obtained from the expressions

$$\theta_x = \cos^{-1} \frac{R_x}{R} = \cos^{-1} \frac{+0.533}{35.35} = 89.1° \qquad \text{Ans.}$$

$$\theta_y = \cos^{-1} \frac{R_y}{R} = \cos^{-1} \frac{+26.175}{35.35} = 42.2°$$

$$\theta_z = \cos^{-1} \frac{R_z}{R} = \cos^{-1} \frac{+23.754}{35.35} = 47.8°$$

Alternatively, the solution can be obtained using vector methods. Unit vectors $\mathbf{e}_1$, $\mathbf{e}_2$, and $\mathbf{e}_3$, along the lines of action of forces $\mathbf{F}_1$, $\mathbf{F}_2$, and $\mathbf{F}_3$, respectively, are

$$\mathbf{e}_1 = (\cos 26° \cos 120°)\mathbf{i} + (\cos 26° \cos 30°)\mathbf{j} + (\cos 64°)\mathbf{k} = -0.4494\mathbf{i} + 0.7784\mathbf{j} + 0.4384\mathbf{k}$$
$$\mathbf{e}_2 = (\cos 60° \cos 60°)\mathbf{i} + (\cos 60° \cos 150°)\mathbf{j} + (\cos 30°)\mathbf{k} = 0.2500\mathbf{i} - 0.4330\mathbf{j} + 0.8660\mathbf{k}$$
$$\mathbf{e}_3 = (\cos 16° \cos 50°)\mathbf{i} + (\cos 16° \cos 40°)\mathbf{j} + (\cos 74°)\mathbf{k} = 0.6179\mathbf{i} + 0.7364\mathbf{j} + 0.2756\mathbf{k}$$

The forces are then written in Cartesian vector form as

$$\mathbf{F}_1 = F_1\mathbf{e}_1 = 25(-0.4494\mathbf{i} + 0.7784\mathbf{j} + 0.4384\mathbf{k}) = -11.235\mathbf{i} + 19.460\mathbf{j} + 10.960\mathbf{k} \text{ kN}$$
$$\mathbf{F}_2 = F_2\mathbf{e}_2 = 10(0.2500\mathbf{i} - 0.4330\mathbf{j} + 0.8660\mathbf{k}) = 2.500\mathbf{i} - 4.330\mathbf{j} + 8.660\mathbf{k} \text{ kN}$$
$$\mathbf{F}_3 = F_3\mathbf{e}_3 = 15(0.6179\mathbf{i} + 0.7364\mathbf{j} + 0.2756\mathbf{k}) = 9.269\mathbf{i} + 11.046\mathbf{j} + 4.134\mathbf{k} \text{ kN}$$

The resultant $\mathbf{R}$ of the three forces is

$$\mathbf{R} = \mathbf{F}_1 + \mathbf{F}_2 + \mathbf{F}_3 = \Sigma F_x\mathbf{i} + \Sigma F_y\mathbf{j} + \Sigma F_z\mathbf{k} = 0.534\mathbf{i} + 26.176\mathbf{j} + 23.754\mathbf{k} \text{ kN}$$

Once the scalar components of the resultant are determined, the magnitude and the direction angles are determined as in the first part of the example.

Determine the magnitude R of the resultant of the three forces shown in Fig. 2-31 and the angles θ_x, θ_y, and θ_z between the line of action of the resultant and the positive x-, y-, and z-coordinate axes.

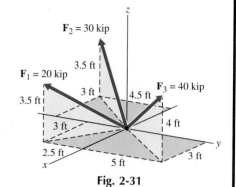

Fig. 2-31

SOLUTION

In addition to the origin of coordinates, the lines of action of forces $\mathbf{F}_1$, $\mathbf{F}_2$, and $\mathbf{F}_3$ pass through points $(3, -2.5, 3.5)$, $(-3, -4.5, 3.5)$, and $(3, 5, 4)$, respectively. Since the coordinates of these points in space are known, it is easy to determine the unit vectors associated with each of the forces:

$$\mathbf{e}_1 = \frac{3\mathbf{i} - 2.5\mathbf{j} + 3.5\mathbf{k}}{\sqrt{(3)^2 + (-2.5)^2 + (3.5)^2}} = +0.5721\mathbf{i} - 0.4767\mathbf{j} + 0.6674\mathbf{k}$$

$$\mathbf{e}_2 = \frac{-3\mathbf{i} - 4.5\mathbf{j} + 3.5\mathbf{k}}{\sqrt{(-3)^2 + (-4.5)^2 + (3.5)^2}} = -0.4657\mathbf{i} - 0.6985\mathbf{j} + 0.5433\mathbf{k}$$

$$\mathbf{e}_3 = \frac{3\mathbf{i} + 5\mathbf{j} + 4\mathbf{k}}{\sqrt{(3)^2 + (5)^2 + (4)^2}} = +0.4243\mathbf{i} + 0.7071\mathbf{j} + 0.5657\mathbf{k}$$

Once the unit vectors $\mathbf{e}_1$, $\mathbf{e}_2$, and $\mathbf{e}_3$ are known, the three forces can be expressed in Cartesian vector form as

$$\begin{aligned}
\mathbf{F}_1 = F_1\mathbf{e}_1 &= 20(+0.5721\mathbf{i} - 0.4767\mathbf{j} + 0.6674\mathbf{k}) \\
&= +11.442\mathbf{i} - 9.534\mathbf{j} + 13.348\mathbf{k} \text{ kip} \\
\mathbf{F}_2 = F_2\mathbf{e}_2 &= 30(-0.4657\mathbf{i} - 0.6985\mathbf{j} + 0.5433\mathbf{k}) \\
&= -13.971\mathbf{i} - 20.955\mathbf{j} + 16.299\mathbf{k} \text{ kip} \\
\mathbf{F}_3 = F_3\mathbf{e}_3 &= 40(+0.4243\mathbf{i} + 0.7071\mathbf{j} + 0.5657\mathbf{k}) \\
&= +16.972\mathbf{i} + 28.284\mathbf{j} + 22.628\mathbf{k} \text{ kip}
\end{aligned}$$

The resultant $\mathbf{R}$ of the three forces is

$$\mathbf{R} = \mathbf{F}_1 + \mathbf{F}_2 + \mathbf{F}_3 = R_x\mathbf{i} + R_y\mathbf{j} + R_z\mathbf{k} \text{ kip}$$

where

$$\begin{aligned}
R_x = \Sigma F_x = F_{1x} + F_{2x} + F_{3x} \\
= +11.442 - 13.971 + 16.972 = +14.443 \text{ kip} \\
R_y = \Sigma F_y = F_{1y} + F_{2y} + F_{3y} \\
= -9.534 - 20.955 + 28.284 = -2.205 \text{ kip} \\
R_z = \Sigma F_z = F_{1z} + F_{2z} + F_{3z} \\
= +13.348 + 16.299 + 22.628 = +52.28 \text{ kip}
\end{aligned}$$

Thus

$$\mathbf{R} = +14.443\mathbf{i} - 2.205\mathbf{j} + 52.28\mathbf{k} \text{ kip}$$

The magnitude R of the resultant is

$$\begin{aligned}
R &= \sqrt{R_x^2 + R_y^2 + R_z^2} \\
&= \sqrt{(+14.443)^2 + (-2.205)^2 + (+52.28)^2} = 54.28 = 54.3 \text{ kip} \qquad \text{Ans.}
\end{aligned}$$

The angles θ_x, θ_y, and θ_z are obtained from the expressions

$$\theta_x = \cos^{-1}\frac{R_x}{R} = \cos^{-1}\frac{+14.443}{54.28} = 74.6° \qquad \text{Ans.}$$

$$\theta_y = \cos^{-1}\frac{R_y}{R} = \cos^{-1}\frac{-2.205}{54.28} = 92.3° \qquad \text{Ans.}$$

$$\theta_z = \cos^{-1}\frac{R_z}{R} = \cos^{-1}\frac{+52.28}{54.28} = 15.60°$$

PROBLEMS

Use the rectangular component method to solve the following problems. Determine the magnitude R of the resultant and the angle θ_x between the line of action of the resultant and the x-axis for

2-47*   The three forces shown in Fig. P2-47.

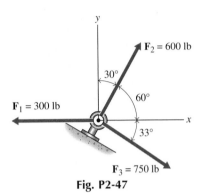

Fig. P2-47

2-48*   The three forces shown in Fig. P2-48.

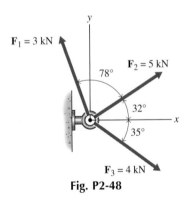

Fig. P2-48

2-49   The three forces shown in Fig. P2-49.

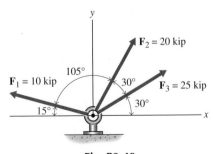

Fig. P2-49

2-50   The three forces shown in Fig. P2-50.

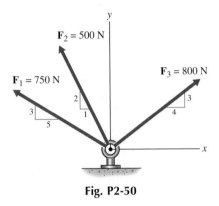

Fig. P2-50

2-51*   The three forces shown in Fig. P2-51.

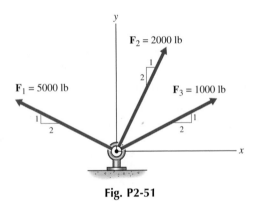

Fig. P2-51

2-52*   The four forces shown in Fig. P2-52.

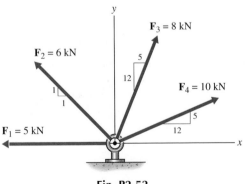

Fig. P2-52

2-53 The four forces shown in Fig. P2-53.

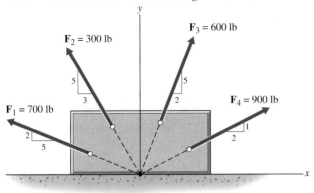

Fig. P2-53

2-54 The five forces shown in Fig. P2-54.

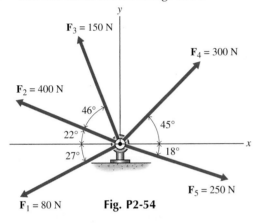

$F_1 = 80$ N **Fig. P2-54**

Use the rectangular component method to solve the following problems. Determine the magnitude R of the resultant and the angles θ_x, θ_y, and θ_z between the line of action of the resultant and the positive x-, y-, and z-coordinate axes for

2-55* The three forces shown in Fig. P2-55.

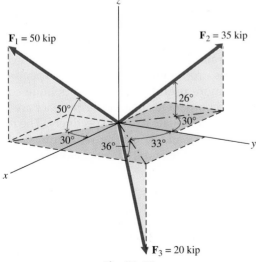

Fig. P2-55

2-56 The three forces shown in Fig. P2-56.

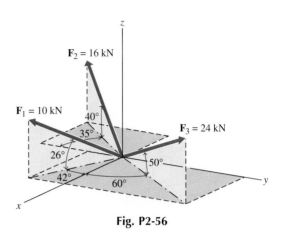

Fig. P2-56

2-57* The three forces shown in Fig. P2-57.

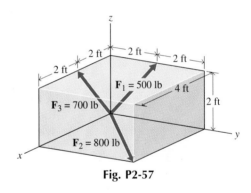

Fig. P2-57

2-58* The three forces shown in Fig. P2-58.

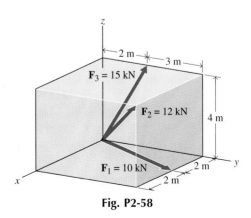

Fig. P2-58

55

Use the rectangular component method to solve the following problems. Determine the magnitude R of the resultant and the angles θ_x, θ_y, and θ_z between the line of action of the resultant and the positive x-, y-, and z-coordinate axes for

2-59 The three forces shown in Fig. P2-59.

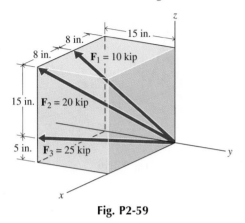

Fig. P2-59

2-60 The three forces shown in Fig. P2-60.

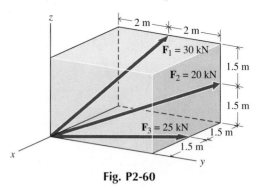

Fig. P2-60

SUMMARY

A force is defined as the action of one physical body on another. Since the interaction can occur when there is contact between the bodies or when the bodies are physically separated, forces are classified as either surface forces (a push or a pull produced by mechanical means) or body forces (the gravitational pull of the earth). The characteristics of a force are its magnitude, its direction (slope and sense), and its point of application. Since both a magnitude and a direction are needed to characterize a force and forces add according to the parallelogram law of addition, forces are vector quantities. A number of forces treated as a group constitute a force system.

The components of a force are any system of forces that can be combined by the parallelogram law to produce the original force. For most engineering problems, rectangular (mutually perpendicular) components of a force are more useful than general oblique components. The three rectangular components of a force $\mathbf{F}$ in space are $\mathbf{F}_x$, $\mathbf{F}_y$, and $\mathbf{F}_z$ along the x-, y-, and z-coordinate axes, respectively. The force $\mathbf{F}$ and its scalar components F_x, F_y, and F_z can be written in Cartesian vector form by using unit vectors $\mathbf{i}$, $\mathbf{j}$, and $\mathbf{k}$ directed along the positive x-, y-, and z-coordinate axes as

$$\mathbf{F} = \mathbf{F}_x + \mathbf{F}_y + \mathbf{F}_z = F_x\mathbf{i} + F_y\mathbf{j} + F_z\mathbf{k}$$
$$= F\cos\theta_x\mathbf{i} + F\cos\theta_y\mathbf{j} + F\cos\theta_z\mathbf{k} \qquad (2\text{-}3)$$

where the scalar components F_x, F_y, and F_z are related to the magnitude F and the direction of the force $\mathbf{F}$ by the expressions

$$F_x = F \cos \theta_x \qquad F_y = F \cos \theta_y \qquad F_z = F \cos \theta_z$$

$$F = \sqrt{F_x^2 + F_y^2 + F_z^2} \qquad (2\text{-}4)$$

$$\theta_x = \cos^{-1}\frac{F_x}{F} \qquad \theta_y = \cos^{-1}\frac{F_y}{F} \qquad \theta_z = \cos^{-1}\frac{F_z}{F}$$

$$\cos^2 \theta_x + \cos^2 \theta_y + \cos^2 \theta_z = 1$$

The rectangular component of a force $\mathbf{F}$ along an arbitrary direction n can be obtained by using the vector dot-product. Thus, if $\mathbf{e}_n$ is a unit vector in the specified direction n, the magnitude of the rectangular component $\mathbf{F}_n$ of the force $\mathbf{F}$ is

$$F_n = \mathbf{F} \cdot \mathbf{e}_n = (F_x\mathbf{i} + F_y\mathbf{j} + F_z\mathbf{k}) \cdot \mathbf{e}_n$$

If the angles between the direction n and the x-, y-, and z-axes are θ_x', θ_y', and θ_z', the unit vector $\mathbf{e}_n$ can be written in Cartesian vector form as

$$\mathbf{e}_n = \cos \theta_x'\mathbf{i} + \cos \theta_y'\mathbf{j} + \cos \theta_z'\mathbf{k}$$

Thus, the magnitude of the force $\mathbf{F}_n$ is

$$F_n = \mathbf{F} \cdot \mathbf{e}_n = F_x \cos \theta_x' + F_y \cos \theta_y' + F_z \cos \theta_z' \qquad (2\text{-}5)$$

The rectangular component $\mathbf{F}_n$ of the force $\mathbf{F}$ is expressed in Cartesian vector form as

$$\mathbf{F}_n = (\mathbf{F} \cdot \mathbf{e}_n)\mathbf{e}_n = F_n\mathbf{e}_n = F_n(\cos \theta_x'\mathbf{i} + \cos \theta_y'\mathbf{j} + \cos \theta_z'\mathbf{k}) \qquad (2\text{-}7)$$

The angle α between the line of action of the force $\mathbf{F}$ and the direction n is determined by using the definition of a rectangular component of a force ($F_n = F \cos \alpha = \mathbf{F} \cdot \mathbf{e}_n$). Thus,

$$\alpha = \cos^{-1}\frac{\mathbf{F} \cdot \mathbf{e}_n}{F} = \cos^{-1}\frac{F_n}{F} \qquad (2\text{-}8)$$

A single force, called the resultant $\mathbf{R}$, will produce the same effect on a body as a system of concurrent forces. The resultant can be determined by adding the forces using the parallelogram law; however, this procedure is time-consuming and tedious when more than two forces comprise the system. Resultants are easily obtained, however, by using rectangular components of the forces. For the general case of two or more concurrent forces in space,

$$\mathbf{R}_x = \Sigma\mathbf{F}_x = R_x\mathbf{i} \qquad \mathbf{R}_y = \Sigma\mathbf{F}_y = R_y\mathbf{j} \qquad \mathbf{R}_z = \Sigma\mathbf{F}_z = R_z\mathbf{k}$$

The magnitude R of the resultant and the angles θ_x, θ_y, and θ_z between the line of action of the resultant and the positive coordinate axes are

$$R = \sqrt{R_x^2 + R_y^2 + R_z^2}$$

$$\theta_x = \cos^{-1}\frac{R_x}{R} \qquad \theta_y = \cos^{-1}\frac{R_y}{R} \qquad \theta_z = \cos^{-1}\frac{R_z}{R}$$

REVIEW PROBLEMS

2-61* Three forces are applied to a stalled automobile as shown in Fig. P2-61. Determine the magnitude of the force F_3 and the magnitude of the resultant R if the line of action of the resultant is along the x-axis.

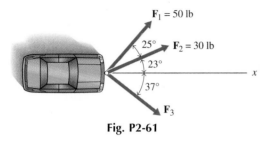

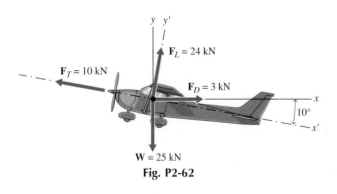

Fig. P2-61

2-63 A plate is supported by rods A and B as shown in Fig. P2-63. A 600-lb force F_1 and an 800-lb force F_2 are applied to the plate through a smooth pin.

a. Determine the magnitudes of forces F_A and F_B if the resultant R of the four forces acting on the plate is zero.

b. If rod A breaks, determine the magnitude of force F_B and the angle of inclination of rod B with respect to the x-axis if the resultant R of the three remaining forces equals zero.

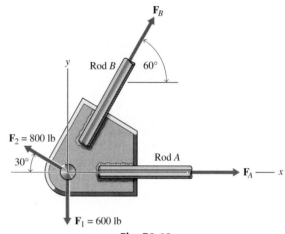

Fig. P2-63

2-62* Four forces act on a small airplane in flight, as shown in Fig. P2-62: its weight, the thrust provided by the engine, the lift provided by the wings, and the drag resulting from its motion through the air. Determine the resultant of the four forces and its line of action with respect to the axis of the plane.

2-64 Three cables are used to drag a heavy crate on a horizontal surface as shown in Fig. P2-64. The resultant R of the forces has a magnitude of 2800 N and its line of action is directed along the x-axis. Determine the magnitudes of forces F_1 and F_3.

Fig. P2-62

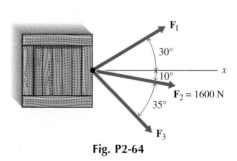

Fig. P2-64

2-65* A 500-lb force is applied to the post shown in Fig. P2-65. Determine

a. The magnitudes of the x and y components of the force.

b. The magnitudes of the u and v components of the force.

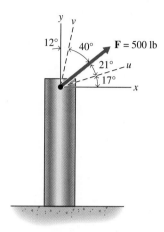

Fig. P2-65

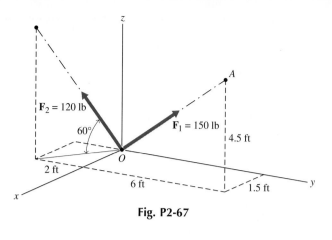

Fig. P2-67

2-66* Two forces F_1 and F_2 are applied to an eyebolt as shown in Fig. P2-66. Determine

a. The magnitude and direction (angle θ_x) of the resultant R of the two forces.

b. The magnitudes of two other forces F_u and F_v that would have the same resultant.

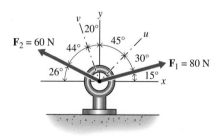

Fig. P2-66

2-67 Two forces are applied at a point in a body as shown in Fig. P2-67. Determine

a. The magnitude and direction (angles θ_x, θ_y, and θ_z) of the resultant R of the two forces.

b. The magnitude of the rectangular component of force F_1 along the line of action of force F_2.

c. The angle α between forces F_1 and F_2.

2-68 Three forces are applied with cables to the anchor block shown in Fig. P2-68. Determine

a. The magnitude and direction (angles θ_x, θ_y, and θ_z) of the resultant R of the three forces.

b. The magnitude of the rectangular component of force F_1 along the line of action of force F_2.

c. The angle α between forces F_1 and F_3.

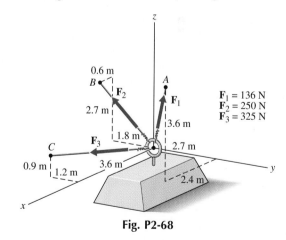

Fig. P2-68

3

STATICS OF PARTICLES

Particle equilibrium principles are sufficient to determine the forces in the cables supporting the weights of the traffic lights.

3-1   INTRODUCTION

Statics was defined in Chapter 1 as the branch of rigid-body mechanics concerned with bodies that are acted on by a balanced system of forces (the resultant of all forces acting on the body is zero) and hence are at rest or moving with a constant velocity in a straight line.

A body with negligible dimensions is commonly referred to as a particle. In mechanics, either large bodies or small bodies can be referred to as a particle when the size and shape of the body have no effect on the response of the body to a system of forces. Under these conditions, the mass of the body can be assumed to be concentrated at a point. For example, the Earth can be modeled as a particle for orbital motion studies since the size of the Earth is insignificant when compared to the size of its orbit and the shape of the Earth does not influence the description of its position or the action of forces applied to it.

Since it is assumed that the mass of a particle is concentrated at a point and that the size and shape of a particle can be neglected, a particle can be subjected only to a system of concurrent forces. Newton's first law of motion states that "in the absence of external forces ($\mathbf{R} = \mathbf{0}$), a particle originally at rest or moving with a constant velocity (in equilibrium) will remain at rest or continue to move with a constant velocity along a straight line. Thus, a necessary condition for equilibrium of a particle is

$$\mathbf{R} = \Sigma\mathbf{F} = \mathbf{0} \tag{3-1}$$

A particle in equilibrium must also satisfy Newton's second law of motion, which can be expressed in equation form (Eq. 1-1) as

$$\mathbf{R} = \Sigma\mathbf{F} = m\mathbf{a} \tag{1-1}$$

In order to satisfy both Eqs. 1-1 and 3-1

$$m\mathbf{a} = \mathbf{0}$$

Since the mass of the particle is not zero, the acceleration of a particle in equilibrium is zero ($\mathbf{a} = \mathbf{0}$). Thus, a particle initially at rest will remain at rest, and a particle moving with a constant velocity will maintain that velocity. Therefore, Eq. 3-1 is both a necessary condition and a sufficient condition for equilibrium.

The particle assumption is valid for many practical applications and thus provides a means for introducing the student to some interesting engineering problems early in a statics course. For this reason, this short chapter on statics (equilibrium) of a particle has been introduced before the more difficult problems associated with equilibrium of a rigid body (which involves the concepts of moments and distributed loads) are considered.

The force system acting on a body in a typical statics problem consists of known forces and unknown forces. Both must be clearly identified before a solution to a specific problem is attempted. A method commonly used to identify all forces acting on a body in a given situation is described in the following section.

3-2   FREE-BODY DIAGRAMS

A carefully prepared drawing or sketch that shows a "body of interest" separated from all interacting bodies is known as a free-body diagram (FBD). Once the body of interest is selected, the forces exerted by all other bodies on the one being considered must be determined and shown on the diagram. It is important that "all" forces acting "on" the body of interest be shown. Recall also that "a force cannot exist unless there is a body to exert the force." Frequently, the student will overlook and omit a force from the free-body diagram or show a force on the free-body diagram when there is no body present to exert the force.

The number of forces on a free-body diagram is determined by noting the number of bodies that exert forces on the body of interest. These forces may be either forces of contact or body forces. An important body force is the Earth-pull on (or weight of) a body.

Each known force should be shown on a free-body diagram with its correct magnitude, slope, and sense. Letter symbols are used for the magnitudes of unknown forces. If a force has a known line of action but an unknown magnitude and sense, the sense of the force can be assumed. The correct sense will become apparent after solving for the unknown magnitude. By definition, the magnitude of a force is always positive; therefore, if the solution yields a negative magnitude, the minus sign indicates that the sense of the force is opposite to that assumed on the free-body diagram.

If both magnitude and direction of a force acting on the body of interest is unknown (such as a pin reaction in a pin-connected structure), it is frequently convenient to show the two rectangular components of the force on the free-body diagram instead of the actual force. In this way, one deals with two forces of unknown magnitude but known direction. After the two rectangular components of the force are determined, the magnitude and direction of the actual force can easily be found. However, do not show *both* the force of unknown magnitude and its rectangular components on the same diagram.

The word *free* in the name "free-body diagram" emphasizes the idea that all bodies exerting forces on the body of interest are removed or withdrawn and are replaced by the forces they exert. Do not show both the bodies removed and the forces exerted by them on the free-body diagram. Sometimes it may be convenient to indicate, by light-weight dotted lines, the faint outlines of the bodies removed, in order to visualize the geometry and specify dimensions required for solution of the problem.

In drawing a free-body diagram of a given body, certain assumptions are made regarding the nature of the forces (reactions) exerted by other bodies on the body of interest. Two common assumptions are the following:

1. If a surface of contact at which a force is applied by one body to another body has only a small degree of roughness, it may be assumed to be smooth (frictionless), and hence the action (or reaction) of the one body on the other is directed normal to the surface of contact.
2. A body that possesses only a small degree of bending stiffness (resistance to bending), such as a cord, rope, chain, or the like, may be considered to be perfectly flexible, and hence the pull of such a body on any other body is directed along the axis of the flexible body.

A large number of additional assumptions will be discussed in

Section 6-2-2 in the chapter on equilibrium of rigid bodies, where the topic of idealization of supports and connections is considered.

The term "body of interest" used in the definition of a free-body diagram may mean any definite part of a structure or machine such as an eye-bar in a bridge truss or a connecting rod in an automobile engine. The body of interest may also be taken as a group of physical bodies joined together (considered as one body), such as an entire bridge or a complete engine.

The importance of drawing a free-body diagram before attempting to solve a mechanics problem cannot be overemphasized. A procedure that can be followed to construct a complete and correct free-body diagram, contains the following four steps.

Constructing a Free-body Diagram

Step 1.   Decide which body or combination of bodies is to be shown on the free-body diagram.
Step 2.   Prepare a drawing or sketch of the outline of this isolated or free body.
Step 3.   Carefully trace around the boundary of the free body and identify all the forces exerted by contacting or attracting bodies that were removed during the isolation process.
Step 4.   Choose the set of coordinate axes to be used in solving the problem and indicate these directions on the free-body diagram. Place any dimensions required for solution of the problem on the diagram.

Application of these four steps to any statics or dynamics problem should produce a complete and correct free-body diagram, which is an essential first step for the solution of any problem.

The free-body diagram is the "road map" for writing the equations of equilibrium. Every equation of equilibrium must be supported by a properly drawn, complete, free-body diagram. The symbols used in the equations of equilibrium must match the symbols used on the free-body diagram. For example, use $A \cos 30°$ rather than A_x if A is used to represent a force of known direction (30° with respect to the x-axis) on the free-body diagram.

3-3   EQUILIBRIUM OF A PARTICLE

In Section 3-1 it was noted that the term "particle" is used in statics to describe a body when the size and shape of the body will not significantly affect the solution of the problem being considered and when the mass of the body can be assumed to be concentrated at a point. As a result, we observed that a particle can be subjected only to a system of concurrent forces and that the necessary and sufficient conditions for equilibrium can be expressed mathematically as

$$\mathbf{R} = \Sigma\mathbf{F} = \mathbf{0} \tag{3-1}$$

where $\Sigma\mathbf{F}$ is the vector sum of all forces acting on the particle.

3-3-1   Two-Dimensional Problems

For a system of coplanar (say the xy plane), concurrent forces, Eq. 3-1 can be written as

$$\begin{aligned}
\mathbf{R} = \mathbf{R}_x + \mathbf{R}_y &= \mathbf{R}_n + \mathbf{R}_t = \mathbf{0} \\
&= R_x\mathbf{i} + R_y\mathbf{j} = R_n\mathbf{e}_n + R_t\mathbf{e}_t = \mathbf{0} \\
&= \Sigma F_x\mathbf{i} + \Sigma F_y\mathbf{j} = \Sigma F_n\mathbf{e}_n + \Sigma F_t\mathbf{e}_t = \mathbf{0}
\end{aligned} \tag{3-2}$$

Equation 3-2 is satisfied only if

$$\begin{aligned}
\mathbf{R}_x &= R_x\mathbf{i} = \Sigma F_x\mathbf{i} = \mathbf{0} \\
\mathbf{R}_y &= R_y\mathbf{j} = \Sigma F_y\mathbf{j} = \mathbf{0} \\
\mathbf{R}_n &= R_n\mathbf{e}_n = \Sigma F_n\mathbf{e}_n = \mathbf{0} \\
\mathbf{R}_t &= R_t\mathbf{e}_t = \Sigma F_t\mathbf{e}_t = \mathbf{0}
\end{aligned}$$

In scalar form, these equations become

$$\begin{aligned}
R_x &= \Sigma F_x = 0 \\
R_y &= \Sigma F_y = 0 \\
R_n &= \Sigma F_n = 0 \\
R_t &= \Sigma F_t = 0
\end{aligned} \tag{3-3}$$

That is, the sum of the rectangular components in *any* direction must be zero. Though this would appear to give an infinite number of equations, no more than two of the equations are independent. The remaining equations could be obtained from combinations of the two independent equations. However, it is sometimes convenient to use $\Sigma F_x = 0$ and $\Sigma F_n = 0$ as the two independent equations rather than $\Sigma F_x = 0$ and $\Sigma F_y = 0$ (see Example 3-1).

Equations 3-3 can be used to determine two unknown quantities (two magnitudes, two slopes, or a magnitude and a slope). The procedure is illustrated in the following examples.

A free-body diagram of a particle subjected to the action of four forces is shown in Fig. 3-1a. Determine the magnitudes of forces $\mathbf{F}_1$ and $\mathbf{F}_2$ so that the particle is in equilibrium.

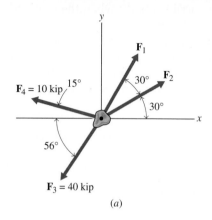

(a)

SOLUTION

The particle is subjected to a system of coplanar, concurrent forces. The necessary and sufficient conditions for equilibrium are given by Eqs. 3-3 as $\Sigma F_x = 0$ and $\Sigma F_y = 0$. Applying these equations, by using the free-body diagram shown in Fig. 3-1a, yields

$$+ \rightarrow \Sigma F_x = F_{1x} + F_{2x} + F_{3x} + F_{4x} = 0$$
$$= F_1 \cos 60° + F_2 \cos 30° - 40 \cos 56° - 10 \cos 15° = 0$$
$$= 0.5000F_1 + 0.8660F_2 - 22.37 - 9.659 = 0$$

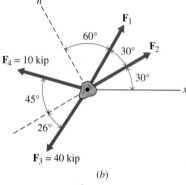

(b)

Fig. 3-1

from which

$$F_1 + 1.732F_2 = 64.06 \qquad (a)$$

$$+\uparrow\Sigma F_y = F_{1y} + F_{2y} + F_{3y} + F_{4y} = 0$$
$$= F_1 \sin 60° + F_2 \sin 30° - 40 \sin 56° + 10 \sin 15° = 0$$
$$= 0.8660F_1 + 0.5000F_2 - 33.16 + 2.588 = 0$$

from which

$$F_1 + 0.5774F_2 = 35.30 \qquad (b)$$

Solving Eqs. a and b simultaneously yields

$$F_1 = 20.9 \text{ kip} \qquad \text{Ans.}$$
$$F_2 = 24.9 \text{ kip} \qquad \text{Ans.}$$

Alternatively, summing forces in a direction n perpendicular to the line of action of force $\mathbf{F}_2$, as shown in Fig. 3-1b, yields

$$+\nwarrow\Sigma F_n = F_{1n} + F_{3n} + F_{4n} = 0$$
$$= F_1 \sin 30° - 40 \sin 26° + 10 \sin 45° = 0$$

from which

$$F_1 = 20.93 = 20.9 \text{ kip} \qquad \text{Ans.}$$

Once $\mathbf{F}_1$ is known, forces can be summed in any other direction to obtain $\mathbf{F}_2$. Thus, summing forces in the x-direction yields

$$+ \rightarrow \Sigma F_x = F_{1x} + F_{2x} + F_{3x} + F_{4x} = 0$$
$$= 20.93 \cos 60° + F_2 \cos 30° - 40 \cos 56° - 10 \cos 15° = 0$$

from which

$$F_2 = 24.90 = 24.9 \text{ kip} \qquad \text{Ans.}$$

Summing forces in a direction perpendicular to one of the unknown forces eliminates the need to solve simultaneous equations in two-dimensional problems.

A free-body diagram of a particle subjected to the action of four forces is shown in Fig. 3-2a. Determine the magnitude and direction (angle θ) of force $\mathbf{F}_4$ so that the particle is in equilibrium.

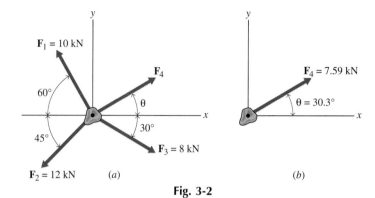

Fig. 3-2

SOLUTION

The particle is subjected to a system of coplanar, concurrent forces. The necessary and sufficient conditions for equilibrium are given by Eqs. 3-3 as $\Sigma F_x = 0$ and $\Sigma F_y = 0$. Applying these equations, by using the free-body diagram shown in Fig. 3-2a, yields

$$+ \rightarrow \Sigma F_x = F_{1x} + F_{2x} + F_{3x} + F_{4x} = 0$$
$$= -10 \cos 60° - 12 \cos 45° + 8 \cos 30° + F_4 \cos \theta = 0$$

from which

$$F_{4x} = F_4 \cos \theta = 6.557 \text{ kN} \qquad (a)$$

$$+\uparrow\Sigma F_y = F_{1y} + F_{2y} + F_{3y} + F_{4y} = 0$$
$$= 10 \sin 60° - 12 \sin 45° - 8 \sin 30° + F_4 \sin \theta = 0$$

from which

$$F_{4y} = F_4 \sin \theta = 3.825 \text{ kN} \qquad (b)$$

Once the components F_{4x} and F_{4y} are known, Eqs. 2-1 are used to determine F_4 and θ:

$$F_4 = \sqrt{F_{4x}^2 + F_{4y}^2} = \sqrt{(6.557)^2 + (3.825)^2} = 7.59 \text{ kN} \qquad \text{Ans.}$$
$$\theta = \tan^{-1}\frac{F_{4y}}{F_{4x}} = \tan^{-1}\frac{3.825}{6.557} = 30.3° \qquad \text{Ans.}$$

The results are shown in Fig. 3-2b.

A homogeneous steel cylinder weighing 500 lb is supported by a flexible cable and a smooth inclined plane as shown in Fig. 3-3a. Determine the tension T in the cable and the force R exerted by the inclined plane on the cylinder.

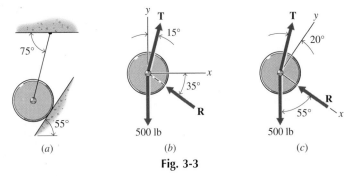

(a)   (b)   (c)

Fig. 3-3

SOLUTION

The tension T in the cable and the force R exerted by the inclined plane on the cylinder can be determined by considering the free-body diagram of the cylinder shown in Fig. 3-3b. From this free-body diagram it is obvious that the cylinder is subjected to a system of coplanar, concurrent forces. Thus, the necessary and sufficient conditions for equilibrium are given by Eqs. 3-3 as $\Sigma F_x = 0$ and $\Sigma F_y = 0$. Applying these equations yields

$$+ \rightarrow \Sigma F_x = T \sin 15° - R \cos 35° = 0.2588T - 0.8192R = 0$$

from which

$$T = 3.165R \qquad\qquad (a)$$

$$+\uparrow\Sigma F_y = T \cos 15° + R \sin 35° - 500 = 0.9659T + 0.5736R - 500 = 0$$

from which

$$R = 871.7 - 1.684T \qquad\qquad (b)$$

Solving Eqs. a and b simultaneously yields

$$T = 436 \text{ lb} \qquad \angle 75° \qquad\qquad \text{Ans.}$$
$$R = 137.7 \text{ lb} \qquad 35° \qquad\qquad \text{Ans.}$$

The solution to a problem can frequently be simplified by proper selection of the coordinate system. For example, if the x- and y-axes are chosen parallel and perpendicular to the inclined surface as shown on the free-body diagram of Fig. 3-3c, the need to solve simultaneous equations is eliminated. Thus,

$$+ \nearrow \Sigma F_y = T \cos 20° - 500 \sin 55° = 0.9397T - 409.6 = 0$$

from which

$$T = 436 \text{ lb} \qquad \angle 75° \qquad\qquad \text{Ans.}$$

Once the force T is known, $\Sigma F_x = 0$ yields R directly. Thus,

$$+ \searrow \Sigma F_x = -R - T \sin 20° + 500 \cos 55° = 0$$

from which

$$R = 286.8 - 0.3420T$$
$$= 286.8 - 0.3420(436) = 137.7 \text{ lb} \qquad 35° \qquad\qquad \text{Ans.}$$

The cable system shown in Fig. 3-4a is being used to lift body A. The cable system is in equilibrium at the cable positions shown in the figure when a 500-N force is applied at joint 1. Determine the tensions in cables A, B, C, and D and the mass of body A that is being lifted.

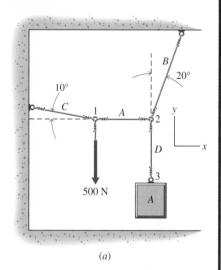

(a)

SOLUTION

The tensions in the cables can be determined by using the free-body diagrams of joints 1 and 2 as shown in Figs. 3-4b and 3-4c, respectively. Each joint is subjected to a system of coplanar, concurrent forces. Thus, the necessary and sufficient conditions for equilibrium are given by Eqs. 3-3 as $\Sigma F_x = 0$ and $\Sigma F_y = 0$.

From the free-body diagram for joint 1 (see Fig. 3-4b),

$$+\uparrow \Sigma F_y = T_C \sin 10° - 500 = 0$$

from which

$$T_C = \frac{500}{\sin 10°} = 2879 \text{ N}$$

$$+ \rightarrow \Sigma F_x = T_A - T_C \cos 10° = 0$$

from which

$$T_A = T_C \cos 10° = 2879 \cos 10° = 2835 \text{ N}$$

From the free-body diagram for joint 2 (see Fig. 3-4c),

$$+ \rightarrow \Sigma F_x = T_B \sin 20° - T_A = 0$$

from which

$$T_B = \frac{T_A}{\sin 20°} = \frac{2835}{\sin 20°} = 8289 \text{ N}$$

$$+\uparrow \Sigma F_y = T_B \cos 20° - T_D = 0$$

from which

$$T_D = T_B \cos 20° = 8289 \cos 20° = 7789 \text{ N}$$

The cable tensions rounded to three significant figures are

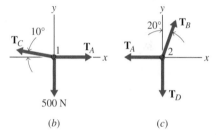

(b)   (c)

$$T_A = 2840 \text{ N} \qquad \text{Ans.}$$
$$T_B = 8290 \text{ N} \qquad \text{Ans.}$$
$$T_C = 2880 \text{ N} \qquad \text{Ans.}$$
$$T_D = 7790 \text{ N} \qquad \text{Ans.}$$

Finally, from the free-body diagram for body A (see Fig. 3-4d),

$$+\uparrow \Sigma F_y = T_D - W_A = 0$$

from which

$$W_A = T_D = 7789 \text{ N}$$

Thus,

$$m_A = \frac{W_A}{g} = \frac{7789}{9.81} = 794 \text{ kg} \qquad \text{Ans.}$$

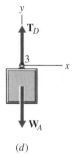

(d)

Fig. 3-4

3-3-2   Three-Dimensional Problems

For a three-dimensional system of concurrent forces, Eq. 3-1 can be written as

$$\begin{aligned}
\mathbf{R} = \Sigma\mathbf{F} &= \mathbf{0}\\
&= \mathbf{R}_x + \mathbf{R}_y + \mathbf{R}_z = \mathbf{0}\\
&= R_x\mathbf{i} + R_y\mathbf{j} + R_z\mathbf{k} = \mathbf{0}\\
&= \Sigma F_x\mathbf{i} + \Sigma F_y\mathbf{j} + \Sigma F_z\mathbf{k} = \mathbf{0}
\end{aligned} \qquad (3\text{-}4)$$

Equation 3-4 is satisfied only if

$$\begin{aligned}
\mathbf{R}_x &= R_x\mathbf{i} = \Sigma F_x\mathbf{i} = \mathbf{0}\\
\mathbf{R}_y &= R_y\mathbf{j} = \Sigma F_y\mathbf{j} = \mathbf{0}\\
\mathbf{R}_z &= R_z\mathbf{k} = \Sigma F_z\mathbf{k} = \mathbf{0}
\end{aligned} \qquad (3\text{-}5)$$

In scalar form, these equations become

$$\begin{aligned}
R_x &= \Sigma F_x = 0\\
R_y &= \Sigma F_y = 0\\
R_z &= \Sigma F_z = 0
\end{aligned} \qquad (3\text{-}6)$$

Equations 3-5 and 3-6 can be used to determine three unknown quantities (three magnitudes, three slopes, or any combination of three magnitudes and slopes). The procedure is illustrated in the following examples. Example 3-5 illustrates the scalar method of solution for a three-dimensional problem. Example 3-6 illustrates the vector method of solution for a similar problem.

EXAMPLE PROBLEM 3-5

A free-body diagram of a particle subjected to the action of four forces is shown in Fig. 3-5. Determine the magnitude and the coordinate direction angles of the unknown force $\mathbf{F}_4$ so that the particle is in equilibrium.

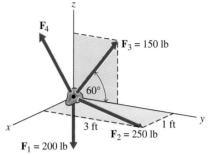

Fig. 3-5

SOLUTION

The necessary and sufficient conditions for equilibrium of a particle subjected to a three-dimensional system of concurrent forces are given by Eqs. 3-6 as

$$R_x = \Sigma F_x = 0 \qquad R_y = \Sigma F_y = 0 \qquad R_z = \Sigma F_z = 0$$

The scalar components for each of the forces shown on the free-body diagram are:

For $\mathbf{F}_1$:
$$F_{1x} = 0$$
$$F_{1y} = 0$$
$$F_{1z} = -200 \text{ lb}$$

For $\mathbf{F}_2$:
$$F_{2x} = (1/\sqrt{10})(250) = 79.06 \text{ lb}$$
$$F_{2y} = (3/\sqrt{10})(250) = 237.2 \text{ lb}$$
$$F_{2z} = 0$$

For $\mathbf{F}_3$:
$$F_{3x} = 0$$
$$F_{3y} = 150 \cos 60° = 75.0 \text{ lb}$$
$$F_{3z} = 150 \sin 60° = 129.9 \text{ lb}$$

Thus

$$R_x = \Sigma F_x = F_{1x} + F_{2x} + F_{3x} + F_{4x} = 0 + 79.06 + 0 + F_{4x} = 0$$
$$F_{4x} = -79.06 \text{ lb}$$
$$R_y = \Sigma F_y = F_{1y} + F_{2y} + F_{3y} + F_{4y} = 0 + 237.2 + 75 + F_{4y} = 0$$
$$F_{4y} = -312.2 \text{ lb}$$
$$R_z = \Sigma F_z = F_{1z} + F_{2z} + F_{3z} + F_{4z} = -200 + 0 + 129.9 + F_{4z} = 0$$
$$F_{4z} = 70.1 \text{ lb}$$

Once the rectangular components of the force $\mathbf{F}_4$ are known, Eqs. 2-2 can be used to determine its magnitude and coordinate direction angles. Thus

$$F_4 = \sqrt{F_{4x}^2 + F_{4y}^2 + F_{4z}^2}$$
$$= \sqrt{(-79.06)^2 + (-312.2)^2 + (70.1)^2}$$
$$= 329.6 = 330 \text{ lb} \qquad \text{Ans.}$$
$$\theta_x = \cos^{-1} \frac{F_{4x}}{F_4} = \cos^{-1} \frac{-79.06}{329.6} = 103.9° \qquad \text{Ans.}$$
$$\theta_y = \cos^{-1} \frac{F_{4y}}{F_4} = \cos^{-1} \frac{-312.2}{329.6} = 161.3° \qquad \text{Ans.}$$
$$\theta_z = \cos^{-1} \frac{F_{4z}}{F_4} = \cos^{-1} \frac{70.1}{329.6} = 77.7° \qquad \text{Ans.}$$

A block is supported by a system of cables as shown in Fig. 3-6a. The weight of the block is 500 N. Determine the tensions in cables A, B, and C.

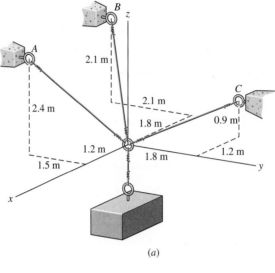

(a)

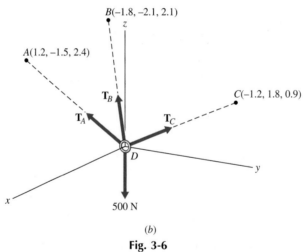

(b)

Fig. 3-6

SOLUTION

A free-body diagram for joint D of the cable system is shown in Fig. 3-6b. This diagram shows that joint D is subjected to a three-dimensional system of concurrent forces with three unknown cable tensions T_A, T_B, and T_C. The coordinates of the support points for each of the cables is shown on the free-body diagram in (x, y, z) format as an aid in writing vector equations for the cable tensions. The necessary and sufficient conditions for equilibrium of the joint are given by Eq. 3-4 as

$$\mathbf{R} = \Sigma \mathbf{F} = \mathbf{T}_A + \mathbf{T}_B + \mathbf{T}_C + \mathbf{W} = 0$$

The cable tensions and the weight of the block can be expressed in Cartesian vector form as

$$\mathbf{T}_A = T_A\left[\frac{1.2\mathbf{i} - 1.5\mathbf{j} + 2.4\mathbf{k}}{\sqrt{9.45}}\right] = T_A(0.3904\mathbf{i} - 0.4880\mathbf{j} + 0.7807\mathbf{k})$$

$$\mathbf{T}_B = T_B\left[\frac{-1.8\mathbf{i} - 2.1\mathbf{j} + 2.1\mathbf{k}}{\sqrt{12.06}}\right] = T_B(-0.5183\mathbf{i} - 0.6047\mathbf{j} + 0.6047\mathbf{k}) \quad (a)$$

$$\mathbf{T}_C = T_C\left[\frac{-1.2\mathbf{i} + 1.8\mathbf{j} + 0.9\mathbf{k}}{\sqrt{5.49}}\right] = T_C(-0.5121\mathbf{i} + 0.7682\mathbf{j} + 0.3841\mathbf{k})$$

$$\mathbf{W} = -500\mathbf{k}$$

Substituting Eqs. a into Eq. 3-4 yields

$$\begin{aligned}\mathbf{R} = &(0.3904T_A - 0.5183T_B - 0.5121T_C)\mathbf{i} \\ &+ (-0.4880T_A - 0.6047T_B + 0.7682T_C)\mathbf{j} \\ &+ (0.7807T_A + 0.6047T_B + 0.3841T_C - 500)\mathbf{k} = 0 \quad (b)\end{aligned}$$

Since each of the components of Eq. b must equal zero if the resultant $\mathbf{R}$ is to be zero, the following equations must be satisfied.

$$0.3904T_A - 0.5183T_B - 0.5121T_C = 0$$
$$-0.4880T_A - 0.6047T_B + 0.7682T_C = 0$$
$$0.7807T_A + 0.6047T_B + 0.3841T_C = 500$$

The simultaneous solution of these three linear equations gives

$$T_A = 459 \text{ N} \qquad \text{Ans.}$$
$$T_B = 32.4 \text{ N} \qquad \text{Ans.}$$
$$T_C = 317 \text{ N} \qquad \text{Ans.}$$

PROBLEMS

3-1* Determine the magnitudes of forces F_2 and F_3 so that the particle shown in Fig. P3-1 is in equilibrium.

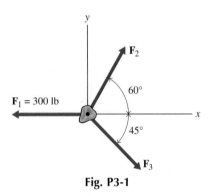

Fig. P3-1

3-2* Determine the magnitudes of forces F_3 and F_4 so that the particle shown in Fig. P3-2 is in equilibrium.

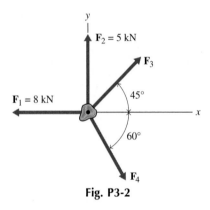

Fig. P3-2

3-3 Determine the magnitudes of forces F_1 and F_2 so that the particle shown in Fig. P3-3 is in equilibrium.

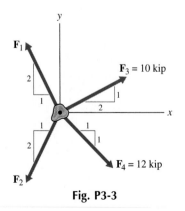

Fig. P3-3

3-4 Determine the magnitudes of forces F_1 and F_2 so that the particle shown in Fig. P3-4 is in equilibrium.

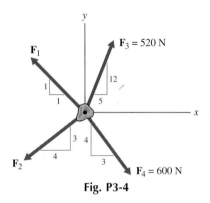

Fig. P3-4

3-5* Determine the magnitude and direction angle θ of force F_4 so that the particle shown in Fig. P3-5 is in equilibrium.

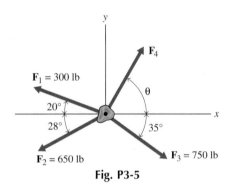

Fig. P3-5

3-6* Determine the magnitude and direction angle θ of force F_4 so that the particle shown in Fig. P3-6 is in equilibrium.

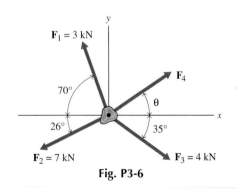

Fig. P3-6

3-7 Determine the magnitude and direction angle θ of force $\mathbf{F}_4$ so that the particle shown in Fig. P3-7 is in equilibrium.

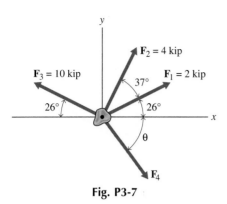

Fig. P3-7

3-8 Determine the magnitude and direction angle θ of force $\mathbf{F}_4$ so that the particle shown in Fig. P3-8 is in equilibrium.

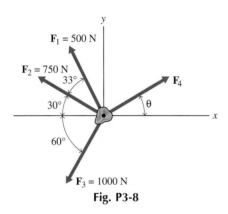

Fig. P3-8

3-9* A homogeneous sphere weighing 50 lb rests against two smooth planes that form a V-shaped trough as shown in Fig. P3-9. Determine the forces exerted on the sphere by the planes at contact points A and B.

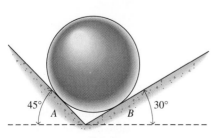

Fig. P3-9

3-10* A block with a mass of 10 kg is held in equilibrium on a smooth horizontal surface by two flexible cables as shown in Fig. P3-10. Determine the force exerted on the block by the horizontal surface and the angle θ between the inclined cable and the horizontal.

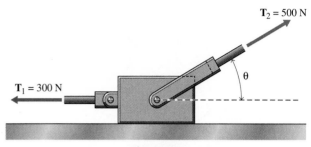

Fig. P3-10

3-11 Two flexible cables A and B are used to support a 220-lb traffic light as shown in Fig. P3-11. Determine the tension in each of the cables.

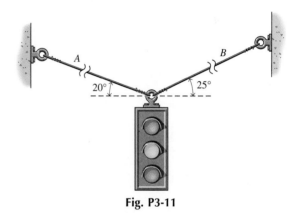

Fig. P3-11

3-12 Three smooth homogeneous cylinders A, B, and C are stacked in a box as shown in Fig. P3-12. Each cylinder has a diameter of 250 mm and a mass of 245 kg. Determine:

a. The force exerted by cylinder B on cylinder A.
b. The forces exerted on cylinder B by the vertical and horizontal surfaces at D and E.

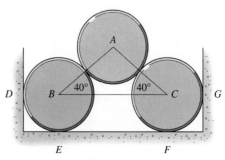

Fig. P3-12

75

3-13* Three smooth homogeneous cylinders A, B, and C are stacked in a V-shaped trough as shown in Fig. P3-13. Each cylinder weighs 100 lb and has a diameter of 5 in. Determine the minimum angle θ for equilibrium.

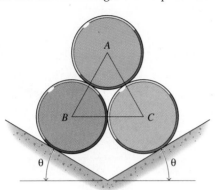

Fig. P3-13

3-14* The 200-mm diameter pipes shown in Fig. P3-14 each have a mass of 200 kg. Determine the forces exerted by the supports on the pipes at contact surfaces A, B, and C. Assume all surfaces to be smooth.

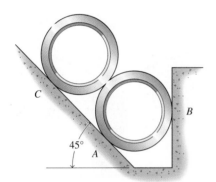

Fig. P3-14

3-15 Two 10-in. diameter pipes and a 6-in. diameter pipe are supported in a pipe rack as shown in Fig. P3-15. The 10-in. diameter pipes each weigh 300 lb and the 6-in. diam-

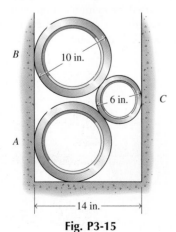

Fig. P3-15

eter pipe weighs 175 lb. Determine the forces exerted on the pipes by the supports at contact surfaces A, B, and C. Assume all surfaces to be smooth.

3-16 A body with a mass of 250 kg is supported by the flexible cable system shown in Fig. P3-16. Determine the tensions in cables A, B, C, and D.

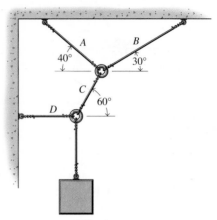

Fig. P3-16

3-17* Two bodies weighing 150 lb and 200 lb, respectively, rest on a cylinder and are connected by a rope as shown in Fig. P3-17. Find the reactions of the cylinder on the bodies, the tension in the rope, and the angle θ. Assume all surfaces to be smooth.

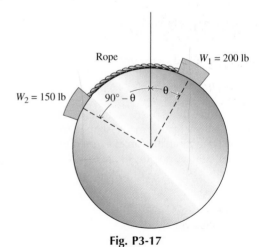

Fig. P3-17

3-18 Two bodies A and B weighing 800 N and 200 N, respectively, are held in equilibrium on perpendicular surfaces by a connecting flexible cable that makes an angle θ with the horizontal as shown in Fig. P3-18. Find the reactions of the surfaces on the bodies, the tension in the cable, and the angle θ. Assume all surfaces to be smooth.

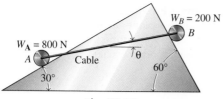
Fig. P3-18

3-19* The particle shown in Fig. P3-19 is in equilibrium under the action of the four forces shown on the free-body diagram. Determine the magnitude and the coordinate direction angles of the unknown force $\mathbf{F}_4$.

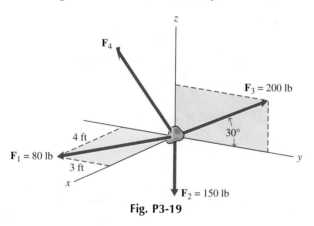
Fig. P3-19

3-20* The particle shown in Fig. P3-20 is in equilibrium under the action of the four forces shown on the free-body diagram. Determine the magnitude and the coordinate direction angles of the unknown force $\mathbf{F}_4$.

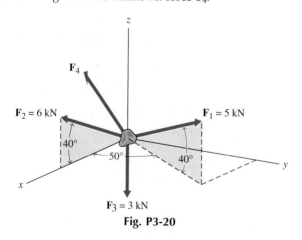
Fig. P3-20

3-21 The particle shown in Fig. P3-21 is in equilibrium under the action of the four forces shown on the free-body diagram. Determine the magnitude and the coordinate direction angles of the unknown force $\mathbf{F}_4$.

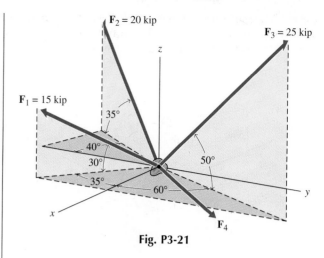
Fig. P3-21

3-22 The particle shown in Fig. P3-22 is in equilibrium under the action of the four forces shown on the free-body diagram. Determine the magnitude and the coordinate direction angles of the unknown force $\mathbf{F}_4$.

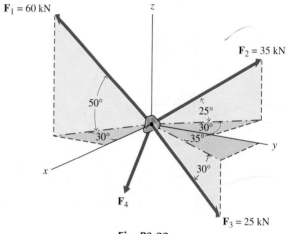
Fig. P3-22

3-23* A 3000-lb cylinder is supported by a system of cables as shown in Fig. P3-23. Determine the tensions in cables A, B, and C.

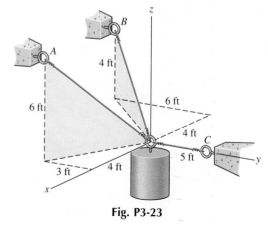
Fig. P3-23

3-24* The traffic light shown in Fig. P3-24 is supported by a system of cables. Determine the tensions in cables A, B, and C if the traffic light has a mass of 75 kg.

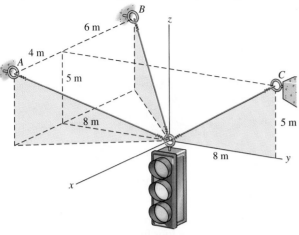

Fig. P3-24

3-25 The circular disk shown in Fig. P3-25 weighs 500 lb. Determine the tensions in cables A, B, and C.

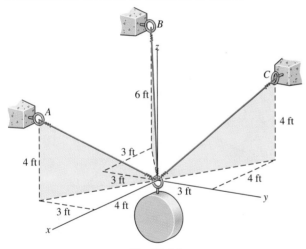

Fig. P3-25

3-26 The particle shown in Fig. P3-26 is in equilibrium

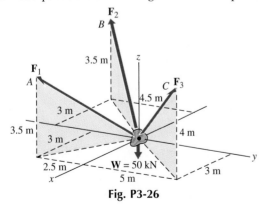

Fig. P3-26

under the action of the four forces shown on the free-body diagram. Determine the tensions in cables A, B, and C.

3-27* The force **F** required to support the 2500-lb concrete slab in the xy plane, as shown in Fig. P3-27, is the same as its weight. Determine the tensions in cables A, B, and C used to support the slab.

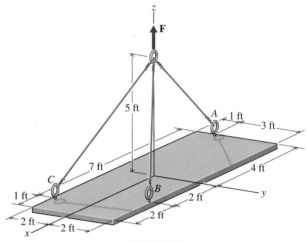

Fig. P3-27

3-28* The crate shown in Fig. P3-28 has a mass of 500 kg. Determine the tensions in cables A, B, and C used to support the crate.

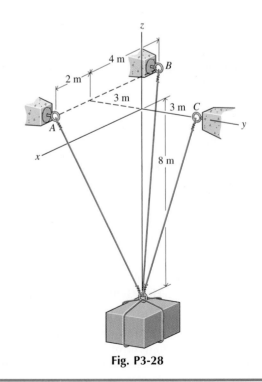

Fig. P3-28

78

SUMMARY

The term "particle" is used in statics to describe a body when the size and shape of the body will not significantly affect the solution of the problem being considered and when the mass of the body can be assumed to be concentrated at a point. As a result, a particle can be subjected only to a system of concurrent forces and the necessary and sufficient conditions for equilibrium can be expressed mathematically as

$$\mathbf{R} = \Sigma\mathbf{F} = \mathbf{0} \qquad (3\text{-}1)$$

The force system acting on a body in a typical problem consists of known forces and unknown forces. Both must be clearly identified before a solution to a specific problem is attempted. A carefully prepared drawing that shows a "body of interest" separated from all other interacting bodies and with all external forces applied is known as a free-body diagram (FBD). The importance of drawing a free-body diagram before attempting to solve a mechanics problem cannot be overemphasized. A procedure that can be followed to construct a complete and correct free-body diagram, contains the following four steps.

1. Decide which body or combination of bodies is to be shown on the free-body diagram.
2. Prepare a drawing or sketch of the outline of this isolated or free body.
3. Carefully trace around the boundary of the free body and identify all of the forces exerted by contacting or attracting bodies that were removed during the isolation process.
4. Choose the set of coordinate axes to be used in solving the problem and indicate these directions on the free-body diagram. Place any dimensions required for solution of the problem on the diagram.

Each known force should be shown on a free-body diagram with its correct magnitude, slope, and sense. Letter symbols can be used for the magnitudes of unknown forces. If a force has a known line of action but an unknown magnitude and sense, the sense of the force can be assumed. By definition, the magnitude of a force is always positive; therefore, if the solution yields a negative magnitude, the minus sign indicates that the sense of the force is opposite to that assumed on the free-body diagram. Most engineers consider an appropriate free-body diagram to be the single most important tool for the solution of mechanics problems.

For a three-dimensional system of concurrent forces, Eq. 3-1 can be written as

$$\begin{aligned}
\mathbf{R} = \Sigma\mathbf{F} &= \mathbf{R}_x + \mathbf{R}_y + \mathbf{R}_z \\
&= R_x\mathbf{i} + R_y\mathbf{j} + R_z\mathbf{k} \\
&= \Sigma F_x\mathbf{i} + \Sigma F_y\mathbf{j} + \Sigma F_z\mathbf{k} = \mathbf{0} \qquad (3\text{-}4)
\end{aligned}$$

Equation 3-4 is satisfied only if

$$\Sigma F_x = 0 \qquad \Sigma F_y = 0 \qquad \Sigma F_z = 0 \qquad (3\text{-}6)$$

Equations 3-6 can be used to determine three unknown quantities (three magnitudes, three slopes, or any combination of three magnitudes and slopes).

REVIEW PROBLEMS

3-29* The weights of cylinders A, B, and C of Fig. P3-29 are 175 lb, 275 lb, and 700 lb, respectively. Determine the forces exerted on cylinders A and B by the horizontal and vertical surfaces. Assume that all surfaces are smooth.

3-31 A continuous cable is used to support blocks A and B as shown in Fig. P3-31. Block A is supported by a small wheel that is free to roll on the cable. Determine the displacement y of block A for equilibrium if the weights of blocks A and B are 50 lb and 75 lb, respectively.

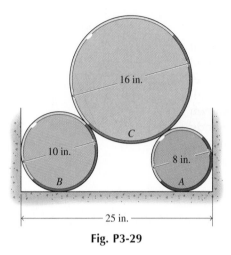

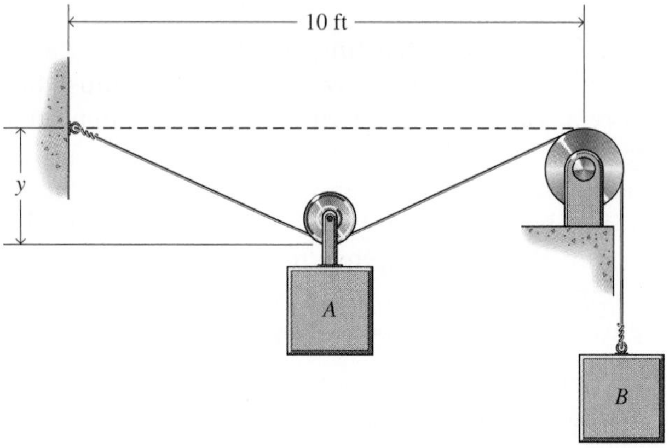

Fig. P3-29

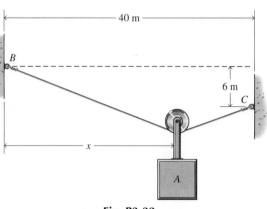

Fig. P3-31

3-30* The masses of cylinders A and B of Fig. P3-30 are 40 kg and 90 kg, respectively. Determine the forces exerted on the cylinders by the inclined surfaces and the magnitude and direction of the force exerted by cylinder A on cylinder B when the cylinders are in equilibrium. Assume that all surfaces are smooth.

3-32 The mass of block A in Fig. P3-32 is 200 kg. Block A is supported by a small wheel that is free to roll on the continuous cable between supports B and C. The length of the cable is 43 m. Determine the distance x and the tension T in the cable when the system is in equilibrium.

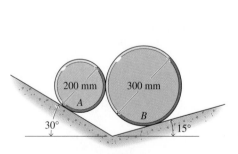

Fig. P3-30

Fig. P3-32

3-33* The hot-air balloon shown in Fig. P3-33 is tethered with three mooring cables. If the net lift of the balloon is 750 lb, determine the force exerted on the balloon by each of the cables.

3-34 Two forces are applied in a horizontal plane to a loading ring at the top of a post as shown in Fig. P3-34. The post can transmit only an axial compressive force. Two guy wires AC and BC are used to hold the loading ring in equilibrium. Determine the force transmitted by the post and the forces in the two guy wires.

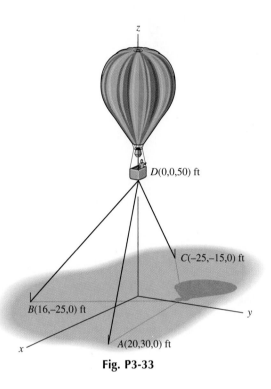

Fig. P3-33

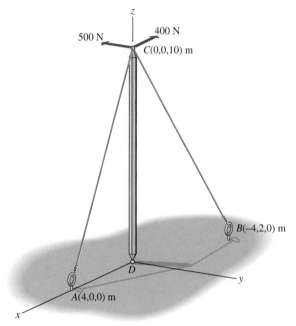

Fig. P3-34

Computer Problems

C3-35 A 75-lb stop light is suspended between two poles as shown in Fig. P3-35. Neglect the weight of the flexible cables and plot the tension in both cables as a function of the sag distance d ($0 \le d \le 8$ ft). Determine the minimum sag d for which both tensions are less than:

a. 100 lb.
b. 250 lb.
c. 500 lb.

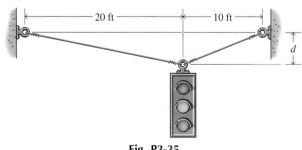

Fig. P3-35

C3-36 A 50-kg load is suspended from a pulley as shown in Fig. P3-36a. The tension in the flexible cable does not change as it passes around the small frictionless pulleys, and the weight of the cable may be neglected. Plot the force P required for equilibrium as a function of the sag distance d ($0 \le d \le 1$ m). Determine the minimum sag d for which P is less than:

a. Twice the weight of the load.
b. Four times the weight of the load.
c. Eight times the weight of the load.

Repeat the problem if the load is securely fastened to the hoisting rope as shown in Fig. P3-36b. (Plot both the force P and the tension T_{AB} in segment AB of the cable on the same graph.)

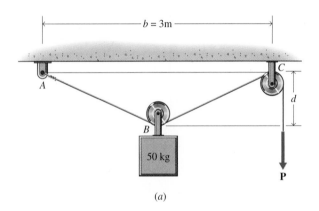

(a)

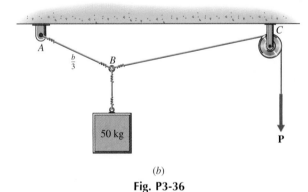

(b)

Fig. P3-36

C3-37 Three identical steel disks are stacked in a box as shown in Fig. P3-37. The weight and diameter of the smooth disks are 50 lb and 12 in., respectively. Plot the three forces exerted on disk C (by disk A, by the side wall, and by the floor) as a function of the distance b between the walls of the box (24 in. $\le b \le$ 36 in.). Determine the range of b for which:

a. The force at the floor is larger than the other two forces.
b. None of the three forces exceeds 100 lb.
c. None of the three forces exceeds 200 lb.

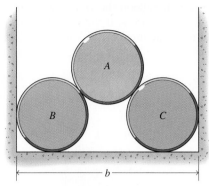

Fig. P3-37

C3-38 A pair of steel disks are stacked in a box as shown in Fig. P3-38. The masses and diameters of the smooth disks are $m_A = 5$ kg, $m_b = 20$ kg, $d_A = 100$ mm, and $d_B = 200$ mm. Plot the two forces exerted on disk A (by disk B and by the side wall) as a function of the distance b between the walls of the box (200 mm $\le b \le$ 300 mm). Determine the range of b for which:

a. The force at the side wall is less than W_A, the weight of disk A.
b. Neither of the two forces exceeds $2W_A$.
c. Neither of the two forces exceeds $4W_A$.

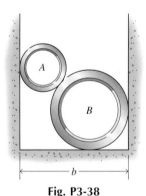

Fig. P3-38

C3-39 Two small blocks are connected by a flexible cord as shown in Fig. P3-39. The radius of the smooth cylindrical surface is 18 in., and the length of the rope is such that the angle between the two blocks is 90°. Plot the angle θ (between block 2 and the vertical) as a function of the weight W_2 ($W_2 \le 150$ lb). Determine the weight W_2 for which:

82

a. $\theta = 10°$.
b. $\theta = 80°$.

Do you think the equilibrium positions of parts a and b are stable? (That is, if the blocks were disturbed slightly, do you think they would return to the equilibrium position or do you think they would slide off of the cylindrical surface?)

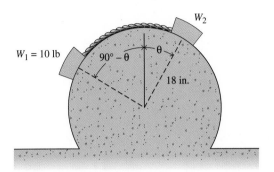

Fig. P3-39

C3-40 Two small wheels are connected by a light-weight, rigid rod as shown in Fig. P3-40. Plot the angle θ (between the rod and the horizontal) as a function of the weight W_1 ($W_1 \leq 10\ W_2$). Determine the weight W_1 for which:

a. $\theta = -50°$.
b. $\theta = 10°$.
c. $\theta = 25°$.

Do you think the equilibrium positions of parts a, b, and c are stable? (That is, if the wheels were disturbed slightly, do you think they would return to the equilibrium position or do you think they would slide off of the triangular surface?)

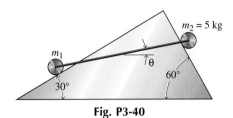

Fig. P3-40

4

RIGID BODIES: EQUIVALENT FORCE/ MOMENT SYSTEMS

The force exerted by the tugboat on the container ship is equivalent to a force and a couple at the center of mass of the ship. A small force applied to the bow of a long ship has a large effect in controlling the direction of the ship.

4-1   INTRODUCTION

The resultant force $\mathbf{R}$ of a system of two or more concurrent forces $\mathbf{F}_1$, $\mathbf{F}_2, \ldots, \mathbf{F}_n$ was defined in Chapter 2 as the single force that will produce the same effect on a body as the original system of forces. When the resultant force $\mathbf{R}$ of a concurrent force system is zero, the body on which the system of forces acts is in equilibrium, and the force system is said to be balanced. Methods to determine resultants of coplanar and noncoplanar concurrent force systems were discussed in Chapter 2 and applied to the equilibrium of a particle in Chapter 3.

For the case of a three-dimensional body that has a definite size and shape, the particle idealization discussed in Chapter 3 is no longer valid, in general, since the forces acting on the body are usually not concurrent. For these more general force systems, the condition $\mathbf{R} = 0$ is a necessary but not a sufficient condition for equilibrium of the body. A second restriction related to the tendency of a force to produce rotation of a body must also be satisfied and gives rise to the concept of a moment. In this chapter, the moment of a force about a point and the moment of a force about a line (axis) will be defined and methods will be developed for finding the resultant forces and the resultant moments (couples) for four other force systems that can be applied to a rigid body; namely,

1. Coplanar, parallel force systems
2. Coplanar, nonparallel, nonconcurrent force systems
3. Noncoplanar, parallel force systems
4. Noncoplanar, nonparallel, nonconcurrent force systems

The resultants of these force systems will be used in later chapters dealing with equilibrium (Chapter 6) and motion (Chapter 16) of rigid bodies.

4-2   MOMENTS AND THEIR CHARACTERISTICS

The moment of a force about a point or axis is a measure of the tendency of the force to rotate a body about that point or axis. For example, the moment of force $\mathbf{F}$ about point O in Fig. 4-1a is a measure of the tendency of the force to rotate the body about line AA. Line AA is perpendicular to the plane containing force $\mathbf{F}$ and point O.

A moment has both a magnitude and a direction, and adds according to the parallelogram law of addition; therefore, it is a vector quantity. The magnitude of a moment $|\mathbf{M}|$ is defined as the product of the magnitude of a force $|\mathbf{F}|$ and the perpendicular distance d from the line of action of the force to the axis. Thus, in Fig. 4-1b, the magnitude of the moment of the force $\mathbf{F}$ about point O (actually about axis AA, which is perpendicular to the page and passes through point O) is

$$M_O = |\mathbf{M}_O| = |\mathbf{F}|d \qquad (4\text{-}1)$$

Point O is called the moment center, distance d is called the moment arm, and line AA is called the axis of the moment.

The direction (sense) of a moment in a two-dimensional problem can be specified by using a small curved arrow about the point as shown in Fig. 4-1b. If the force tends to produce a counterclockwise

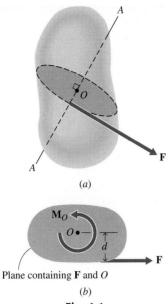

(a)

Plane containing **F** and O

(b)

Fig. 4-1

rotation, the moment is defined to be positive. In a similar manner, if
the force tends to produce a clockwise rotation, the moment is defined
to be negative.

Since the magnitude of a moment of a force is the product of a
force and a length, the dimensional expression for a moment is FL. In
the U. S. Customary System, the units commonly used for moments
are lb · ft and lb · in. or in. · lb, ft · lb, and ft · kip. In the SI System, the
units commonly used for moments are N · m, kN · m, and so on. It is
immaterial whether the unit of force or the unit of length is stated first.

The procedure for determining moments in simple two-dimensional problems (coplanar forces in, say, the xy-plane) is illustrated in
the following examples.

Three forces are applied to the bar shown in Fig. 4-2. Determine

a. The moment of force $\mathbf{F}_A$ about point E.
b. The moment of force $\mathbf{F}_E$ about point A.
c. The moment of force $\mathbf{F}_D$ about point B.

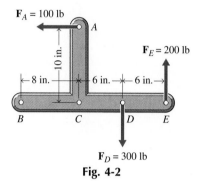

Fig. 4-2

SOLUTION

The magnitude of a moment about an arbitrary point O can be determined by using Eq. 4-1. Thus

$$M_O = |\mathbf{M}_O| = |\mathbf{F}|d$$

a.
$$M_E = |\mathbf{F}_A|d_{E/A} = 100(10) = 1000 \text{ in.} \cdot \text{lb}$$
$$\mathbf{M}_E = 1000 \text{ in.} \cdot \text{lb}\downarrow \qquad \text{Ans.}$$

b.
$$M_A = |\mathbf{F}_E|d_{A/E} = 200(12) = 2400 \text{ in.} \cdot \text{lb}$$
$$\mathbf{M}_A = 2400 \text{ in.} \cdot \text{lb}\downarrow \qquad \text{Ans.}$$

c.
$$M_B = |\mathbf{F}_D|d_{B/D} = 300(14) = 4200 \text{ in.} \cdot \text{lb}$$
$$\mathbf{M}_B = 4200 \text{ in.} \cdot \text{lb}\downarrow \qquad \text{Ans.}$$

Four forces are applied to a plate as shown in Fig. 4-3a. Determine

a. The moment of force $\mathbf{F}_B$ about point A.
b. The moment of force $\mathbf{F}_C$ about point B.
c. The moment of force $\mathbf{F}_C$ about point A.

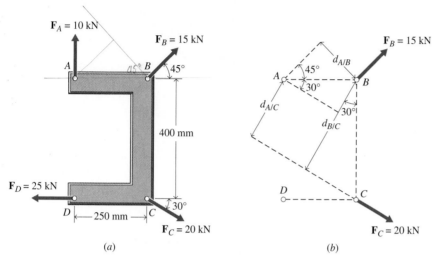

Fig. 4-3

SOLUTION

The magnitude of a moment about an arbitrary point O can be determined by using Eq. 4-1. Thus

$$M_O = |\mathbf{M}_O| = |\mathbf{F}|d$$

In this example, the problem is slightly more involved because of the geometry. As shown in Fig. 4-3b, the perpendicular distances from the forces to the points of interest are as follows:

$$d_{A/B} = 250 \sin 45° = 176.8 \text{ mm}$$
$$d_{B/C} = 400 \cos 30° = 346.4 \text{ mm}$$
$$d_{A/C} = 400 \cos 30° - 250 \sin 30° = 221.4 \text{ mm}$$

a. $M_A = |\mathbf{F}_B|d_{A/B} = 15(10^3)(176.8)(10^{-3}) = 2.65(10^3) \text{ N} \cdot \text{m} = 2.65 \text{ kN} \cdot \text{m}$
$$\mathbf{M}_A = 2.65 \text{ kN} \cdot \text{m} \downarrow \qquad \text{Ans.}$$

b. $M_B = |\mathbf{F}_C|d_{B/C} = 20(10^3)(346.4)(10^{-3}) = 6.93(10^3) \text{ N} \cdot \text{m} = 6.93 \text{ kN} \cdot \text{m}$
$$\mathbf{M}_B = 6.93 \text{ kN} \cdot \text{m} \downarrow \qquad \text{Ans.}$$

c. $M_A = |\mathbf{F}_C|d_{A/C} = 20(10^3)(221.4)(10^{-3}) = 4.43(10^3) \text{ N} \cdot \text{m} = 4.43 \text{ kN} \cdot \text{m}$
$$\mathbf{M}_A = 4.43 \text{ kN} \cdot \text{m} \downarrow \qquad \text{Ans.}$$

PROBLEMS

4-1* Two forces are applied at a point in a plane as shown in Fig. P4-1. Determine

a. The moments of force F_1 about points O, A, B, and C.
b. The moments of force F_2 about points O, A, B, and C.

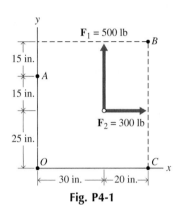

Fig. P4-1

4-2* Two forces are applied at a point in a plane as shown in Fig. P4-2. Determine

a. The moments of force F_1 about points O, A, B, and C.
b. The moments of force F_2 about points O, A, B, and C.

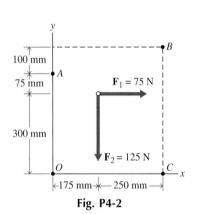

Fig. P4-2

4-3* Two forces are applied to a bracket as shown in Fig. P4-3. Determine

a. The moment of force F_1 about point O.
b. The moment of force F_2 about point O.

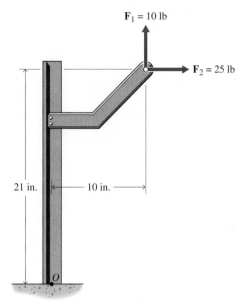

Fig. P4-3

4-4* Three forces are applied to a plate as shown in Fig. P4-4. Determine

a. The moment of force F_1 about point B.
b. The moment of force F_3 about point A.
c. The moment of force F_2 about point B.

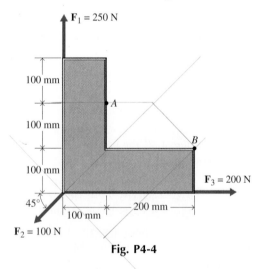

Fig. P4-4

4-5 Three forces are applied to a circular plate as shown in Fig. P4-5. Determine

a. The moment of force F_1 about point O.

b. The moment of force $\mathbf{F}_3$ about point O.
c. The moment of force $\mathbf{F}_2$ about point A.

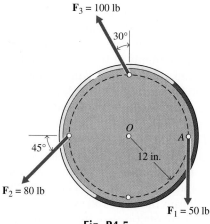

Fig. P4-5

4-6 Two forces are applied to a beam as shown in Fig. P4-6. Determine

a. The moment of force $\mathbf{F}_1$ about point B.
b. The moment of force $\mathbf{F}_2$ about point A.

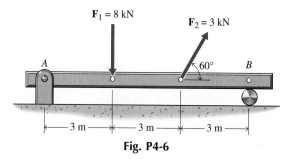

Fig. P4-6

4-7* Three forces are applied to a triangular plate as shown in Fig. P4-7. Determine

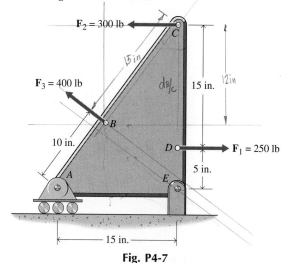

Fig. P4-7

a. The moment of force $\mathbf{F}_3$ about point C.
b. The moment of force $\mathbf{F}_2$ about point B.
c. The moment of force $\mathbf{F}_1$ about point B.
d. The moment of force $\mathbf{F}_3$ about point E.

4-8* Three forces are applied to a bracket as shown in Fig. P4-8. Determine

a. The moment of force $\mathbf{F}_1$ about point B.
b. The moment of force $\mathbf{F}_2$ about point A.
c. The moment of force $\mathbf{F}_3$ about point C.
d. The moment of force $\mathbf{F}_3$ about point E.

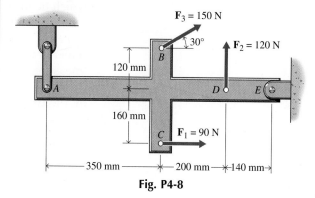

Fig. P4-8

4-9 Determine the moments of the 600-lb force shown in Fig. P4-9 about points A, B, and O.

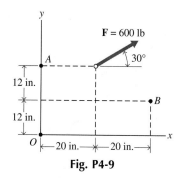

Fig. P4-9

4-10 Determine the moments of the 16-kN force shown in Fig. P4-10 about points A, B, and O.

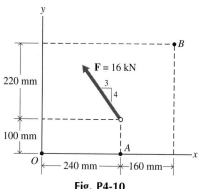

Fig. P4-10

4-2-1   Principle of Moments: Varignon's Theorem

A concept often used in solving mechanics (statics, dynamics, mechanics of materials) problems is the principle of moments. This principle, when applied to a system of forces, states that the moment **M** of the resultant **R** of a system of forces with respect to any axis or point is equal to the vector sum of the moments of the individual forces of the system with respect to the same axis or point. Application of this principle to a pair of concurrent forces is known as Varignon's theorem. Varignon's theorem can be illustrated by using the concurrent force system, shown in Fig. 4-4, where **R** is the resultant of forces **A** and **B**, which lie in the xy-plane. The point of concurrency A and the moment center O have been arbitrarily selected to lie on the y-axis. The distances d, a, and b are the perpendicular distances from the moment center O to the forces **R**, **A**, and **B**, respectively. The angles γ, α, and β (measured from the x-axis) locate the forces **R**, **A**, and **B**, respectively.

The magnitudes of the moments produced by the resultant **R** and by the two forces **A** and **B** with respect to point O are

$$M_R = Rd = R(h \cos \gamma)$$
$$M_A = Aa = A(h \cos \alpha) \tag{a}$$
$$M_B = Bb = B(h \cos \beta)$$

From Fig. 4-4 note also that

$$R \cos \gamma = A \cos \alpha + B \cos \beta \tag{b}$$

Substituting Eqs. a into Eq. b and multiplying both sides of the equation by h yields

$$M_R = M_A + M_B \tag{4-2}$$

Equation 4-2 indicates that the moment of the resultant **R** with respect to a point O is equal to the sum of the moments of the forces **A** and **B** with respect to the same point O.

The following examples illustrate the use of the principle of moments.

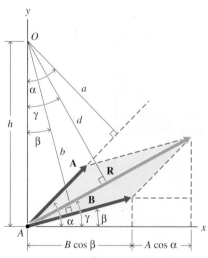

Fig. 4-4

Determine the moment about point O (actually the z-axis through point O) of the 500-lb force shown in Fig. 4-5a.

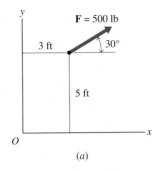

(a)

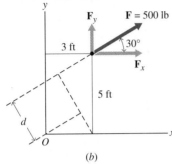

(b)

Fig. 4-5

SOLUTION

Inspection of Fig. 4-5b shows that the perpendicular distance d from the line of action of the force $\mathbf{F}$ to point O is

$$d = 5 \cos 30° - 3 \sin 30° = 2.830 \text{ ft}$$

Thus, from Eq. 4-1,

$$\curvearrowleft + M_O = -Fd = -500(2.830) = -1415 \text{ ft} \cdot \text{lb}$$
$$\mathbf{M}_O = 1415 \text{ ft} \cdot \text{lb}\curvearrowright \qquad \text{Ans.}$$

Alternatively, the moment of force $\mathbf{F}$ about point O also can be determined by using the principle of moments. The magnitudes of the rectangular components of the 500-lb force are

$$F_x = F \cos 30° = 500 \cos 30° = 433.0 \text{ lb}$$
$$F_y = F \sin 30° = 500 \sin 30° = 250.0 \text{ lb}$$

Once the forces F_x and F_y are known, the moment M_O is

$$\curvearrowleft + M_O = F_y(3) - F_x(5) = 250.0(3) - 433.0(5) = -1415 \text{ ft} \cdot \text{lb}$$
$$\mathbf{M}_O = 1415 \text{ ft} \cdot \text{lb}\curvearrowright \qquad \text{Ans.}$$

Use the principle of moments to determine the moment about point B of the 300-N force shown in Fig. 4-6a.

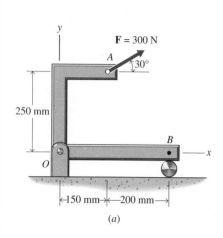

(a)

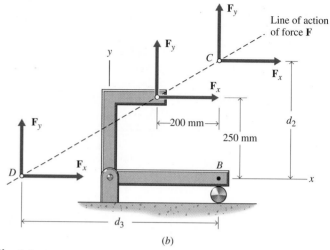

(b)

Fig. 4-6

SOLUTION

The magnitudes of the rectangular components of the 300-N force are

$$F_x = F \cos 30° = 300 \cos 30° = 259.8 \text{ N}$$
$$F_y = F \sin 30° = 300 \sin 30° = 150.0 \text{ N}$$

Once the forces F_x and F_y are known, the moment M_B is

$$\llcorner + M_B = -F_x(0.250) - F_y(0.200)$$
$$= -259.8(0.250) - 150.0(0.200) = -95.0 \text{ N} \cdot \text{m}$$
$$\mathbf{M}_B = 95.0 \text{ N} \cdot \text{m} \downarrow \qquad \text{Ans.}$$

The moment about point B can also be determined by moving the force $\mathbf{F}$ along its line of action (principle of transmissibility) to points C or D as shown in Fig. 4-6b. For both of these points, one component of the force produces no moment about point B. Thus, for point C

$$d_2 = 250 + 200 \tan 30° = 365.5 \text{ mm}$$
$$\llcorner + M_B = -F_x d_2 = -259.8(0.3655) = -95.0 \text{ N} \cdot \text{m}$$
$$\mathbf{M}_B = 95.0 \text{ N} \cdot \text{m} \downarrow \qquad \text{Ans.}$$

For point D

$$d_3 = 200 + 250 \cot 30° = 633.0 \text{ mm}$$
$$\llcorner + M_B = -F_y d_3 = -150.0(0.633) = -95.0 \text{ N} \cdot \text{m}$$
$$\mathbf{M}_B = 95.0 \text{ N} \cdot \text{m} \downarrow \qquad \text{Ans.}$$

Three forces $\mathbf{F}_1$, $\mathbf{F}_2$, and $\mathbf{F}_3$ are applied to a beam as shown in Fig. 4-7a. Determine the moments $\mathbf{M}_{A1}$, $\mathbf{M}_{A2}$, and $\mathbf{M}_{A3}$ about point A produced by each of the forces.

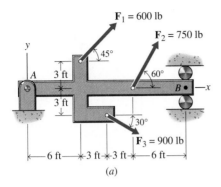

(a)

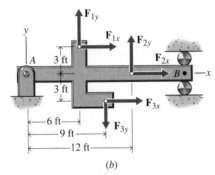

(b)

Fig. 4-7

SOLUTION

The magnitudes of the rectangular components (see Fig. 4-7b) of the three forces are

$$F_{1x} = F_1 \cos 45° = 600 \cos 45° = 424.3 \text{ lb}$$
$$F_{1y} = F_1 \sin 45° = 600 \sin 45° = 424.3 \text{ lb}$$
$$F_{2x} = F_2 \cos 60° = 750 \cos 60° = 375.0 \text{ lb}$$
$$F_{2y} = F_2 \sin 60° = 750 \sin 60° = 649.5 \text{ lb}$$
$$F_{3x} = F_3 \cos 30° = 900 \cos 30° = 779.4 \text{ lb}$$
$$F_{3y} = F_3 \sin 30° = 900 \sin 30° = 450.0 \text{ lb}$$

The moments $\mathbf{M}_{A1}$, $\mathbf{M}_{A2}$, and $\mathbf{M}_{A3}$ are determined by using the principle of moments. Thus

$$\curvearrowright + M_{A1} = F_{1y}(6) - F_{1x}(3) = 424.3(6) - 424.3(3) = 1273 \text{ ft} \cdot \text{lb}$$
$$\mathbf{M}_{A1} = 1273 \text{ ft} \cdot \text{lb} \curvearrowright \qquad \text{Ans.}$$
$$\curvearrowright + M_{A2} = F_{2y}(12) - F_{2x}(0) = 649.5(12) - 375.0(0) = 7794 \text{ ft} \cdot \text{lb}$$
$$\mathbf{M}_{A2} = 7794 \text{ ft} \cdot \text{lb} \curvearrowright \qquad \text{Ans.}$$
$$\curvearrowright + M_{A3} = F_{3x}(3) - F_{3y}(9) = 779.4(3) - 450.0(9) = -1712 \text{ ft} \cdot \text{lb}$$
$$\mathbf{M}_{A3} = 1712 \text{ ft} \cdot \text{lb} \curvearrowleft \qquad \text{Ans.}$$

PROBLEMS

4-11* A 300-lb force is applied to a bracket as shown in Fig. P4-11. Determine the moment of the force about point A.

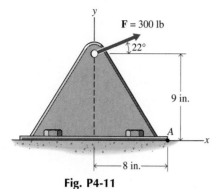

Fig. P4-11

4-12* A 250-N force is applied to a bracket as shown in Fig. P4-12. Determine the moment of the force about point A.

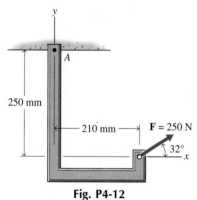

Fig. P4-12

4-13 Determine the moment of the 100-lb force shown in Fig. P4-13

a. About point O.
b. About point A.

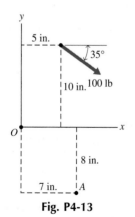

Fig. P4-13

4-14 Determine the moment of the 150-N force shown in Fig. P4-14

a. About point A.
b. About point B.

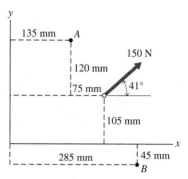

Fig. P4-14

4-15* Two forces F_1 and F_2 are applied to a plate as shown in Fig. P4-15. Determine

a. The moment of force F_1 about point A.
b. The moment of force F_2 about point B.

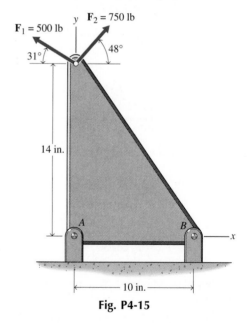

Fig. P4-15

4-16* Two forces F_1 and F_2 are applied to a bracket as shown in Fig. P4-16. Determine

a. The moment of force F_1 about point O.
b. The moment of force F_2 about point A.

96

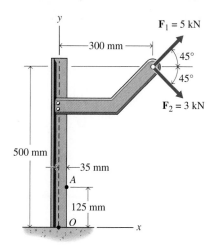

Fig. P4-16

4-17 Two forces F_1 and F_2 are applied to a bracket as shown in Fig. P4-17. Determine

a. The moment of force F_1 about point B.
b. The moment of force F_2 about point A.

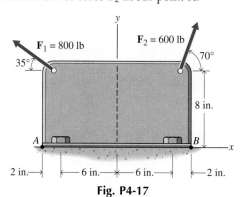

Fig. P4-17

4-18 Two forces are applied to a ring as shown in Fig. P4-18. Determine

a. The moment of force F_1 about point A.
b. The moment of force F_2 about point A.

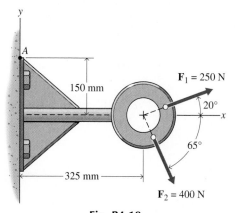

Fig. P4-18

4-19* Two forces F_1 and F_2 are applied to a gusset plate as shown in Fig. P4-19. Determine

a. The moment of force F_1 about point A.
b. The moment of force F_2 about point B.

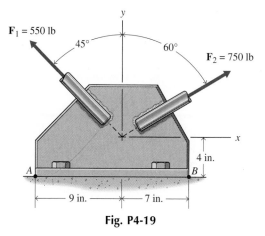

Fig. P4-19

4-20* Two forces F_1 and F_2 are applied to a beam as shown in Fig. P4-20. Determine

a. The moment of force F_1 about point A.
b. The moment of force F_2 about point B.

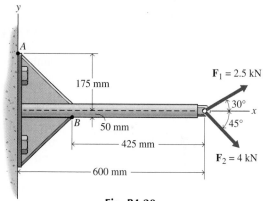

Fig. P4-20

4-21 Three forces F_A, F_B, and F_C are applied to a beam as shown in Fig. P4-21. Determine

a. The moments of forces F_A and F_C about point O.
b. The moment of force F_B about point D.

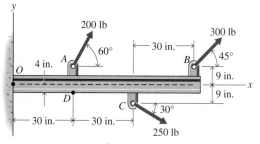

Fig. P4-21

97

4-22 Three forces F_1, F_2, and F_3 are applied to a bracket as shown in Fig. P4-22. Determine the moments of each of the forces about point B.

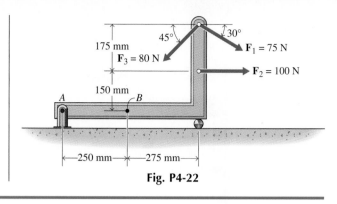

Fig. P4-22

4-3 VECTOR REPRESENTATION OF A MOMENT

For some two-dimensional problems and for most three-dimensional problems, use of Eq. 4-1 for moment determinations is not convenient owing to difficulties in determining the perpendicular distance d between the line of action of the force and the moment center O. For these types of problems, a vector approach simplifies moment calculations.

In Fig. 4-1, the moment of the force $\mathbf{F}$ about point O can be represented by the expression

$$\mathbf{M}_O = \mathbf{r} \times \mathbf{F} \qquad (4\text{-}3)$$

where $\mathbf{r}$ is a position vector (see Appendix A) from point O to a point A on the line of action of the force $\mathbf{F}$, as shown in Fig. 4-8. By definition, the cross-product (see Appendix A) of the two intersecting vectors $\mathbf{r}$ and $\mathbf{F}$ is

$$\mathbf{M}_O = \mathbf{r} \times \mathbf{F} = |\mathbf{r}||\mathbf{F}| \sin \alpha \mathbf{e} \qquad (4\text{-}4)$$

where α is the angle ($0 \le \alpha \le 180°$) between the two intersecting vectors $\mathbf{r}$ and $\mathbf{F}$, and $\mathbf{e}$ is a unit vector perpendicular to the plane containing vectors $\mathbf{r}$ and $\mathbf{F}$. It is obvious from Fig. 4-8 that the term $|\mathbf{r}| \sin \alpha$ equals the perpendicular distance d from the line of action of the force

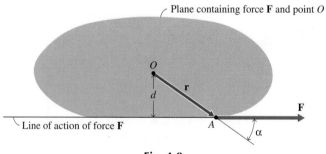

Fig. 4-8

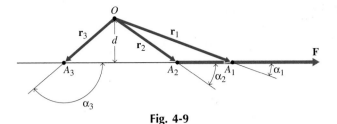

Fig. 4-9

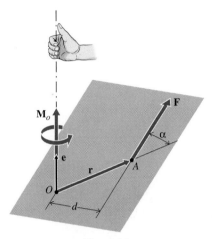

to the moment center O. Note also from Fig. 4-9 that the distance d is independent of the position A along the line of action of the force since

$$|\mathbf{r_1}| \sin \alpha_1 = |\mathbf{r_2}| \sin \alpha_2 = |\mathbf{r_3}| \sin \alpha_3 = d$$

Thus, Eq. 4-4 can be written

$$\mathbf{M}_O = |\mathbf{F}|d\ \mathbf{e} = Fd\ \mathbf{e} = M_O\ \mathbf{e} \qquad (4\text{-}5)$$

In Eq. 4-5 the direction of the unit vector $\mathbf{e}$ is determined (see Fig. 4-10) by using the right-hand rule (fingers of the right hand curl from positive $\mathbf{r}$ to positive $\mathbf{F}$ and the thumb points in the direction of positive $\mathbf{M}_O$). Thus, Eq. 4-5 yields both the magnitude M_O and the direction $\mathbf{e}$ of the moment $\mathbf{M}_O$. It is important to note that the sequence $\mathbf{r} \times \mathbf{F}$ must be maintained in calculating moments since the sequence $\mathbf{F} \times \mathbf{r}$ will produce a moment with the opposite sense. The vector product is *not* a commutative operation.

Fig. 4-10

4-3-1 Moment of a Force About a Point

The vector $\mathbf{r}$ from the point about which the moment is to be determined (say point B) to any point on the line of action of the force $\mathbf{F}$ (say point A) can be expressed in terms of the unit vectors $\mathbf{i}$, $\mathbf{j}$, and $\mathbf{k}$, and the coordinates (x_A, y_A, z_A) and (x_B, y_B, z_B) of points A and B, respectively. Thus, as illustrated in Fig. 4-11,

$$\mathbf{r} = \mathbf{r}_{A/B} = \mathbf{r}_A - \mathbf{r}_B = (x_A - x_B)\mathbf{i} + (y_A - y_B)\mathbf{j} + (z_A - z_B)\mathbf{k} \quad (4\text{-}6)$$

where the subscript A/B indicates A with respect to B.

Equation 4-3 is applicable for both the two-dimensional case (forces in, say, the xy-plane) and the three-dimensional case (forces with arbitrary space orientations).

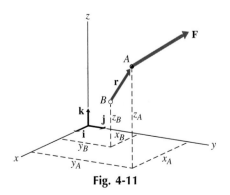

Fig. 4-11

The Two-dimensional Case   Consider first the moment $\mathbf{M}_O$ about the origin of coordinates (see Fig. 4-12a) produced by a force $\mathbf{F}$ in the xy-plane. The line of action of the force passes through point A. For this special case (see Fig. 4-12b)

$$\mathbf{F} = F_x\mathbf{i} + F_y\mathbf{j}$$

and the position vector $\mathbf{r}$ from the origin O to point A is

$$\mathbf{r} = r_x\mathbf{i} + r_y\mathbf{j}$$

The vector product $\mathbf{r} \times \mathbf{F}$ for this two-dimensional case can be written in determinant form as

$$\mathbf{M}_O = \mathbf{r} \times \mathbf{F} = \begin{vmatrix} \mathbf{i} & \mathbf{j} & \mathbf{k} \\ r_x & r_y & 0 \\ F_x & F_y & 0 \end{vmatrix} = (r_x F_y - r_y F_x)\mathbf{k} = M_z\mathbf{k} \qquad (4\text{-}7)$$

Thus, for the two-dimensional case, the moment $\mathbf{M}_O$ about point O due to a force $\mathbf{F}$ in the xy-plane is perpendicular to the plane (directed along the z-axis). The moment is completely defined by the scalar quantity

$$M_O = M_z = r_x F_y - r_y F_x \qquad (4\text{-}8)$$

since a positive value for M_O indicates a tendency to rotate the body in a counterclockwise direction, which by the right-hand rule is along the positive z-axis. Similarly, a negative value indicates a tendency to rotate the body in a clockwise direction, which requires a moment in the negative z-direction.

The following examples illustrate the use of vector algebra for determining moments about a point in two-dimensional problems.

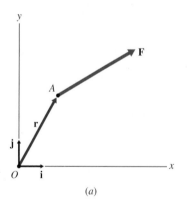

(a)

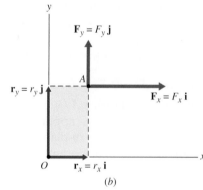

(b)

Fig. 4-12

A 1000-N force is applied to a beam cross section as shown in Fig. 4-13. Determine

a. The moment of the force about point O.
b. The perpendicular distance d from point B to the line of action of the force.

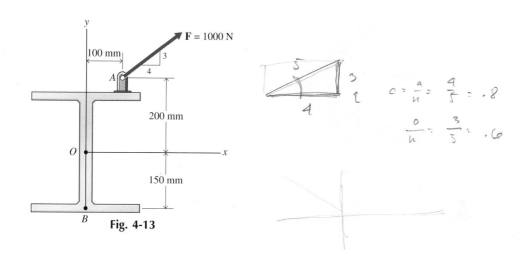

Fig. 4-13

SOLUTION

a. The force **F** and the position vector **r** from point O to point A can be expressed in Cartesian vector form as

$$\mathbf{F} = 1000\,(0.80\mathbf{i} + 0.60\mathbf{j}) = (800\mathbf{i} + 600\mathbf{j})\ \text{N}$$
$$\mathbf{r} = \mathbf{r}_{A/O} = (0.100\mathbf{i} + 0.200\mathbf{j})\ \text{m}$$

From Eq. 4-7

$$\mathbf{M}_O = \mathbf{r} \times \mathbf{F} = \begin{vmatrix} \mathbf{i} & \mathbf{j} & \mathbf{k} \\ r_x & r_y & 0 \\ F_x & F_y & 0 \end{vmatrix} = (r_x F_y - r_y F_x)\mathbf{k} = M_z \mathbf{k}$$

$$\mathbf{M}_O = (r_x F_y - r_y F_x)\mathbf{k} = [(0.100)(600) - (0.200)(800)]\mathbf{k}$$
$$= -100\mathbf{k}\ \text{N} \cdot \text{m} = 100\ \text{N} \cdot \text{m} \downarrow \qquad \text{Ans.}$$

b. The position vector **r** from point B to point A is

$$\mathbf{r} = \mathbf{r}_{A/B} = (0.100\mathbf{i} + 0.350\mathbf{j})\ \text{m}$$
$$\mathbf{M}_B = (r_x F_y - r_y F_x)\mathbf{k} = [(0.100)(600) - (0.350)(800)]\mathbf{k}$$
$$= -220\mathbf{k}\ \text{N} \cdot \text{m} = 200\ \text{N} \cdot \text{m} \downarrow$$
$$d = \frac{|\mathbf{M}_B|}{|\mathbf{F}|} = \frac{220}{1000} = 0.220\ \text{m} = 220\ \text{mm} \qquad \text{Ans.}$$

Four forces are applied to a square plate as shown in Fig. 4-14. Determine the moments produced by each of the forces about the origin O of the xy-coordinate system.

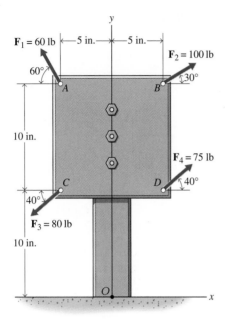

Fig. 4-14

SOLUTION

The four forces and the four position vectors in Cartesian vector form are as follows:

$$\mathbf{F}_1 = (-30.0\mathbf{i} + 52.0\mathbf{j}) \text{ lb} \qquad \mathbf{r}_{A/O} = (-5.00\mathbf{i} + 20.0\mathbf{j}) \text{ in.}$$
$$\mathbf{F}_2 = (86.6\mathbf{i} + 50.0\mathbf{j}) \text{ lb} \qquad \mathbf{r}_{B/O} = (5.00\mathbf{i} + 20.0\mathbf{j}) \text{ in.}$$
$$\mathbf{F}_3 = (-61.3\mathbf{i} - 51.4\mathbf{j}) \text{ lb} \qquad \mathbf{r}_{C/O} = (-5.00\mathbf{i} + 10.0\mathbf{j}) \text{ in.}$$
$$\mathbf{F}_4 = (57.5\mathbf{i} + 48.2\mathbf{j}) \text{ lb} \qquad \mathbf{r}_{D/O} = (5.00\mathbf{i} + 10.0\mathbf{j}) \text{ in.}$$

From Eq. 4-7

$$\mathbf{M}_O = \mathbf{r} \times \mathbf{F} = \begin{vmatrix} \mathbf{i} & \mathbf{j} & \mathbf{k} \\ r_x & r_y & 0 \\ F_x & F_y & 0 \end{vmatrix} = (r_x F_y - r_y F_x)\mathbf{k} = M_z\mathbf{k}$$

$$\mathbf{M}_{O1} = (r_{1x}F_{1y} - r_{1y}F_{1x})\mathbf{k} = [(-5.00)(52.0) - (20.0)(-30.0)]\mathbf{k}$$
$$= 340\mathbf{k} \text{ in} \cdot \text{lb} = 340 \text{ in} \cdot \text{lb} \curvearrowleft \qquad \text{Ans.}$$

$$\mathbf{M}_{O2} = (r_{2x}F_{2y} - r_{2y}F_{2x})\mathbf{k} = [(5.00)(50.0) - (20.0)(86.6)]\mathbf{k}$$
$$= -1482\mathbf{k} \text{ in} \cdot \text{lb} = 1482 \text{ in} \cdot \text{lb} \curvearrowright \qquad \text{Ans.}$$

$$\mathbf{M}_{O3} = (r_{3x}F_{3y} - r_{3y}F_{3x})\mathbf{k} = [(-5.00)(-51.4) - (10.0)(-61.3)]\mathbf{k}$$
$$= 870\mathbf{k} \text{ in} \cdot \text{lb} = 870 \text{ in} \cdot \text{lb} \curvearrowleft \qquad \text{Ans.}$$

$$\mathbf{M}_{O4} = (r_{4x}F_{4y} - r_{4y}F_{4x})\mathbf{k} = [(5.00)(48.2) - (10.0)(57.5)]\mathbf{k}$$
$$= -334\mathbf{k} \text{ in} \cdot \text{lb} = 334 \text{ in} \cdot \text{lb} \curvearrowright \qquad \text{Ans.}$$

PROBLEMS

Use the vector definition $\mathbf{M} = \mathbf{r} \times \mathbf{F}$ in the solution of the following problems.

4-23* Determine the moment of the 910-lb force shown in Fig. P4-23 about point O.

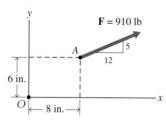

Fig. P4-23

4-24* Determine the moment of the 375-N force shown in Fig. P4-24 about point O.

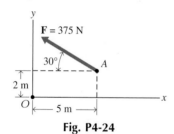

Fig. P4-24

4-25 Determine the moment of the 650-lb force shown in Fig. P4-25

a. About point O.
b. About point B.

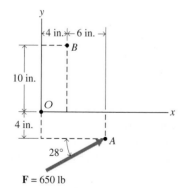

Fig. P4-25

4-26 Determine the moment of the 725-N force shown in Fig. P4-26

a. About point O.
b. About point B.

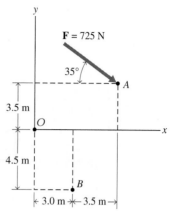

Fig. P4-26

4-27* Use the vector method to solve Problem 4-15 (p. 96).

4-28* Use the vector method to solve Problem 4-16 (p. 96).

4-29 Use the vector method to solve Problem 4-17 (p. 97).

4-30 Use the vector method to solve Problem 4-18 (p. 97).

4-31* Use the vector method to solve Problem 4-19 (p. 97).

4-32* Use the vector method to solve Problem 4-20 (p. 97).

4-33 A 100-lb force is applied to a curved bar as shown in Fig. P4-33. Determine the moment of the force about point B.

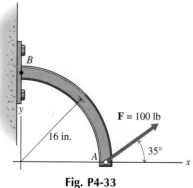

Fig. P4-33

4-34 A 450-N force is applied to a bracket as shown in Fig. P4-34. Determine the moment of the force

a. About point B.
b. About point C.

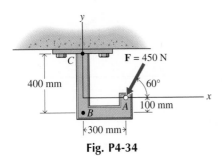

Fig. P4-34

4-35* Use the vector method to solve Problem 4-21 (p. 97).

4-36* Use the vector method to solve Problem 4-22 (p. 98).

4-37 A 583-lb force is applied to a bracket as shown in Fig. P4-37. Determine the moment of the force

a. About point D.
b. About point E.

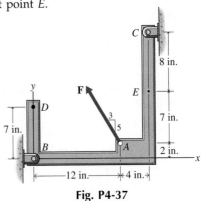

Fig. P4-37

4-38 A 650-N force is applied to a bracket as shown in Fig. P4-38. Determine the moment of the force

a. About point D.
b. About point E.

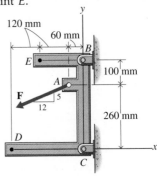

Fig. P4-38

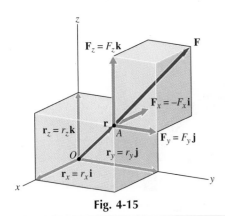

Fig. 4-15

The Three-dimensional Case The moment M_O about the origin of coordinates O produced by a force F with a space (three-dimensional) orientation can also be determined by using Eq. 4-3. For this general case (see Fig. 4-15), the force F can be expressed in Cartesian vector form as

$$F = F_x i + F_y j + F_z k$$

and the position vector r from the origin O to an arbitrary point A on the line of action of the force as

$$r = r_x i + r_y j + r_z k$$

The vector product $\mathbf{r} \times \mathbf{F}$ for this three-dimensional case can be written in determinant form as

$$\mathbf{M}_O = \mathbf{r} \times \mathbf{F} = \begin{vmatrix} \mathbf{i} & \mathbf{j} & \mathbf{k} \\ r_x & r_y & r_z \\ F_x & F_y & F_z \end{vmatrix}$$

$$= (r_y F_z - r_z F_y)\mathbf{i} + (r_z F_x - r_x F_z)\mathbf{j} + (r_x F_y - r_y F_x)\mathbf{k}$$
$$= M_x\mathbf{i} + M_y\mathbf{j} + M_z\mathbf{k} \tag{4-9}$$

where

$$\begin{aligned} M_x &= r_y F_z - r_z F_y \\ M_y &= r_z F_x - r_x F_z \\ M_z &= r_x F_y - r_y F_x \end{aligned} \tag{4-10}$$

are the three scalar components of the moment of force $\mathbf{F}$ about point O. The magnitude of the moment $|\mathbf{M}_O|$ (see Fig. 4-16) is

$$|\mathbf{M}_O| = \sqrt{M_x^2 + M_y^2 + M_z^2} \tag{4-11}$$

Alternatively, the moment $\mathbf{M}_O$ can be written as

$$\mathbf{M}_O = M_O\mathbf{e} \tag{4-12}$$

where

$$\mathbf{e} = \cos\theta_x\,\mathbf{i} + \cos\theta_y\,\mathbf{j} + \cos\theta_z\,\mathbf{k} \tag{4-13}$$

The direction cosines associated with the unit vector $\mathbf{e}$ are

$$\cos\theta_x = \frac{M_x}{|\mathbf{M}_O|} \qquad \cos\theta_y = \frac{M_y}{|\mathbf{M}_O|} \qquad \cos\theta_z = \frac{M_z}{|\mathbf{M}_O|} \tag{4-14}$$

A moment obeys all the rules of vector combination and can be considered a sliding vector with a line of action coinciding with the moment axis.

The principle of moments discussed in Section 4-2-1 is not restricted to two concurrent forces but may be extended to any force system. The proof for an arbitrary number of forces follows from the distributive property of the vector product. Thus,

$$\mathbf{M}_O = \mathbf{r} \times \mathbf{R}$$

but

$$\mathbf{R} = \mathbf{F}_1 + \mathbf{F}_2 + \cdots + \mathbf{F}_n$$

therefore

$$\begin{aligned} \mathbf{M}_O &= \mathbf{r} \times (\mathbf{F}_1 + \mathbf{F}_2 + \cdots + \mathbf{F}_n) \\ &= (\mathbf{r} \times \mathbf{F}_1) + (\mathbf{r} \times \mathbf{F}_2) + \cdots + (\mathbf{r} \times \mathbf{F}_n) \end{aligned}$$

Thus

$$\mathbf{M}_O = \mathbf{M}_R = \mathbf{M}_1 + \mathbf{M}_2 + \cdots + \mathbf{M}_n \tag{4-15}$$

Equation 4-15 indicates that the moment of the resultant of any number of forces is equal to the sum of the moments of the individual forces.

The following example illustrates the use of vector algebra for determining moments about a point in three-dimensional problems.

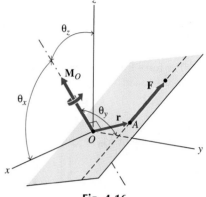

Fig. 4-16

A force with a magnitude of 840 N acts at a point in a body as shown in Fig. 4-17. Determine

a. The moment of the force about point B.
b. The direction angles associated with the unit vector $\mathbf{e}$ along the axis of the moment.
c. The perpendicular distance d from point B to the line of action of the force.

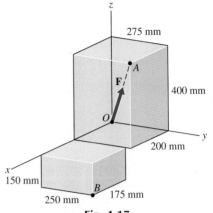

Fig. 4-17

SOLUTION

a. The force $\mathbf{F}$ and the position vector $\mathbf{r}$ from point B to point A can be written in Cartesian vector form as

$$\mathbf{F} = (320\mathbf{i} + 440\mathbf{j} + 640\mathbf{k})\,\text{N}$$
$$\mathbf{r}_{A/B} = (-0.175\mathbf{i} + 0.025\mathbf{j} + 0.550\mathbf{k})\,\text{m}$$

The moment $\mathbf{M}_B$ is given by Eq. 4-9 as

$$\mathbf{M}_B = \mathbf{r}_{A/B} \times \mathbf{F} = \begin{vmatrix} \mathbf{i} & \mathbf{j} & \mathbf{k} \\ r_x & r_y & r_z \\ F_x & F_y & F_z \end{vmatrix} = \begin{vmatrix} \mathbf{i} & \mathbf{j} & \mathbf{k} \\ -0.175 & 0.025 & 0.550 \\ 320 & 440 & 640 \end{vmatrix}$$
$$= (-226\mathbf{i} + 288\mathbf{j} - 85\mathbf{k})\,\text{N}\cdot\text{m} \quad \text{Ans.}$$

The position vector $\mathbf{r}$ can also be written from point B to point O as

$$\mathbf{r}_{O/B} = (-0.375\mathbf{i} - 0.250\mathbf{j} + 0.150\mathbf{k})\,\text{m}$$

The moment $\mathbf{M}_B$ is then given by Eq. 4-9 as

$$\mathbf{M}_B = \mathbf{r}_{O/B} \times \mathbf{F} = \begin{vmatrix} \mathbf{i} & \mathbf{j} & \mathbf{k} \\ r_x & r_y & r_z \\ F_x & F_y & F_z \end{vmatrix} = \begin{vmatrix} \mathbf{i} & \mathbf{j} & \mathbf{k} \\ -0.375 & -0.250 & 0.150 \\ 320 & 440 & 640 \end{vmatrix}$$
$$= (-226\mathbf{i} + 288\mathbf{j} - 85\mathbf{k})\,\text{N}\cdot\text{m} \quad \text{Ans.}$$

b. The magnitude of moment $\mathbf{M}_B$ is obtained by using Eq. 4-11. Thus

$$|\mathbf{M}_B| = \sqrt{M_x^2 + M_y^2 + M_z^2} = \sqrt{(-226)^2 + (288)^2 + (-85)^2} = 375.8\,\text{N}\cdot\text{m}$$

The direction angles are obtained by using Eqs. 4-14. Thus

$$\theta_x = \cos^{-1}\frac{M_x}{|\mathbf{M}_O|} = \cos^{-1}\frac{-226}{375.8} = 127.0° \qquad \text{Ans.}$$

$$\theta_y = \cos^{-1}\frac{M_y}{|\mathbf{M}_O|} = \cos^{-1}\frac{288}{375.8} = 40.0° \qquad \text{Ans.}$$

$$\theta_z = \cos^{-1}\frac{M_z}{|\mathbf{M}_O|} = \cos^{-1}\frac{-85}{375.8} = 103.1° \qquad \text{Ans.}$$

c. The distance d is obtained by using the definition of a moment. Thus

$$d = \frac{M}{F} = \frac{375.8}{840} = 0.447\,\text{m} = 447\,\text{mm} \qquad \text{Ans.}$$

PROBLEMS

4-39* A force with a magnitude of 600 lb acts at a point in a body as shown in Fig. P4-39. Determine the moment of the force about point B.

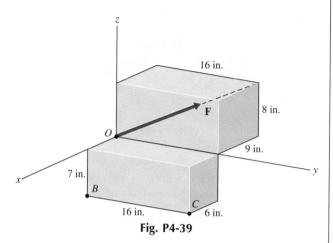

Fig. P4-39

4-40* A force with a magnitude of 850 N acts at a point in a body as shown in Fig. P4-40. Determine the moment of the force about point B.

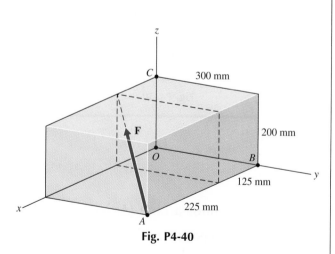

Fig. P4-40

4-41 A force with a magnitude of 480 lb acts at a point in a body as shown in Fig. P4-39. Determine the moment of the force about point C.

4-42 A force with a magnitude of 680 N acts at a point in a body as shown in Fig. P4-40. Determine the moment of the force about point C.

4-43* A force with a magnitude of 580 lb acts at a point in a body as shown in Fig. P4-43. Determine

a. The moment of the force about point B.
b. The direction angles associated with the unit vector **e** along the axis of the moment.

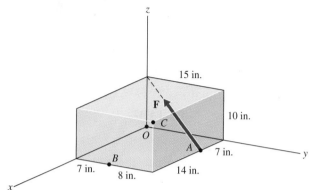

Fig. P4-43

4-44* A force with a magnitude of 730 N acts at a point in a body as shown in Fig. P4-44. Determine

a. The moment of the force about point B.
b. The direction angles associated with the unit vector **e** along the axis of the moment.

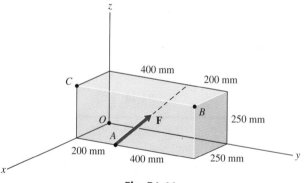

Fig. P4-44

4-45 A force with a magnitude of 870 lb acts at a point in a body as shown in Fig. P4-43. Determine

a. The moment of the force about point C.
b. The direction angles associated with the unit vector **e** along the axis of the moment.

4-46 A force with a magnitude of 585 N acts at a point in a body as shown in Fig. P4-44. Determine

a. The moment of the force about point C.
b. The direction angles associated with the unit vector **e** along the axis of the moment.

107

4-47* Determine the moment of the 580-lb force shown in Fig. P4-47 about point B.

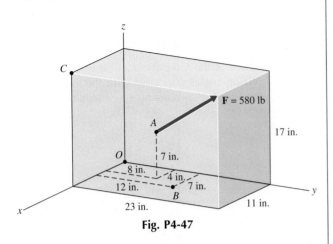

Fig. P4-47

4-49 A pipe bracket is loaded as shown in Fig. P4-49. Determine the moment of the force F about point B.

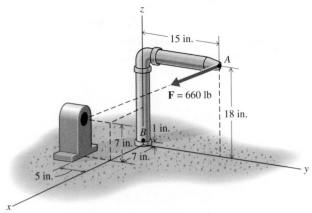

Fig. P4-49

4-48* Determine the moment of the 760-N force shown in Fig. P4-48 about point B.

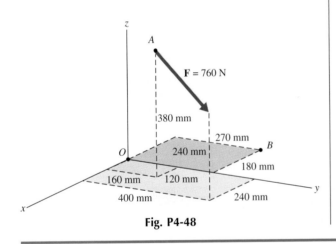

Fig. P4-48

4-50 A bar is bent and loaded as shown in Fig. P4-50. Determine the moment of force F about point O.

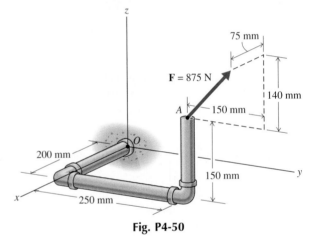

Fig. P4-50

4-3-2 Moment of a Force About a Line (Axis)

The moment M_O of a force F about a point O was defined as the vector product

$$M_O = r \times F \tag{4-3}$$

Although it is possible, mathematically, to define the moment of a force about a point, the quantity has no physical significance in mechanics since bodies rotate about axes (as illustrated in Fig. 4-1) and not points. The vector definition of a moment about a point (Eq. 4-3) is only an intermediate step in a process that allows us to find the moment about an axis that passes through the point.

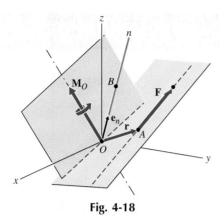

Fig. 4-18

The moment $\mathbf{M}_{OB}$ of a force with respect to a line (say line OB in Fig. 4-18) can be determined by first calculating the moment $\mathbf{M}_O$ about point O on the line (or about any other point on the line). Then, the moment vector $\mathbf{M}_O$ can be resolved into components that are parallel $\mathbf{M}_{\parallel}$ and perpendicular $\mathbf{M}_{\perp}$ to the line OB, as shown in Fig. 4-19. If $\mathbf{e}_n$ is a unit vector in the n-direction along line OB, as shown in Fig. 4-18, then

$$\mathbf{M}_{OB} = \mathbf{M}_{\parallel} = (\mathbf{M}_O \cdot \mathbf{e}_n)\,\mathbf{e}_n$$
$$= [(\mathbf{r} \times \mathbf{F}) \cdot \mathbf{e}_n]\,\mathbf{e}_n = M_{OB}\,\mathbf{e}_n \qquad (4\text{-}16)$$

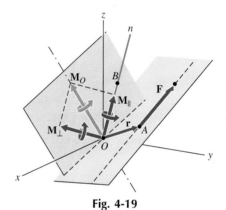

Fig. 4-19

These two operations, the vector product $\mathbf{r} \times \mathbf{F}$ of the position vector $\mathbf{r}$ and the force $\mathbf{F}$ to obtain the moment $\mathbf{M}_O$ about point O, followed by the scalar product $\mathbf{M}_O \cdot \mathbf{e}_n$ of the moment $\mathbf{M}_O$ about point O and the unit vector $\mathbf{e}_n$ along the desired moment axis, to obtain the moment M_{OB} can be performed in sequence or combined into one operation. The quantity inside the brackets is called the triple scalar product. The triple scalar product can be written in determinant form as

$$M_{OB} = \mathbf{M}_O \cdot \mathbf{e}_n = (\mathbf{r} \times \mathbf{F}) \cdot \mathbf{e}_n = \begin{vmatrix} \mathbf{i} & \mathbf{j} & \mathbf{k} \\ r_x & r_y & r_z \\ F_x & F_y & F_z \end{vmatrix} \cdot \mathbf{e}_n \qquad (4\text{-}17)$$

or alternatively as

$$M_{OB} = \mathbf{M}_O \cdot \mathbf{e}_n = (\mathbf{r} \times \mathbf{F}) \cdot \mathbf{e}_n = \begin{vmatrix} e_{nx} & e_{ny} & e_{nz} \\ r_x & r_y & r_z \\ F_x & F_y & F_z \end{vmatrix} \qquad (4\text{-}18)$$

where e_{nx}, e_{ny}, and e_{nz} are the Cartesian components (direction cosines) of the unit vector $\mathbf{e}_n$. The unit vector $\mathbf{e}_n$ is usually selected in the direction from O toward B. A positive coefficient of $\mathbf{e}_n$ in the expression $\mathbf{M}_{OB}$ means that the moment vector has the same sense as that selected for $\mathbf{e}_n$, whereas a negative sign indicates that $\mathbf{M}_{OB}$ is opposite to the sense of $\mathbf{e}_n$.

The following examples illustrate the use of vector algebra for determining moments about lines in three-dimensional problems.

The force **F** in Fig. 4-20 has a magnitude of 440 lb. Determine

a. The moment $\mathbf{M}_B$ of the force about point B.
b. The component of moment $\mathbf{M}_B$ parallel to line BC.
c. The component of moment $\mathbf{M}_B$ perpendicular to line BC.
d. The unit vector associated with the component of moment $\mathbf{M}_B$ perpendicular to line BC.

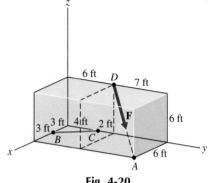

Fig. 4-20

SOLUTION

a. The force **F** and the position vector **r** from point B to point A can be written in Cartesian vector form as

$$\mathbf{F} = 440[(6/11)\mathbf{i} + (7/11)\mathbf{j} - (6/11)\mathbf{k}] = (240\mathbf{i} + 280\mathbf{j} - 240\mathbf{k}) \text{ lb}$$
$$\mathbf{r}_{A/B} = (3\mathbf{i} + 13\mathbf{j}) \text{ ft}$$

The moment $\mathbf{M}_B$ is given by Eq. 4-9 as

$$\mathbf{M}_B = \mathbf{r}_{A/B} \times \mathbf{F} = \begin{vmatrix} \mathbf{i} & \mathbf{j} & \mathbf{k} \\ r_x & r_y & r_z \\ F_x & F_y & F_z \end{vmatrix} = \begin{vmatrix} \mathbf{i} & \mathbf{j} & \mathbf{k} \\ 3 & 13 & 0 \\ 240 & 280 & -240 \end{vmatrix}$$
$$= (-3120\mathbf{i} + 720\mathbf{j} - 2280\mathbf{k}) \text{ ft} \cdot \text{lb} \qquad \text{Ans.}$$

b. The unit vector $\mathbf{e}_{BC}$ associated with line BC is

$$\mathbf{e}_{BC} = -0.60\mathbf{i} + 0.80\mathbf{j}$$

The component of moment $\mathbf{M}_B$ parallel to line BC is given by Eq. 4-16 as

$$\mathbf{M}_{BC} = \mathbf{M}_\| = (\mathbf{M}_B \cdot \mathbf{e}_{BC}) \, \mathbf{e}_{BC} = M_{BC} \, \mathbf{e}_{BC}$$
$$M_{BC} = \mathbf{M}_B \cdot \mathbf{e}_{BC} = (-3120\mathbf{i} + 720\mathbf{j} - 2280\mathbf{k}) \cdot (-0.60\mathbf{i} + 0.80\mathbf{j})$$
$$= (-3120)(-0.60) + (720)(0.80) = 2448 \text{ ft} \cdot \text{lb}$$
$$\mathbf{M}_{BC} = M_{BC} \, \mathbf{e}_{BC} = 2448(-0.60\mathbf{i} + 0.80\mathbf{j}) = (-1469\mathbf{i} + 1958\mathbf{j}) \text{ ft} \cdot \text{lb} \quad \text{Ans.}$$

c. The moment $\mathbf{M}_\perp$ is obtained as the difference between $\mathbf{M}_B$ and $\mathbf{M}_\|$ since $\mathbf{M}_\|$ and $\mathbf{M}_\perp$ are the two rectangular components of $\mathbf{M}_B$. Thus,

$$\mathbf{M}_\perp = \mathbf{M}_B - \mathbf{M}_\| = \mathbf{M}_B - \mathbf{M}_{BC}$$
$$= (-3120\mathbf{i} + 720\mathbf{j} - 2280\mathbf{k}) - (-1469\mathbf{i} + 1958\mathbf{j})$$
$$= (-1651\mathbf{i} - 1238\mathbf{j} - 2280\mathbf{k}) \text{ ft} \cdot \text{lb} \qquad \text{Ans.}$$

d. The magnitude of moment $\mathbf{M}_\perp$ is

$$|\mathbf{M}_\perp| = \sqrt{(-1651)^2 + (-1238)^2 + (-2280)^2} = 3075 \text{ ft} \cdot \text{lb}$$

Therefore

$$\mathbf{e}_\perp = (-1651/3075)\mathbf{i} + (-1238/3075)\mathbf{j} + (-2280/3075)\mathbf{k}$$
$$= -0.537\mathbf{i} - 0.403\mathbf{j} - 0.741\mathbf{k} \qquad \text{Ans.}$$

As a check:

$$\mathbf{e}_\| \cdot \mathbf{e}_\perp = (-0.60\mathbf{i} + 0.80\mathbf{j}) \cdot (-0.5369\mathbf{i} - 0.4026\mathbf{j} - 0.7415\mathbf{k})$$
$$= 0.00006 \cong 0$$

which verifies, except for round-off in the expression for $\mathbf{e}_\perp$, that the moment components $\mathbf{M}_\|$ and $\mathbf{M}_\perp$ are perpendicular.

The force **F** in Fig. 4-21 has a magnitude of 721 N.

a. Determine the moment M_{CD} of the force about line CD.
b. Determine the moment M_{CE} of the force about line CE.

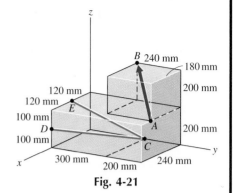

Fig. 4-21

SOLUTION

The force **F** can be expressed in Cartesian vector form as

$$\mathbf{F} = 721[(-180/360.6)\mathbf{i} + (-240/360.6)\mathbf{j} + (200/360.6)\mathbf{k}]$$
$$= (-360\mathbf{i} - 480\mathbf{j} + 400\mathbf{k}) \text{ N}$$

a. For line CD, the unit vector $\mathbf{e}_{CD}$ and the position vector $\mathbf{r}_{A/C}$ are

$$\mathbf{e}_{CD} = (240/397)\mathbf{i} - (300/397)\mathbf{j} + (100/397)\mathbf{k}$$
$$= 0.605\mathbf{i} - 0.756\mathbf{j} + 0.252\mathbf{k}$$
$$\mathbf{r}_{A/C} = (0.180\mathbf{i} + 0.200\mathbf{j} + 0.200\mathbf{k}) \text{ m}$$

Thus, from Eq. 4-18, the moment M_{CD} is

$$M_{CD} = \begin{vmatrix} e_{nx} & e_{ny} & e_{nz} \\ r_x & r_y & r_z \\ F_x & F_y & F_z \end{vmatrix} = \begin{vmatrix} 0.605 & -0.756 & 0.252 \\ 0.180 & 0.200 & 0.200 \\ -360 & -480 & 400 \end{vmatrix}$$

$$= 0.605(176) + 0.756(144) + 0.252(-14.4) = +212 \text{ N} \cdot \text{m} \quad \text{Ans.}$$

Similarly, from the position vector $\mathbf{r}_{B/D}$ and Eq. 4-18, the moment M_{CD} is

$$\mathbf{r}_{B/D} = (-0.240\mathbf{i} + 0.260\mathbf{j} + 0.300\mathbf{k}) \text{ m}$$

$$M_{CD} = \begin{vmatrix} e_{nx} & e_{ny} & e_{nz} \\ r_x & r_y & r_z \\ F_x & F_y & F_z \end{vmatrix} = \begin{vmatrix} 0.605 & -0.756 & 0.252 \\ -0.240 & 0.260 & 0.300 \\ -360 & -480 & 400 \end{vmatrix}$$

$$= 0.605(248) + 0.756(12) + 0.252(208.8) = +212 \text{ N} \cdot \text{m} \quad \text{Ans.}$$

b. For line CE, the unit vector $\mathbf{e}_{CE}$ is

$$\mathbf{e}_{CE} = (120/380)\mathbf{i} - (300/380)\mathbf{j} + (200/380)\mathbf{k}$$
$$= 0.316\mathbf{i} - 0.789\mathbf{j} + 0.526\mathbf{k}$$

From the position vector $\mathbf{r}_{A/C}$ and Eq. 4-18, the moment M_{CE} is

$$M_{CE} = \begin{vmatrix} e_{nx} & e_{ny} & e_{nz} \\ r_x & r_y & r_z \\ F_x & F_y & F_z \end{vmatrix} = \begin{vmatrix} 0.316 & -0.789 & 0.526 \\ 0.180 & 0.200 & 0.200 \\ -360 & -480 & 400 \end{vmatrix}$$

$$= 0.316(176) + 0.789(144) + 0.526(-14.4) = +161.7 \text{ N} \cdot \text{m} \quad \text{Ans.}$$

Similarly, from the position vector $\mathbf{r}_{B/E}$ and Eq. 4-18, the moment M_{CE} is

$$\mathbf{r}_{B/E} = (-0.120\mathbf{i} + 0.260\mathbf{j} + 0.200\mathbf{k}) \text{ m}$$

$$M_{CE} = \begin{vmatrix} e_{nx} & e_{ny} & e_{nz} \\ r_x & r_y & r_z \\ F_x & F_y & F_z \end{vmatrix} = \begin{vmatrix} 0.316 & -0.789 & 0.526 \\ -0.120 & 0.260 & 0.200 \\ -360 & -480 & 400 \end{vmatrix}$$

$$= 0.316(200) + 0.789(24) + 0.526(151.2) = +161.7 \text{ N} \cdot \text{m} \quad \text{Ans.}$$

PROBLEMS

4-51* The force **F** in Fig. P4-51 can be expressed in Cartesian vector form as **F** = (60**i** + 100**j** + 120**k**) lb. Determine the scalar component of the moment at point *B* about line *BC*.

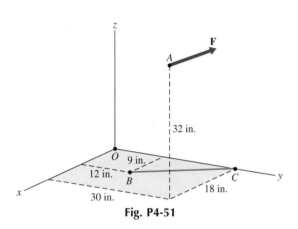

Fig. P4-51

4-52* The force **F** in Fig. P4-52 can be expressed in Cartesian vector form as **F** = (−120**i** + 300**j** + 150**k**) N. Determine the scalar component of the moment at point *B* about line *BC*.

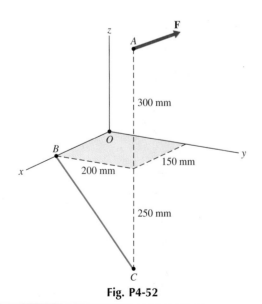

Fig. P4-52

4-53 The magnitude of the force **F** in Fig. P4-53 is 450 lb. Determine the moment of the force about line *BC*. Express the result in Cartesian vector form.

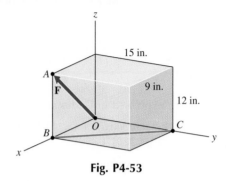

Fig. P4-53

4-54 The magnitude of the force **F** in Fig. P4-54 is 735 N. Determine the moment of the force about line *CD*. Express the result in Cartesian vector form.

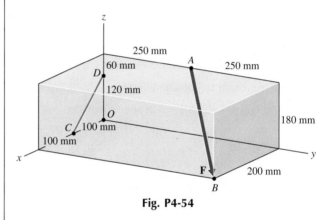

Fig. P4-54

4-55* The magnitude of the force **F** in Fig. P4-55 is 680 lb. Determine

a. The scalar component of the moment at point *O* about line *OC*.
b. The scalar component of the moment at point *D* about line *DE*.

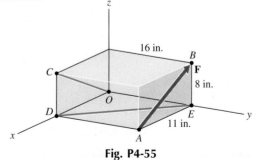

Fig. P4-55

4-56* The magnitude of the force **F** in Fig. P4-56 is 635 N. Determine

a. The scalar component of the moment at point O about line OC.
b. The scalar component of the moment at point D about line DE.

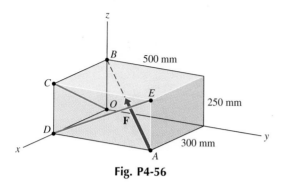

Fig. P4-56

4-57 The magnitude of the force **F** in Fig. P4-57 is 781 lb. Determine

a. The scalar component of the moment at point C about line CD.
b. The component of the moment at point C parallel to line CE.
c. The component of the moment at point C perpendicular to line CE.
d. The unit vector associated with the component of the moment at point C perpendicular to line CE.

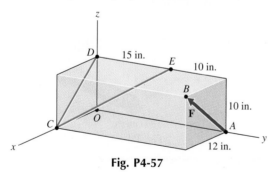

Fig. P4-57

4-58 The magnitude of the force **F** in Fig. P4-58 is 976 N. Determine

a. The scalar component of the moment at point C about line CD.
b. The component of the moment at point C parallel to line CE.
c. The component of the moment at point C perpendicular to line CE.
d. The unit vector associated with the component of the moment at point C perpendicular to line CE.

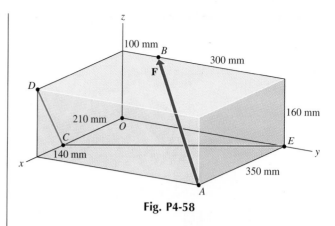

Fig. P4-58

4-59* The magnitude of the force **F** in Fig. P4-59 is 107 lb. Determine the scalar component of the moment at point O about line OC.

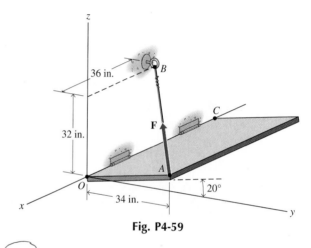

Fig. P4-59

4-60* A 534-N force is applied to a lever-shaft assembly as shown in Fig. P4-60. Determine the scalar component of the moment at point O about line OB.

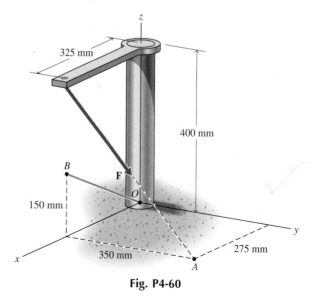

Fig. P4-60

113

4-61 A curved bar is subjected to a 660-lb force as shown in Fig. P4-61. Determine the moment of the force about line *BC*. Express the results in Cartesian vector form.

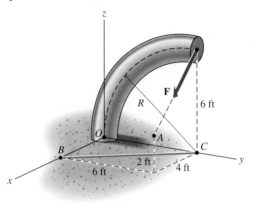

Fig. P4-61

4-63 A 200-lb force is applied to a lever-shaft assembly as shown in Fig. P4-63. Determine the moment of the force about line *OC*. Express the results in Cartesian vector form.

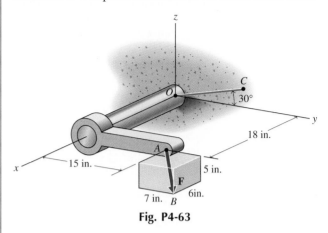

Fig. P4-63

4-62 A bracket is subjected to an 825-N force as shown in Fig. P4-62. Determine the moment of the force about line *OB*. Express the results in Cartesian vector form.

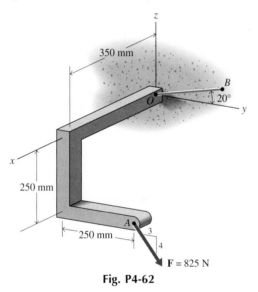

Fig. P4-62

4-64 A bracket is subjected to a 384-N force as shown in Fig. P4-64. Determine the moment of the force about line *OC*. Express the results in Cartesian vector form.

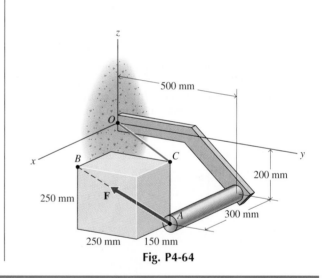

Fig. P4-64

4-4 COUPLES

Two equal, noncollinear, parallel forces of opposite sense (see Fig. 4-22) are called a couple. Since the two forces are equal, parallel, and of opposite sense, the sum of the forces in any direction is zero. Therefore, a couple tends only to rotate the body on which it acts. The moment of a couple is defined as the sum of the moments of the pair of forces that comprise the couple. For the two forces $\mathbf{F}_1$ and $\mathbf{F}_2$ shown in

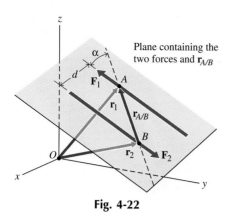

Fig. 4-22

Fig. 4-22, the moments of the couple about points A and B in the plane of the couple are

$$M_A = |\mathbf{F}_2|d \qquad M_B = |\mathbf{F}_1|d$$

However,

$$|\mathbf{F}_1| = |\mathbf{F}_2| = F$$

therefore,

$$M_A = M_B = Fd$$

which indicates that the magnitude of the moment of a couple about a point in the plane of the couple is equal to the magnitude of one of the forces multiplied by the perpendicular distance between the forces.

Other characteristics of a couple can be determined by considering two parallel forces in space such as those shown in Fig. 4-22. The sum of the moments of the two forces about any point O is

$$\mathbf{M}_O = \mathbf{r}_1 \times \mathbf{F}_1 + \mathbf{r}_2 \times \mathbf{F}_2$$

or since $\mathbf{F}_2$ equals $-\mathbf{F}_1$

$$\mathbf{M}_O = \mathbf{r}_1 \times \mathbf{F}_1 + \mathbf{r}_2 \times (-\mathbf{F}_1)$$
$$= (\mathbf{r}_1 - \mathbf{r}_2) \times \mathbf{F}_1 = \mathbf{r}_{A/B} \times \mathbf{F}_1$$

where $\mathbf{r}_{A/B}$ is the position vector from any point B on $\mathbf{F}_2$ to any point A on $\mathbf{F}_1$. Therefore, from the definition of the vector cross-product

$$\mathbf{M}_O = \mathbf{r}_{A/B} \times \mathbf{F}_1$$
$$= |\mathbf{r}_{A/B}||\mathbf{F}_1| \sin \alpha \; \mathbf{e} = F_1 d \; \mathbf{e} \qquad (4\text{-}19)$$

where d is the perpendicular distance between the forces of the couple and $\mathbf{e}$ is a unit vector perpendicular to the plane of the couple with its sense in the direction specified for moments by the right-hand rule. It is obvious from Eq. 4-19 that the moment of a couple does not depend on the location of the moment center O. Thus, the moment of a couple is a free vector.

The characteristics of a couple, which control its "external effect" on a rigid body are as follows:

1.   The magnitude of the moment of the couple.
2.   The sense (direction of rotation) of the couple.
3.   The aspect of the plane of the couple, that is, the direction or slope of the plane (not its location) as defined by a normal n to the plane.

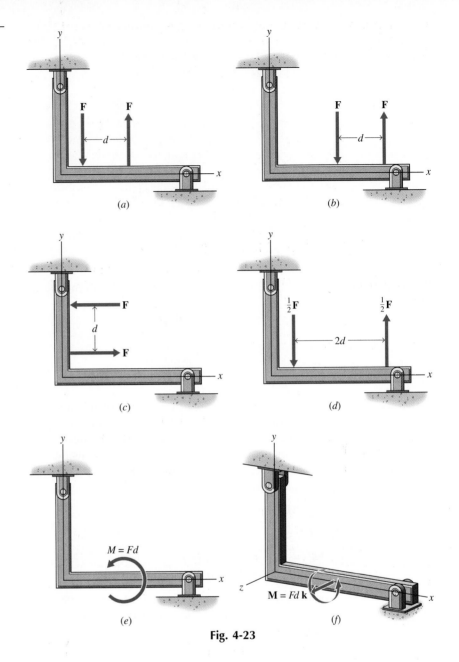

Fig. 4-23

Equation 4-19 indicates that several transformations of a couple can be made without changing any of the external effects of the couple on the body. For example,

1. A couple can be translated to a parallel position in its plane (see Figs. 4-23a and 4-23b) or to any parallel plane (since position vectors $\mathbf{r}_1$ and $\mathbf{r}_2$ do not appear in Eq. 4-19).
2. A couple can be rotated in its plane (see Figs. 4-23a and 4-23c).
3. The magnitude of the two forces of a couple and the distance between them can be changed provided the product Fd remains constant (see Figs. 4-23a and 4-23d).

For two-dimensional problems, a couple is frequently represented by a curved arrow on a sketch of the body as shown in Fig. 4-23e. The magnitude of the moment of the couple $|\mathbf{M}| = M = Fd$ is provided, and the curved arrow indicates the sense of the couple. A couple can also be represented formally as a vector as indicated in Fig. 4-23f.

Any number of couples $\mathbf{C}_1, \mathbf{C}_2, \ldots, \mathbf{C}_n$ in a plane can be combined to yield a resultant couple $\mathbf{C}$ equal to the algebraic sum of the individual couples. A system of couples in space (see Fig. 4-24a) can be combined into a single resultant couple $\mathbf{C}$, by representing each couple of the system (since a couple is a free vector) by a vector, drawn for convenience, from the origin of a set of rectangular axes. Each couple can then be resolved into components $\mathbf{C}_x$, $\mathbf{C}_y$, and $\mathbf{C}_z$ along the coordinate axes. These vector components represent couples lying in the yz-, zx-, and xy-planes, respectively. The x-, y-, and z-components of the resultant couple $\mathbf{C}$ are obtained as $\Sigma\mathbf{C}_x$, $\Sigma\mathbf{C}_y$, and $\Sigma\mathbf{C}_z$ as shown in Fig. 4-24b. The original system of couples is thus reduced to three couples lying in the coordinate planes. The resultant couple $\mathbf{C}$ for the system (see Fig. 4-24c) can be written in vector form as

$$\mathbf{C} = \Sigma\mathbf{C}_x + \Sigma\mathbf{C}_y + \Sigma\mathbf{C}_z = \Sigma C_x\mathbf{i} + \Sigma C_y\mathbf{j} + \Sigma C_z\mathbf{k} \qquad (4\text{-}20)$$

The magnitude of the couple $\mathbf{C}$ is

$$|\mathbf{C}| = \sqrt{(\Sigma C_x)^2 + (\Sigma C_y)^2 + (\Sigma C_z)^2} \qquad (4\text{-}21)$$

Alternatively, the couple $\mathbf{C}$ can be written as

$$\mathbf{C} = C\mathbf{e} \qquad (4\text{-}22)$$

where

$$\mathbf{e} = \cos\theta_x\mathbf{i} + \cos\theta_y\mathbf{j} + \cos\theta_z\mathbf{k}$$

The direction cosines associated with the unit vector $\mathbf{e}$ are

$$\theta_x = \cos^{-1}\frac{\Sigma C_x}{|\mathbf{C}|} \qquad \theta_y = \cos^{-1}\frac{\Sigma C_y}{|\mathbf{C}|} \qquad \theta_z = \cos^{-1}\frac{\Sigma C_z}{|\mathbf{C}|} \qquad (4\text{-}23)$$

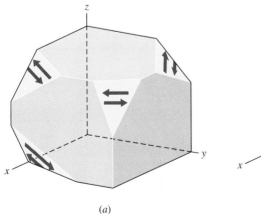
(a)

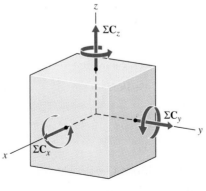
(b)

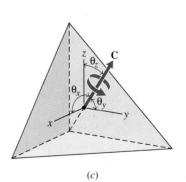
(c)

Fig. 4-24

A beam is loaded with a system of forces as shown in Fig. 4-25. Express the resultant of the force system in Cartesian vector form.

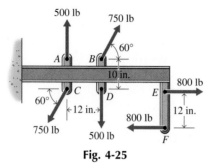

Fig. 4-25

SOLUTION

An examination of Fig. 4-25 indicates that the force system consists of a system of three couples in the same plane. A scalar analysis yields, for forces A and D:

$$M_1 = F_A d_1 = 500(12) = 6000 \text{ in.} \cdot \text{lb} \downarrow = -6000 \text{ in.} \cdot \text{lb}$$

For forces B and C, the perpendicular distance between forces B and C is not obvious from Fig. 4-25; therefore components of the forces will be used.

$$F_{Bx} = F_B \cos 60° = 750 \cos 60° = 375 \text{ lb}$$

$$F_{By} = F_B \sin 60° = 750 \sin 60° = 649.5 \text{ lb}$$

$$M_2 = F_{Bx} d_2 = 375(10) = 3750 \text{ in.} \cdot \text{lb} \downarrow = -3750 \text{ in.} \cdot \text{lb}$$

$$M_3 = F_{By} d_3 = 649.5(12) = 7794 \text{ in.} \cdot \text{lb} \downarrow = +7794 \text{ in.} \cdot \text{lb}$$

For forces E and F:

$$M_4 = F_E d_4 = 800(12) = 9600 \text{ in.} \cdot \text{lb} \downarrow = -9600 \text{ in.} \cdot \text{lb}$$

For any number of couples in a plane, the resultant couple $\mathbf{C}$ is equal to the algebraic sum of the individual couples. Thus,

$$C = \Sigma M = -6000 + 7794 - 3750 - 9600 = -11,556 \text{ in.} \cdot \text{lb} = -963 \text{ ft} \cdot \text{lb}$$

$$\mathbf{C} = -963\mathbf{k} \text{ ft} \cdot \text{lb} \qquad \text{Ans.}$$

Alternatively, a vector analysis yields:

$$\mathbf{C} = (\mathbf{r}_{D/A} \times \mathbf{F}_D) + (\mathbf{r}_{B/C} \times \mathbf{F}_B) + (\mathbf{r}_{E/F} \times \mathbf{F}_E)$$

$$= [(12\mathbf{i} - 10\mathbf{j}) \times (-500\mathbf{j})] + [(12\mathbf{i} + 10\mathbf{j}) \times (375\mathbf{i} + 649.5\mathbf{j})]$$

$$+ [(12\mathbf{j}) \times (800\mathbf{i})] = -11,556\mathbf{k} \text{ in} \cdot \text{lb} = -963\mathbf{k} \text{ ft} \cdot \text{lb} \qquad \text{Ans.}$$

The magnitudes of the four couples applied to the block shown in Fig. 4-26 are $|\mathbf{C}_1| = 75$ N·m, $|\mathbf{C}_2| = 50$ N·m, $|\mathbf{C}_3| = 60$ N·m, and $|\mathbf{C}_4| = 90$ N·m. Determine the magnitude of the resultant couple $\mathbf{C}$ and the direction angles associated with the unit vector $\mathbf{e}$ used to describe the normal to the plane of the resultant couple $\mathbf{C}$.

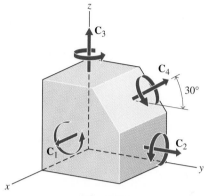

Fig. 4-26

SOLUTION

The x-, y-, and z-components of the resultant couple $\mathbf{C}$ are

$$\Sigma C_x = C_1 = 75 \text{ N·m}$$

$$\Sigma C_y = C_2 + C_4 \cos 30° = 50 + 90 \cos 30° = 127.9 \text{ N·m}$$

$$\Sigma C_z = C_3 + C_4 \sin 30° = 60 + 90 \sin 30° = 105.0 \text{ N·m}$$

The resultant couple $\mathbf{C}$ for the system can be written in vector form as

$$\mathbf{C} = \Sigma \mathbf{C}_x + \Sigma \mathbf{C}_y + \Sigma \mathbf{C}_z = (75\mathbf{i} + 127.9\mathbf{j} + 105.0\mathbf{k}) \text{ N·m}$$

The magnitude of the couple $\mathbf{C}$ is

$$|\mathbf{C}| = \sqrt{(\Sigma C_x)^2 + (\Sigma C_y)^2 + (\Sigma C_z)^2}$$
$$= \sqrt{(75.0)^2 + (127.9)^2 + (105.0)^2} = 181.7 \text{ N·m} \qquad \text{Ans.}$$

The direction angles are

$$\theta_x = \cos^{-1} \frac{\Sigma C_x}{|\mathbf{C}|} = \cos^{-1} \frac{75.0}{181.7} = 65.6° \qquad \text{Ans.}$$

$$\theta_y = \cos^{-1} \frac{\Sigma C_y}{|\mathbf{C}|} = \cos^{-1} \frac{127.9}{181.7} = 45.3° \qquad \text{Ans.}$$

$$\theta_z = \cos^{-1} \frac{\Sigma C_z}{|\mathbf{C}|} = \cos^{-1} \frac{105.0}{181.7} = 54.7° \qquad \text{Ans.}$$

PROBLEMS

4-65* Determine the moment of the couple shown in Fig. P4-65 and the perpendicular distance between the two forces.

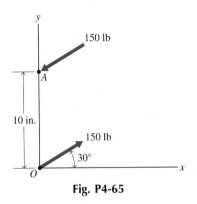

Fig. P4-65

4-66* Determine the moment of the couple shown in Fig. P4-66 and the perpendicular distance between the two forces.

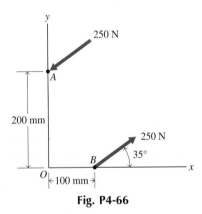

Fig. P4-66

4-67* Two parallel forces of opposite sense $\mathbf{F}_1 = (-30\mathbf{i} - 60\mathbf{j} + 50\mathbf{k})$ lb and $\mathbf{F}_2 = (30\mathbf{i} + 60\mathbf{j} - 50\mathbf{k})$ lb act at points A and O of a body as shown in Fig. P4-67. Determine the moment of the couple and the perpendicular distance between the two forces.

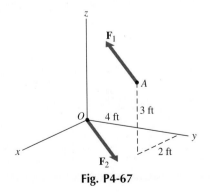

Fig. P4-67

4-68* Two parallel forces of opposite sense $\mathbf{F}_1 = (-75\mathbf{i} + 100\mathbf{j} - 200\mathbf{k})$ N and $\mathbf{F}_2 = (75\mathbf{i} - 100\mathbf{j} + 200\mathbf{k})$ N act at points A and O of a body as shown in Fig. P4-68. Determine the moment of the couple and the perpendicular distance between the two forces.

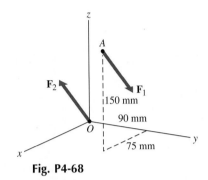

Fig. P4-68

4-69 Two parallel forces of opposite sense $\mathbf{F}_1 = (-70\mathbf{i} - 120\mathbf{j} - 80\mathbf{k})$ lb and $\mathbf{F}_2 = (70\mathbf{i} + 120\mathbf{j} + 80\mathbf{k})$ lb act at points B and A of a body as shown in Fig. P4-69. Determine the moment of the couple and the perpendicular distance between the two forces.

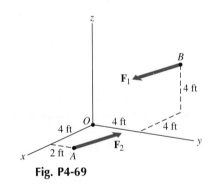

Fig. P4-69

4-70 Two parallel forces of opposite sense $\mathbf{F}_1 = (125\mathbf{i} + 200\mathbf{j} + 250\mathbf{k})$ N and $\mathbf{F}_2 = (-125\mathbf{i} - 200\mathbf{j} - 250\mathbf{k})$ N act at points A and B of a body as shown in Fig. P4-70. Determine the moment of the couple and the perpendicular distance between the two forces.

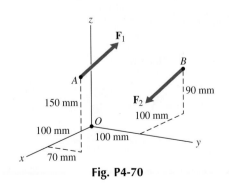

Fig. P4-70

4-71* A bracket is loaded with a system of forces as shown in Fig. P4-71. Express the resultant of the force system in Cartesian vector form.

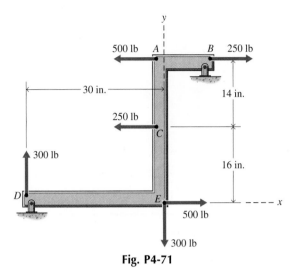

Fig. P4-71

4-73 A bracket is loaded with a system of forces as shown in Fig. P4-73. Express the resultant of the force system in Cartesian vector form.

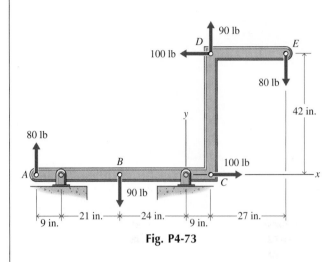

Fig. P4-73

4-72* A plate is loaded with a system of forces as shown in Fig. P4-72. Express the resultant of the force system in Cartesian vector form.

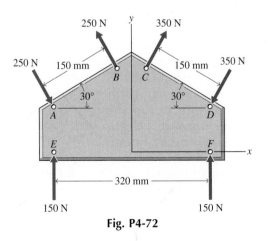

Fig. P4-72

4-74 A bracket is loaded with a system of forces as shown in Fig. P4-74. Express the resultant of the force system in Cartesian vector form.

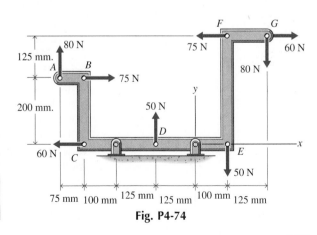

Fig. P4-74

4-75* Three couples are applied to a bent bar as shown in Fig. P4-75. Determine the magnitude of the resultant couple **C** and the direction angles associated with the unit vector **e** used to describe the normal to the plane of the resultant couple **C**.

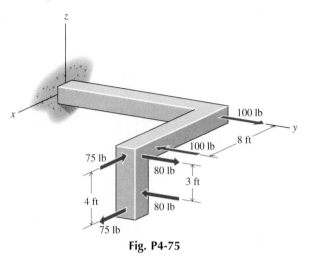

Fig. P4-75

4-76 Three couples are applied to a rectangular block as shown in Fig. P4-76. Determine the magnitude of the resultant couple **C** and the direction angles associated with the unit vector **e** used to describe the normal to the plane of the resultant couple **C**.

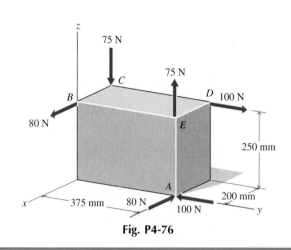

Fig. P4-76

4-5 RESOLUTION OF A FORCE INTO A FORCE AND A COUPLE

In many problems in mechanics it is convenient to resolve a force **F** into a parallel force **F** and a couple **C**. Thus, in Fig. 4-27a, let **F** represent a force acting on a body at point A. An arbitrary point O in the body and the plane containing both force **F** and point O are shown shaded in Fig. 4-27a. The aspect of the shaded plane in the body can be described by using its outer normal n and a unit vector **e** along the outer normal. A two-dimensional representation of the shaded plane is shown in Fig. 4-27b. If two equal and opposite collinear forces **F**, parallel to the original force, are introduced at point O, as shown in Fig. 4-27c, the three forces have the same external effect on the body as the original force **F**, since the effects of the two equal, opposite, and collinear forces cancel. The new three-force system can be considered to be a force **F** acting at point O (parallel to the original force and of the same magnitude and sense), and a couple **C** that has the same moment as the moment of the original force about point O, as shown in Fig. 4-27d. The magnitudes and action lines of the forces of this couple, however, may be changed in accordance with the transformations of a couple discussed in Section 4-4.

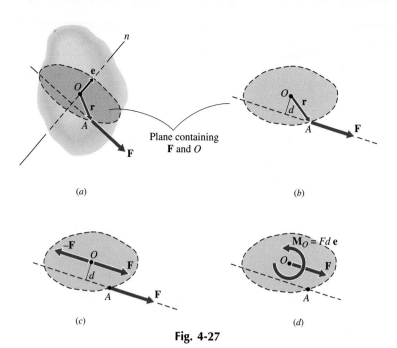

Fig. 4-27

Since a force can be resolved into a force and a couple lying in the same plane, it follows, conversely, that a force and a couple lying in the same plane can be combined into a single force in the plane by reversing the procedure illustrated in Fig. 4-27. Thus, the sole external effect of combining a couple with a force is to move the action line of the force into a parallel position. The magnitude and sense of the force remain unchanged.

The procedure for transforming a force into a force $\mathbf{F}$ and a couple $\mathbf{C}$ is illustrated in the following examples.

A 300-lb force $\mathbf{F}_A$ is applied to a bracket at point A as shown in Fig. 4-28a. Replace the force $\mathbf{F}_A$ by a force $\mathbf{F}_O$ and a couple $\mathbf{C}$ at point O.

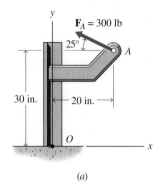

(a)

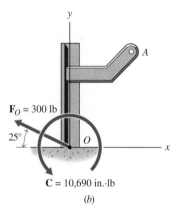

(b)

Fig. 4-28

SOLUTION

Force $\mathbf{F}_O$ has the same magnitude and sense as force $\mathbf{F}_A$. Couple $\mathbf{C}$ has the same magnitude and sense as the moment of force $\mathbf{F}_A$ about point O. Thus,

$$F_O = F_A = 300 \text{ lb}$$
$$C = |\mathbf{F}_A| d_{A/O} = 300 \ (20 \sin 25° + 30 \cos 25°) = 10{,}690 \text{ in.} \cdot \text{lb}$$

therefore, as shown in Fig. 4-28b,

$$\mathbf{F}_O = 300 \text{ lb } 25° \diagup \qquad \text{Ans.}$$
$$\mathbf{C} = 10{,}690 \text{ in.} \cdot \text{lb} \downharpoonleft \qquad \text{Ans.}$$

Alternatively, the force $\mathbf{F}_O$ and couple $\mathbf{C}$ can be expressed in Cartesian vector form as

$$\mathbf{F}_O = \mathbf{F}_A = F_A \mathbf{e}_A = 300(\cos 155° \ \mathbf{i} + \sin 155° \mathbf{j})$$
$$= -271.9\mathbf{i} + 126.8\mathbf{j} \text{ lb} \qquad \text{Ans.}$$
$$\mathbf{C} = \mathbf{r}_{A/O} \times \mathbf{F}_A = (20\mathbf{i} + 30\mathbf{j}) \times (-271.9 + 126.8\mathbf{j})$$
$$= 10{,}690\mathbf{k} \text{ in.} \cdot \text{lb} \qquad \text{Ans.}$$

The force **F** shown in Fig. 4-29a has a magnitude of 763 N. Replace the force **F** by a force **F**$_O$ at point O and a couple **C**.

a. Express the force **F**$_O$ and the couple **C** in Cartesian vector form.
b. Determine the direction angles θ_x, θ_y, and θ_z associated with the unit vector **e** that describes the aspect of the plane of the couple.

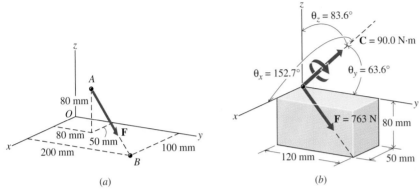

(a)

(b)

Fig. 4-29

SOLUTION

a. The force **F**$_A$ that is equal to force **F** shown in Fig. 4-29a can be expressed in Cartesian vector form as

$$\mathbf{F}_A = \mathbf{F} = F\,\mathbf{e}_F = 763\left[\frac{100-50}{\sqrt{23300}}\,\mathbf{i} + \frac{200-80}{\sqrt{23300}}\,\mathbf{j} + \frac{0-80}{\sqrt{23300}}\,\mathbf{k}\right]$$
$$= (250\mathbf{i} + 600\mathbf{j} - 400\mathbf{k})\text{ N} \qquad \text{Ans.}$$

The moment of the force **F** about point O is

$$\mathbf{C} = \mathbf{r}_{A/O} \times \mathbf{F} = \begin{vmatrix} \mathbf{i} & \mathbf{j} & \mathbf{k} \\ r_x & r_y & r_z \\ F_x & F_y & F_z \end{vmatrix} = \begin{vmatrix} \mathbf{i} & \mathbf{j} & \mathbf{k} \\ 0.05 & 0.08 & 0.08 \\ 250 & 600 & -400 \end{vmatrix}$$
$$= (-80.0\mathbf{i} + 40.0\mathbf{j} + 10.0\mathbf{k})\text{ N}\cdot\text{m} \qquad \text{Ans.}$$

b. The magnitude of the couple **C** is

$$|\mathbf{C}| = \sqrt{M_x^2 + M_y^2 + M_z^2} = \sqrt{(-80.0)^2 + (40.0)^2 + (10.0)^2} = 90.0\text{ N}\cdot\text{m}$$

The direction angles are obtained by using Eqs. 4-14. Thus

$$\theta_x = \cos^{-1}\frac{M_x}{|\mathbf{C}|} = \cos^{-1}\frac{-80.0}{90.0} = 152.7° \qquad \text{Ans.}$$

$$\theta_y = \cos^{-1}\frac{M_y}{|\mathbf{C}|} = \cos^{-1}\frac{40.0}{90.0} = 63.6° \qquad \text{Ans.}$$

$$\theta_z = \cos^{-1}\frac{M_z}{|\mathbf{C}|} = \cos^{-1}\frac{10.0}{90.0} = 83.6° \qquad \text{Ans.}$$

The force **F** and the couple **C** are shown in Fig. 4-29b.

PROBLEMS

4-77* A 250-lb force is applied to a beam as shown in Fig. P4-77. Replace the force by a force at point A and a couple. Express your answer in Cartesian vector form.

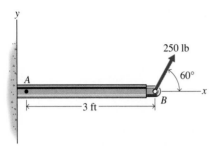

Fig. P4-77

4-78* A 500-N force is applied to a beam as shown in Fig. P4-78. Replace the force by a force at point C and a couple. Express your answer in Cartesian vector form.

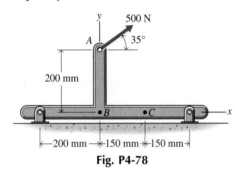

Fig. P4-78

4-79 A 350-lb force is applied to a bracket as shown in Fig. P4-79. Replace the force by a force at point A and a couple. Express your answer in Cartesian vector form.

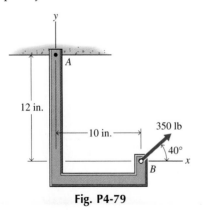

Fig. P4-79

4-80 An 800-N force is applied to a bracket as shown in Fig. P4-80. Replace the force by a force at point B and a couple. Express your answer in Cartesian vector form.

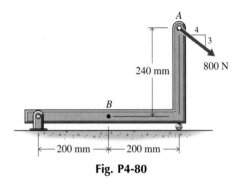

Fig. P4-80

4-81* The force **F** shown in Fig. P4-81 has a magnitude of 780 lb. Replace the force **F** by a force $\mathbf{F}_O$ at point O and a couple **C**.

a. Express the force $\mathbf{F}_O$ and the couple **C** in Cartesian vector form.
b. Determine the direction angles associated with the unit vector **e** that describes the aspect of the plane of the couple.

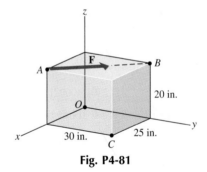

Fig. P4-81

4-82* The force **F** shown in Fig. P4-82 has a magnitude of 595 N. Replace the force **F** by a force $\mathbf{F}_O$ at point O and a couple **C**.

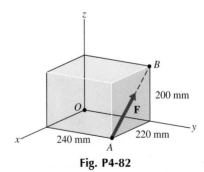

Fig. P4-82

a. Express the force $\mathbf{F}_O$ and the couple $\mathbf{C}$ in Cartesian vector form.
b. Determine the direction angles associated with the unit vector $\mathbf{e}$ that describes the aspect of the plane of the couple.

4-83 The force $\mathbf{F}$ shown in Fig. P4-83 has a magnitude of 928 lb. Replace the force $\mathbf{F}$ by a force $\mathbf{F}_O$ at point O and a couple $\mathbf{C}$.

a. Express the force $\mathbf{F}_O$ and the couple $\mathbf{C}$ in Cartesian vector form.
b. Determine the direction angles associated with the unit vector $\mathbf{e}$ that describes the aspect of the plane of the couple.

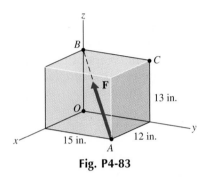

Fig. P4-83

4-84 The force $\mathbf{F}$ shown in Fig. P4-84 has a magnitude of 494 N. Replace the force $\mathbf{F}$ by a force $\mathbf{F}_O$ at point O and a couple $\mathbf{C}$.

a. Express the force $\mathbf{F}_O$ and the couple $\mathbf{C}$ in Cartesian vector form.
b. Determine the direction angles associated with the unit vector $\mathbf{e}$ that describes the aspect of the plane of the couple.

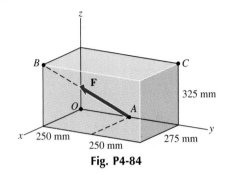

Fig. P4-84

4-85 Solve Problem 4-83 using point C instead of point O.

4-86 Solve Problem 4-84 using point C instead of point O.

4-6 SIMPLIFICATION OF A FORCE SYSTEM: RESULTANTS

Two force systems are equivalent if they produce the same external effect when applied to a rigid body. The "resultant" of any force system is the simplest equivalent system to which the given system will reduce. For some systems, the resultant is a single force. For other systems, the simplest equivalent system is a couple. Still other force systems reduce to a force and a couple as the simplest equivalent system.

4-6-1 Coplanar Force Systems

The resultant of a system of coplanar forces $\mathbf{F}_1, \mathbf{F}_2, \mathbf{F}_3, \ldots, \mathbf{F}_n$ can be determined by using the rectangular components of the forces in any two convenient perpendicular directions. Thus, for a system of forces in the xy-plane, such as the one shown in Fig. 4-30a, the resultant force $\mathbf{R}$ can be expressed as

$$\mathbf{R} = \mathbf{R}_x + \mathbf{R}_y = R_x\mathbf{i} + R_y\mathbf{j} = R\mathbf{e}$$

where

$$R_x = \Sigma F_x \qquad R_y = \Sigma F_y$$
$$R = |\mathbf{R}| = \sqrt{(\Sigma F_x)^2 + (\Sigma F_y)^2} \qquad \text{(4-24)}$$
$$\mathbf{e} = \cos\theta_x\,\mathbf{i} + \cos\theta_y\,\mathbf{j}$$
$$\cos\theta_x = \frac{\Sigma F_x}{R} \qquad \cos\theta_y = \frac{\Sigma F_y}{R}$$

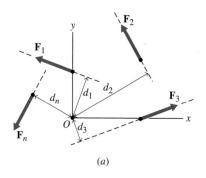

(a)

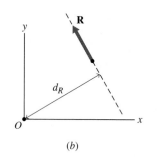

(b)

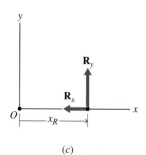

(c)

Fig. 4-30

The location of the line of action of $\mathbf{R}$ with respect to an arbitrary point O (say the origin of the xy-coordinate system) can be computed by using the principle of moments. The moment of $\mathbf{R}$ about point O must equal the sum of the moments of the forces $\mathbf{F}_1$, $\mathbf{F}_2$, $\mathbf{F}_3$, . . . , $\mathbf{F}_n$ of the original system about the same point O, as shown in Fig. 4-30b. Thus,

$$Rd_R = F_1 d_1 + F_2 d_2 + F_3 d_3 + \cdots + F_n d_n = \Sigma M_O$$

Therefore

$$d_R = \frac{\Sigma M_O}{R} \tag{4-25}$$

where ΣM_O stands for the algebraic sum of the moments of the forces of the original system about point O. The direction of d_R is selected so that the product Rd_R produces a moment about point O with the same sense (clockwise or counterclockwise) as the algebraic sum of the moments of the forces of the original system about point O.

The location of the line of action of $\mathbf{R}$ with respect to point O can also be specified by determining the intercept of the line of action of the force with one of the coordinate axes. For example in Fig. 4-30c, the intercept x_R is determined from the equation

$$x_R = \frac{\Sigma M_O}{R_y} \tag{4-26}$$

since the component R_x of the resultant force $\mathbf{R}$ does not produce a moment about point O. The special case for a system of coplanar parallel forces is illustrated in Fig. 4-31.

In the event that the resultant force $\mathbf{R}$ of a system of coplanar forces $\mathbf{F}_1$, $\mathbf{F}_2$, . . . , $\mathbf{F}_n$ is zero but the moment ΣM_O is not zero, the resultant is a couple $\mathbf{C}$ whose vector is perpendicular to the plane containing the forces (the xy-plane for the case being discussed). Thus, the resultant of a coplanar system of forces may be either a force $\mathbf{R}$ or a couple $\mathbf{C}$.

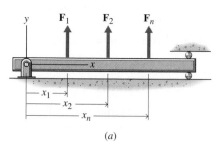

(a)

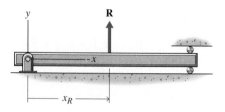

(b)

Fig. 4-31

EXAMPLE PROBLEM 4-15

Four forces are applied to a rectangular plate as shown in Fig. 4-32a. Determine the magnitude and direction of the resultant of the four forces.

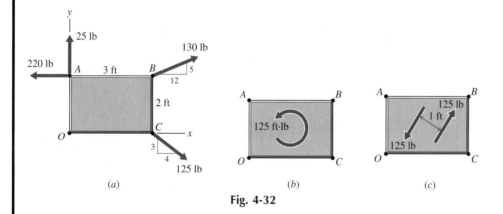

Fig. 4-32

SOLUTION

The resultant of a system of coplanar forces is either a force **R** or a couple **C**. The resultant force is obtained by using Eqs. 4-24. Thus,

$$R_x = \Sigma F_x = -220 + (12/13)(130) + (4/5)(125) = 0$$
$$R_y = \Sigma F_y = 25 + (5/13)(130) - (3/5)(125) = 0$$
$$\mathbf{R} = R_x\mathbf{i} + R_y\mathbf{j} = \mathbf{0} \qquad \text{Ans.}$$

The couple **C**, if it exists, can be determined by summing moments about any convenient point in the plate. An examination of the force system indicates that the determination can be simplified by summing moments about point B. Thus,

$$\begin{aligned} \wr + \Sigma M_B &= -F_{Ay}d_A + F_{Cx}d_C \\ &= -25(3) + (4/5)(125)(2) = 125 \text{ ft} \cdot \text{lb} \qquad \text{Ans.} \end{aligned}$$

Alternatively, the moments can be summed about point O. Thus,

$$\begin{aligned} \wr + \Sigma M_O &= F_{Ax}d_A - F_{Bx}d_{Bx} + F_{By}d_{By} - F_{Cy}\,d_{Cy} \\ &= 220(2) - (12/13)(130)(2) + (5/13)(130)(3) \\ &\quad - (3/5)(125)(3) = 125 \text{ ft} \cdot \text{lb} \qquad \text{Ans.} \end{aligned}$$

Since the resultant of the force system is a couple, it can be represented by a curved arrow at any point on the plate, as illustrated in Fig. 4-32b, or by a pair of equal but opposite parallel forces, as illustrated in Fig. 4-32c.

Three forces and a couple are applied to a bracket as shown in Fig. 4-33a. Determine

a. The magnitude and direction of the resultant.
b. The perpendicular distance d_R from point O to the line of action of the resultant.
c. The distance x_R from point O to the intercept of the line of action of the resultant with the x-axis.

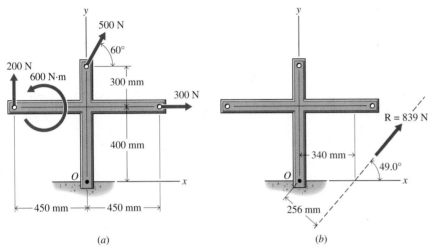

(a)   (b)

Fig. 4-33

SOLUTION

a. The resultant of a system of coplanar forces is either a force **R** or a couple **C**. The resultant force is obtained by using Eqs. 4-24. Thus,

$$R_x = \Sigma F_x = 500 \cos 60° + 300 = 550 \text{ N}$$
$$R_y = \Sigma F_y = 500 \sin 60° + 200 = 633 \text{ N}$$
$$R = |\mathbf{R}| = \sqrt{(\Sigma F_x)^2 + (\Sigma F_y)^2}$$
$$= \sqrt{(550)^2 + (633)^2} = 838.6 \text{ N} = 839 \text{ N} \qquad \text{Ans.}$$
$$\theta_x = \cos^{-1} \frac{\Sigma F_x}{R} = \cos^{-1} \frac{550}{838.6} = 49.0° \qquad \text{Ans.}$$

b. From Eq. 4-25

$$(\text{↓}+\Sigma M_O = -300(0.400) - 500 \cos 60° (0.700)$$
$$- 200(0.450) + 600 = 215 \text{ N} \cdot \text{m}$$
$$d_R = \frac{\Sigma M_O}{R} = \frac{215}{838.6} = 0.2564 \text{ m} = 256 \text{ mm} \qquad \text{Ans.}$$

c. From Eq. 4-26

$$x_R = \frac{\Sigma M_O}{R_y} = \frac{215}{633} = 0.3397 \text{ m} = 340 \text{ mm} \qquad \text{Ans.}$$

The results are illustrated in Fig. 4-33b.

PROBLEMS

4-87* Three forces are applied to a beam as shown in Fig. P4-87. Replace the three forces by an equivalent force-couple system at point C.

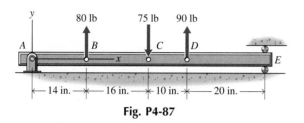

Fig. P4-87

4-88* Four forces are applied to a post as shown in Fig. P4-88. Replace the four forces by an equivalent force-couple system at point C.

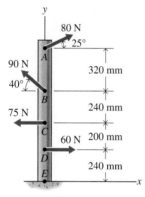

Fig. P4-88

4-89 Three forces are applied to a beam as shown in Fig. P4-87. Determine the resultant **R** of the three forces and the location of its line of action with respect to support A.

4-90 Four forces are applied to a post as shown in Fig. P4-88. Determine the resultant **R** of the four forces and the location of its line of action with respect to support E.

4-91* Four forces are applied to the cross section of a beam as shown in Fig. P4-91. Determine the magnitude and direction of the resultant of the four forces and the distance x_R from point O to the intercept of the line of action of the resultant with the x-axis.

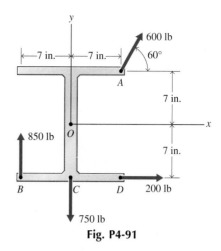

Fig. P4-91

4-92* Three forces are applied to an angle bracket as shown in Fig. P4-92. Determine the magnitude and direction of the resultant of the three forces and the perpendicular distance d_R from point O to the line of action of the resultant.

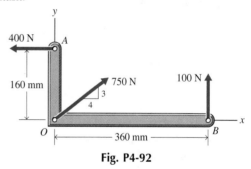

Fig. P4-92

4-93 Four forces are applied to a circular disk as shown in Fig. P4-93. Determine the magnitude and direction of the resultant of the four forces and the distance x_R from point O to the intercept of the line of action of the resultant with the x-axis.

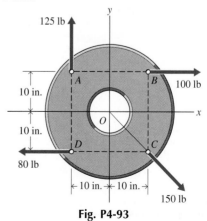

Fig. P4-93

131

4-94 Four forces and a couple are applied to a rectangular plate as shown in Fig. P4-94. Determine the magnitude and direction of the resultant of the force-couple system and the distance x_R from point O to the intercept of the line of action of the resultant with the x-axis.

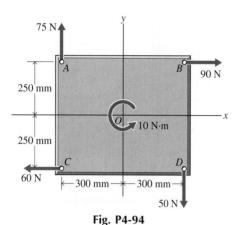

Fig. P4-94

4-95* Four forces are applied to a circular disk as shown in Fig. P4-93. Replace the four forces by an equivalent force-couple system at point C.

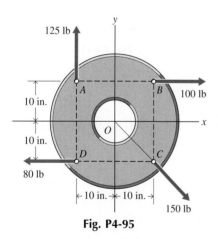

Fig. P4-95

4-96* Four forces and a couple are applied to a rectangular plate as shown in Fig. P4-94. Replace the four forces by an equivalent force-couple system at point C.

4-97 Four forces and a couple are applied to a frame as shown in Fig. P4-97. Determine

a. The magnitude and direction of the resultant.
b. The perpendicular distance d_R from point A to the line of action of the resultant.

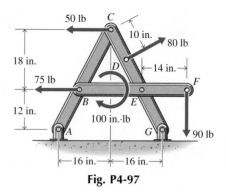

Fig. P4-97

4-98 Four forces are applied to a bracket as shown in Fig. P4-98. Determine

a. The magnitude and direction of the resultant.
b. The perpendicular distance d_R from point A to the line of action of the resultant.

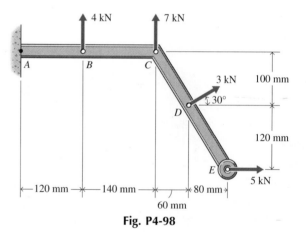

Fig. P4-98

4-99* Four forces are applied to a truss as shown in Fig. P4-99. Determine

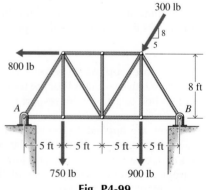

Fig. P4-99

a. The magnitude and direction of the resultant.
b. The perpendicular distance d_R from support A to the line of action of the resultant.

4-100* Four forces are applied to a truss as shown in Fig. P4-100. Determine

a. The magnitude and direction of the resultant.
b. The perpendicular distance d_R from point A to the line of action of the resultant.

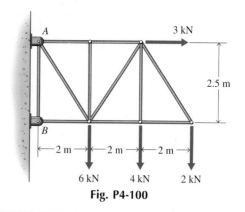

Fig. P4-100

4-6-2 Noncoplanar, Parallel Force Systems

If all forces of a three-dimensional system are parallel, the resultant force $\mathbf{R}$ is the algebraic sum of the forces of the system. The line of action of the resultant is determined by using the principle of moments. Thus, as shown in Fig. 4-34a, for a system of forces $\mathbf{F}_1, \mathbf{F}_2, \ldots, \mathbf{F}_n$, perpendicular to the xy-plane,

$$\mathbf{R} = \mathbf{F}_1 + \mathbf{F}_2 + \cdots + \mathbf{F}_n = R\mathbf{k} = \Sigma F\, \mathbf{k} \qquad (4\text{-}27)$$
$$\mathbf{M}_O = \mathbf{r} \times \mathbf{R} = \mathbf{r}_1 \times \mathbf{F}_1 + \mathbf{r}_2 \times \mathbf{F}_2 + \mathbf{r}_3 \times \mathbf{F}_3 + \cdots + \mathbf{r}_n \times \mathbf{F}_n \qquad (4\text{-}28)$$

The intersection of the line of action of the resultant force $\mathbf{R}$ with the xy-plane (see Fig. 4-34b) is located by equating the moments of $\mathbf{R}$ about the x- and y-axes to the sums of the moments of the forces of the system $\mathbf{F}_1, \mathbf{F}_2, \ldots, \mathbf{F}_n$ about the x- and y-axes. It is important to note from Eqs. 4-10 that $M_x = F_z r_y$ while $M_y = -F_z r_x$. Therefore,

$$Rx_R = F_1 x_1 + F_2 x_2 + \cdots + F_n x_n = -\Sigma M_y$$
$$Ry_R = F_1 y_1 + F_2 y_2 + \cdots + F_n y_n = \Sigma M_x$$

which gives

$$x_R = -\frac{\Sigma M_y}{R} \qquad y_R = \frac{\Sigma M_x}{R} \qquad (4\text{-}29)$$

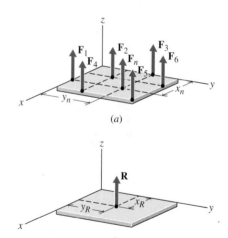

Fig. 4-34

In the event that the resultant force $\mathbf{R}$ of the system of parallel forces is zero but the moments ΣM_x and ΣM_y are not zero, the resultant is a couple $\mathbf{C}$ whose vector lies in a plane perpendicular to the forces (in the xy-plane for the system of forces illustrated in Fig. 4-34). Thus, the resultant of a noncoplanar system of parallel forces may be either a force $\mathbf{R}$ or a couple $\mathbf{C}$. The following example illustrates the procedure for determining resultants of noncoplanar, parallel force systems.

Determine the resultant of the parallel force system shown in Fig. 4-35 and locate the intersection of the line of action of the resultant with the xy-plane.

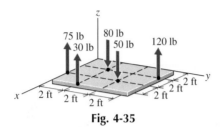

Fig. 4-35

SOLUTION

The resultant **R** of the force system is determined by using Eq. 4-27:

$$\mathbf{R} = \Sigma F \, \mathbf{k}$$
$$= (75 + 30 - 80 - 50 + 120)\mathbf{k} = 95\mathbf{k} \text{ lb} \qquad \text{Ans.}$$

The intersection (x_R, y_R) of the line of action of the resultant with the xy-plane is determined by using Eqs. 4-29. With positive moments defined by using the right-hand rule

$$\Sigma M_y = -75(4) - 30(6) + 80(2) + 50(4) - 120(2) = -360 \text{ ft} \cdot \text{lb}$$
$$\Sigma M_x = 75(0) + 30(2) - 80(2) - 50(4) + 120(6) = 420 \text{ ft} \cdot \text{lb}$$

Therefore,

$$x_R = -\frac{\Sigma M_y}{R} = -\frac{-360}{95} = 3.79 \text{ ft} \qquad \text{Ans.}$$

$$y_R = \frac{\Sigma M_x}{R} = \frac{420}{95} = 4.42 \text{ ft} \qquad \text{Ans.}$$

The coordinates (x_R, y_R) can also be determined by using a vector analysis:

$$\mathbf{M}_O = \mathbf{r} \times \mathbf{R} = \mathbf{r}_1 \times \mathbf{F}_1 + \mathbf{r}_2 \times \mathbf{F}_2 + \mathbf{r}_3 \times \mathbf{F}_3 + \mathbf{r}_4 \times \mathbf{F}_4 + \mathbf{r}_5 \times \mathbf{F}_5$$
$$= [(x_R\mathbf{i} + y_R\mathbf{j}) \times (95\mathbf{k})]$$
$$= [(4\mathbf{i}) \times (75\mathbf{k})] + [(6\mathbf{i} + 2\mathbf{j}) \times (30\mathbf{k})] + [(2\mathbf{i} + 2\mathbf{j}) \times (-80\mathbf{k})]$$
$$\quad + [(4\mathbf{i} + 4\mathbf{j}) \times (-50\mathbf{k})] + [(2\mathbf{i} + 6\mathbf{j}) \times (120\mathbf{k})]$$
$$= -95x_R\mathbf{j} + 95y_R\mathbf{i} = 420\mathbf{i} - 360\mathbf{j}$$

Solving for x_R and y_R yields

$$x_R = \frac{360}{95} = 3.79 \text{ ft} \qquad \text{Ans.}$$

$$y_R = \frac{420}{95} = 4.42 \text{ ft} \qquad \text{Ans.}$$

PROBLEMS

4-101* Determine the resultant of the parallel force system shown in Fig. P4-101 and locate the intersection of the line of action of the resultant with the *xy*-plane.

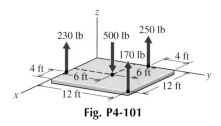

Fig. P4-101

4-102* Determine the resultant of the parallel force system shown in Fig. P4-102 and locate the intersection of the line of action of the resultant with the *xz*-plane.

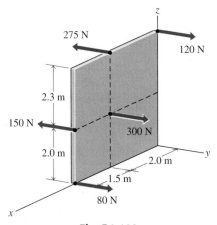

Fig. P4-102

4-103 Determine the resultant of the parallel force system shown in Fig. P4-103 and locate the intersection of the line of action of the resultant with the *yz*-plane.

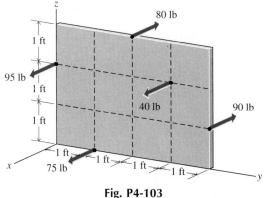

Fig. P4-103

4-104 Determine the resultant of the parallel force system shown in Fig. P4-104 and locate the intersection of the line of action of the resultant with the *xz*-plane.

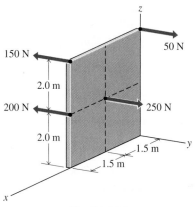

Fig. P4-104

4-105* The resultant of the parallel force system shown in Fig. P4-105 is a couple that can be expressed in Cartesian vector form as

$$\mathbf{C} = (-1160\mathbf{j} + 2250\mathbf{k}) \text{ in.} \cdot \text{lb}$$

Determine the magnitudes of forces $\mathbf{F}_1$, $\mathbf{F}_2$, and $\mathbf{F}_3$.

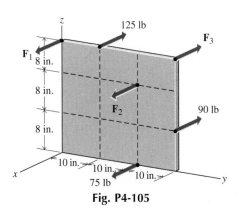

Fig. P4-105

135

4-106* The resultant of the parallel force system shown in Fig. P4-106 is a couple that can be expressed in Cartesian vector form as

$$C = (-180i + 435j) \text{ N} \cdot \text{m}$$

Determine the magnitudes of forces F_1, F_2, and F_3.

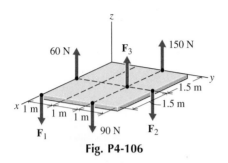

Fig. P4-106

4-107 The resultant of the parallel force system shown in Fig. P4-107 is a couple that can be expressed in Cartesian vector form as

$$C = (850j - 950k) \text{ in.} \cdot \text{lb}$$

Determine the magnitude and sense of each of the unknown forces (F_1, F_2, and F_3).

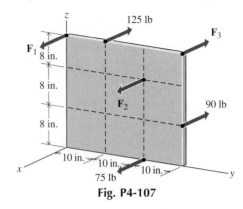

Fig. P4-107

4-108 The resultant of the parallel force system shown in Fig. P4-106 is a couple that can be expressed in Cartesian vector form as

$$C = (200i + 350j) \text{ N} \cdot \text{m}$$

Determine the magnitude and sense of each of the unknown forces (F_1, F_2, and F_3).

4-6-3 General Force Systems

The resultant **R** of a general, three-dimensional system of forces F_1, F_2, F_3, . . . , F_n, such as the one shown in Fig. 4-36a, can be determined by resolving each force of the system into an equal parallel force through any point (taken for convenience as the origin O of a system of coordinate axes) and a couple as shown in Fig. 4-36b. Thus, as shown in Fig. 4-36c, the given system is replaced by two systems:

1. A system of noncoplanar, concurrent forces through the origin O that have the same magnitudes and directions as the forces of the original system.
2. A system of noncoplanar couples.

Each force and couple of the two systems can be resolved into components along the coordinate axes as shown in Figs. 4-36d and 4-36e. The resultant of the concurrent force system is a force through the origin O that can be expressed as

$$R = R_x + R_y + R_z = R_x i + R_y j + R_z k = Re \qquad (4\text{-}30)$$

where

$$R_x = \Sigma F_x \qquad R_y = \Sigma F_y \qquad R_z = \Sigma F_z$$
$$R = |R| = \sqrt{(\Sigma F_x)^2 + (\Sigma F_y)^2 + (\Sigma F_z)^2}$$
$$e = \cos \theta_x \, i + \cos \theta_y \, j + \cos \theta_z \, k$$
$$\cos \theta_x = \frac{\Sigma F_x}{|R|} \qquad \cos \theta_y = \frac{\Sigma F_y}{|R|} \qquad \cos \theta_z = \frac{\Sigma F_z}{|R|}$$

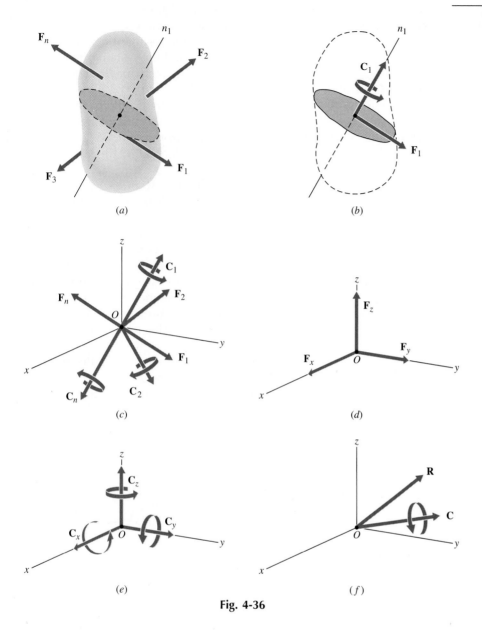

Fig. 4-36

The resultant of the system of noncoplanar couples is a couple **C** that can be expressed as

$$\mathbf{C} = \Sigma\mathbf{C}_x + \Sigma\mathbf{C}_y + \Sigma\mathbf{C}_z = \Sigma C_x\,\mathbf{i} + \Sigma C_y\,\mathbf{j} + \Sigma C_z\,\mathbf{k} = C\mathbf{e} \quad \text{(4-31)}$$

where

$$C = |\mathbf{C}| = \sqrt{(\Sigma C_x)^2 + (\Sigma C_y)^2 + (\Sigma C_z)^2}$$
$$\mathbf{e} = \cos\theta_x\,\mathbf{i} + \cos\theta_y\,\mathbf{j} + \cos\theta_z\,\mathbf{k}$$
$$\cos\theta_x = \frac{\Sigma C_x}{|\mathbf{C}|} \qquad \cos\theta_y = \frac{\Sigma C_y}{|\mathbf{C}|} \qquad \cos\theta_z = \frac{\Sigma C_z}{|\mathbf{C}|}$$

The resultant force **R** and the resultant couple **C** shown in Fig. 4-36*f* together constitute the resultant of the system with respect to point O. In special cases, the resultant couple **C** may vanish, leaving

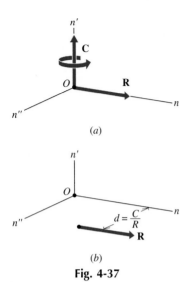

(a)

(b)

Fig. 4-37

the force $\mathbf{R}$ as the resultant of the system. Again in special cases, the resultant force $\mathbf{R}$ may vanish, leaving the couple $\mathbf{C}$ as the resultant of the system. If the resultant force $\mathbf{R}$ and the resultant couple $\mathbf{C}$ both vanish, the resultant of the system is zero and the system is in equilibrium. Thus, the resultant of a general force system may be a force $\mathbf{R}$, a couple $\mathbf{C}$, or both a force $\mathbf{R}$ and a couple $\mathbf{C}$.

When the couple $\mathbf{C}$ is perpendicular to the resultant force $\mathbf{R}$, as shown in Fig. 4-37, the two can be combined to form a single force $\mathbf{R}$ whose line of action is a distance $d = C/R$ from point O in a direction that makes the direction of the moment of R about O the same as that of C.

Another transformation of the resultant ($\mathbf{R}$ and $\mathbf{C}$) of a general force system is illustrated in Figs. 4-38 and 4-39. In this case, the couple $\mathbf{C}$ is resolved into components parallel and perpendicular to the resultant force $\mathbf{R}$, as shown in Fig. 4-38b. The resultant force $\mathbf{R}$ and the perpendicular component of the couple $\mathbf{C}_\perp$ can be combined as illustrated in Fig. 4-37. In addition, the parallel component of the couple $\mathbf{C}_\parallel$ can be translated to coincide with the line of action of the resultant force $\mathbf{R}$, as shown in Fig. 4-38c. The combination of couple $\mathbf{C}_\parallel$ and resultant force $\mathbf{R}$ is known as a wrench. The action may be described as a push (or pull) and a twist about an axis parallel to the push (or pull). When the force and moment vectors have the same sense, as shown in Fig. 4-38c, the wrench is positive. When the vectors have the opposite sense, as shown in Fig. 4-39c, the wrench is negative.

The analysis of general, three-dimensional force systems is illustrated in the following examples.

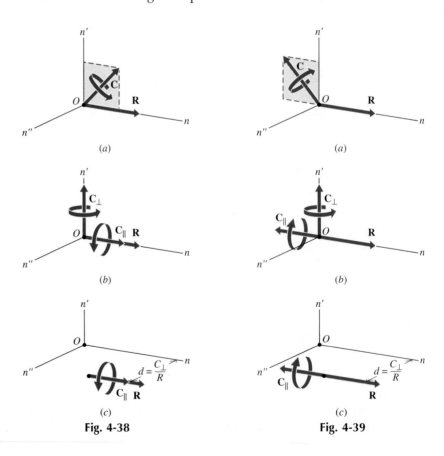

(a)

(b)

(c)

Fig. 4-38

(a)

(b)

(c)

Fig. 4-39

Replace the force system shown in Fig. 4-40a with a force **R** through point O and a couple **C**.

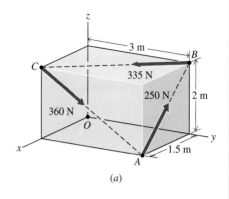

(a)

SOLUTION

The three forces and their positions with respect to point O can be written in Cartesian vector form as:

$$\mathbf{F}_A = 250\left[-\frac{1.5}{2.5}\mathbf{i} + \frac{2.0}{2.5}\mathbf{k}\right] = (-150\mathbf{i} + 200\mathbf{k})\ \text{N} \qquad \mathbf{r}_{A/O} = (1.5\mathbf{i} + 3.0\mathbf{j})\ \text{m}$$

$$\mathbf{F}_B = 335\left[\frac{1.5}{3.354}\mathbf{i} - \frac{3.0}{3.354}\mathbf{j}\right] = (150\mathbf{i} - 300\mathbf{j})\ \text{N} \qquad \mathbf{r}_{B/O} = (3.0\mathbf{j} + 2.0\mathbf{k})\ \text{m}$$

$$\mathbf{F}_C = 360\left[\frac{3.0}{3.606}\mathbf{j} - \frac{2.0}{3.606}\mathbf{k}\right] = (300\mathbf{j} - 200\mathbf{k})\ \text{N} \qquad \mathbf{r}_{C/O} = (1.5\mathbf{i} + 2.0\mathbf{k})\ \text{m}$$

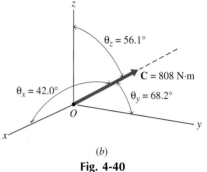

(b)

Fig. 4-40

Each of the three forces can be replaced by an equal force through point O and a couple. The vector sum of the concurrent forces is

$$\mathbf{R} = \mathbf{F}_A + \mathbf{F}_B + \mathbf{F}_C$$
$$= -150\mathbf{i} + 200\mathbf{k} + 150\mathbf{i} - 300\mathbf{j} + 300\mathbf{j} - 200\mathbf{k} = 0 \qquad \text{Ans.}$$

The moment $\mathbf{M}_O$ of the resultant couple is

$$\mathbf{C} = \mathbf{M}_O = \mathbf{r}_{A/O} \times \mathbf{F}_A + \mathbf{r}_{B/O} \times \mathbf{F}_B + \mathbf{r}_{C/O} \times \mathbf{F}_C$$

$$= \begin{vmatrix} \mathbf{i} & \mathbf{j} & \mathbf{k} \\ 1.5 & 3.0 & 0 \\ -150 & 0 & 200 \end{vmatrix} + \begin{vmatrix} \mathbf{i} & \mathbf{j} & \mathbf{k} \\ 0 & 3.0 & 2.0 \\ 150 & -300 & 0 \end{vmatrix} + \begin{vmatrix} \mathbf{i} & \mathbf{j} & \mathbf{k} \\ 1.5 & 0 & 2.0 \\ 0 & 300 & -200 \end{vmatrix}$$

$$= (600\mathbf{i} + 300\mathbf{j} + 450\mathbf{k})\ \text{N} \cdot \text{m} \qquad \text{Ans.}$$

The magnitude of the resultant couple is

$$|\mathbf{C}| = \sqrt{C_x^2 + C_y^2 + C_z^2}$$
$$= \sqrt{(600)^2 + (300)^2 + (450)^2} = 807.8 = 808\ \text{N} \cdot \text{m} \qquad \text{Ans.}$$

Finally, the direction angles that locate the axis of the couple are

$$\theta_x = \cos^{-1}\frac{C_x}{|\mathbf{C}|} = \cos^{-1}\frac{600}{807.8} = 42.0° \qquad \text{Ans.}$$

$$\theta_y = \cos^{-1}\frac{C_y}{|\mathbf{C}|} = \cos^{-1}\frac{300}{807.8} = 68.2° \qquad \text{Ans.}$$

$$\theta_z = \cos^{-1}\frac{C_z}{|\mathbf{C}|} = \cos^{-1}\frac{450}{807.8} = 56.1° \qquad \text{Ans.}$$

The resultant of the force system is the couple shown in Fig. 4.40b.

Three forces are applied to a rigid body as shown in Fig. 4-41a.

a. Reduce the forces to a wrench.
b. Determine the intersection of the wrench with the xy-plane.

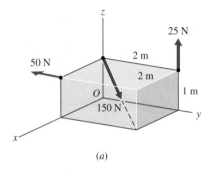

(a)

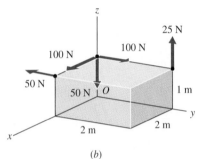

(b)

Fig. 4-41

SOLUTION

a. The rectangular components of the three forces are shown in Fig. 4-41b. From these data, the equivalent force **R** and the couple **C** at point O can be written in Cartesian vector form as

$$\mathbf{R} = (100\mathbf{i} + 50\mathbf{j} - 25\mathbf{k}) \text{ N}$$
$$\mathbf{C} = (100\mathbf{j} - 100\mathbf{k}) \text{ N} \cdot \text{m}$$

The magnitude of the resultant force is

$$|\mathbf{R}| = \sqrt{R_x^2 + R_y^2 + R_z^2}$$
$$= \sqrt{(100)^2 + (50)^2 + (-25)^2} = 114.56 = 114.6 \text{ N} \cdot \text{m} \qquad \text{Ans.}$$

The direction angles that locate the line of action of the resultant force are

$$\theta_x = \cos^{-1} \frac{R_x}{|\mathbf{R}|} = \cos^{-1} \frac{100}{114.56} = 29.2° \qquad \text{Ans.}$$

$$\theta_y = \cos^{-1} \frac{R_y}{|\mathbf{R}|} = \cos^{-1} \frac{50}{114.56} = 64.1° \qquad \text{Ans.}$$

$$\theta_z = \cos^{-1} \frac{R_z}{|\mathbf{R}|} = \cos^{-1} \frac{-25}{114.56} = 102.6° \qquad \text{Ans.}$$

A unit vector $\mathbf{e}_R$ along the line of action of the resultant force **R** is

$$\mathbf{e} = \cos \theta_x \, \mathbf{i} + \cos \theta_y \, \mathbf{j} + \cos \theta_z \, \mathbf{k}$$
$$= 0.8729\mathbf{i} + 0.4365\mathbf{j} - 0.2182\mathbf{k}$$

The components of the couple parallel and perpendicular to the line of action of the resultant force $\mathbf{R}$ are

$$|\mathbf{C}_\parallel| = \mathbf{C} \cdot \mathbf{e}_R = (100\mathbf{j} - 100\mathbf{k}) \cdot (0.8729\mathbf{i} + 0.4365\mathbf{j} - 0.2182\mathbf{k})$$
$$= 65.46 = 65.5 \text{ N} \cdot \text{m} \qquad \text{Ans.}$$
$$\mathbf{C}_\parallel = |\mathbf{C}_\parallel| \, \mathbf{e}_R = 65.46(0.8729\mathbf{i} + 0.4365\mathbf{j} - 0.2182\mathbf{k})$$
$$= (57.14\mathbf{i} + 28.57\mathbf{j} - 14.28\mathbf{k}) \text{ N} \cdot \text{m}$$
$$\mathbf{C}_\perp = \mathbf{C} - \mathbf{C}_\parallel = (-57.14\mathbf{i} + 71.43\mathbf{j} - 85.72\mathbf{k}) \text{ N} \cdot \text{m}$$

Thus, the force $\mathbf{R}$ and the couple $\mathbf{C}$ that comprise the wrench are

$$\mathbf{R} = (100\mathbf{i} + 50\mathbf{j} - 25\mathbf{k}) \text{ N} \qquad \text{Ans.}$$
$$\mathbf{C}_\parallel = (57.1\mathbf{i} + 28.6\mathbf{j} - 14.28\mathbf{k}) \text{ N} \cdot \text{m} \qquad \text{Ans.}$$

b. The resultant force $\mathbf{R}$ and the couple $\mathbf{C}_\perp$ can be combined by translating force $\mathbf{R}$ such that $\mathbf{r} \times \mathbf{R} = \mathbf{C}_\perp$ as illustrated schematically in Fig. 4-42. Thus,

$$\begin{vmatrix} \mathbf{i} & \mathbf{j} & \mathbf{k} \\ x_R & y_R & 0 \\ 100 & 50 & -25 \end{vmatrix} = -57.14\mathbf{i} + 71.43\mathbf{j} - 85.72\mathbf{k}$$

$$-25y_R\mathbf{i} + 25x_R\mathbf{j} + (50x_R - 100y_R)\mathbf{k} = -57.14\mathbf{i} + 71.43\mathbf{j} - 85.72\mathbf{k}$$

From which

$$-25y_R = -57.14 \qquad y_R = 2.29 \text{ m} \qquad \text{Ans.}$$
$$25x_R = 71.43 \qquad x_R = 2.86 \text{ m} \qquad \text{Ans.}$$

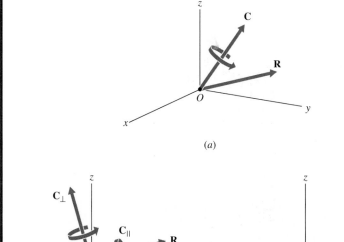

(a)

(b)   (c)

Fig. 4-42

Forces are applied at points A, B, and C of the bar shown in Fig. 4-43a. Replace this system of forces with a force $\mathbf{R}$ through point O and a couple $\mathbf{C}$.

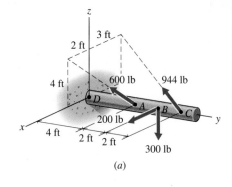

(a)

SOLUTION

The forces and their positions with respect to point O are

$$\mathbf{F}_A = 600\left[\frac{2}{6}\mathbf{i} - \frac{4}{6}\mathbf{j} + \frac{4}{6}\mathbf{k}\right] = (200\mathbf{i} - 400\mathbf{j} + 400\mathbf{k})\text{ lb}$$

$$\mathbf{F}_B = (200\mathbf{i} - 300\mathbf{k})\text{ lb}$$

$$\mathbf{F}_C = 944\left[-\frac{3}{9.434}\mathbf{i} - \frac{8}{9.434}\mathbf{j} + \frac{4}{9.434}\mathbf{k}\right] = (-300\mathbf{i} - 800\mathbf{j} + 400\mathbf{k})\text{ lb}$$

$$\mathbf{r}_{A/O} = 4\mathbf{j}\text{ ft} \qquad \mathbf{r}_{B/O} = 6\mathbf{j}\text{ ft} \qquad \mathbf{r}_{C/O} = 8\mathbf{j}\text{ ft}$$

Each of the three forces can be replaced by an equal force through point O and a couple. The vector sum of these concurrent forces is the resultant force $\mathbf{R}$.

$$\begin{aligned}
\mathbf{R} &= \mathbf{F}_A + \mathbf{F}_B + \mathbf{F}_C \\
&= 200\mathbf{i} - 400\mathbf{j} + 400\mathbf{k} + 200\mathbf{i} - 300\mathbf{k} - 300\mathbf{i} - 800\mathbf{j} + 400\mathbf{k} \\
&= (100\mathbf{i} - 1200\mathbf{j} + 500\mathbf{k})\text{ lb} \qquad\qquad \text{Ans.}
\end{aligned}$$

The magnitude of the resultant force is

$$|\mathbf{R}| = \sqrt{R_x^2 + R_y^2 + R_z^2} = \sqrt{(100)^2 + (-1200)^2 + (500)^2} = 1304\text{ lb} \quad \text{Ans.}$$

The direction angles that locate the line of action of the resultant force are

$$\theta_x = \cos^{-1}\frac{R_x}{|\mathbf{R}|} = \cos^{-1}\frac{100}{1304} = 85.6° \qquad\qquad \text{Ans.}$$

$$\theta_y = \cos^{-1}\frac{R_y}{|\mathbf{R}|} = \cos^{-1}\frac{-1200}{1304} = 157.0° \qquad\qquad \text{Ans.}$$

$$\theta_z = \cos^{-1}\frac{R_z}{|\mathbf{R}|} = \cos^{-1}\frac{500}{1304} = 67.5° \qquad\qquad \text{Ans.}$$

The moment $\mathbf{M}_O$ of the resultant couple is

$$\begin{aligned}
\mathbf{C} = \mathbf{M}_O &= \mathbf{r}_{A/O} \times \mathbf{F}_A + \mathbf{r}_{B/O} \times \mathbf{F}_B + \mathbf{r}_{C/O} \times \mathbf{F}_C \\
&= 1600\mathbf{i} - 800\mathbf{k} - 1800\mathbf{i} - 1200\mathbf{k} + 3200\mathbf{i} + 2400\mathbf{k} \\
&= (3000\boldsymbol{i} + 400\boldsymbol{k})\text{ ft}\cdot\text{lb} \qquad\qquad \text{Ans.}
\end{aligned}$$

The magnitude of the resultant couple is

$$|\mathbf{C}| = \sqrt{C_x^2 + C_y^2 + C_z^2} = \sqrt{(3000)^2 + (400)^2} = 3027 = 3030\text{ ft}\cdot\text{lb} \quad \text{Ans.}$$

The direction angles that locate the axis of the couple are

$$\theta_x = \cos^{-1}\frac{C_x}{|\mathbf{C}|} = \cos^{-1}\frac{3000}{3027} = 7.66° \qquad\qquad \text{Ans.}$$

$$\theta_y = \cos^{-1}\frac{C_y}{|\mathbf{C}|} = \cos^{-1}\frac{0}{3027} = 90.0° \qquad\qquad \text{Ans.}$$

$$\theta_z = \cos^{-1}\frac{C_z}{|\mathbf{C}|} = \cos^{-1}\frac{400}{3027} = 82.4° \qquad\qquad \text{Ans.}$$

The resultant of the force system is the force $\mathbf{R}$ and the couple $\mathbf{C}$ shown in Fig. 4.43b.

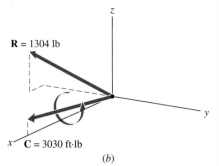

(b)

Fig. 4-43

PROBLEMS

4-109* Replace the force system shown in Fig. P4-109 with a force **R** through point D and a couple **C**.

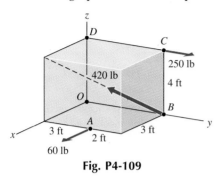

Fig. P4-109

4-110* Replace the force system shown in Fig. P4-110 with a force **R** through point C and a couple **C**.

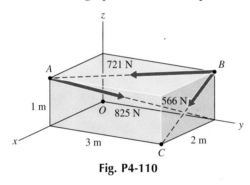

Fig. P4-110

4-111 Replace the force system shown in Fig. P4-109 with a force **R** through point O and a couple **C**.

4-112 Replace the force system shown in Fig. P4-110 with a force **R** through point O and a couple **C**.

4-113* Replace the force system shown in Fig. P4-113 with a force **R** through point O and a couple **C**.

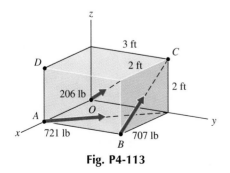

Fig. P4-113

4-114* Replace the force system shown in Fig. P4-114 with a force **R** through point D and a couple **C**.

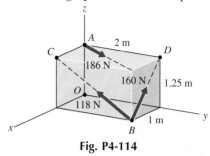

Fig. P4-114

4-115 Replace the force system shown in Fig. P4-113 with a force **R** through point D and a couple **C**.

4-116 Replace the force system shown in Fig. P4-114 with a force **R** through point O and a couple **C**.

4-117* Forces are applied at points A, B, and C of the bar shown in Fig. P4-117. Replace this system of forces with a force **R** through point O and a couple **C**.

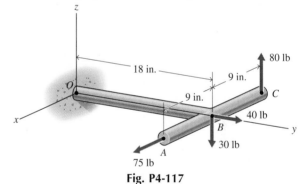

Fig. P4-117

4-118* Forces are applied at points A, B, and C of the bar shown in Fig. P4-118. Replace this system of forces with a force **R** through point O and a couple **C**.

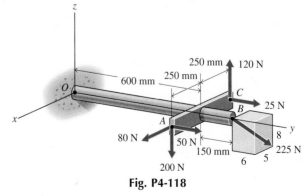

Fig. P4-118

4-119 The three forces shown in Fig. P4-119 can be written in Cartesian vector form as

$$\mathbf{F}_A = (-250\mathbf{i} - 200\mathbf{j} + 300\mathbf{k})\text{ lb}$$
$$\mathbf{F}_B = (-125\mathbf{i} + 250\mathbf{j} + 100\mathbf{k})\text{ lb}$$
$$\mathbf{F}_C = (-200\mathbf{i} - 200\mathbf{j} - 300\mathbf{k})\text{ lb}$$

Replace this system of forces with a force $\mathbf{R}$ through point O and a couple $\mathbf{C}$.

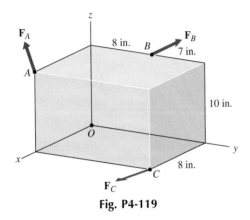

Fig. P4-119

4-120 Three forces are applied to a V-bracket as shown in Fig. P4-120. The three forces can be written in Cartesian vector form as

$$\mathbf{F}_A = (-500\mathbf{i} - 400\mathbf{j} + 300\mathbf{k})\text{ N}$$
$$\mathbf{F}_B = (600\mathbf{i} - 500\mathbf{j} - 400\mathbf{k})\text{ N}$$
$$\mathbf{F}_C = (300\mathbf{i} + 450\mathbf{j} - 250\mathbf{k})\text{ N}$$

Replace this system of forces with a force $\mathbf{R}$ through point O and a couple $\mathbf{C}$.

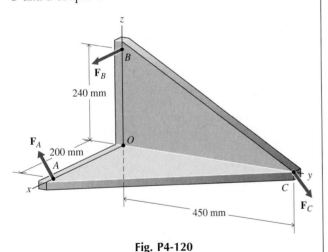

Fig. P4-120

4-121* Reduce the forces shown in Fig. P4-121 to a wrench and locate the intersection of the wrench with the xy-plane.

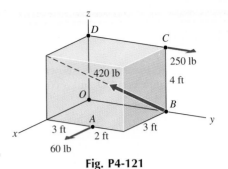

Fig. P4-121

4-122* Reduce the forces shown in Fig. P4-122 to a wrench and locate the intersection of the wrench with the xy-plane.

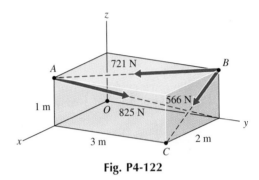

Fig. P4-122

4-123 Reduce the forces shown in Fig. P4-123 to a wrench and locate the intersection of the wrench with the xy-plane.

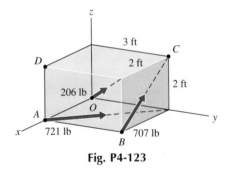

Fig. P4-123

4-124 Reduce the forces shown in Fig. P4-124 to a wrench and locate the intersection of the wrench with the xy-plane.

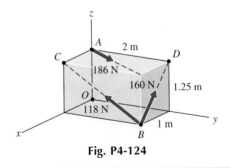

Fig. P4-124

SUMMARY

For a three-dimensional body, the particle idealization discussed in Chapter 3 is not valid, in general, since the forces acting on the body are usually not concurrent. For a general force system, $\mathbf{R} = \mathbf{0}$ is a necessary but not a sufficient condition for equilibrium of the body. A second restriction related to the tendency of a force to produce rotation of a body must also be satisfied and gives rise to the concept of a moment. In this chapter, the moment of a force about a point and the moment of a force about a line (axis) were defined and methods were developed for finding the resultant forces and the resultant moments (couples) for any general force system that may be applied to the body.

A moment is a vector quantity since it has both a magnitude and a direction, and adds according to the parallelogram law of addition. The moment of a force $\mathbf{F}$ about a point O can be represented by the vector cross-product:

$$\mathbf{M}_O = \mathbf{r} \times \mathbf{F} \tag{4-3}$$

where $\mathbf{r}$ is a position vector from point O to any point on the line of action of the force $\mathbf{F}$. The cross-product of the two intersecting vectors $\mathbf{r}$ and $\mathbf{F}$ is

$$\mathbf{M}_O = \mathbf{r} \times \mathbf{F} = M_x\mathbf{i} + M_y\mathbf{j} + M_z\mathbf{k} = M_O\mathbf{e} \tag{4-9}$$

where

$$M_x = r_y F_z - r_z F_y \qquad M_y = r_z F_x - r_x F_z \qquad M_z = r_x F_y - r_y F_x$$

are the three scalar components of the moment. The magnitude of the moment $\mathbf{M}_O$ is

$$M_O = |\mathbf{M}_O| = \sqrt{M_x^2 + M_y^2 + M_z^2} \tag{4-11}$$

The direction cosines associated with the unit vector $\mathbf{e}$ are

$$\cos\theta_x = \frac{M_x}{|\mathbf{M}_O|} \qquad \cos\theta_y = \frac{M_y}{|\mathbf{M}_O|} \qquad \cos\theta_z = \frac{M_z}{|\mathbf{M}_O|} \tag{4-14}$$

The moment $\mathbf{M}_{OB}$ of a force $\mathbf{F}$ about a line OB in a direction n specified by the unit vector $\mathbf{e}_n$ is

$$\mathbf{M}_{OB} = (\mathbf{M}_O \cdot \mathbf{e}_n)\mathbf{e}_n = [(\mathbf{r} \times \mathbf{F}) \cdot \mathbf{e}_n]\mathbf{e}_n = M_{OB}\mathbf{e}_n \tag{4-16}$$

Two equal parallel forces of opposite sense are called a couple. A couple has no tendency to translate a body in any direction but tends only to rotate the body on which it acts. Several transformations of a couple can be made without changing any of the external effects of the couple on the body. A couple (1) can be translated to a parallel position in its plane or to any parallel plane, and (2) can be rotated in its plane. Also, (3) the magnitude F of the two forces of a couple and the distance d between them can be changed provided the product Fd remains constant. Any system of couples in a plane or in space can be combined into a single resultant couple $\mathbf{C}$.

Any force $\mathbf{F}$ can be resolved into a parallel force $\mathbf{F}$ and a couple $\mathbf{C}$. Alternatively, a force and a couple in the same plane can be combined into a single force. The sole effect of combining a couple with a force is to move the action line of the force into a parallel position. The magnitude and sense of the force remain unchanged.

The resultant of a force system acting on a rigid body is the simplest force system that can replace the original system without altering the external effect of the system on the body. For a coplanar system of forces, the resultant is either a force **R** or a couple **C**. For a three-dimensional system of forces, the resultant may be a force **R**, a couple **C**, or both a force **R** and a couple **C**. The combination of a force **R** and a couple **C** (a push or pull together with a twist about the line of action of the force) is known as a wrench.

REVIEW PROBLEMS

4-125* A 2500-lb jet engine is suspended from the wing of an airplane as shown in Fig. P4-125. Determine the moment produced by the engine at point *A* in the wing when the plane is

a. On the ground with the engine not operating.
b. In flight with the engine developing a thrust **T** of 15,000 lb.

4-126* A 200-N force is applied at corner *B* of a rectangular plate as shown in Fig. P4-126. Determine

a. The moment of the force about point *O*.
b. The moment of the force about line *OD*.

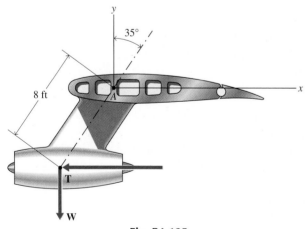

Fig. P4-125

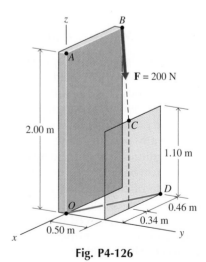

Fig. P4-126

4-127 A force **F** with a magnitude of 500 lb acts at a point D in a body as shown in Fig. P4-127. Determine

a. The moment of the force about point B.
b. The direction angles associated with the unit vector **e** along the axis of the moment.
c. The moment of the force about line BC.

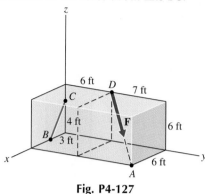

Fig. P4-127

4-128 A bent rod supports two forces as shown in Fig. P4-128. Determine

a. The moment of the two forces about point O.
b. The moment of the two forces about line OA.

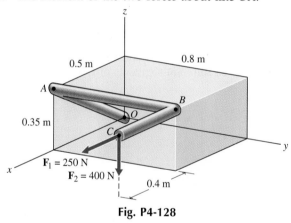

Fig. P4-128

4-129* Determine the resultant of the parallel force system shown in Fig. P4-129 and locate the intersection of its line of action with the xy-plane.

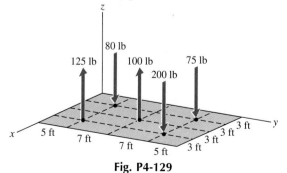

Fig. P4-129

4-130* The input and output torques from a gear box are shown in Fig. P4-130. Determine the magnitude and direction of the resultant torque.

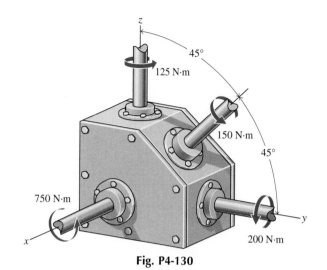

Fig. P4-130

4-131 The driving wheel of a truck is subjected to the force-couple system shown in Fig. P4-131. Replace this system by an equivalent single force and determine the point of application of the force along the vertical diameter of the wheel.

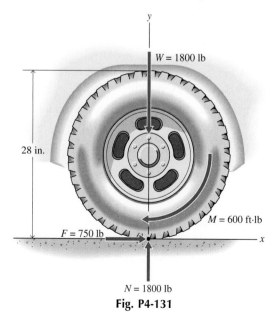

Fig. P4-131

147

4-132 A bracket is subjected to the force-couple system shown in Fig. P4-132. Determine

a. The magnitude and direction of the resultant force **R**.
b. The perpendicular distance from support O to the line of action of the resultant.

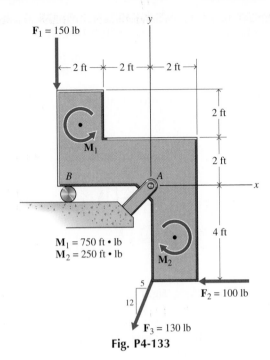

Fig. P4-133

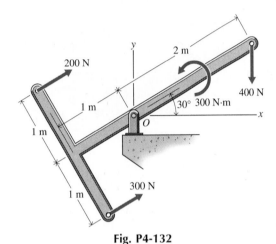

Fig. P4-132

4-133* A bracket is subjected to the force-couple system shown in Fig. P4-133. Determine

a. The magnitude and direction of the resultant force **R**.
b. The perpendicular distance from support A to the line of action of the resultant.

4-134* A bent rod supports a 450-N force as shown in Fig. P4-134.

a. Replace the 450-N force with a force **R** through point O and a couple **C**.
b. Determine the twisting moments produced by force **F** in the three different segments of the rod.

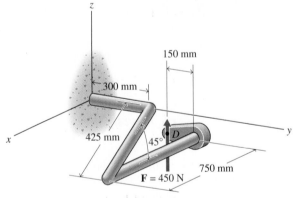

Fig. P4-134

5

DISTRIBUTED FORCES: CENTROIDS AND CENTER OF GRAVITY

A high-wire walker uses a balance pole to maintain his center of mass over the wire.

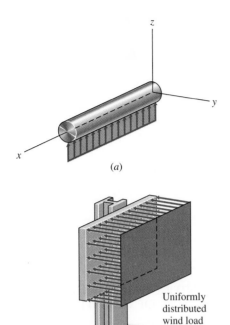

(a)

Uniformly
distributed
wind load

(b)

Fig. 5-1

Fig. 5-2

5-1 INTRODUCTION

In previous chapters, we have dealt primarily with concentrated forces each of which could be represented as a simple vector quantity that has a magnitude, a direction, a line of action, and in some instances a point of application. In many instances, the loads are not concentrated at a point but are distributed along a line (such as the contact line of a roller bearing illustrated in Fig. 5-1a) or over a surface (such as the wind load on a sign illustrated in Fig. 5-1b). In these cases, the loads are known as distributed loads. The distribution can be uniform or nonuniform. Other forces, known as body forces, which result from gravitational, electrical, and magnetic effects, are distributed over the mass of the body. When the areas over which the loads are applied become significant with respect to the size of the body, the assumption of a concentrated force is no longer valid, and the actual distribution of the load must be included in the analysis.

A distributed force at any point is characterized by its intensity and its direction. A force distributed over an area and acting normal to the surface (usually due to the action of a liquid or gas) is known as a pressure. Internal distributed forces in solids (known as stresses) may or may not act normal to the surface of interest. The units for both pressure and stress are force per unit area (lb/in.^{2} or N/m^2). Force distributed over the volume of a body (body forces) are measured in units of force per unit volume (lb./in.^{3} or N/m^3).

In the preceding chapters, moments of forces about points or axes have been considered. In the analysis of many problems in engineering, expressions are encountered that represent moments of masses, forces, volumes, areas, or lines with respect to axes or planes. For example, consider the moment of an area A (in the xy-plane) about the y-axis as shown in Fig. 5-2. Since an area is a distributed quantity, its moment about an axis cannot be defined as the product of the area and the perpendicular distance of the area from the axis (similar to the manner in which the moment of a concentrated force was defined), since the different parts of the area are at different distances from the axis; and, hence, the distance of the area from the axis is indefinite and meaningless. The area can, however, be considered to be composed of a large number of very small (differential) elements of area dA, and the moment dM of an element about an axis can then be defined as the product of the area of the element and the distance of the element from the axis. Thus,

$$dM_i = x_i \, dA_i \tag{5-1}$$

where the subscript i denotes the ith element. The moment of the area A about the y-axis is then defined as the algebraic sum of the moments of the n elements of area about the y-axis. Thus,

$$M_y = \sum_{i=1}^{n} x_i \, dA_i \tag{5-2}$$

or in integral form

$$M_y = \int_A x \, dA \tag{5-3}$$

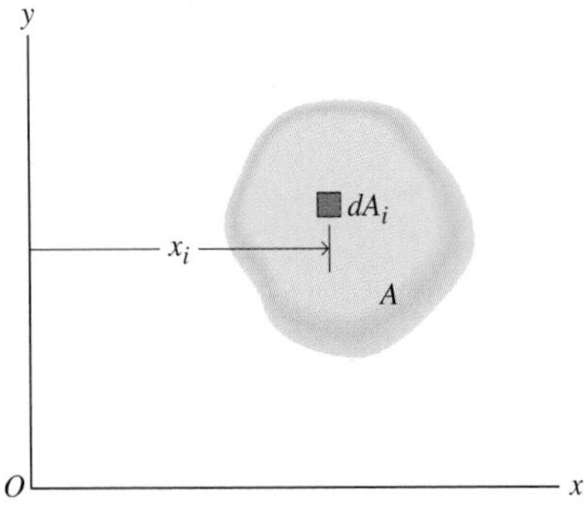

The moment of a mass, force, volume, area, or line with respect to a line or a plane can be defined in a similar way.

The moment thus defined is called the first moment of the quantity being considered since the first power of the distance (x in the above case) is used in the expression. Later, in Chapter 10, integrals of the form $\int_A x^2\, dA$ will be introduced. Such integrals are known as second moments since the second power of the distance appears in the expression.

The sign of the moment of an element about an axis may be positive or negative since the coordinate of the element may be positive or negative whereas the masses, forces, volumes, areas, and lengths are always positive. Likewise, the moment of the quantity (mass, force, volume, area, or length) about an axis or plane may be positive, negative, or zero since the sum of the positive moments of the elements may be larger than, smaller than, or equal to the sum of the negative moments of the elements.

The dimensional expression for the moment of a line is length squared (L^2); therefore, the moment of a line about an axis or plane will have the units in.^{2}, ft^2, mm^2, m^2, and so on. Similarly, the dimensions of the moments of an area and of a volume are length cubed (L^3) and length to the fourth power (L^4), respectively.

Concepts required for the treatment of distributed forces include center of mass, center of gravity, and centroid. These topics are developed in the next two sections of this chapter.

5-2   CENTER OF MASS AND CENTER OF GRAVITY

5-2-1   Center of Mass

The term "center of mass" or "mass center" is used to denote the point in a system of particles or physical body where the mass can be conceived to be concentrated so that the moment of the concentrated mass with respect to any axis or plane equals to the moment of the distributed mass with respect to the same axis or plane.

For example, consider a set of n particles as shown in Fig. 5-3. The coordinates of the ith particle of mass m_i are (x_i, y_i, z_i) and the distances from the coordinate planes to the mass center G of the set of particles are $(\bar{x}_i, \bar{y}_i, \bar{z}_i)$. By definition,

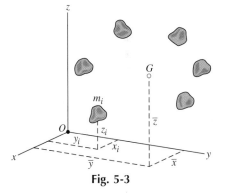

Fig. 5-3

$$M_{yz} = m\bar{x} = \sum_{i=1}^{n} m_i x_i \quad \text{or} \quad \bar{x} = \frac{1}{m} \sum_{i=1}^{n} m_i x_i$$

$$M_{zx} = m\bar{y} = \sum_{i=1}^{n} m_i y_i \quad \text{or} \quad \bar{y} = \frac{1}{m} \sum_{i=1}^{n} m_i y_i \qquad (5\text{-}4)$$

$$M_{xy} = m\bar{z} = \sum_{i=1}^{n} m_i z_i \quad \text{or} \quad \bar{z} = \frac{1}{m} \sum_{i=1}^{n} m_i z_i$$

where

$$m = \sum_{i=1}^{n} m_i$$

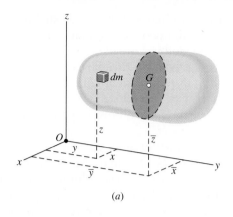

(a)

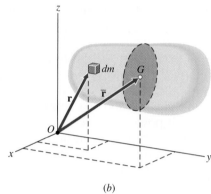

(b)

Fig. 5-4

If the particles form a continuous body, as shown in Fig. 5-4a, the summations can be replaced by integrals over the mass of the body to give

$$M_{yz} = m\bar{x} = \int x \, dm \qquad \text{or} \qquad \bar{x} = \frac{1}{m} \int x \, dm$$

$$M_{zx} = m\bar{y} = \int y \, dm \qquad \text{or} \qquad \bar{y} = \frac{1}{m} \int y \, dm \qquad (5\text{-}5)$$

$$M_{xy} = m\bar{z} = \int z \, dm \qquad \text{or} \qquad \bar{z} = \frac{1}{m} \int z \, dm$$

where

$$m = \int dm$$

Equations 5-4 can be combined into a single vector equation by multiplying the first, second, and third equations by $\mathbf{i}$, $\mathbf{j}$, and $\mathbf{k}$, respectively, and adding. Thus

$$m\bar{x}\mathbf{i} + m\bar{y}\mathbf{j} + m\bar{z}\mathbf{k} = \sum_{i=1}^{n} m_i x_i \mathbf{i} + \sum_{i=1}^{n} m_i y_i \mathbf{j} + \sum_{i=1}^{n} m_i z_i \mathbf{k}$$

from which

$$m(\bar{x}\mathbf{i} + \bar{y}\mathbf{j} + \bar{z}\mathbf{k}) = \sum_{i=1}^{n} m_i(x_i\mathbf{i} + y_i\mathbf{j} + z_i\mathbf{k})$$

which reduces to

$$\mathbf{M}_O = m\mathbf{r} = \sum_{i=1}^{n} m_i \mathbf{r}_i \qquad \text{or} \qquad \mathbf{r} = \frac{1}{m} \sum_{i=1}^{n} m_i \mathbf{r}_i \qquad (5\text{-}6)$$

since the position vector from the origin to the ith particle, as shown in Fig. 5-4b, is

$$\mathbf{r}_i = x_i \mathbf{i} + y_i \mathbf{j} + z_i \mathbf{k}$$

and the position vector from the origin to the mass center is

$$\bar{\mathbf{r}} = \bar{x}\mathbf{i} + \bar{y}\mathbf{j} + \bar{z}\mathbf{k}$$

If the particles form a continuous body, the summation can be replaced by an integral over the mass of the body to give

$$m\bar{\mathbf{r}} = \int_m \mathbf{r} \, dm = \int_V \mathbf{r} \rho \, dV$$

or   (5-7)

$$\bar{\mathbf{r}} = \frac{1}{m} \int_m \mathbf{r} \, dm = \frac{1}{m} \int_V \mathbf{r} \rho \, dV$$

where $\mathbf{r}$ is the position vector from the origin to the element dm of the body, ρ is the density of the element, and dV is its volume.

5-2-2   Center of Gravity

The weight W of a body is the resultant of the distributed body forces exerted on the individual particles of the body by the Earth. The point G in the body through which the weight acts is defined as the "center of gravity" of the body. The magnitude of the force exerted on a particular particle by the Earth depends on the mass of the particle and the distance between the particle and the center of the Earth (Newton's law of gravitation). For practical engineering work, where the size of the body is small in comparison to the size of the Earth, all particles can be assumed to be at the same distance from the center of the Earth (experience the same gravitational acceleration g). Also, because of the size of the Earth, the lines of action of the individual particle forces that are concurrent at the center of the Earth can be assumed to be parallel. These two assumptions yield a center of gravity that coincides with the center of mass since

$$W = mg$$

where g is the gravitational attraction exerted on the body by the Earth. Approximate values for the gravitational constant that are suitable for most engineering computations are $g = 32.17$ ft/s^2 = 9.807 m/s^2. If both sides of Eqs. 5-5 are multiplied by g, they can be expressed in terms of the weight W of a body as

$$M_{yz} = W\bar{x} = \int x \, dW \qquad \text{or} \qquad \bar{x} = \frac{1}{W} \int x \, dW$$

$$M_{zx} = W\bar{y} = \int y \, dW \qquad \text{or} \qquad \bar{y} = \frac{1}{W} \int y \, dW \qquad (5\text{-}8)$$

$$M_{xy} = W\bar{z} = \int z \, dW \qquad \text{or} \qquad \bar{z} = \frac{1}{W} \int z \, dW$$

where

$$W = \int dW$$

For a body with a specific shape, the center of gravity can be located by considering the body to be made of an infinite number of small elements each of which have a weight dW given by the expression

$$dW = \gamma \, dV$$

where γ is the specific weight of the material (weight per unit volume) and dV is the volume of the element. The total weight W of the body is

$$W = \int_V \gamma \, dV$$

If an xyz-coordinate system is selected such that the line of action of the weight W is parallel to the z-axis, the moment of the weight dW of an element about the y-axis is

$$dM_y = x \, dW = x \, (\gamma \, dV)$$

From the definition of the center of gravity,

$$M_y = \bar{x} \, W = \bar{x} \int_V \gamma \, dV = \int_V x \, (\gamma \, dV)$$

Thus, the x-coordinate of a point on the action line of the weight W is

$$\bar{x} = \frac{\int_V x \, (\gamma \, dV)}{\int_V \gamma \, dV}$$

Similarly,

$$\bar{y} = \frac{\int_V y \, (\gamma \, dV)}{\int_V \gamma \, dV} \qquad \bar{z} = \frac{\int_V z \, (\gamma \, dV)}{\int_V \gamma \, dV} \tag{5-9}$$

In the event that the specific weight γ is not constant, but can be expressed as a function of the coordinates, it will be necessary to account for this variation in the calculations of $\bar{x}$, $\bar{y}$, and $\bar{z}$.

The following example illustrates the procedure used to locate the "center of mass" or the "center of gravity" of a system of particles.

Four bodies A, B, C, and D (which can be treated as particles) are attached to a shaft as shown in Fig. 5-5. The masses of the bodies are 0.2, 0.4, 0.6, and 0.8 slug, respectively, and the distances from the axis of the shaft to their mass centers are 1.50, 2.50, 2.00, and 1.25 ft, respectively. Find the mass center for the four bodies.

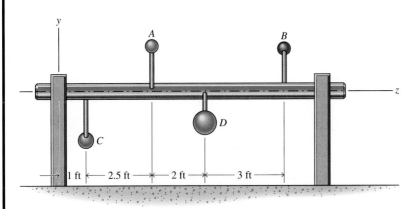

Front View

Fig. 5-5

End View

SOLUTION

A typical equation from Eqs. 5-4 that is used to locate the mass center of a system of particles is

$$\bar{x} = \frac{1}{m} \sum_{i=1}^{n} m_i x_i \quad \text{where} \quad m = \sum_{i=1}^{n} m_i$$

Thus, for the system of four bodies shown in Fig. 5-5:

$$\Sigma m_i = 0.2 + 0.4 + 0.6 + 0.8 = 2.0 \text{ slug}$$

$$\Sigma m_i x_i = m_A x_A + m_B x_B + m_C x_C + m_D x_D$$
$$= 0.2(-1.50 \cos 60°) + 0.4(2.50 \cos 30°) + 0.6(2.00 \cos 45°)$$
$$+ 0.8(-1.25 \cos 45°) = 0.8574 \text{ slug} \cdot \text{ft}$$

$$\Sigma m_i y_i = m_A y_A + m_B y_B + m_C y_C + m_D y_D$$
$$= 0.2(1.50 \sin 60°) + 0.4(2.50 \sin 30°) + 0.6(-2.00 \sin 45°)$$
$$+ 0.8(-1.25 \sin 45°) = -0.7958 \text{ slug} \cdot \text{ft}$$

$$\Sigma m_i z_i = m_A z_A + m_B z_B + m_C z_C + m_D z_D$$
$$= 0.2(3.5) + 0.4(8.5) + 0.6(1.0) + 0.8(5.5) = 9.10 \text{ slug} \cdot \text{ft}$$

$$\bar{x} = \frac{\Sigma m_i x_i}{m} = \frac{0.8574}{2.00} = 0.429 \text{ ft} \qquad \text{Ans.}$$

$$\bar{y} = \frac{\Sigma m_i y_i}{m} = \frac{-0.7958}{2.00} = -0.398 \text{ ft} \qquad \text{Ans.}$$

$$\bar{z} = \frac{\Sigma m_i z_i}{m} = \frac{9.10}{2.00} = 4.55 \text{ ft} \qquad \text{Ans.}$$

PROBLEMS

5-1* Locate the center of gravity for the four particles shown in Fig. P5-1 if $W_A = 20$ lb, $W_B = 25$ lb, $W_C = 30$ lb, and $W_D = 40$ lb.

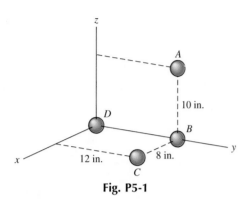

Fig. P5-1

5-2* Locate the center of mass for the four particles shown in Fig. P5-2 if $m_A = 16$ kg, $m_B = 24$ kg, $m_C = 14$ kg, and $m_D = 36$ kg.

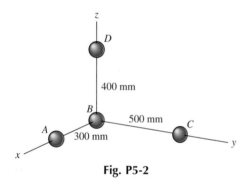

Fig. P5-2

5-3* Locate the center of gravity for the five particles shown in Fig. P5-3 if $W_A = 25$ lb, $W_B = 35$ lb, $W_C = 15$ lb, $W_D = 28$ lb, and $W_E = 16$ lb.

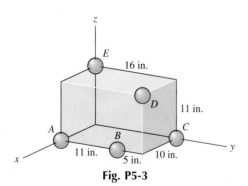

Fig. P5-3

5-4* Locate the center of mass for the five particles shown in Fig. P5-4 if $m_A = 2$ kg, $m_B = 3$ kg, $m_C = 4$ kg, $m_D = 3$ kg, and $m_E = 2$ kg.

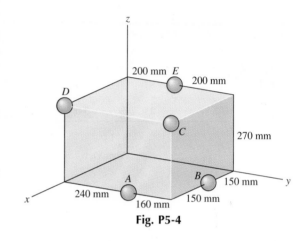

Fig. P5-4

5-5 Locate the center of gravity for the five particles shown in Fig. P5-3 if $W_A = 15$ lb, $W_B = 24$ lb, $W_C = 35$ lb, $W_D = 18$ lb, and $W_E = 26$ lb.

5-6 Locate the center of mass for the five particles shown in Fig. P5-4 if $m_A = 6$ kg, $m_B = 9$ kg, $m_C = 5$ kg, $m_D = 7$ kg, and $m_E = 4$ kg.

5-7* Three bodies with masses of 2, 4, and 6 slugs are located at points (2, 3, 4), (3, −4, 5), and (−3, 4, 6), respectively. Locate the mass center of the system if the distances are measured in feet.

5-8* Three bodies with masses of 3, 6, and 7 kg are located at points (4, −3, 1), (−1, 3, 2), and (2, 2, −4), respectively. Locate the mass center of the system if the distances are measured in meters.

5-3 CENTROIDS OF VOLUMES, AREAS, AND LINES

5-3-1 Centroids of Volumes

If the specific weight γ of a body is constant, Eqs. 5-9 reduce to

$$\bar{x} = \frac{1}{V} \int_V x\, dV \qquad \bar{y} = \frac{1}{V} \int_V y\, dV \qquad \bar{z} = \frac{1}{V} \int_V z\, dV \qquad (5\text{-}10)$$

Equations 5-10 indicate that the coordinates $\bar{x}$, $\bar{y}$, and $\bar{z}$ depend only on the geometry of the body and are independent of the physical properties. The point located by such a set of coordinates is known as the "centroid" C of the volume of the body. The term *centroid* is usually used in connection with geometrical figures (volumes, areas, and lines); whereas the terms *center of mass* and *center of gravity* are used in connection with physical bodies. Note that the centroid C of a volume has the same position as the center of gravity G of the body if the body is homogeneous. If the specific weight is variable, the center of gravity of the body and the centroid of the volume will usually be at different points, as indicated in Fig. 5-6. In this case, since the specific weight of the lower portion of the cone is greater than the specific weight of the upper portion of the cone, the center of gravity G, which depends on the weights of the two parts, will be below the centroid C, which depends only on the volume of the two parts.

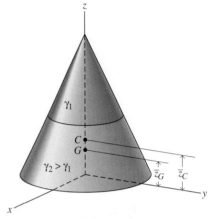

Fig. 5-6

5-3-2 Centroids of Areas

The center of gravity G of a homogeneous thin plate of uniform thickness t and surface area A can be determined by considering a differential element of volume dV that can be expressed in terms of a differential element of the surface area dA of the plate as

$$dV = t\, dA$$

Thus, for a thin plate, Eqs. 5-10 reduce to

$$\bar{x} = \frac{1}{A} \int_A x\, dA \qquad \bar{y} = \frac{1}{A} \int_A y\, dA \qquad \bar{z} = \frac{1}{A} \int_A z\, dA \qquad (5\text{-}11)$$

For the case of a thin three-dimensional shell, three coordinates $\bar{x}$, $\bar{y}$, and $\bar{z}$ are required to specify the location of the center of gravity G of the shell. For a flat plate with one axis of the *xyz*-coordinate system perpendicular to the surface of the plate (say the *z*-axis), only two coordinates in the plane of the plate ($\bar{x}$ and $\bar{y}$) are required to specify the location of the center of gravity G of the plate. The two coordinates ($\bar{x}$ and $\bar{y}$) in the plane of the plate also locate the centroid of the area A of the plate.

5-3-3 Centroids of Lines

The center of gravity of a homogeneous curved wire with a small uniform cross-sectional area A and length L can be determined by considering a small element of volume dV that can be expressed in terms of a differential element of length dL as

$$dV = A\, dL$$

Thus, for a slender rod or wire, Eqs. 5-10 reduce to

$$\bar{x} = \frac{1}{L} \int_L x \, dL \qquad \bar{y} = \frac{1}{L} \int_L y \, dL \qquad \bar{z} = \frac{1}{L} \int_L z \, dL \qquad (5\text{-}12)$$

Two or three coordinates, depending on the shape, are required to specify the location of the center of gravity G of the wire or the centroid C of the line defining the shape of the wire.

5-3-4   Centroid, Center of Mass, or Center of Gravity by Integration

The procedure involved in the determination, by integration, of the coordinates of the centroid, center of mass, or center of gravity of a body can be summarized as follows:

1. Prepare a sketch of the body approximately to scale.
2. Establish a coordinate system. Rectangular coordinates are used with most shapes that have flat planes for boundaries. Polar coordinates are usually used for shapes with circular boundaries. Whenever a line or plane of symmetry exists in a body, a coordinate axis or plane should be chosen to coincide with this line or plane. The centroid, center of mass, or center of gravity will always lie on such a line or plane since the moments of symmetrically located pairs of elements (one with a positive coordinate and the other with an equal negative coordinate) will always cancel.
3. Select an element of volume, area, or length. For center of mass or center of gravity determinations, determine the mass or weight of the element by using the appropriate expression (constant or variable) for the density or specific weight. The element can frequently be selected so that only single integration is required for the complete body or for the several parts into which the body can be divided. Sometimes, however, it may be necessary to use double integration or perhaps triple integration for some shapes. If possible, the element should be chosen so that all parts are the same distance from the reference axis or plane. This distance will be the moment arm for first moment determinations. When the parts of the element are at different distances from the reference axis or plane, the location of the centroid, center of mass, or center of gravity of the element must be known in order to establish the moment arm for moment calculations. Integrate the expression to determine the volume, area, length, mass, or weight of the body.
4. Write an expression for the first moment of the element with respect to one of the reference axes or planes. Integrate the expression to determine the first moment with respect to the reference axis or plane.
5. Use the appropriate equation (Eqs. 5-4, 5-5, etc.) to obtain the coordinate of the centroid, center of mass, or center of gravity with respect to the reference axis or plane.
6. Repeat steps 3 to 5, using different reference axes or planes for the other coordinates of the centroid, center of mass, or center of gravity.
7. Locate the centroid, center of mass, or center of gravity on the sketch. Gross errors are often detected by using this last step.

Locate the centroid of the rectangular area shown in Fig. 5-7a.

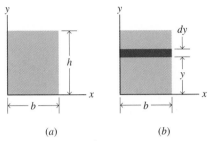

(a) (b)

Fig. 5-7

SOLUTION

Symmetry considerations require that the centroid of a rectangular area be located at the center of the rectangle. Thus, for the rectangular area shown in Fig. 5-7a, $\bar{x} = b/2$ and $\bar{y} = h/2$. These results are established by integration in the following fashion. For the differential element of area shown in Fig. 5-7b, $dA = b\,dy$. The element dA is located a distance y from the x-axis; therefore, the moment of the area about the x-axis is

$$M_x = \int_A y\,dA = \int_0^h y\,(b\,dy) = b\left[\frac{y^2}{2}\right]_0^h = \frac{bh^2}{2}$$

From Eq. 5-11

$$\bar{y} = \frac{M_x}{A} = \frac{bh^2/2}{bh} = \frac{h}{2} \qquad\qquad \text{Ans.}$$

In a similar manner by using an element of area $dA = h\,dx$, the moment of the area about the y-axis is

$$M_y = \int_A x\,dA = \int_0^b x\,(h\,dx) = h\left[\frac{x^2}{2}\right]_0^b = \frac{hb^2}{2}$$

From Eq. 5-11

$$\bar{x} = \frac{M_y}{A} = \frac{hb^2/2}{bh} = \frac{b}{2} \qquad\qquad \text{Ans.}$$

The element of area $dA = b\,dy$, used to calculate M_x, was not used to calculate M_y since all parts of the horizontal strip are located at different distances x from the y-axis. As a result of this example, it is now known that $\bar{x} = b/2$ for the element of area $dA = b\,dy$ shown in Fig. 5-7b. This result will be used frequently in later examples to simplify the integrations.

Locate the y-coordinate of the centroid of the area of the quarter circle shown in Fig. 5-8a.

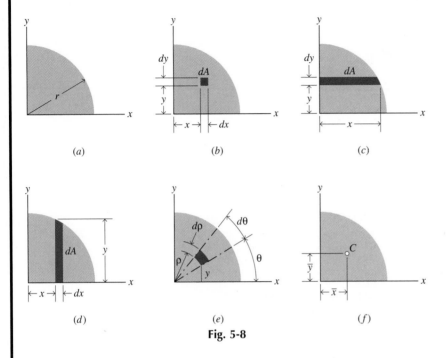

(a)   (b)   (c)

(d)   (e)   (f)

Fig. 5-8

SOLUTION

Four different elements will be used to solve this problem.

METHOD 1: Double integral in rectangular coordinates

For the differential element shown in Fig. 5-8b, $dA = dy\,dx$. The element dA is at a distance y from the x-axis; therefore, the moment of the area about the x-axis is

$$M_x = \int_A y\,dA = \int_0^r \int_0^{\sqrt{r^2-x^2}} y\,dy\,dx$$

$$= \int_0^r \left[\frac{y^2}{2}\right]_0^{\sqrt{r^2-x^2}} dx = \int_0^r \frac{r^2 - x^2}{2}\,dx = \left[\frac{r^2 x}{2} - \frac{x^3}{6}\right]_0^r = \frac{r^3}{3}$$

From Eq. 5-11

$$y = \frac{M_x}{A} = \frac{r^3/3}{\pi r^2/4} = \frac{4r}{3\pi} \qquad\qquad \text{Ans.}$$

METHOD 2: Single integral using a horizontal strip

Alternatively, the element of area can be selected as shown in Fig. 5-8c. For this element, which is located a distance y from the x-axis, $dA = x\, dy = \sqrt{r^2 - y^2}\, dy$. Therefore, the moment of the area about the x-axis is

$$M_x = \int_A y\, dA = \int_0^r y\, \sqrt{r^2 - y^2}\, dy = \left[-\frac{(r^2 - y^2)^{3/2}}{3} \right]_0^r = \frac{r^3}{3}$$

From Eq. 5-11

$$\bar{y} = \frac{M_x}{A} = \frac{r^3/3}{\pi r^2/4} = \frac{4r}{3\pi} \qquad\qquad \text{Ans.}$$

METHOD 3: Single integral using a vertical strip

The element of area could also be selected as shown in Fig. 5-8d. For this element, $dA = y\, dx = \sqrt{r^2 - x^2}\, dx$; however, all parts of the element are at different distances y from the x-axis. For this type of element, the results of Example Problem 5-2 can be used to compute a moment dM_x, which can be integrated to yield moment M_x. Thus

$$dM_x = \frac{y}{2}\, dA = \frac{y}{2}\, y\, dx = \frac{y^2}{2}\, dx = \frac{r^2 - x^2}{2}\, dx$$

$$M_x = \int_A dM_x = \int_0^r \frac{r^2 - x^2}{2}\, dx = \left[\frac{r^2 x}{2} - \frac{x^3}{6} \right]_0^r = \frac{r^3}{3}$$

From Eq. 5-11

$$\bar{y} = \frac{M_x}{A} = \frac{r^3/3}{\pi r^2/4} = \frac{4r}{3\pi} \qquad\qquad \text{Ans.}$$

METHOD 4: Double integral using polar coordinates

Finally, polar coordinates can be used to locate the centroid of the quarter circle. With polar coordinates, the element of area is $dA = \rho\, d\theta\, d\rho$ and the distance from the x-axis to the element is $y = \rho \sin \theta$ as shown in Fig. 5-8e. Thus

$$M_x = \int_A y\, dA = \int_0^r \int_0^{\pi/2} \rho^2 \sin \theta\, d\theta\, d\rho$$

$$= \int_0^r \rho^2 \left[-\cos \theta \right]_0^{\pi/2} d\rho = \int_0^r \rho^2\, d\rho = \left[\frac{\rho^3}{3} \right]_0^r = \frac{r^3}{3}$$

From Eq. 5-11

$$\bar{y} = \frac{M_x}{A} = \frac{r^3/3}{\pi r^2/4} = \frac{4r}{3\pi} \qquad\qquad \text{Ans.}$$

In a completely similar manner, the x-coordinate of the centroid is obtained as

$$\bar{x} = \frac{M_y}{A} = \frac{r^3/3}{\pi r^2/4} = \frac{4r}{3\pi}$$

The results are illustrated in Fig. 5-8f.

Locate the centroid of the triangular area shown in Fig. 5-9a.

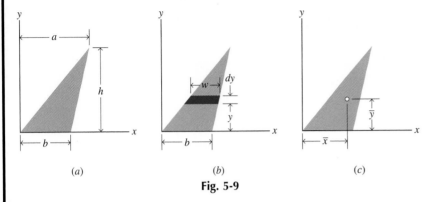

(a) (b) (c)

Fig. 5-9

SOLUTION

If the horizontal differential element of area shown in Fig. 5-9b is selected for the integration, the same element can be used for the complete area. From similar triangles in Fig. 5-9b,

$$\frac{h-y}{w} = \frac{h}{b} \quad \text{or} \quad w = \frac{b}{h}(h-y)$$

Therefore, for this element, $dA = w\, dy = (b/h)(h-y)\, dy$ and the distance from the x-axis to the element is y. Thus, the moment of the area about the x-axis is

$$M_x = \int_A y\, dA = \int_0^h \frac{b}{h}(hy - y^2)\, dy = \frac{b}{h}\left[\frac{hy^2}{2} - \frac{y^3}{3}\right]_0^h = \frac{bh^2}{6}$$

From Eq. 5-11

$$\bar{y} = \frac{M_x}{A} = \frac{bh^2/6}{bh/2} = \frac{h}{3} \qquad \text{Ans.}$$

The distance x from the y-axis to the centroid of the element of area dA is

$$x = \frac{ay}{h} + \frac{w}{2} = \frac{b}{2} + \frac{(2a-b)y}{2h}$$

Thus, the moment of the area about the y-axis is

$$M_y = \int_A x\, dA = \int_0^h \left[\frac{b}{2} + \frac{(2a-b)y}{2h}\right]\left[\frac{b}{h}(h-y)\right] dy$$

$$= \frac{1}{2h^2}\left[b^2h^2y - \frac{b^2hy^2}{2} + \frac{2abhy^2}{2} - \frac{2aby^3}{3} - \frac{b^2hy^2}{2} + \frac{b^2y^3}{3}\right]_0^h$$

$$= \frac{bh(a+b)}{6}$$

From Eq. 5-11

$$\bar{x} = \frac{M_y}{A} = \frac{bh(a+b)/6}{bh/2} = \frac{a+b}{3} \qquad \text{Ans.}$$

The results are shown in Fig. 5-9c.

A circular arc of thin homogeneous wire is shown in Fig. 5-10a.

a. Locate the x- and y-coordinates of the mass center.
b. Use the results of part a to determine the coordinates of the mass center for a half circle.

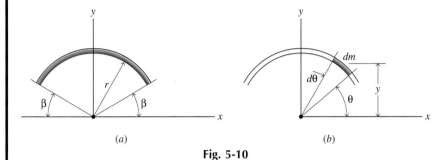

(a)   (b)

Fig. 5-10

SOLUTION

a. The wire can be assumed to consist of a large number of differential elements of length dL as shown in Fig. 5-10b. The mass of each of these elements is

$$dm = \rho \, dV = \rho A \, dL = \rho A(r \, d\theta)$$

Therefore, the total mass of the wire is

$$m = \int dm = \int_{\beta}^{\pi-\beta} \rho Ar \, d\theta = \rho Ar \int_{\beta}^{\pi-\beta} d\theta = \rho Ar(\pi - 2\beta)$$

The distance y from the x-axis to the element dm is $y = r \sin \theta$. Thus,

$$m\bar{y} = \int y \, dm = \int_{\beta}^{\pi-\beta} (r \sin \theta)(\rho Ar \, d\theta)$$
$$= \rho Ar^2 \int_{\beta}^{\pi-\beta} \sin \theta \, d\theta = \rho Ar^2(2 \cos \beta)$$

Therefore,

$$\bar{y} = \frac{2\rho Ar^2 \cos \beta}{\rho Ar(\pi - 2\beta)} = \frac{2r \cos \beta}{\pi - 2\beta} \qquad \text{Ans.}$$

Since the length of wire is symmetric about the y-axis,

$$\bar{x} = 0 \qquad \text{Ans.}$$

b. For the half circle, $\beta = 0$

$$\bar{y} = \frac{2r}{\pi} \qquad \text{Ans.}$$
$$\bar{x} = 0 \qquad \text{Ans.}$$

Locate the center of gravity G of the homogeneous right circular cone shown in Fig. 5-11a, which has an altitude h and radius r and is made of a material with a specific weight γ.

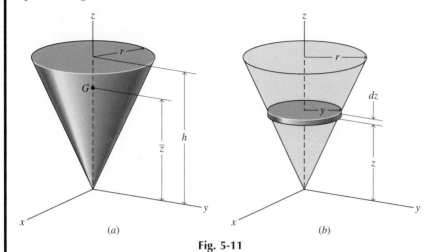

Fig. 5-11

SOLUTION

From symmetry, it is obvious that $\bar{x} = \bar{y} = 0$. The coordinate $\bar{z}$ of the center of gravity G of the cone can be located by using the differential element of volume shown in Fig. 5-11b. The weight dW of the differential element is

$$dW = \gamma \, dV = \gamma \, (\pi y^2) \, dz = \gamma \pi \left(\frac{rz}{h}\right)^2 dz = \frac{\gamma \pi r^2}{h^2} z^2 \, dz$$

From Eq. 5-9

$$\bar{z} = \frac{\int z \, dW}{\int dW} = \frac{\int_V z \, (\gamma \, dV)}{\int_V \gamma \, dV}$$

Thus,

$$W\bar{z} = \int z \, dW = \int_0^h \frac{\gamma \pi r^2}{h^2} z^3 \, dz = \frac{1}{4} \gamma \pi r^2 h^2$$

The weight of the cone is

$$W = \int_V \gamma \, dV = \int_0^h \frac{\gamma \pi r^2}{h^2} z^2 \, dz = \frac{\gamma \pi r^2}{h^2} \left[\frac{z^3}{3}\right]_0^h = \frac{1}{3} \gamma \pi r^2 h$$

Therefore

$$\bar{z} = \frac{W\bar{z}}{W} = \frac{\gamma \pi r^2 h^2/4}{\gamma \pi r^2 h/3} = \frac{3h}{4} \qquad \text{Ans.}$$

Since the xz-plane and the yz-plane are planes of symmetry,

$$\bar{x} = \bar{y} = 0 \qquad \text{Ans.}$$

EXAMPLE PROBLEM 5-7

Locate the centroid of the volume of the hemisphere shown in Fig. 5-12a.

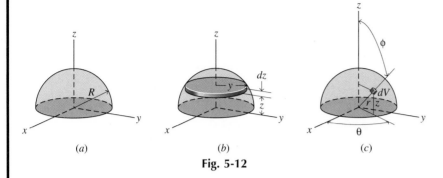

Fig. 5-12

SOLUTION

From symmetry, it is obvious that $\bar{x} = \bar{y} = 0$. The coordinate $\bar{z}$ of the centroid C of the hemisphere will be located by using two different methods.

METHOD 1:   Single integral using rectangular coordinates

For the differential element of volume shown in Fig. 5-12b,

$$dV = \pi y^2\, dz = \pi(R^2 - z^2)\, dz$$
$$V = \int_V dV = \int_0^R \pi(R^2 - z^2)\, dz = \pi\left[R^2 z - \frac{z^3}{3}\right]_0^R = \frac{2}{3}\pi R^3$$

and

$$M_{xy} = \int_V z\, dV = \int_0^R \pi(R^2 z - z^3)\, dz = \left[\frac{\pi R^2 z^2}{2} - \frac{\pi z^4}{2}\right]_0^R = \frac{\pi R^4}{4}$$

From Eq. 5-10

$$\bar{z} = \frac{M_{xy}}{V} = \frac{\pi R^4/4}{2\pi R^3/3} = \frac{3}{8}R \qquad\qquad \text{Ans.}$$

METHOD 2:   Triple integral using spherical coordinates

For the element of volume shown in Fig. 5-12c,

$$z = r \cos\phi$$
$$dV = (r\, d\phi)(r \sin\phi\, d\theta)\, dr = r^2 \sin\phi\, dr\, d\phi\, d\theta$$

Thus,

$$
\begin{aligned}
M_{xy} &= \int_V z\, dV = \int_0^{2\pi}\int_0^{\pi/2}\int_0^R r^3 \sin\phi \cos\phi\, dr\, d\phi\, d\theta \\
&= \frac{R^4}{4}\int_0^{2\pi}\int_0^{\pi/2} \sin\phi \cos\phi\, d\phi\, d\theta \\
&= \frac{R^4}{8}\int_0^{2\pi}\left[\sin^2\phi\right]_0^{\pi/2} d\theta = \frac{R^4}{8}\int_0^{2\pi} d\theta = \frac{\pi R^4}{4}
\end{aligned}
$$

From Eq. 5-10

$$\bar{z} = \frac{M_{xy}}{V} = \frac{\pi R^4/4}{2\pi R^3/3} = \frac{3}{8}R \qquad\qquad \text{Ans.}$$

Locate the centroid of the arc of thin wire shown in Fig. 5-13a.

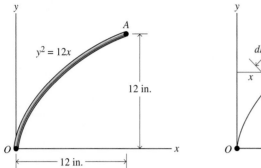

Fig. 5-13

SOLUTION

The wire can be assumed to consist of a large number of differential elements of length dL as shown in Fig. 5-13b. The length dL of each of these elements is given by the expression

$$dL = \sqrt{(dx)^2 + (dy)^2} = \sqrt{(dx/dy)^2 + 1}\, dy$$

For the parabolic arc $y^2 = 12x$:

$$dL = \sqrt{(y/6)^2 + 1}\, dy = \frac{1}{6}\sqrt{y^2 + 36}\, dy$$

$$L = \frac{1}{6}\int_0^{12} \sqrt{y^2 + 36}\, dy$$

$$= \frac{1}{6}\left[\frac{y}{2}\sqrt{y^2 + 36} + 18\ln\left(y + \sqrt{y^2 + 36}\right)\right]_0^{12} = 17.747 \text{ in.}$$

The element dL is located at distances x and y from the coordinate y- and x-axes, respectively. Thus,

$$M_x = \int_L y\, dL = \frac{1}{6}\int_0^{12} y\sqrt{y^2 + 36}\, dy = \frac{1}{6}\left[\frac{1}{3}\sqrt{(y^2 + 36)^3}\right]_0^{12} = 122.16 \text{ in.}^2$$

$$M_y = \int_L x\, dL = \frac{1}{72}\int_0^{12} y^2\sqrt{y^2 + 36}\, dy$$

$$= \frac{1}{72}\left[\frac{y}{4}\sqrt{(y^2 + 36)^3} - \frac{36y}{8}\sqrt{(y^2 + 36)} - 162\ln\left(y + \sqrt{y^2 + 36}\right)\right]_0^{12}$$

$$= 87.31 \text{ in.}^2$$

From Eq. 5-12

$$\bar{y} = \frac{M_x}{L} = \frac{122.16}{17.747} = 6.88 \text{ in.} \qquad \text{Ans.}$$

$$\bar{x} = \frac{M_y}{A} = \frac{87.312}{17.747} = 4.92 \text{ in.} \qquad \text{Ans.}$$

PROBLEMS

5-9–5-25 Locate the centroid of the shaded area shown in each of the following figures.

5-9* Fig. P5-9 if $b = 12$ in. and $h = 8$ in.

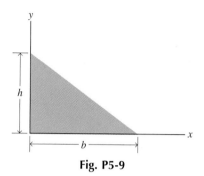

Fig. P5-9

5-10* Fig. P5-10 if $b = 200$ mm and $h = 300$ mm.

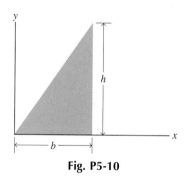

Fig. P5-10

5-11 Fig. P5-11.

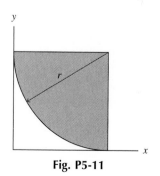

Fig. P5-11

5-12 Fig. P5-12.

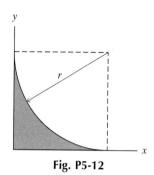

Fig. P5-12

5-13* Fig. P5-13.

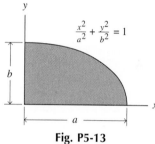

$$\frac{x^2}{a^2} + \frac{y^2}{b^2} = 1$$

Fig. P5-13

5-14* Fig. P5-14.

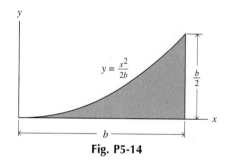

$$y = \frac{x^2}{2b}$$

Fig. P5-14

5-15 Fig. P5-15.

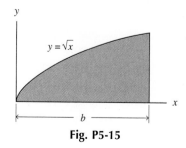

$$y = \sqrt{x}$$

Fig. P5-15

5-16–5-24 Locate the centroid of the shaded area shown in each of the following figures.

5-16 Fig. P5-16.

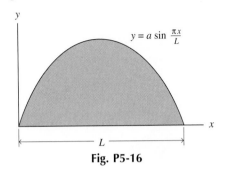

$$y = a \sin \frac{\pi x}{L}$$

Fig. P5-16

5-17* Fig. P5-17.

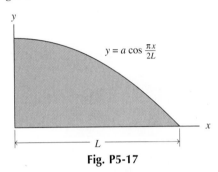

$$y = a \cos \frac{\pi x}{2L}$$

Fig. P5-17

5-18* Fig. P5-18.

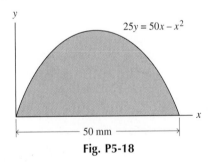

$$25y = 50x - x^2$$

50 mm

Fig. P5-18

5-19 Fig. P5-19.

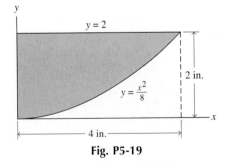

$y = 2$

2 in.

$y = \frac{x^2}{8}$

4 in.

Fig. P5-19

5-20 Fig. P5-20.

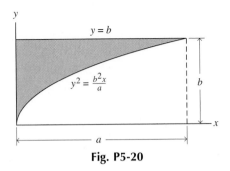

$y = b$

$$y^2 = \frac{b^2 x}{a}$$

b

a

Fig. P5-20

5-21* Fig. P5-21.

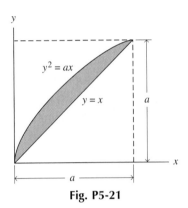

$y^2 = ax$

$y = x$

a

a

Fig. P5-21

5-22* Fig. P5-22.

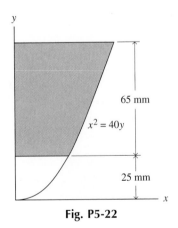

65 mm

$x^2 = 40y$

25 mm

Fig. P5-22

5-23 Fig. P5-23.

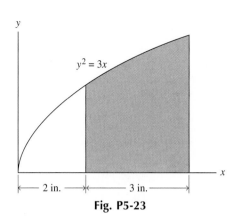

$y^2 = 3x$

← 2 in. → ← 3 in. →

Fig. P5-23

5-24 Fig. P5-24.

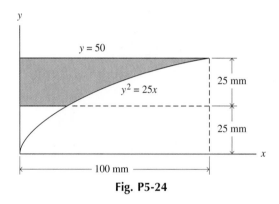

$y = 50$

$y^2 = 25x$

25 mm

25 mm

100 mm

Fig. P5-24

5-25–5-28 Locate the centroid of the curved slender rod shown in the following figures.

5-25* Fig. P5-25.

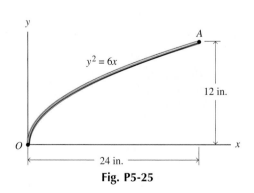

$y^2 = 6x$

A

12 in.

O

24 in.

Fig. P5-25

5-26* Fig. P5-26 if $b = 200$ mm.

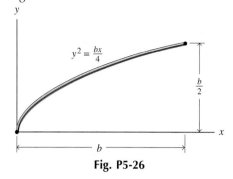

$y^2 = \frac{bx}{4}$

$\frac{b}{2}$

b

Fig. P5-26

5-27 Fig. P5-27 if $b = 36$ in.

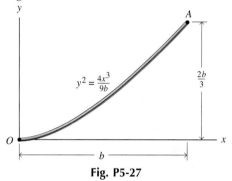

A

$y^2 = \frac{4x^3}{9b}$

$\frac{2b}{3}$

O

b

Fig. P5-27

5-28 Fig. P5-28 if $b = 50$ mm.

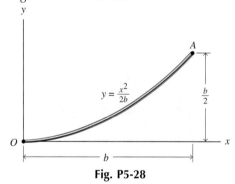

A

$y = \frac{x^2}{2b}$

$\frac{b}{2}$

O

b

Fig. P5-28

5-29* Locate the centroid of the volume of the tetrahedron shown in Fig. P5-29.

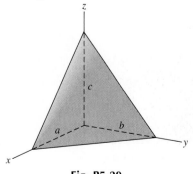

z

c

a

b

y

x

Fig. P5-29

5-30* Locate the centroid of the volume of the portion of a right circular cone shown in Fig. P5-30.

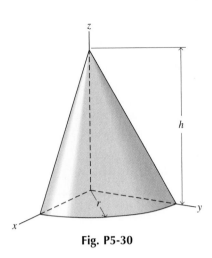

Fig. P5-30

5-31–5-34 Locate the centroid of the volume obtained by revolving the shaded area shown in the following figures about the *x*-axis.

5-31 Fig. P5-31.

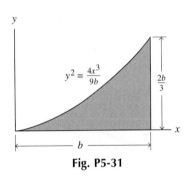

$$y^2 = \frac{4x^3}{9b}$$

$$\frac{2b}{3}$$

b

Fig. P5-31

5-32 Fig. P5-32.

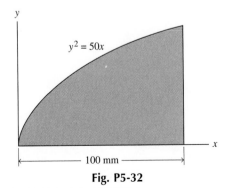

$$y^2 = 50x$$

100 mm

Fig. P5-32

5-33* Fig. P5-33.

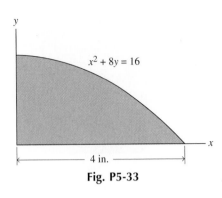

$$x^2 + 8y = 16$$

4 in.

Fig. P5-33

5-34* Fig. P5-34.

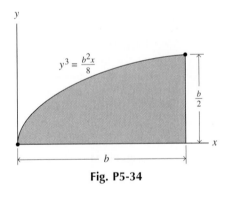

$$y^3 = \frac{b^2 x}{8}$$

$$\frac{b}{2}$$

b

Fig. P5-34

5-35 Locate the centroid of the volume obtained by revolving the shaded area shown in Fig. P5-31 about the *y*-axis.

5-36 Locate the centroid of the volume obtained by revolving the shaded area shown in Fig. P5-34 about the *y*-axis.

5-37* Locate the mass center of a slender straight rod that has a cross-sectional area A and a length L if the density ρ at any point P is proportional to the distance from the left end of the rod to point P.

5-38* Locate the mass center of the right circular cone shown in Fig. 5-11 (p. 164) if the density ρ at any point P is proportional to the distance from the *xy*-plane to point P.

5-39 Locate the mass center of the hemisphere shown in Fig. P5-39 if the density ρ at any point P is proportional to the distance from the xy-plane to point P.

5-40 Locate the mass center of the right circular cone shown in Fig. 5-11 (p. 164) if the density ρ at any point P is proportional to the square of the distance from the xy-plane to point P.

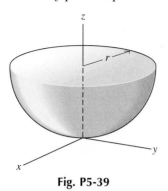

Fig. P5-39

5-4 CENTROIDS OF COMPOSITE BODIES

If the location of the centroid of a line, area, or volume is known, the first moment with respect to an axis or plane is easily found by multiplying the length, area, or volume by the distance from its centroid to the axis or plane. Thus, if a given line, area, or volume can be divided into parts with known locations for their respective centroids, the moment of the total line, area, or volume can be found, without integrating, by obtaining the algebraic sum of the first moments of the parts into which the line, area, or volume is divided. Thus, for example, in the case of a composite area, if $A_1, A_2, \cdots, A_n$ denote the parts into which the area A is divided, and $\bar{x}_1, \bar{x}_2, \cdots, \bar{x}_n$ denote the x-coordinates of the centroids of the respective parts, then

$$M_y = (A_1 + A_2 + \cdots + A_n)\, \bar{x} = A_1\bar{x}_1 + A_2\bar{x}_2 + \cdots + A_n\bar{x}_n$$

or

$$M_y = A\bar{x} = \sum_{i=1}^{n} A_i\bar{x}_i \qquad \text{or} \qquad \bar{x} = \frac{M_y}{A} = \frac{1}{A}\sum_{i=1}^{n} A_i\bar{x}_i$$

Similarly

$$M_x = A\bar{y} = \sum_{i=1}^{n} A_i\bar{y}_i \qquad \text{or} \qquad \bar{y} = \frac{M_x}{A} = \frac{1}{A}\sum_{i=1}^{n} A_i\bar{y}_i \qquad (5\text{-}13)$$

If a hole is considered as one of the component parts of a composite body, the area represented by the hole is considered a negative quantity.

Similar equations can be developed for composite lines, volumes, masses, and weights. The final results would show the A of Eqs. 5-13 replaced with L, V, m, and W, respectively.

Tables 5-1 and 5-2 contain a listing of the centroid locations for some common shapes.

Circular arc

$L = 2r\alpha$

$\bar{x} = \dfrac{r \sin \alpha}{\alpha}$

$\bar{y} = 0$

Circular sector

$A = r^2\alpha$

$\bar{x} = \dfrac{2r \sin \alpha}{3\alpha}$

$\bar{y} = 0$

Quarter circular arc

$L = \dfrac{\pi r}{2}$

$\bar{x} = \dfrac{2r}{\pi}$

$\bar{y} = \dfrac{2r}{\pi}$

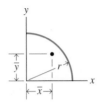

Quadrant of a circle

$A = \dfrac{\pi r^2}{4}$

$\bar{x} = \dfrac{4r}{3\pi}$

$\bar{y} = \dfrac{4r}{3\pi}$

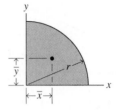

Semicircular arc

$L = \pi r$

$\bar{x} = r$

$\bar{y} = \dfrac{2r}{\pi}$

Semicircular area

$A = \dfrac{\pi r^2}{2}$

$\bar{x} = r$

$\bar{y} = \dfrac{4r}{3\pi}$

Rectangular area

$A = bh$

$\bar{x} = \dfrac{b}{2}$

$\bar{y} = \dfrac{h}{2}$

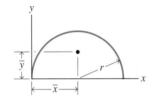

Quadrant of an ellipse

$A = \dfrac{\pi ab}{4}$

$\bar{x} = \dfrac{4a}{3\pi}$

$\bar{y} = \dfrac{4b}{3\pi}$

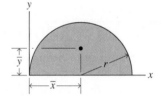

Triangular area

$A = \dfrac{bh}{2}$

$\bar{x} = \dfrac{2b}{3}$

$\bar{y} = \dfrac{h}{3}$

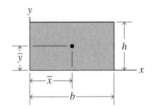

Parabolic spandrel

$A = \dfrac{bh}{3}$

$\bar{x} = \dfrac{3b}{4}$

$\bar{y} = \dfrac{3h}{10}$

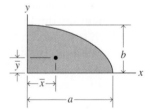

Triangular area

$A = \dfrac{bh}{2}$

$\bar{x} = \dfrac{a + b}{3}$

$\bar{y} = \dfrac{h}{3}$

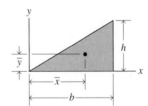

Quadrant of a parabola

$A = \dfrac{2bh}{3}$

$\bar{x} = \dfrac{5b}{8}$

$\bar{y} = \dfrac{2h}{5}$

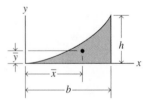

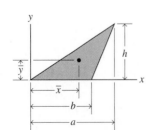

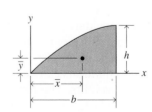

TABLE 5-2 CENTROID LOCATIONS FOR A FEW COMMON VOLUMES

Rectangular parallelepiped

$V = abc$

$\bar{x} = \dfrac{a}{2}$

$\bar{y} = \dfrac{b}{2}$

$\bar{z} = \dfrac{c}{2}$

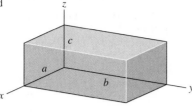

Rectangular tetrahedron

$V = \dfrac{abc}{6}$

$\bar{x} = \dfrac{a}{4}$

$\bar{y} = \dfrac{b}{4}$

$\bar{z} = \dfrac{c}{4}$

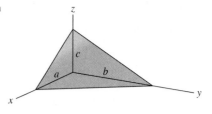

Circular cylinder

$V = \pi r^2 L$

$\bar{x} = 0$

$\bar{y} = \dfrac{L}{2}$

$\bar{z} = 0$

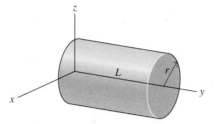

Semicylinder

$V = \dfrac{\pi r^2 L}{2}$

$\bar{x} = 0$

$\bar{y} = \dfrac{L}{2}$

$\bar{z} = \dfrac{4r}{3\pi}$

Hemisphere

$V = \dfrac{2\pi r^3}{3}$

$\bar{x} = 0$

$\bar{y} = 0$

$\bar{z} = \dfrac{3r}{8}$

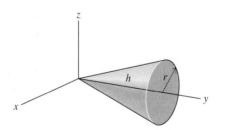

Paraboloid

$V = \dfrac{\pi r^2 h}{2}$

$\bar{x} = 0$

$\bar{y} = \dfrac{2h}{3}$

$\bar{z} = 0$

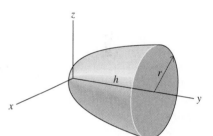

Right circular cone

$V = \dfrac{\pi r^2 h}{3}$

$\bar{x} = 0$

$\bar{y} = \dfrac{3h}{4}$

$\bar{z} = 0$

Half cone

$V = \dfrac{\pi r^2 h}{6}$

$\bar{x} = 0$

$\bar{y} = \dfrac{3h}{4}$

$\bar{z} = \dfrac{r}{\pi}$

The following examples illustrate the procedure for determining the locations of centroids of composite lines, areas, and volumes and centers of mass and centers of gravity for composite bodies.

A slender steel rod is bent into the shape shown in Fig. 5-14a. Locate the centroid of the rod.

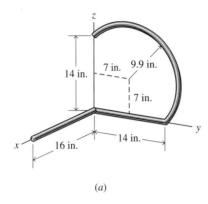

(a)

SOLUTION

The rod can be divided into three parts as shown in Fig. 5-14b. The centroid locations for each of these parts are known or can be determined from the relationships listed in Table 5-1. For the semicircular arc,

$$L_3 = \pi r = \pi(9.90) = 31.1 \text{ in.}$$

$$\bar{y} = 7 + \frac{r \sin \alpha}{\alpha} \cos 45° = 7 + \frac{9.90 \sin (\pi/2)}{\pi/2} \cos 45° = 11.457 \text{ in.}$$

$$\bar{z}_3 = 7 + \frac{r \sin \alpha}{\alpha} \sin 45° = 7 + \frac{9.90 \sin (\pi/2)}{\pi/2} \sin 45° = 11.457 \text{ in.}$$

The centroid for the composite rod can be determined by listing the length, centroid location, and first moments for the individual parts in a table and applying Eqs. 5-14. Thus,

Part	L_i (in.)	$\bar{x}_i$ (in.)	M_{yz} (in.²)	$\bar{y}_i$ (in.)	M_{zx} (in.²)	$\bar{z}_i$ (in.)	M_{xy} (in.²)
1	16.0	8	128	0	0	0	0
2	14.0	0	0	7	98	0	0
3	31.1	0	0	11.457	356.3	11.457	356.3
Σ	61.1		128		454.3		356.3

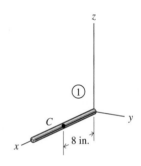

From Eqs. 5-13

$$\bar{x} = \frac{M_{yz}}{L} = \frac{128}{61.1} = 2.09 \text{ in.} \qquad \text{Ans.}$$

$$\bar{y} = \frac{M_{zx}}{L} = \frac{454.3}{61.1} = 7.44 \text{ in.} \qquad \text{Ans.}$$

$$\bar{z} = \frac{M_{xy}}{L} = \frac{356.3}{61.1} = 5.83 \text{ in.} \qquad \text{Ans.}$$

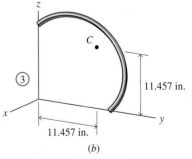

(b)

Fig. 5-14

Locate the centroid of the composite area shown in Fig. 5-15a.

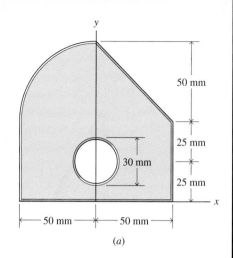

(a)

SOLUTION

The composite area can be divided into four parts, a rectangle, a triangle, a quarter circle, and a circle, as shown in Fig. 5-15b. Recall that the area of the hole is negative since it must be subtracted from the area of the rectangle. The centroid locations for each of these parts can be determined from the relationships listed in Table 5-1.

For the triangle:

$$\bar{x} = \frac{b}{3} = \frac{50}{3} = 16.67 \text{ mm}$$

$$\bar{y} = 50 + \frac{h}{3} = 50 + \frac{50}{3} = 66.67 \text{ mm}$$

For the quarter circle:

$$\bar{x} = -\frac{4r}{3\pi} = -\frac{4(50)}{3\pi} = -21.22 \text{ mm}$$

$$\bar{y} = 50 + \frac{4r}{3\pi} = 50 + \frac{4(50)}{3\pi} = 71.22 \text{ mm}$$

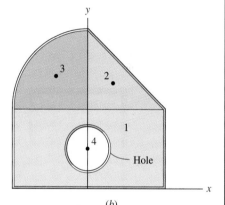

(b)

Fig. 5-15

The centroid for the composite area is determined by listing the area, centroid location, and first moments for the individual parts in a table and applying Eqs. 5-13. Thus,

Part	A_i (mm²)	$\bar{x}_i$ (mm)	M_y (mm³)	$\bar{y}_i$ (mm)	M_x (mm³)
1	5000	0	0	25	125,000
2	1250	16.67	20,838	66.67	83,338
3	1963	−21.22	−41,665	71.22	139,805
4	−707	0	0	25	−17,675
Σ	7506		−20,827		330,468

From Eqs. 5-13

$$\bar{x} = \frac{M_y}{A} = \frac{-20,827}{7506} = -2.77 \text{ mm} \qquad \text{Ans.}$$

$$\bar{y} = \frac{M_x}{A} = \frac{330,468}{7506} = 44.0 \text{ mm} \qquad \text{Ans.}$$

PROBLEMS

5-41–5-46 Locate the centroid of the slender rod shown in the following figures.

5-41* Fig. P5-41.

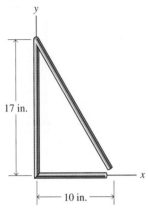

Fig. P5-41

5-42* Fig. P5-42.

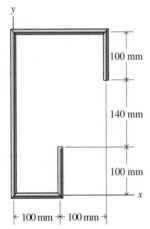

Fig. P5-42

5-43 Fig. P5-43.

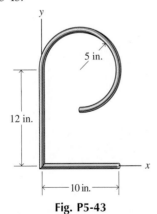

Fig. P5-43

5-44 Fig. P5-44.

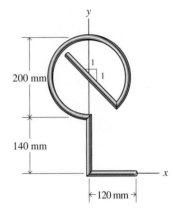

Fig. P5-44

5-45* Fig. P5-45.

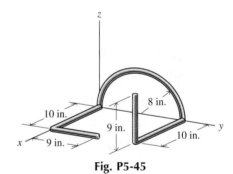

Fig. P5-45

5-46 Fig. P5-46.

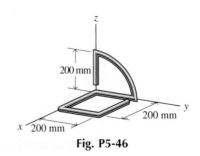

Fig. P5-46

5-47–5-52 Locate the centroid of the shaded area shown in the following figures.

5-47* Fig. P5-47.

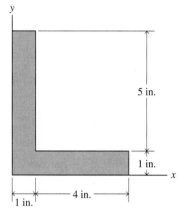

Fig. P5-47

5-48* Fig. P5-48.

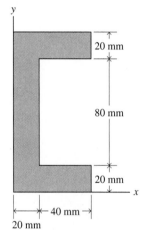

Fig. P5-48

5-49 Fig. P5-49.

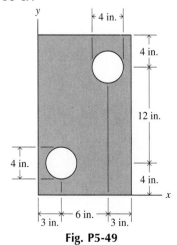

Fig. P5-49

5-50 Fig. P5-50.

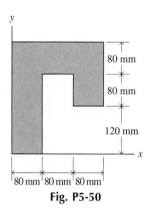

Fig. P5-50

5-51* Fig. P5-51.

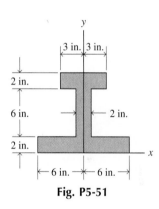

Fig. P5-51

5-52* Fig. P5-52.

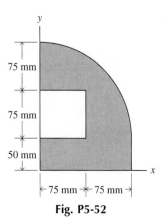

Fig. P5-52

5-53–5-63 Locate the centroid of the slender rod shown in the following figures.

5-53 Fig. P5-53.

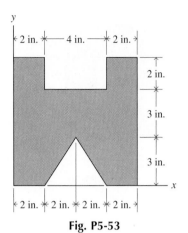

Fig. P5-53

5-54 Fig. P5-54.

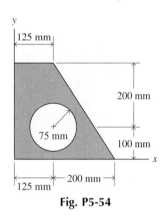

Fig. P5-54

5-55* Fig. P5-55.

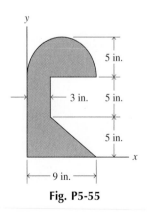

Fig. P5-55

5-56* Fig. P5-56.

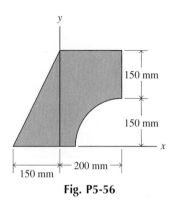

Fig. P5-56

5-57 Fig. P5-57.

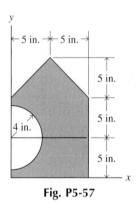

Fig. P5-57

5-58 Fig. P5-58.

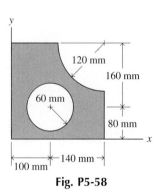

Fig. P5-58

5-59* Fig. P5-59.

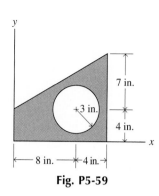

Fig. P5-59

5-60* Fig. P5-60.

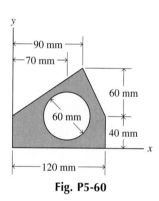

Fig. P5-60

5-61 Fig. P5-61.

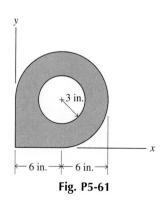

Fig. P5-61

5-62 Fig. P5-62.

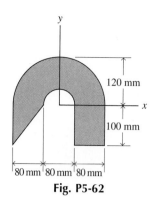

Fig. P5-62

5-63* Fig. P5-63.

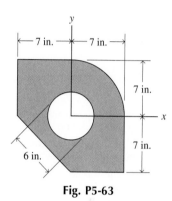

Fig. P5-63

5-64* Two channel sections and a plate are used to form the cross section shown in Fig. P5-64. Each of the channels has a cross-sectional area of 2605 mm². Locate the y-coordinate of the centroid of the composite section with respect to the top surface of the plate.

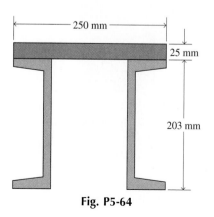

Fig. P5-64

179

5-65* A bracket is made of brass ($\gamma = 0.316$ lb/in.³) and aluminum ($\gamma = 0.100$ lb/in.³) plates as shown in Fig. P5-65.

a. Locate the centroid of the bracket.
b. Locate the center of gravity of the bracket.

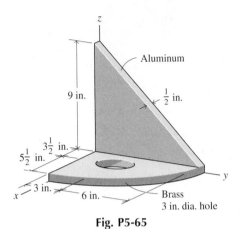

Fig. P5-65

5-66* A cylinder with a hemispherical cavity and a conical cap is shown in Fig. P5-66.

a. Locate the centroid of the composite volume if $R = 140$ mm, $L = 250$ mm, and $h = 300$ mm.
b. Locate the center of mass of the composite volume if the cylinder is made of steel ($\rho = 7870$ kg/m³) and the cap is made of aluminum ($\rho = 2770$ kg/m³).

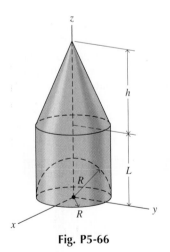

Fig. P5-66

5-67 A bracket is made of steel ($\gamma = 0.284$ lb/in.³) and aluminum ($\gamma = 0.100$ lb/in.³) plates as shown in Fig. P5-67.

a. Locate the centroid of the bracket.
b. Locate the center of gravity of the bracket.

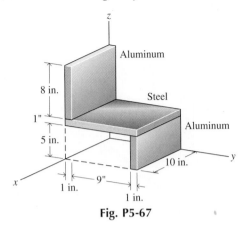

Fig. P5-67

5-68 A cylinder with a conical cavity and a hemispherical cap is shown in Fig. P5-68.

a. Locate the centroid of the composite volume if $R = 200$ mm and $h = 250$ mm.
b. Locate the center of mass of the composite volume if the cylinder is made of brass ($\rho = 8750$ kg/m³) and the cap is made of aluminum ($\rho = 2770$ kg/m³).

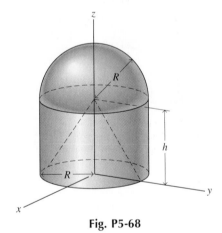

Fig. P5-68

5-69* Locate the center of gravity of the bracket shown in Fig. P5-69 if the holes have 6-in. diameters.

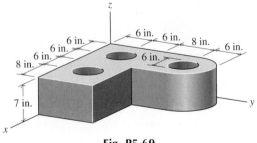

Fig. P5-69

5-70* Locate the center of mass of the machine component shown in Fig. P5-70. The brass ($\rho = 8750$ kg/m³) disk C is mounted on the steel ($\rho = 7870$ kg/m³) shaft B.

5-71 Locate the center of gravity of the machine component shown in Fig. P5-71.

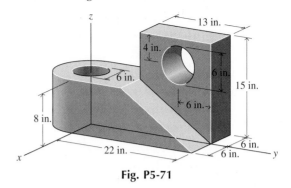

Fig. P5-71

5-72 Locate the center of mass of the machine component shown in Fig. P5-72.

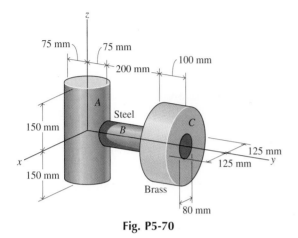

Fig. P5-70

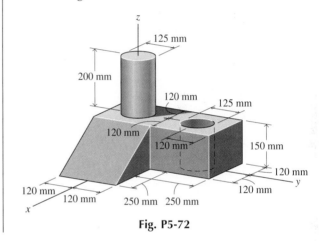

Fig. P5-72

5-5 THEOREMS OF PAPPUS AND GULDINUS

Two theorems stated by Pappus[1] and Guldinus,[2] before the development of calculus, can be used to determine the surface area generated by revolving a plane curve or the volume generated by revolving an area about an axis that does not intersect any part of the plane curve or area. Applications of the theorems require use of the equations previously developed for locating the centroids of lines and areas.

> Theorem 1   The area A of a surface of revolution generated by revolving a plane curve of length L about any nonintersecting axis in its plane is equal to the product of the length of the curve and the length of the path traveled by the centroid of the curve.

[1] Pappus of Alexandria (about A.D. 380), a Greek geometer.

[2] Paul Guldin (1577–1643), a Swiss mathematician.

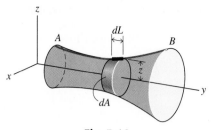

Fig. 5-16

If the curve AB of Fig. 5-16 is revolved about the y-axis, the increment of surface area dA generated by the element of length dL of line AB is

$$dA = 2\pi z \; dL$$

Thus, the total surface area A generated by revolving the line AB about the y-axis is

$$A = 2\pi \int_L z \; dL$$

The coordinate $\bar{z}$ of the centroid C of a line in the yz-plane is given by Eq. 5-12 as

$$\bar{z} = \frac{1}{L} \int_0^L z \; dL$$

Thus, the surface area A generated by revolving the line AB about the y-axis is

$$A = 2\pi \bar{z} \; L \qquad\qquad (5\text{-}14)$$

where $2\pi\bar{z}$ is the distance traveled by the centroid C of the line L in generating the surface area A. The theorem is also valid if the line AB is rotated through an angle θ other than 2π radians. Thus, for any angle of rotation θ $(0 \le \theta \le 2\pi)$, the surface area A generated is

$$A = \theta z \; L \qquad\qquad (5\text{-}15)$$

Surface areas of a wide variety of shapes can be determined by using Theorem 1. Examples include the surface area of the cylinder, cone, sphere, torus, ellipsoid, and any other surface of revolution for which the generating line can be described and the location of its centroid determined.

Theorem 2   The volume V of the solid of revolution generated by revolving a plane area A about any nonintersecting axis in its plane is equal to the product of the area and the length of the path traveled by the centroid of the area.

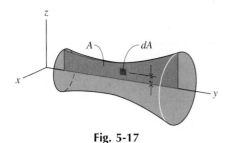

Fig. 5-17

If the plane area A of Fig. 5-17 is revolved about the y-axis, the increment of volume dV generated by the element of area dA is

$$dV = 2\pi z \, dA$$

Thus, the total volume V generated by revolving the area A about the y-axis is

$$V = 2\pi \int_A z \, dA$$

The coordinate $\bar{z}$ of the centroid C of an area in the yz-plane is given by Eq. 5-11 as

$$\bar{z} = \frac{1}{A} \int_A z \, dA$$

Thus, the volume V generated by revolving the area A about the y-axis is

$$V = 2\pi\bar{z} \, A \qquad (5\text{-}16)$$

where $2\pi\bar{z}$ is the distance traveled by the centroid C of the area A in generating the volume V. The theorem is also valid if the area A is rotated through an angle θ other than 2π radians. Thus, for any angle of rotation θ $(0 \le \theta \le 2\pi)$, the volume V generated is

$$V = \theta\bar{z} \, A \qquad (5\text{-}17)$$

The following examples illustrate the procedure for determining surface areas and volumes for solids of revolution by using the theorems of Pappus and Guldinus.

Determine the surface area A and volume V of the solid (torus) generated by revolving the circle shown in Fig. 5-18 through an angle of $360°$ about the y-axis.

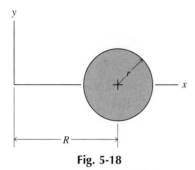

Fig. 5-18

SOLUTION

The circumference L_c and area A_c of the circle shown in Fig. 5-18 are

$$L_c = 2\pi r \qquad A_c = \pi r^2$$

The centroid C_L of the boundary line L_c and the centroid C_A of the area A_c of the circle shown in Fig. 5-18 are both located at the position $x = R$. Thus

$$\bar{x}_L = \bar{x}_A = R$$

From Eqs. 5-14 and 5-16,

$$A = 2\pi\bar{x}_L L_c = 2\pi(R)(2\pi r) = 4\pi^2 Rr \qquad \text{Ans.}$$
$$V = 2\pi\bar{x}_A A_c = 2\pi(R)(\pi r^2) = 2\pi^2 Rr^2 \qquad \text{Ans.}$$

EXAMPLE PROBLEM 5-12

Determine the surface area A and volume V of the solid generated by revolving the shaded area shown in Fig. 5-19 through an angle of $360°$ about the y axis.

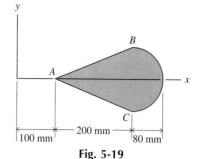

Fig. 5-19

SOLUTION

The length L of the perimeter $ABCA$ of the shaded area of Fig. 5-19 is

$$L = 2L_{AB} + L_{BC} = 2\sqrt{(200)^2 + (80)^2} + \pi(80) = 2(215.4) + 251.3 = 682.1 \text{ mm}$$

For line AB:

$$\bar{x}_{AB} = 100 + \frac{200}{2} = 200 \text{ mm}$$

For semicircle BC:

$$\bar{x}_{BC} = 300 + \frac{2(80)}{\pi} = 350.9 \text{ mm}$$

The first moment about the y-axis of the boundary line $ABCA$ is

$$M_y = 2(215.4)(200) + 251.3(350.9) = 174{,}341 \text{ mm}^2$$

From Eq. 5-13

$$\bar{x}_L = \frac{M_y}{L} = \frac{174{,}341}{682.1} = 255.6 \text{ mm}$$

The surface area A of the solid is given by Eq. 5-14 as

$$A = 2\pi\bar{x}_L L = 2\pi(255.6)(682.1) = 1.095(10^6) \text{ mm}^2 = 1.095 \text{ m}^2 \quad \text{Ans.}$$

The shaded area $A_s = A_T + A_{SC}$

$$= \frac{1}{2}(200)(160) + \frac{\pi(80)^2}{2} = 16{,}000 + 10{,}053 = 26{,}053 \text{ mm}^2$$

For the triangle:

$$\bar{x}_T = 100 + \frac{2}{3}(200) = 233.3 \text{ mm}$$

For the semicircle:

$$\bar{x}_{SC} = 300 + \frac{4(80)}{3\pi} = 334.0 \text{ mm}$$

The first moment about the y-axis of the shaded area is

$$M_y = 16{,}000(233.3) + 10{,}053(334.0) = 7{,}090{,}502 \text{ mm}^3$$

From Eq. 5-13

$$\bar{x}_A = \frac{M_y}{A} = \frac{7{,}090{,}502}{26{,}053} = 272.2 \text{ mm}$$

The volume V of the solid is given by Eq. 5-16 as

$$V = 2\pi\bar{x}_A A = 2\pi(272.2)(26{,}053) = 44.56(10^6) \text{ mm}^3 = 0.0446 \text{ m}^3 \quad \text{Ans.}$$

PROBLEMS

Use the Theorems of Pappus and Guldinus to solve the following problems.

5-73* Determine the surface area A and volume V of the solid body generated by revolving the shaded area shown in Fig. P5-73 through an angle of 360° about the y-axis.

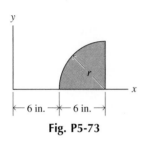

Fig. P5-73

5-74* Determine the surface area A and volume V of the solid body generated by revolving the shaded area shown in Fig. P5-74 through an angle of 360° about the y-axis.

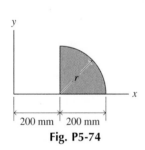

Fig. P5-74

5-75 Determine the surface area A and volume V of the solid body generated by revolving the shaded area shown in Fig. P5-75 through an angle of 360° about the x-axis.

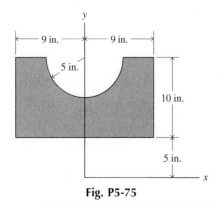

Fig. P5-75

5-76 Determine the surface area A and volume V of the solid body generated by revolving the shaded area shown in Fig. P5-76 through an angle of 360° about the x-axis.

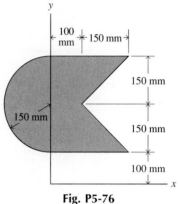

Fig. P5-76

5-77* Determine the volume V of the solid body generated by revolving the shaded area shown in Fig. P5-77 through an angle of 180° about the y-axis.

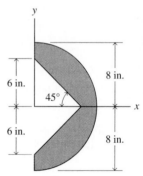

Fig. P5-77

5-78* Determine the volume V of the solid body generated by revolving the shaded area shown in Fig. P5-78 through an angle of 270° about the y-axis.

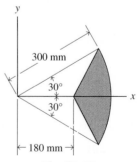

Fig. P5-78

186

5-79 A steel ($\gamma = 0.284$ lb/in.^{3}) flywheel is formed by revolving the shaded area shown in Fig. P5-79 through an angle of 360° about the x-axis. Determine the weight W of the flywheel.

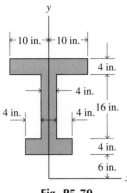

Fig. P5-79

5-80 Determine the volume V of the solid body generated by revolving the shaded area shown in Fig. P5-80 through an angle of 360° about the y-axis.

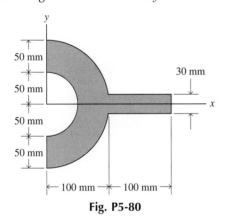

Fig. P5-80

5-81* Determine the volume V of the solid body generated by revolving the shaded area shown in Fig. P5-81 through an angle of 360° about the y-axis.

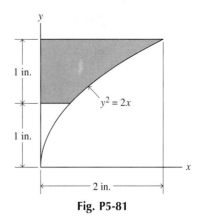

Fig. P5-81

5-82* Determine the volume V of the solid body generated by revolving the shaded area shown in Fig. P5-82 through an angle of 360° about the x-axis.

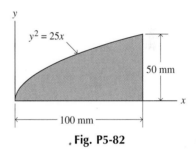

Fig. P5-82

5-83 Determine the volume V of the solid body generated by revolving the shaded area shown in Fig. P5-81 through an angle of 360° about the x-axis.

5-84 Determine the volume V of the solid body generated by revolving the shaded area shown in Fig. P5-82 through an angle of 360° about the y-axis.

5-6   DISTRIBUTED LOADS ON BEAMS

When a load is applied to a rigid body, it is often distributed along a line or over an area A. In many instances, it is convenient to replace this distributed load with a resultant force that is equivalent to the distributed load. Thus, the quantities to be determined are the magnitude of the force and the location of its line of action.

Consider the beam shown in Fig. 5-20a. Since distributed loads on beams usually do not vary across the width of the cross section of the beam, the actual load intensity on the beam can be multiplied by the width of the beam to yield a distributed line load w (N/m or lb/ft) whose magnitude varies only with position x along the beam. The distributed load w versus position x curve is known as a load diagram. The magnitude of the differential force $d\mathbf{R}$ exerted on the beam by the distributed load w in an increment of length dx (see Fig. 5-20b) is

$$dR = w \, dx$$

Therefore, the magnitude of the single concentrated resultant force $\mathbf{R}$ that is equivalent to the distributed load w is

$$R = \int_{L} w \, dx \tag{5-18}$$

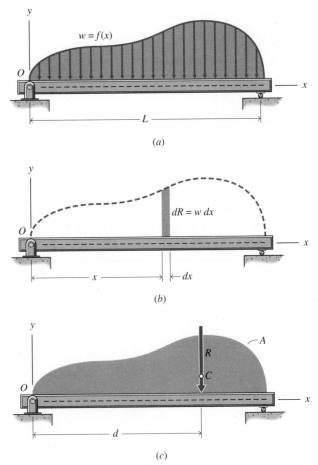

Fig. 5-20

Thus, the magnitude of the resultant force **R** is equal to the area A under the load diagram, as shown in Fig. 5-20c.

The location of the line of action of the resultant force **R** can be determined by equating the moment of the resultant force **R** about an arbitrary point O to the moment of the distributed load about the same point O. The moment produced by the force dR about point O of Fig. 5-20b is

$$dM_o = x \, dR$$

Therefore, the total moment produced by the distributed load w about point O is

$$M_o = \int dM_o = \int x \, dR \qquad (a)$$

The moment produced by the resultant force **R** about point O is

$$M_o = Rd \qquad (b)$$

where d is the distance along the beam from point O to the line of action of the resultant force **R**. Thus, from Eqs. a and b

$$M_o = Rd = \int x \, dR = \bar{x}R \quad \text{or} \quad d = \bar{x} = \frac{M_o}{R} \qquad (5\text{-}19)$$

Equation 5-19 indicates that the line of action of the resultant force **R** passes through the centroid of the area A under the load diagram, as shown in Fig. 5-20c.

The following examples illustrate the procedure for determining the resultant of a distributed load w and the location of its line of action by using the procedures previously developed for locating the centroids of areas.

A beam is subjected to a system of loads that can be represented by the load diagram shown in Fig. 5-21a. Determine the resultant of this system of loads and locate its line of action with respect to the left support of the beam.

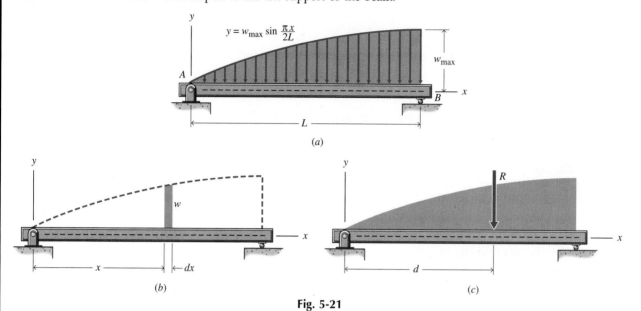

Fig. 5-21

SOLUTION

The magnitude of the resultant **R** of the distributed load shown in Fig. 5-21a is equal to the area under the load diagram, and the line of action of the resultant passes through the centroid of the area. Since the area under this load diagram and the location of its centroid do not normally appear in tables of areas and centroid locations, it is necessary to use integration methods to determine the magnitude of the resultant and the location of its line of action.

The area under the load diagram is determined by using the element of area shown in Fig. 5-21b. Thus,

$$A = \int_A w \, dx = \int_0^L w_{max} \sin \frac{\pi x}{2L} \, dx = \frac{2L w_{max}}{\pi} \left[-\cos \frac{\pi x}{2L} \right]_0^L = \frac{2L w_{max}}{\pi}$$

Thus,

$$R = A = \frac{2L w_{max}}{\pi} = 0.637 w_{max} L \qquad \text{Ans.}$$

The moment of the area about support A is

$$M_A = \int_A x \, (w \, dx) = \int_0^L w_{max} \, x \sin \frac{\pi x}{2L} \, dx$$

$$= w_{max} \left[\frac{4L^2}{\pi^2} \sin \frac{\pi x}{2L} - \frac{2L}{\pi} x \cos \frac{\pi x}{2L} \right]_0^L = \frac{4L^2 w_{max}}{\pi^2}$$

From Eq. 5-11

$$d = \bar{x} = \frac{M_A}{A} = \frac{4L^2 w_{max}/\pi^2}{2L w_{max}/\pi} = \frac{2L}{\pi} = 0.637L \qquad \text{Ans.}$$

The results are shown in Fig. 5-21c.

A beam is subjected to a system of loads that can be represented by the load diagram shown in Fig. 5-22a. Determine the resultant of this system of distributed loads and locate its line of action with respect to the left support of the beam.

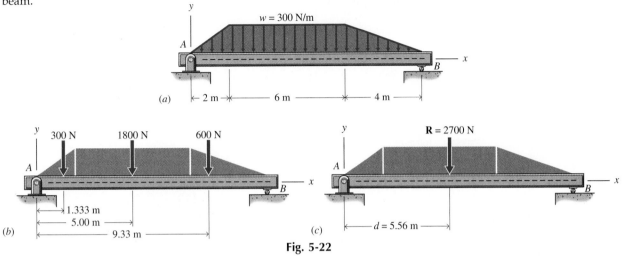

Fig. 5-22

SOLUTION

The magnitude of the resultant **R** of the distributed load shown in Fig. 5-22a is equal to the area under the load diagram. The load diagram can be divided into two triangles and a rectangle. Thus, from Table 5-1:

$$F_1 = A_1 = \frac{1}{2}b_1h_1 = \frac{1}{2}(2)(300) = 300 \text{ N}$$

$$\bar{x}_1 = \frac{2}{3}b_1 = \frac{2}{3}(2) = 1.333 \text{ m}$$

$$F_2 = A_2 = b_2h_2 = (6)(300) = 1800 \text{ N}$$

$$\bar{x}_2 = 2 + \frac{1}{2}b_2 = 2 + \frac{1}{2}(6) = 5.00 \text{ m}$$

$$F_3 = A_3 = \frac{1}{2}b_3h_3 = \frac{1}{2}(4)(300) = 600 \text{ N}$$

$$\bar{x}_3 = 8 + \frac{1}{3}b_3 = 8 + \frac{1}{3}(4) = 9.33 \text{ m}$$

The equivalent forces for the three different areas and the locations of their lines of action are shown in Fig. 5-22b. Thus,

$$R = F_1 + F_2 + F_3 = 300 + 1800 + 600 = 2700 \text{ N} \qquad \text{Ans.}$$

The line of action of the resultant with respect to the left support is located by summing moments about point A. Thus,

$$M_A = Rd = F_1\bar{x}_1 + F_2\bar{x}_2 + F_3\bar{x}_3$$
$$= 300(1.333) + 1800(5.00) + 600(9.33) = 15{,}000 \text{ N} \cdot \text{m}$$

Finally,

$$d = \bar{x} = \frac{M_A}{R} = \frac{15{,}000}{2700} = 5.56\text{m} \qquad \text{Ans.}$$

The resultant force **F** and its line of action are shown in Fig. 5-22c.

PROBLEMS

5-85* A beam is subjected to a system of loads that can be represented by the load diagram shown in Fig. P5-85. Determine the resultant **R** of this system of distributed loads and locate its line of action with respect to the left support of the beam.

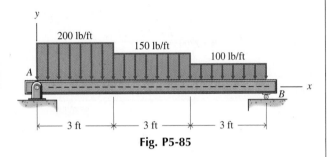

Fig. P5-85

5-86* A beam is subjected to a system of loads that can be represented by the load diagram shown in Fig. P5-86. Determine the resultant **R** of this system of distributed loads and locate its line of action with respect to the left support of the beam.

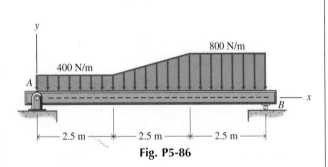

Fig. P5-86

5-87 A vertical beam is subjected to a system of loads that can be represented by the load diagram shown in Fig. P5-87. Determine the resultant **R** of this system of distributed loads and locate its line of action with respect to the support at point O.

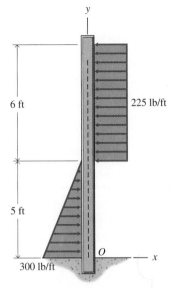

Fig. P5-87

5-88–5-93 A beam is subjected to a system of loads that can be represented by the load diagram shown in the following figures. For each figure determine the resultant **R** of this system of distributed loads and locate its line of action with respect to the left support of the beam.

5-88 Fig. P5-88.

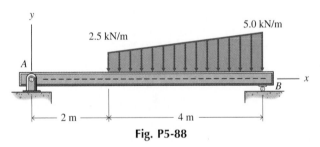

Fig. P5-88

5-89* Fig. P5-89.

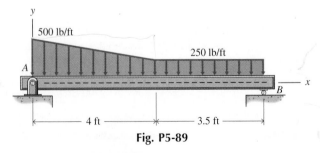

Fig. P5-89

192

5-90* Fig. P5-90.

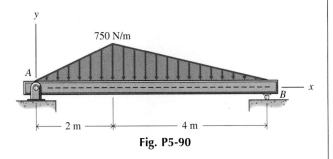

Fig. P5-90

5-91 Fig. P5-91.

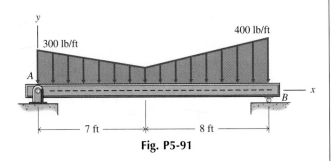

Fig. P5-91

5-92 Fig. P5-92.

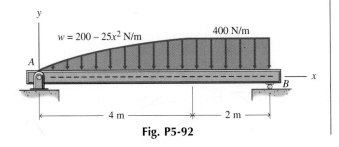

Fig. P5-92

5-93* Fig. P5-93.

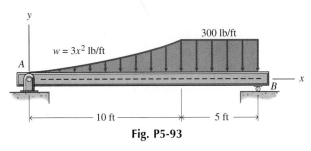

Fig. P5-93

5-94* A vertical beam is subjected to a system of loads that can be represented by the load diagram shown in Fig. P5-94. Determine the resultant **R** of this system of distributed loads and locate its line of action with respect to the support at point O.

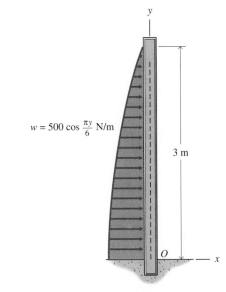

Fig. P5-94

5-95 A beam is subjected to a system of loads that can be represented by the load diagram shown in Fig. P5-95. Determine the resultant **R** of this system of distributed loads and locate its line of action with respect to the left support of the beam.

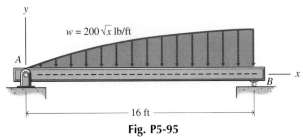

Fig. P5-95

5-96 A vertical beam is subjected to a system of loads that can be represented by the load diagram shown in Fig. P5-96. Determine the resultant **R** of this system of distributed loads and locate its line of action with respect to the support at point *O*.

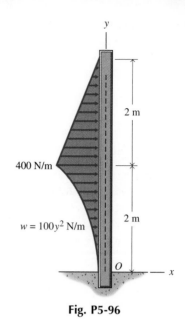

400 N/m

$w = 100y^2$ N/m

2 m

2 m

O

x

y

Fig. P5-96

5-7   FORCES ON SUBMERGED SURFACES

A fluid (either a liquid or a gas) at rest can, by definition, transmit compressive forces but not tensile or shear forces. Since a shear force acts tangent to a surface, a fluid at rest can exert only a compressive normal force (known as a pressure) on a submerged surface. The pressure, called a hydrostatic pressure (equal in all directions), is due to the weight of the fluid above any point on the submerged surface; therefore, fluid pressures vary linearly with depth in fluids with a constant specific weight. The absolute pressure p_A at a depth d is

$$p_A = p_0 + \gamma d = p_0 + \rho g d \qquad (5\text{-}20)$$

where

p_0 = atmospheric pressure at the surface of the fluid
γ = specific weight of the fluid
ρ = density of the fluid
g = gravitational acceleration

In the U. S. Customary system of units, the specific weight γ of fresh water is 62.4 lb/ft³. In the SI system of units, the density ρ of fresh water is 1000 kg/m³. The gravitational acceleration g is 32.2 ft/s² in the U. S. Customary system and 9.81 m/s² in the SI system.

In general, pressure-measuring instruments record pressures above atmospheric pressure. Such pressures are known as "gage pressures," and it is obvious from Eq. 5-20 that the gage pressure p_g is

$$p_g = p_A - p_0 = \gamma d = \rho g d \qquad (5\text{-}21)$$

For the analysis of many engineering problems involving fluid forces, it is necessary to determine the resultant force **R** due to the distribution of pressure on a submerged surface and the location of the

194

intersection of the line of action of the resultant force with the submerged surface. The point P on the submerged surface where the line of action of the resultant force $\mathbf{R}$ intersects the submerged surface is known as the "center of pressure."

5-7-1 Forces on Submerged Plane Surfaces

For the case of fluid pressures on submerged plane surfaces, the load diagram (area) introduced in Section 5-6 for a distributed load along a line becomes a pressure solid (volume), as shown in Fig. 5-23a, since the intensity of a distributed load (pressure) on the submerged surface varies over an area instead of a length. When the distributed pressure p is applied to an area in the xy-plane, the ordinate $p(x,y)$ along the z-axis represents the intensity of the force (force per unit area). The magnitude of the increment of force $d\mathbf{R}$ on an element of area dA is

$$dR = p\, dA = dV_{ps}$$

where dV_{ps} is an element of volume of the pressure solid, as shown in Fig.5-23a. The magnitude of the resultant force $\mathbf{R}$ acting on the submerged surface is

$$R = \int_A p\, dA = \int_V dV_{ps} = V_{ps} \tag{5-22}$$

where V_{ps} is the volume of the pressure solid.

The line of action of the resultant force $\mathbf{R}$ with respect to the x- and y-axes (called the center of pressure) can be located by using the principle of moments.

For moments about the y-axis:

$$Rd_x = \int x\, dR = \int_A x\, p\, dA = \int_V x\, dV_{ps} = \overline{x}_{ps} V_{ps} \tag{5-23a}$$

For moments about the x-axis:

$$Rd_y = \int y\, dR = \int_A y\, p\, dA = \int_V y\, dV_{ps} = \overline{y}_{ps} V_{ps} \tag{5.23b}$$

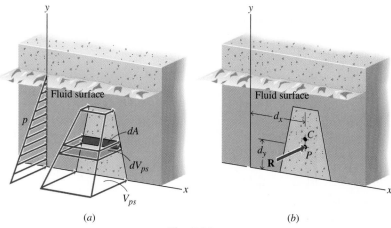

(a)   (b)

Fig. 5-23

Equations 5-23 indicate that the line of action of the resultant force $\mathbf{R}$ passes through the centroid C_V of the volume of the pressure solid. If the pressure is uniformly distributed over the area, the center of pressure P will coincide with the centroid C_A of the area. If the pressure is not uniformly distributed over the area, the center of pressure P and the centroid C_A of the area will be at different points, as shown in Fig. 5-23b.

5-7-2   Forces on Submerged Curved Surfaces

Equations 5-22 and 5-23 apply only to submerged plane surfaces; however, many surfaces of interest in engineering applications are curved such as those associated with pipes, dams, and tanks. For such problems, the resultant force $\mathbf{R}$ and the intersection of its line of action with the curved surface can be determined by integration for each individual problem, but general formulas applicable to a broad class of problems cannot be developed. To overcome this difficulty, the procedure illustrated in Fig. 5-24 has been developed.

In Fig. 5-24a, a cyclindrical gate with a radius a and a length L is being used to close an opening in the wall of a tank containing a fluid. The pressure distribution on the gate is shown in Fig. 5-24b. From such a distribution, horizontal and vertical components of the resultant force can be determined by integration and combined to yield the resultant force $\mathbf{R}$. The pressure-solid approach can also be used to determine the resultant force $\mathbf{R}$ if horizontal and vertical planes are used to isolate the gate and a volume of fluid in contact with the gate, as

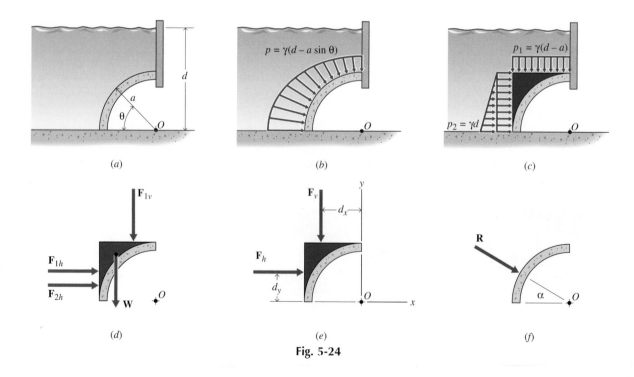

Fig. 5-24

shown in Fig. 5-24c. The force exerted on the horizontal fluid surface by the fluid pressure is

$$F_{1v} = p_1 A_h = \gamma(d - a)(aL)$$

Similarly on the vertical surface,

$$F_{1h} = p_1 A_v = \gamma(d - a)(aL)$$
$$F_{2h} = (p_2 - p_1)A_v = \gamma a(aL)$$

The volume of fluid V_f has a weight W, which is given by the expression

$$W = \gamma V_f = \gamma(a^2 - \frac{1}{4}\pi a^2)L$$

The four forces F_{1v}, F_{1h}, F_{2h}, and W together with their lines of action are shown in Fig. 5-24d. The two vertical forces and the two horizontal forces can be combined to give

$$F_v = F_{1v} + W$$
$$F_h = F_{1h} + F_{2h}$$

where F_v and F_h are the rectangular components of a resultant force **R**. That is, **R** is the resultant of F_{1v}, F_{1h}, F_{2h}, and W, which are the forces exerted by the adjoining water and the earth on the volume of water in contact with the gate. This force is the same as the force exerted by the water on the gate because the volume of water in contact with the gate is in equilibrium and the force exerted on the water by the gate is equal in magnitude and opposite in direction to the force exerted on the gate by the water. The magnitude of the resultant is

$$R = \sqrt{(F_h)^2 + (F_v)^2}$$

The slope of the line of action of the resultant is given by the expression

$$\alpha = \tan^{-1}\frac{F_v}{F_h}$$

Finally, the location of the line of action of the resultant with respect to an arbitrary point can be determined by summing moments about the point. For point O shown in Fig. 5-24e,

$$Rd = F_v d_x - F_h d_y$$

For the cylindrical gate, the line of action of the resultant passes through point O, as shown in Fig. 5-24f. This results from the fact that the pressure always acts normal to the surface; therefore, for the cylindrical gate, the line of action of each increment $d\mathbf{R}$ of the resultant passes through point O. In other words, the increments form a system of concurrent forces.

The following examples illustrate the procedure for determining the resultant force and locating the center of pressure for submerged surfaces by using both integration and pressure-solid approaches.

EXAMPLE PROBLEM 5-15

The water behind a dam is 100 ft deep as shown in Fig. 5-25a. Determine

a. The magnitude of the resultant force **R** exerted on a 30-ft length of the dam by the water pressure.
b. The distance from the water surface to the center of pressure.

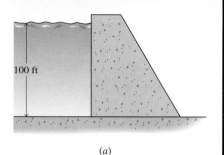

(a)

SOLUTION

A cross section through the pressure solid is shown in Fig. 5-25b. At the base of the dam, the pressure is

$$p = \gamma d = 62.4(100) = 6240 \text{ lb/ft}^2$$

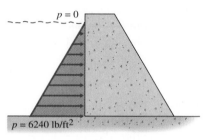

(b)

a. Thus, for the 30-ft length of dam, the volume of the pressure solid is

$$V_{ps} = \frac{1}{2}pA = \frac{1}{2}(6240)(100)(30) = 9,360,000 \text{ lb} = 9360 \text{ kip} \quad \text{Ans.}$$

$$R = V_{ps} = 9360 \text{ kip}$$

b. Since the width of the pressure solid is constant and the cross section is a triangle, the distance from the water surface to the centroid of the solid is

$$d_P = \frac{2}{3}d = \frac{2}{3}(100) = 66.7 \text{ ft} \quad \text{Ans.}$$

The results are shown in Fig. 5-25c.

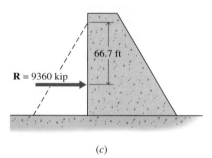

(c)

EXAMPLE PROBLEM 5-16

Determine the magnitude of the resultant of the pressure forces exerted on the rectangular plate shown in Fig. 5-26a and locate the center of pressure with respect to the fluid surface.

SOLUTION

For the element of area dA shown in Fig. 5-26a,

$$p = \rho g(d_C - y) \qquad dA = b\, dy$$

Therefore, the magnitude of the increment of force $d\mathbf{R}$ on the element of area dA is

$$dR = p\, dA = \rho g(d_C - y)(b\, dy)$$

The magnitude of the resultant force **R** is

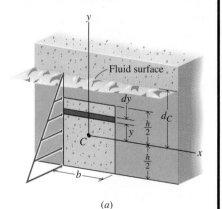

(a)

Fig. 5-26a

$$R = \int_A dR = \int_{-h/2}^{+h/2} \rho g(d_C - y)(b \; dy) = \rho g b \left[d_C y - \frac{1}{2} y^2 \right]_{-h/2}^{+h/2} = \rho g d_C bh$$

The line of action of the resultant force $\mathbf{R}$ is located by using the principle of moments. Summing moments about an x-axis at the fluid surface yields

$$M_x = R d_p = \int_A (d_C - y) dR$$

$$= \int_{-h/2}^{+h/2} \rho g (d_C - y)^2 (b \; dy)$$

$$= \rho g b \left[d_C^2 y - d_C y^2 + \frac{1}{3} y^3 \right]_{-h/2}^{+h/2} = \frac{1}{12} \rho g b h \left(12 d_C^2 + h^2 \right)$$

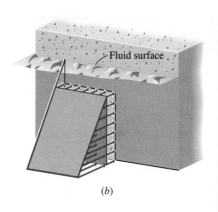

(b)

Therefore, the distance d_p from the fluid surface to the center of pressure is

$$d_P = \frac{M_x}{R} = \frac{\dfrac{1}{12} \rho g b h (12 d_C^2 + h^2)}{\rho g d_C b h} = \frac{12 d_C^2 + h^2}{12 d_C} \qquad \text{Ans.}$$

This problem can also be solved by using the pressure solid shown in Fig. 5-26b since the centroid locations are known for the rectangular prism and triangular wedge shown in the figure. At the top, centroid, and bottom, respectively, of the plate the pressures are

$$p_T = \rho g \left(d_C - \frac{h}{2} \right) \qquad p_C = \rho g d_C \qquad p_B = \rho g \left(d_C + \frac{h}{2} \right)$$

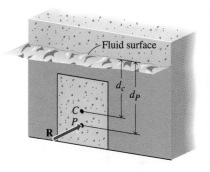

(c)

Fig. 5-26b,c

For the rectangular prism:

$$V_{rp} = p_T bh = \rho g \left(d_C - \frac{h}{2} \right) bh \qquad d_{rp} = d_C$$

For the triangular wedge:

$$V_{tw} = \frac{1}{2} (p_B - p_T) bh = \frac{1}{2} (\rho g h) bh \qquad d_{tw} = d_C + \frac{1}{6} h$$

For the pressure solid:

$$V_{ps} = V_{rp} + V_{tw} = \rho g d_C bh \qquad d_{ps} = d_P$$

Therefore:

$$R = V_{ps} = \rho g d_C bh \qquad \text{Ans.}$$
$$V_{ps} d_{ps} = V_{rp} d_{rp} + V_{tw} d_{tw}$$
$$\rho g d_C b h d_P = \rho g \left(d_C - \frac{h}{2} \right) bh d_C + \frac{1}{2} (\rho g h)(bh) \left(d_C + \frac{h}{6} \right)$$
$$= \frac{1}{12} \rho g b h (12 d_C^2 + h^2)$$

Solving for d_P yields

$$d_P = \frac{\dfrac{1}{12} \rho g b h (12 d_C^2 + h^2)}{\rho g d_C b h} = \frac{12 d_C^2 + h^2}{12 d_C} \qquad \text{Ans.}$$

The distance between the center of pressure and the centroid is

$$d_P - d_C = \frac{12 d_C^2 + h^2}{12 d_C} - d_C = \frac{h^2}{12 d_C}$$

The resultant force $\mathbf{R}$ and the locations of the centroid C and the center of pressure P are shown in Fig. 5-26c.

Determine the magnitude of the resultant of the pressure forces exerted on the circular plate shown in Fig. 5-27a and locate the center of pressure with respect to the fluid surface.

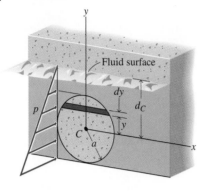

(a)

Fig. 5-27

SOLUTION

For the element of area dA shown in Fig. 5-27a,

$$p = \rho g(d_C - y) \qquad dA = 2(a^2 - y^2)^{1/2} \, dy$$

Therefore, the magnitude of the increment of force $d\mathbf{R}$ on the element of area dA is

$$dR = p \, dA = \rho g(d_C - y)(2)(a^2 - y^2)^{1/2} \, dy$$

The magnitude of the resultant force $\mathbf{R}$ acting on the submerged circular area is

$$R = \int_A dR = \int_{-a}^{+a} 2\rho g(d_C - y)(a^2 - y^2)^{1/2} \, dy$$

$$= 2\rho g \left[\frac{1}{2} d_C y (a^2 - y^2)^{1/2} + \frac{1}{2} d_C a^2 \sin^{-1} \frac{y}{a} + \frac{1}{3}(a^2 - y^2)^{3/2} \right]_{-a}^{+a}$$

$$= \rho g \pi a^2 d_C \qquad \text{Ans.}$$

Note that the area of the circular plate is $A = \pi a^2$ and that the pressure at the centroid of the plate is $p_C = \rho g d_C$. Therefore,

$$R = p_C A$$

The line of action of the resultant force $\mathbf{R}$ is located by using the principle of moments. Summing moments about an x-axis at the fluid surface yields

$$M_x = R d_P = \int_A (d_C - y) dR$$

$$= \int_{-a}^{+a} 2\rho g(d_C - y)^2 (a^2 - y^2)^{1/2} \, dy$$

$$= 2\rho g \left[\frac{1}{2} d_C^2 y (a^2 - y^2)^{1/2} + \frac{1}{2} d_C^2 a^2 \sin^{-1} \frac{y}{a} + \frac{2}{3} d_C (a^2 - y^2)^{1/2} \right.$$

$$\left. - \frac{y}{4}(a^2 - y^2)^{3/2} + \frac{1}{8} a^2 y (a^2 - y^2)^{1/2} + \frac{1}{8} a^4 \sin^{-1} \frac{y}{a} \right]_{-a}^{+a}$$

$$= \frac{1}{4} \rho g \pi a^2 (4 d_C^2 + a^2)$$

Therefore, the distance d_P from the fluid surface to the center of pressure is

$$d_P = \frac{M_y}{R} = \frac{\rho g \pi a^2 \, (4 d_C^2 + a^2)/4}{\rho g \pi a^2 d} = \frac{4 d_C^2 + a^2}{4d} \qquad \text{Ans.}$$

Frequently, when circular shapes are involved, the integration is simplified by using polar coordinates as shown in Fig. 5-27b. For the element of area dA shown in Fig. 5-27b,

$$p = \rho g(d_C - y) = \rho g(d_C - r \sin \theta) \qquad dA = r \, dr \, d\theta$$

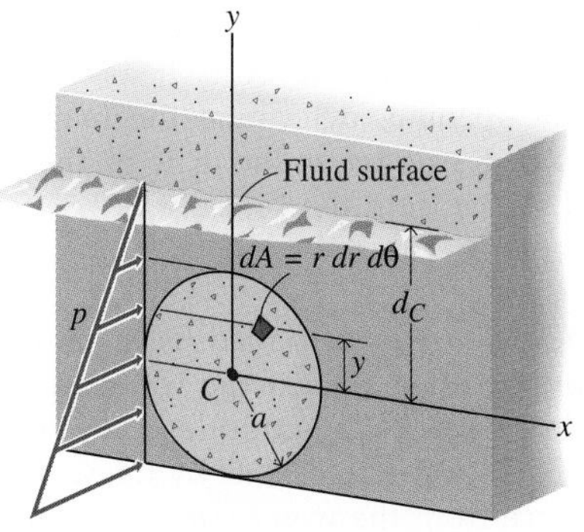

(b)

Fig. 5-27

Therefore, the magnitude of the increment of force $d\mathbf{R}$ on the element of area dA is

$$dR = p \, dA = \rho g(d_C - r \sin \theta)(r \, dr \, d\theta)$$

The magnitude of the resultant force $\mathbf{R}$ acting on the submerged circular area is

$$R = \int_A dR = \int_0^{2\pi} \int_0^a \rho g(d_C - r \sin \theta) \, (r \, dr \, d\theta)$$

$$= \int_0^{2\pi} \rho g\left(\frac{1}{2}d_C a^2 - \frac{1}{3}a^3 \sin \theta\right) d\theta$$

$$= \rho g a^2 \left[\frac{1}{2}d_C \theta + \frac{1}{3}a \cos \theta\right]_0^{2\pi} = \rho g \pi a^2 \, d_C \qquad \text{Ans.}$$

$$M_x = Rd_P = \int_A (d_C - y) \, dR$$

$$= \int_0^{2\pi} \int_0^a \rho g(d_C - r \sin \theta)^2 (r \, dr \, d\theta$$

$$= \int_0^{2\pi} \rho g\left(\frac{1}{2}d_C^2 a^2 - \frac{2}{3}d_C a^3 \sin \theta + \frac{1}{4}a^4 \sin^2 \theta\right) d\theta$$

$$= \rho g\left[\frac{1}{2}d_C^2 a^2 \theta + \frac{2}{3}d_C a^3 \cos \theta + \frac{1}{4}a^4 \left(\frac{1}{2}\theta - \frac{1}{4} \sin 2\theta\right)\right]_0^{2\pi}$$

$$= \frac{1}{4}\rho g \pi a^2 \, (4d_C^2 + a^2)$$

Therefore, the distance d_P from the fluid surface to the center of pressure is

$$d_P = \frac{M_x}{R} = \frac{\rho g \pi a^2 \, (4d_C^2 + a^2)/4}{\rho g \pi a^2 d_C} = \frac{4d_C^2 + a^2}{4d_C} \qquad \text{Ans.}$$

For the special case of the top of the circular area at the fluid surface, the distance $d_C = a$ and

$$R = \rho g \pi a^2 d_C = \rho g \pi a^3$$

$$d_P = \frac{4d_C^2 + a^2}{4d_C} = \frac{5a^2}{4a} = \frac{5}{4}a$$

Note that the pressure-solid technique is of no value here since the shape of the pressure solid (circular cylinder with one end perpendicular and the other end inclined to the axis) is not one generally found in tables.

PROBLEMS

In the following problems use 62.4 lb/ft³ for the specific weight γ of water and 1000 kg/m³ for the density ρ of water.

5-97* If the dam shown in Fig. P5-97 is 200 ft wide, determine the magnitude of the resultant force **R** exerted on the dam by the water pressure.

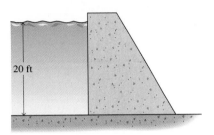

20 ft

Fig. P5-97

5-98* If the dam shown in Fig. P5-98 in 50 m wide, determine the magnitude of the result force **R** exerted on the dam by the water pressure.

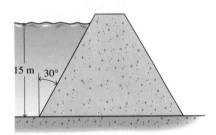

15 m   30°

Fig. P5-98

5-99 The width of the rectangular gate shown in Fig. P5-99 is 8 ft. Determine the magnitude of the resultant force **R** exerted on the gate by the water pressure and the location of the center of pressure with respect to the hinge at the top of the gate.

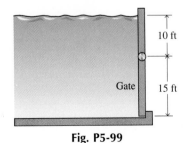

10 ft

Gate   15 ft

Fig. P5-99

5-100 A flat steel plate is used to seal an opening 1 m wide by 2 m high in the vertical wall of a large water tank. When the water level in the tank is 15 m above the top of the opening, determine the magnitude of the resultant force **R** exerted on the plate by the water pressure and the distance from the centroid of the area of the plate to the center of pressure.

5-101* A glass-walled fish tank is 2 ft wide by 6 ft long by 3 ft deep. When the water in the tank is 2.5 ft deep, determine the magnitude of the resultant force **R** exerted on a 2-by-3-ft end plate by the water pressure and the distance from the water surface to the center of pressure.

5-102* The width of the rectangular gate shown in Fig. P5-102 is 4 m. Determine the magnitude of the resultant force **R** exerted on the gate by the water pressure and the location of the center of pressure with respect to the hinge at the bottom of the gate.

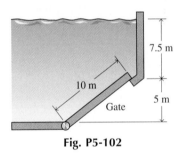

7.5 m

10 m

5 m

Gate

Fig. P5-102

5-103 A water trough 6 ft wide by 10 ft long by 3 ft deep has semicircular ends. When the trough is full of water, determine the magnitude of the resultant force **R** exerted on an end of the trough by the water pressure and the location of the center of pressure with respect to the water surface.

5-104 A flat plate is used to seal an opening in a large water tank as shown in Fig. P5-104. If the opening has the cross section shown in Fig. P5-104*b*, determine the magnitude of the resultant force **R** exerted on the plate by the water pressure and the location of the center of pressure with respect to the bottom of the opening if $d = 5$ m, $h = 2$ m, and $b = 2$ m.

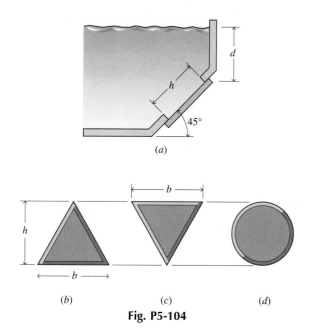

(a)

(b)   *(c)*   *(d)*

Fig. P5-104

5-105* A flat plate is used to seal an opening in a large water tank as shown in Fig. P5-104. If the opening has the cross section shown in Fig. P5-104c, determine the magnitude of the resultant force **R** exerted on the plate by the water pressure and the location of the center of pressure with respect to the bottom of the opening if $d = 15$ ft, $h = 5$ ft, and $b = 5$ ft.

5-106* A flat plate is used to seal an opening in a large water tank as shown in Fig. P5-104. If the opening has the cross section shown in Fig. P5-104c, determine the magnitude of the resultant force **R** exerted on the plate by the water pressure and the location of the center of pressure with respect to the bottom of the opening if $d = 10$ m and $h = 1$ m.

5-107 The width of the cylindrical gate shown in Fig. P5-107 is 4 ft. Determine the magnitude and the slope of the line of action of the resultant force **R** exerted on the gate by the water pressure if $d = 10$ ft and $a = 4$ ft.

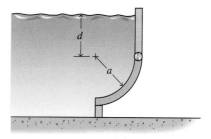

Fig. P5-107

5-108 A 4-m diameter water pipe is filled to the level shown in Fig. P5-108. Determine the magnitude of the resultant force **R** exerted on a 2-m length of the curved section AB of the pipe by the water pressure and locate the line of action of the resultant.

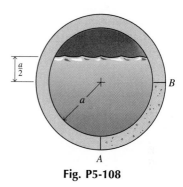

Fig. P5-108

SUMMARY

Previously, we dealt with concentrated forces, which we represented with a vector having a magnitude, a direction, a line of action, and in some cases a point of application. In many instances, surface loads on a body are not concentrated at a point but are distributed along a length or over an area. Other forces, known as body forces, are distributed over the volume of the body. A distributed force at any point is characterized by its intensity and its direction. A force distributed over an area and acting normal to the area is known as a pressure.

Previously, moments of forces about points or axes were considered. In engineering analysis, equations are also encountered that represent moments of masses, forces, volumes, areas, or lines with respect to axes or planes. Such moments are called first moments of the quantity being considered since the first power of a distance is used in the expression.

The term "center of mass" is used to denote the point in a physical body where the mass can be conceived to be concentrated so that the moment of the concentrated mass with respect to an axis or plane equals the moment of the distributed mass with respect to the same axis or plane. The term "center of gravity" is used to denote the point in the body through which the weight of the body acts, regardless of the position (or orientation) of the body. The location of the center of mass in a body is determined by using equations of the form

$$M_{yz} = m\bar{x} = \int_m x \, dm \quad \text{or} \quad \bar{x} = \frac{1}{m} \int_m x \, dm$$

If the density ρ of a body is constant, Eqs. 5-5 reduce to

$$\bar{x} = \frac{1}{V} \int_V x \, dV \qquad \bar{y} = \frac{1}{V} \int_V y \, dV \qquad \bar{z} = \frac{1}{V} \int_V z \, dV \qquad (5\text{-}10)$$

Equations 5-10 indicate that the coordinates $\bar{x}$, $\bar{y}$, and $\bar{z}$ depend only on the geometry of the body and are independent of the physical properties. The point located by such a set of coordinates is known as the "centroid" of the volume of the body. The term "centroid" is usually used in connection with geometrical figures (volumes, areas, and lines), whereas the terms "center of mass" and "center of gravity" are used in connection with physical bodies. The centroid of a volume has the same position as the center of gravity of the body if the body is homogeneous. If the specific weight is variable, the center of gravity of the body and the centroid of the volume will be at different points.

When a load, applied to a rigid body, is distributed along a line or over an area A, it is often convenient for purposes of static analysis to replace this distributed load with a resultant force $\mathbf{R}$ that is equivalent to the distributed load w. For a beam with a distributed load along its length, the magnitude of the resultant force is determined from the expression

$$R = \int_L w \, dx \tag{5-18}$$

which indicates that the magnitude of the resultant force is equal to the area under the load diagram used to represent the distributed load. The line of action of the resultant force passes through the centroid of the area under the load diagram. For the case of fluid pressures on submerged plane surfaces, the load diagram (area) for a distributed load along a line becomes a pressure solid (volume) since the intensity of the pressure on the submerged surface varies over an area instead of a length. The magnitude of the resultant force acting on the submerged surface is

$$R = \int_A p \, dA = \int_V dV_{ps} = V_{ps} \tag{5-22}$$

which indicates that the magnitude of the resultant force is equal to the volume of the pressure solid used to represent the distributed load. The line of action of the resultant force passes through the centroid of the volume of the pressure solid. If the pressure is uniformly distrib-uted over the area, the center of pressure (point of intersection of the line of action of the resultant force with the area) will coincide with the centroid of the area. If the pressure is not uniformly distributed over the area, the center of pressure and the centroid of the area will be at different points.

REVIEW PROBLEMS

5-109–5-112 Locate the centroid of the shaded area shown in the following figures.

5-109* Fig. P5-109.

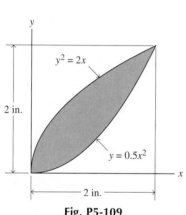

Fig. P5-109

5-111* Fig. P5-111.

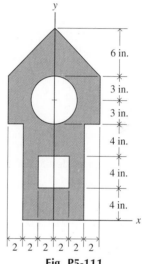

Fig. P5-111

5-110* Fig. P5-110.

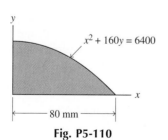

Fig. P5-110

5-112 Fig. P5-112.

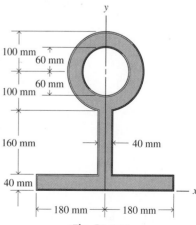

Fig. P5-112

5-113 Locate the centroid of the volume shown in Fig. P5-113 if $R = 10$ in. and $h = 32$ in.

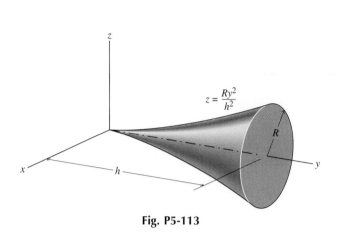

$$z = \frac{Ry^2}{h^2}$$

Fig. P5-113

5-114* Determine the surface area and volume of the solid body generated by revolving the shaded area shown in Fig. P5-114 through an angle of 360° about the x-axis.

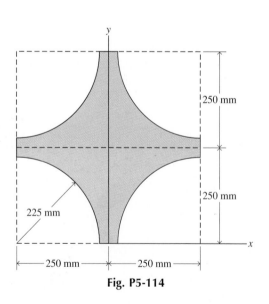

250 mm

250 mm

250 mm

225 mm

250 mm — 250 mm

Fig. P5-114

5-115* Determine the volume of material removed when the groove is cut in the circular shaft shown in Fig. P5-115.

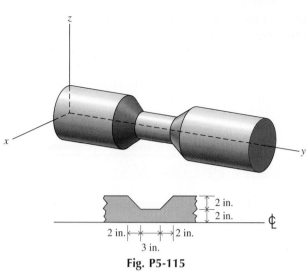

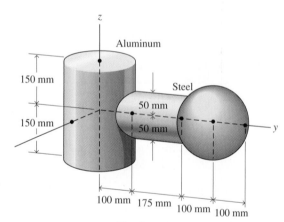

2 in.
2 in.

2 in. — 2 in.
3 in.

Fig. P5-115

5-116 Locate the centroid and the mass center of the volume shown in Fig. P5-116 that consists of an aluminum cylinder ($\rho = 2770$ kg/m³) and a steel ($\rho = 7870$ kg/m³) cylinder and sphere.

Aluminum

150 mm

Steel

50 mm

150 mm

50 mm

100 mm 175 mm
100 mm 100 mm

Fig. P5-116

5-117 Determine the resultant **R** of the system of distributed loads on the beam of Fig. P5-117 and locate its line of action with respect to the left support of the beam.

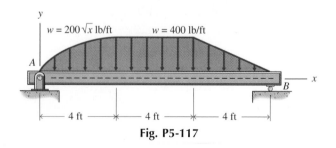

$w = 200\sqrt{x}$ lb/ft $w = 400$ lb/ft

A

B

4 ft — 4 ft — 4 ft

Fig. P5-117

5-118* Determine the resultant **R** of the system of distributed loads on the beam of Fig. P5-118 and locate its line of action with respect to the left support of the beam.

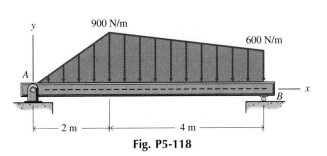

Fig. P5-118

5-119 Determine the magnitude and locate the line of action of the resultant force **R** exerted by the water pressure on a 5-ft length of the dam shown in Fig. P5-119. Use 62.4 lb/ft^3 for the specific weight γ of water.

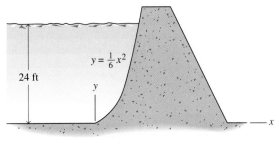

Fig. P5-119

Computer Problems

C5-120 Write a computer program to compute centroid locations using numerical integration and use the program to verify the answers for Problems 5-18, 5-22, 5-25, and 5-26 (use $b = 500$ mm).

C5-121 A rectangular gate holds back water ($\gamma = 62.4$ lb/ft^3) as shown in Fig. P5-121. The gate is 10 ft high and 8 ft wide and is pivoted at a point 4.5 ft from its bottom edge. When the water level is sufficiently low, the gate rests against the stop at C and does not touch the stop at A. When the water level is sufficiently high, the gate presses against the stop at A and does not touch the stop at C.

a. Plot F_A, F_{Bx}, and F_C, the horizontal components of the forces at the two stops and the pivot, as a function of the water depth h ($5 \le h \le 35$ ft).
b. Plot d, the location of the center of pressure relative to the pivot B as a function of the water depth h ($5 \le h \le 35$ ft).

c. If the stop at A were removed, at what depth of water would the gate rotate and allow the water to drain out? Where is the location of the center of pressure for this depth?

C5-122 A water tank at the aquarium has a circular window 2 m in diameter in a vertical wall as shown in Fig. P5-122.

a. Plot R, the resultant of the water ($\rho = 1000$ kg/m^3) pressure on the glass window, as a function of the water depth h ($0.5 \le h \le 5$ m).
b. Plot d, the location of the center of pressure relative to the center of the circular window, as a function of the water depth h ($0.5 \le h \le 5$ m).

Fig. P5-121

Fig. P5-122

207

C5-123 A wine barrel is modeled as a circular cylinder 4 ft in diameter and 6 ft long lying on its side as shown in Fig. P5-123.

a. Plot R, the resultant of the wine ($\gamma = 56$ lb/ft^3) pressure on the end of the barrel, as a function of the wine depth h ($0 \le h \le 4$ ft).

b. Plot d, the location of the center of pressure relative to the center of the circle, as a function of the wine depth h ($0 \le h \le 4$ ft).

C5-124 The tank shown in Fig. P5-124 is basically rectangular (1 m square by 2 m long). It is being raised from its side to its end. If the tank is one-fourth full of oil ($\rho = 850$ kg/m^3):

a. Plot h, the depth of oil in the tank, as a function of the angle θ ($0° \le \theta \le 90°$).

b. Plot R, the resultant of the oil pressure on the end of the tank, as a function of θ ($0° \le \theta \le 90°$).

c. Plot d, the location of the center of pressure relative to the center of the square, as a function of the angle θ ($0° \le \theta \le 90°$).

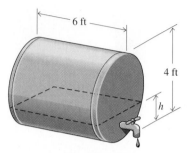

Fig. P5-123

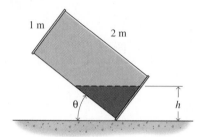

Fig. P5-124

6

EQUILIBRIUM OF RIGID BODIES

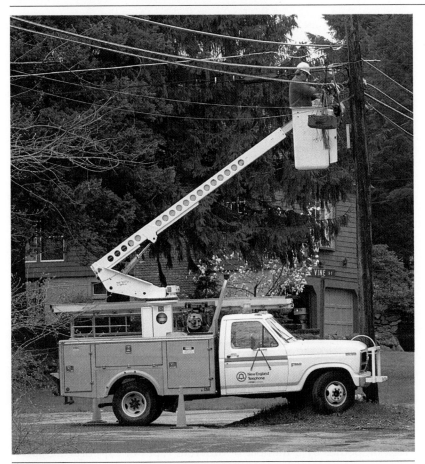

Rigid body principles that involve both force equilibrium and moment equilibrium are required to determine the forces in the hinge and hydraulic cylinder supporting the arm and basket holding the telephone worker.

6-1 INTRODUCTION

The concept of equilibrium was introduced in Chapter 3 and applied to a system of forces acting on a particle. Since any system of forces acting on a particle is a concurrent force system, a particle is in equilibrium if the resultant **R** of the force system acting on the particle is zero. For the case of a rigid body, it was shown in Chapter 4 that the most general force system can be expressed in terms of a resultant force **R** and a resultant couple **C**. Therefore, for a rigid body to be in equilibrium, both the resultant force **R** and the resultant couple **C** must vanish. These two conditions can be expressed by the two vector equations

$$\mathbf{R} = \Sigma F_x\mathbf{i} + \Sigma F_y\mathbf{j} + \Sigma F_z\mathbf{k} = \mathbf{0}$$
$$\mathbf{C} = \Sigma M_x\mathbf{i} + \Sigma M_y\mathbf{j} + \Sigma M_z\mathbf{k} = \mathbf{0}$$

(6-1)

Equations 6-1 can be expressed in scalar form as

$$\Sigma F_x = 0 \qquad \Sigma F_y = 0 \qquad \Sigma F_z = 0$$
$$\Sigma M_x = 0 \qquad \Sigma M_y = 0 \qquad \Sigma M_z = 0$$

(6-2)

Equations 6-2 are the necessary conditions for equilibrium of a rigid body. If all the forces acting on a body can be determined from these equations, then they are also the sufficient conditions for equilibrium.

The forces and moments that act on a rigid body are either external or internal. Forces applied to a rigid body by another body or by the earth are external forces. Fluid pressure on the wall of a tank or the force applied by a truck wheel to a road surface are examples of external forces. The weight of a body is another example of an external force. Internal forces hold the particles forming the rigid body together. If the body of interest is composed of several parts, the forces holding the parts together are also defined as internal forces.

External forces can be divided into applied forces and reaction forces. Forces exerted on the body by external sources are the applied forces. Forces exerted on the body by supports or connections are the reaction forces. Our concern in this chapter is only with external forces and the moments these external forces produce. Since internal forces occur as equal and opposite pairs, they have no effect on equilibrium of the overall rigid body. Internal forces will be considered in Chapter 8.

The best way to identify all forces acting on a body of interest is to use the free-body diagram approach introduced in Chapter 3 when equilibrium of a particle was considered. This free-body diagram of the body of interest must show all the applied forces and all the reaction forces exerted on the body at the supports. In Section 6-2 free-body diagrams for rigid bodies are considered and the reaction forces exerted on bodies by a number of different types of supports are specified.

6-2 FREE-BODY DIAGRAMS

The concept of a free-body diagram was introduced in Chapter 3 and used to solve equilibrium problems involving bodies that could be idealized as a particle. In all these problems, the external forces (applied forces and reaction forces) could be represented as a concurrent force system. In the more general case of a rigid body, systems of forces other than concurrent force systems are encountered and the free-body diagrams become much more complicated. The basic procedure for drawing the diagram, however, remains the same and consists essentially of the following four steps:

Step 1. Decide which body or combination of bodies is to be isolated or cut free from its surroundings.

Step 2. Prepare a drawing or sketch of the outline of this isolated or free body.

Step 3. Carefully trace around the boundary of the free body and identify all the forces exerted by contacting or interacting bodies that were removed during the isolation process.

Step 4. Choose the set of coordinate axes to be used in solving the problem and indicate these directions on the free-body diagram.

Application of these four steps to any statics or dynamics problem will produce a complete and correct free-body diagram, which is *an essential first step for the solution of the problem.*

Forces that are known should be added to the diagram and labeled with their proper magnitudes and directions. Letter symbols can be used to represent magnitudes of forces that are unknown. If the correct sense of an unknown force is not obvious, the sense can be arbitrarily assumed. The algebraic sign of the calculated value of the unknown force will indicate the sense of the force. A plus sign indicates that the force is in the direction assumed, and a minus sign indicates that the force is in a direction opposite from that assumed.

When connections or supports are removed from the isolated body, the actions of these connections or supports must be represented by forces and/or moments on the free-body diagram. The forces and moments used to represent the actions of common connections and supports used with bodies subjected to two-dimensional force systems are identified and discussed in Section 6-2-1. A similar discussion for connections and supports used with bodies subjected to three-dimensional force systems is presented in Section 6-2-2.

6-2-1 Idealization of Two-Dimensional Supports and Connections

Common types of supports and connections used with rigid bodies subjected to two-dimensional force systems, together with the forces and moments used to represent the actions of these supports and connections on a free body, are listed in Table 6-1.

TABLE 6-1   TWO-DIMENSIONAL REACTIONS AT SUPPORTS AND CONNECTIONS

1. Gravitational attraction

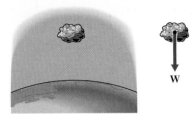

Fig. 6-1

The gravitational attraction of the Earth on a body (see Fig. 6-1) is the weight **W** of the body. The line of action of the force **W** passes through the center of gravity of the body and is directed toward the center of the Earth.

2. Flexible cord, rope, chain, or cable

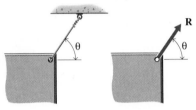

Fig. 6-2

A flexible cord, rope, chain, or cable (see Fig. 6-2) always exerts a tensile force **R** on the body. The line of action of the force **R** is known; it is tangent to the cord, rope, chain, or cable at the point of attachment.

3. Rigid link

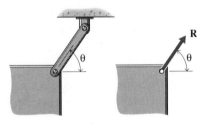

Fig. 6-3

A rigid link (see Fig. 6-3) can exert either a tensile or a compressive force **R** on the body. The line of action of the force **R** is known; it must be directed along the axis of the link (see Section 6-3-1 for proof).

4. Ball, roller, or rocker

Fig. 6-4

A ball, roller, or rocker (see Fig. 6-4) can exert a compressive force **R** on the body. The line of action of the force **R** is perpendicular to the surface supporting the ball, roller, or rocker.

TABLE 6-1   (continued)

5. Smooth surface

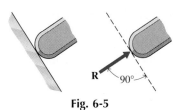

Fig. 6-5

A smooth surface, either flat or curved (see Fig. 6-5), can exert a compressive force **R** on the body. The line of action of the force **R** is perpendicular to the smooth surface at the point of contact between the body and the smooth surface.

6. Smooth pin

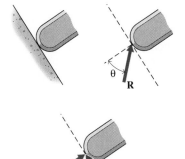

Fig. 6-6

A smooth pin (see Fig. 6-6) can exert a force **R** of unknown magnitude R and direction θ on the body. As a result, the force **R** is usually represented on a free-body diagram by its rectangular components $\mathbf{R}_x$ and $\mathbf{R}_y$.

7. Rough surface

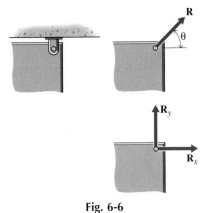

Fig. 6-7

Rough surfaces (see Fig. 6-7) are capable of supporting a tangential frictional force $\mathbf{R}_t$ as well as a compressive normal force $\mathbf{R}_n$. As a result, the force **R** exerted on a body by a rough surface is a compressive force **R** at an unknown angle θ. The force **R** is usually represented on a free-body diagram by its rectangular components $\mathbf{R}_n$ and $\mathbf{R}_t$.

8. Pin in a smooth guide

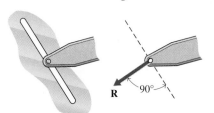

Fig. 6-8

A pin in a smooth guide (see Fig. 6-8) can transmit only a force **R** perpendicular to the surfaces of the guide. The sense of **R** is assumed on the figure and may be either downward to the left or upward to the right.

TABLE 6-1 **(continued)**

9. Collar on a smooth shaft

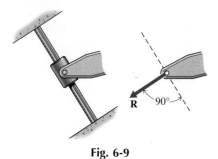

Fig. 6-9

A collar on a smooth shaft (see Fig. 6-9) that is pin-connected to a body can transmit only a force **R** perpendicular to the axis of the shaft. When the connection between the collar and the body is fixed (see Fig. 6-10), the collar can transmit both a force **R** and a moment **M** perpendicular to the axis of the shaft. If the shaft is not smooth, a tangential frictional force $\mathbf{R}_t$ as well as a normal force $\mathbf{R}_n$ can be transmitted.

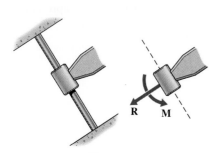

Fig. 6-10

10. Fixed support

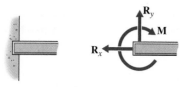

Fig. 6-11

A fixed support (see Fig. 6-11) can exert both a force **R** and a couple **C** on the body. The magnitude R and the direction θ of the force **R** are not known. Therefore, the force **R** is usually represented on a free-body diagram by its rectangular components $\mathbf{R}_x$ and $\mathbf{R}_y$ and the couple **C** by its moment **M**.

11. Linear elastic spring

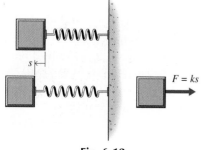

Fig. 6-12

The force **R** exerted on a body by a linear elastic spring (see Fig. 6-12) is proportional to the change in length of the spring. The spring will exert a tensile force if lengthened and a compressive force if shortened. The line of action of the force is along the axis of the spring.

TABLE 6-1   (continued)

12. Ideal pulley

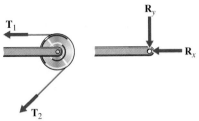

Fig. 6-13

Pulleys (see Fig. 6-13) are used to change the direction of a rope or cable. The pin connecting an ideal pulley to a member can exert a force **R** of unknown magnitude R and direction θ on the body. The force **R** is usually represented on a free-body diagram by its rectangular components $\mathbf{R}_x$ and $\mathbf{R}_y$. Also, since the pin is smooth (frictionless), the tension **T** in the cable must remain constant to satisfy moment equilibrium about the axis of the pulley.

6-2-2 Idealization of Three-Dimensional Supports and Connections

Common types of supports and connections used with rigid bodies subjected to three-dimensional force systems, together with the forces and moments used to represent the actions of these supports and connections on a free-body diagram, are listed in Table 6-2.

TABLE 6-2   THREE-DIMENSIONAL REACTIONS AT SUPPORTS AND CONNECTIONS

1. Ball and socket

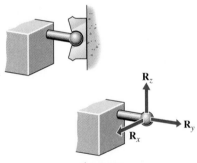

Fig. 6-14

A ball and socket joint (see Fig. 6-14) can transmit a force **R** but no moment. The force **R** is usually represented on a free-body diagram by its three rectangular components $\mathbf{R}_x$, $\mathbf{R}_y$, and $\mathbf{R}_z$.

2. Hinge

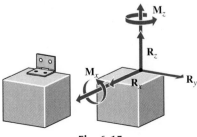

Fig. 6-15

A hinge (see Fig. 6-15) is normally designed to transmit a force **R** in a direction perpendicular to the axis of the hinge pin. The design may also permit a force component to be transmitted along the axis of the pin. Individual hinges have the ability to transmit small moments about axes perpendicular to the axis of the pin. However, properly aligned pairs of hinges transmit only forces under normal conditions of use. Thus, the action of a hinge is represented on a free-body diagram by the force components $\mathbf{R}_x$, $\mathbf{R}_y$, and $\mathbf{R}_z$ and the moments $\mathbf{M}_x$ and $\mathbf{M}_z$ when the axis of the pin is in the y-direction.

TABLE 6-2 **(continued)**

3. Ball bearing

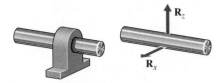

Ideal (smooth) ball bearings (see Fig. 6-16) are designed to transmit a force **R** in a direction perpendicular to the axis of the bearing. The action of the bearing is represented on a free-body diagram by the force components **R**$_x$ and **R**$_z$ when the axis of the bearing is in the y-direction.

Fig. 6-16

4. Journal bearing

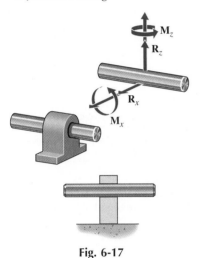

Journal bearings (see Fig. 6-17) are designed to transmit a force **R** in a direction perpendicular to the axis of the bearing. Individual journal bearings have the ability to transmit small moments about axes perpendicular to the axis of the shaft. However, properly aligned pairs of bearings transmit only forces perpendicular to the axis of the shaft under normal conditions of use. Therefore, the action of a journal bearing is represented on a free-body diagram by the force components **R**$_x$ and **R**$_z$ and the couple moments **M**$_x$ and **M**$_z$ when the axis of the bearing is in the y-direction.

Fig. 6-17

5. Thrust bearing

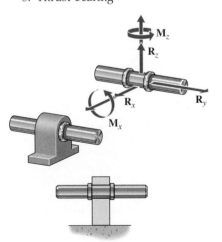

A thrust bearing (see Fig. 6-18), as the name implies, is designed to transmit force components both perpendicular and parallel (thrust) to the axis of the bearing. Individual thrust bearings have the ability to transmit small moments about axes perpendicular to the axis of the shaft. However, properly aligned pairs of bearings transmit only forces under normal conditions of use. Therefore, the action of a thrust bearing is represented on a free-body diagram by the force components **R**$_x$, **R**$_y$, and **R**$_z$ and the couple moments **M**$_x$ and **M**$_z$ when the axis of the bearing is in the y-direction.

Fig. 6-18

TABLE 6-2   (continued)

6. Smooth pin and bracket

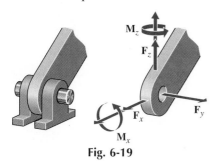

Fig. 6-19

A pin and bracket (see Fig. 6-19) is designed to transmit a force $\mathbf{R}$ in a direction perpendicular to the axis of the pin but may also transmit a force component along the axis of the pin. The unit also has the ability to transmit small moments about axes perpendicular to the axis of the pin. Therefore, the action of a smooth pin and bracket is represented on a free-body diagram by the force components $\mathbf{R}_x$, $\mathbf{R}_y$, and $\mathbf{R}_z$ and the couple moments $\mathbf{M}_x$ and $\mathbf{M}_z$ when the axis of the pin is in the y-direction.

7. Fixed support

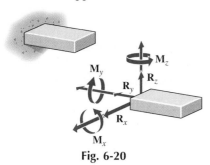

Fig. 6-20

A fixed support (see Fig. 6-20) can resist both a force $\mathbf{R}$ and a couple $\mathbf{C}$. The magnitudes and directions of the force and couple are not known. Thus, the action of a fixed support is represented on a free-body diagram by the force components $\mathbf{R}_x$, $\mathbf{R}_y$, and $\mathbf{R}_z$ and the moments $\mathbf{M}_x$, $\mathbf{M}_y$, and $\mathbf{M}_z$.

Free-body diagrams for a number of structures and machine components are presented in the following examples.

Draw the free-body diagram for the beam shown in Fig. 6-21a.

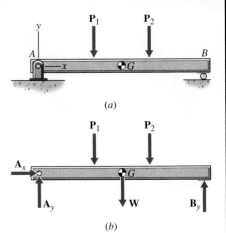

(a)

(b)

Fig. 6-21

SOLUTION

Two concentrated forces P_1 and P_2 are applied to the beam. The weight of the beam is represented by the force W, which has a line of action that passes through the center of gravity G of the beam. The beam is supported at the left end with a smooth pin and bracket and at the right end with a roller. The action of the left support is represented by the forces A_x and A_y. The action of the roller is represented by the force B_y, which acts normal to the surface of the bearing plate. A complete free-body diagram for the beam is shown in Fig. 6-21b.

Draw the free-body diagram for the beam shown in Fig. 6-22a. Neglect the weight of the beam.

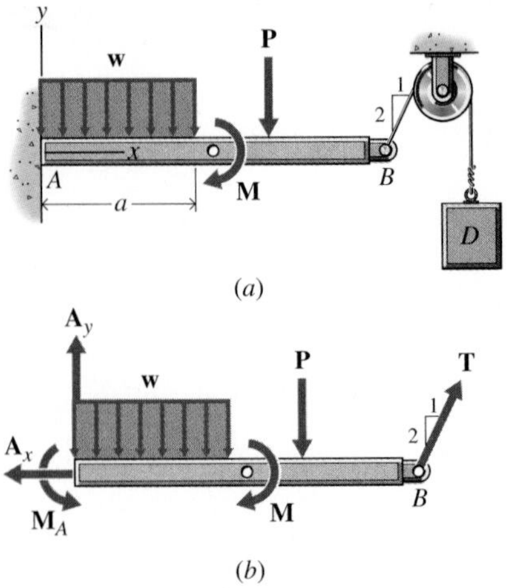

(a)

SOLUTION

A couple M and a concentrated load P are applied to the beam. Also, the beam supports a uniformly distributed load over a length a at the left end of the beam and a body D through a pulley and cable system at the right end of the beam. The support at the left end of the beam is fixed and its action on the beam is represented by the forces A_x and A_y and the couple M_A. The action of the cable at the right end of the beam is represented by a tensile force T with a line of action in the direction of the cable. A complete free-body diagram for the beam is shown in Fig. 6-22b. Note that the pulley and body D are not a part of the free-body diagram for the beam.

(b)

Fig. 6-22

A cylinder is supported on a smooth inclined surface by a two-bar frame as shown in Fig. 6-23a. Draw the free-body diagram (a) for the cylinder, (b) for the two-bar frame, and (c) for the pin at C.

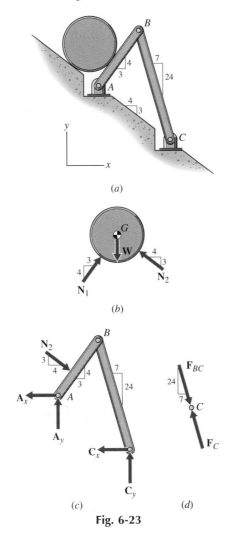

Fig. 6-23

SOLUTION

a. The free-body diagram for the cylinder is shown in Fig. 6-23b. The weight **W** of the cylinder acts through the center of gravity G. The forces $\mathbf{N}_1$ and $\mathbf{N}_2$ act normal to the smooth surfaces at the points of contact.

b. The free-body diagram for the two-bar frame is shown in Fig. 6-23c. The action of the smooth pin and bracket supports at points A and C are represented by forces $\mathbf{A}_x$ and $\mathbf{A}_y$ and $\mathbf{C}_x$ and $\mathbf{C}_y$, respectively. Since bar BC is a link, the resultant $\mathbf{F}_C$ of forces $\mathbf{C}_x$ and $\mathbf{C}_y$ must have a line of action along the axis of the link. As a result, the free-body diagram for pin C can be drawn as shown in Fig. 6-23d. Note that the pin forces at B are internal and do not appear on the free-body diagram.

Draw free-body diagrams (a) for the pulley, (b) for the post AB, and (c) for the beam CD shown in Fig. 6-24a.

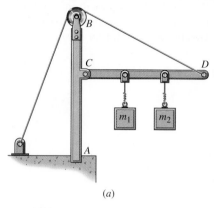

(a)

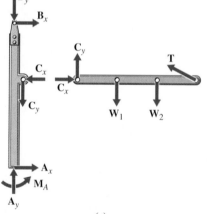

(b)

SOLUTION

a. The free-body diagram for the pulley with the cable removed is shown in Fig. 6-24b. The action of the cable on the pulley is represented by a distributed force as shown in the first part of the figure. The action of the post on the pulley (through the pin at B) is a force $\mathbf{R}$, which is equal in magnitude but opposite in sense to the resultant of the distributed force exerted by the cable. For the solution of problems, a free-body diagram consisting of the pulley and the contacting portion of the cable, as shown in the last part of Fig. 6-24b, is used so that the distributed load becomes an internal force, which does not appear in the analysis.

b. The post is fixed at the base and loaded by the pulley and beam through pins at points B and C, respectively. The free-body diagram is shown in Fig. 6-24c. Note that the forces $\mathbf{B}_x$ and $\mathbf{B}_y$ exerted on the post by the pulley are equal in magnitude but opposite in direction to the forces $\mathbf{B}_x$ and $\mathbf{B}_y$ exerted on the pulley by the post.

c. The free-body diagram for the beam is also shown in Fig. 6-24c. The masses m_1 and m_2 must be converted to forces ($\mathbf{W} = m\mathbf{g}$) before their actions can be represented on the free-body diagram.

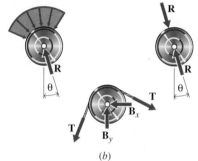

(c)

Fig. 6-24

Draw the free-body diagram for the curved bar AC shown in Fig. 6-25a, which is supported by a ball-and-socket joint at A, a flexible cable at B, and a pin and bracket at C. Neglect the weight of the bar.

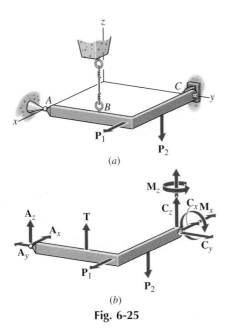

(a)

(b)

Fig. 6-25

SOLUTION

The action of the ball-and-socket joint at support A is represented by three rectangular force components $\mathbf{A}_x$, $\mathbf{A}_y$, and $\mathbf{A}_z$. The action of the pin and bracket at support C can be represented by force components $\mathbf{C}_x$, $\mathbf{C}_y$, and $\mathbf{C}_z$ and moment components $\mathbf{M}_x$ and $\mathbf{M}_z$. The action of the cable is represented by the cable tension $\mathbf{T}$. A complete free-body diagram for bar AC is shown in Fig. 6-25b.

PROBLEMS

Draw complete free-body diagrams for the bodies specified in the following problems. Include the weight of the
member on the diagram except where the problem statement indicates that the weight of the member is to be neglected. Assume that all surfaces are smooth unless indicated otherwise.

6-1 The cantilever beam shown in Fig. P6-1.

Fig. P6-1

6-2 The cylinder shown in Fig. P6-2.

Fig. P6-2

6-3 The bar shown in Fig. P6-3.

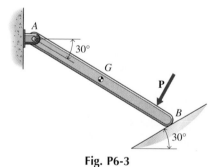

Fig. P6-3

6-4 The beam shown in Fig. P6-4.

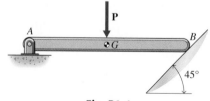

Fig. P6-4

6-5 The beam shown in Fig. P6-5.

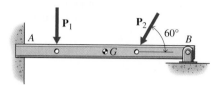

Fig. P6-5

6-6 (a) The cylinder and (b) the bar shown in Fig. P6-6.

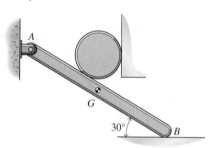

Fig. P6-6

6-7 Beam *AD* shown in Fig. P6-7.

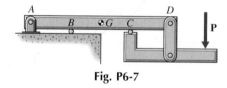

Fig. P6-7

6-8 (a) The bar *AB* and (b) the post *CD* shown in Fig.
P6-8.

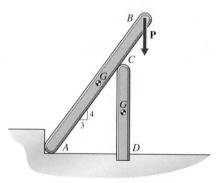

Fig. P6-8

6-9 (a) The bar AB and (b) the post CD shown in Fig. P6-9.

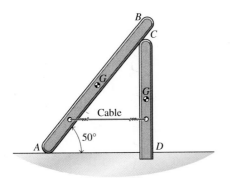

Fig. P6-9

6-10 The ladder shown in Fig. P6-10. The horizontal surface is rough and the vertical surface is smooth.

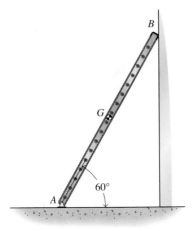

Fig. P6-10

6-11 (a) The cylinder and (b) the frame shown in Fig. P6-11. Neglect the weight of the frame.

Fig. P6-11

6-12 (a) The cylinder and (b) the bar shown in Fig. P6-12.

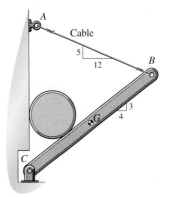

Fig. P6-12

6-13 (a) The bar AB and (b) the bar CB shown in Fig. P6-13. Neglect the weights of the bars.

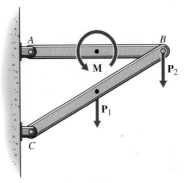

Fig. P6-13

6-14 (a) The bar AB and (b) the bar BD shown in Fig. P6-14. Neglect the weights of the bars.

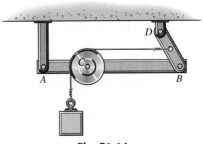

Fig. P6-14

6-15 (a) The bar *BE* and (b) the bar *DF* shown in Fig. P6-15. Neglect the weights of the bars.

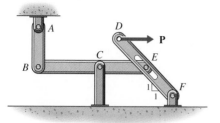

Fig. P6-15

6-16 The bar *CE* shown in Fig. P6-16.

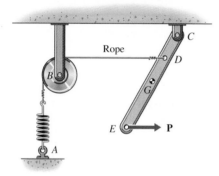

Fig. P6-16

6-17 (a) The bar *AC* and (b) the bar *DF* shown in Fig. P6-17.

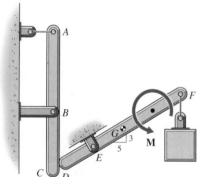

Fig. P6-17

6-18 (a) The bar *AC* and (b) the bar *DE* shown in Fig. P6-18. Neglect the weight of bar *DE*.

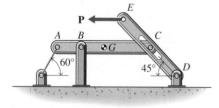

Fig. P6-18

6-19 (a) The bar *AB* and (b) body *C* shown in Fig. P6-19.

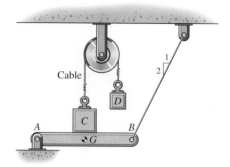

Fig. P6-19

6-20 (a) The bar *AD* and (b) the bar *CF* shown in Fig. P6-20. Neglect the weights of the bars.

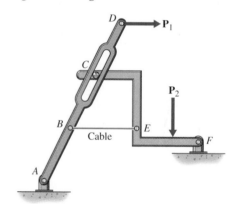

Fig. P6-20

6-21 The block shown in Fig. P6-21. The support at *A* is a ball and socket. The support at *B* is a pin and bracket.

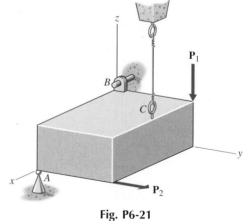

Fig. P6-21

6-22 The door shown in Fig. P6-22. Assume that the hinges at supports C and D are properly aligned so that they are not required to transmit moment components.

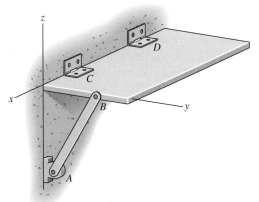

Fig. P6-22

6-23 The door shown in Fig. P6-23.

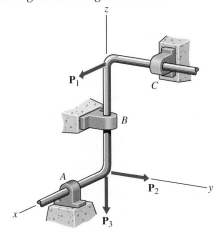

Fig. P6-23

6-24 The bent bar shown in Fig. P6-24. The support at A is a journal bearing and the supports at B and C are ball bearings. Neglect the weight of the bar.

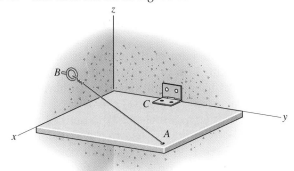

Fig. P6-24

6-25 The shaft shown in Fig. P6-25. The bearing at A is a thrust bearing and the bearing at D is a ball bearing. Neglect the weights of the shaft and the levers.

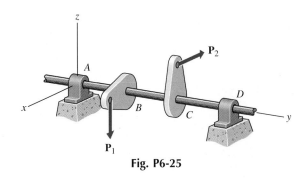

Fig. P6-25

6-26 The bar shown in Fig. P6-26. The bar rests against a smooth vertical wall at end D. The support at A is a ball and socket. Neglect the weight of the bar.

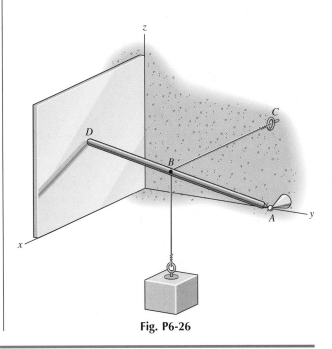

Fig. P6-26

6-3   EQUILIBRIUM IN TWO DIMENSIONS

The term "two dimensional" is frequently used to describe problems in which the forces involved are contained in a plane (say the xy-plane) and the axes of all couples are perpendicular to the plane containing the forces. For two-dimensional problems, since a force in the xy-plane has no z-component and produces no moments about the x- or y-axes, Eqs. 6-1 reduce to

$$\mathbf{R} = \Sigma F_x\mathbf{i} + \Sigma F_y\mathbf{j} = \mathbf{0}$$
$$\mathbf{C} = \Sigma M_z\mathbf{k} = \mathbf{0} \tag{6-3}$$

Thus, three of the six independent scalar equations of equilibrium (Eqs. 6-2) are automatically satisfied; namely,

$$\Sigma F_z = 0 \qquad \Sigma M_x = 0 \qquad \Sigma M_y = 0$$

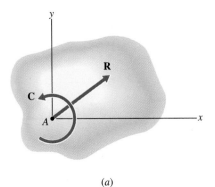

(a)

Therefore, there are only three independent scalar equations of equilibrium for a rigid body subjected to a two-dimensional system of forces. The three equations can be expressed as

$$\Sigma F_x = 0 \qquad \Sigma F_y = 0 \qquad \Sigma M_A = 0 \tag{6-4}$$

The third equation represents the sum of the moments of all forces about a z-axis through any point A on or off the body. Equations 6-4 are both the necessary and sufficient conditions for equilibrium of a body subjected to a two-dimensional system of forces.

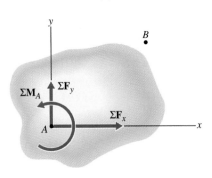

(b)

There are two additional ways in which the equations of equilibrium can be expressed for a body subjected to a two-dimensional system of forces. The resultant force $\mathbf{R}$ and the resultant couple $\mathbf{C}$ of a general two-dimensional force system on a rigid body are shown in Fig. 6-26a. The resultant can be represented in terms of its scalar components as shown in Fig. 6-26b. If the condition $\Sigma M_A = 0$ is satisfied, $\mathbf{C} = \mathbf{0}$. If, in addition, the condition $\Sigma F_x = 0$ is satisfied, $\mathbf{R} = \Sigma F_y\mathbf{j}$. For any point B on or off the body that does not lie on the y-axis, the equation $\Sigma M_B = 0$ can be satisfied only if $\Sigma F_y = 0$. Thus, an alternative set of scalar equilibrium equations for two-dimensional problems is

$$\Sigma F_x = 0 \qquad \Sigma M_A = 0 \qquad \Sigma M_B = 0 \tag{6-5}$$

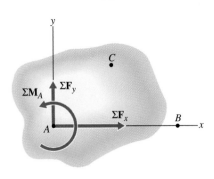

(c)

Fig. 6-26

where points A and B must have different x-coordinates.

The conditions of equilibrium for a two-dimensional force system, can also be expressed by using three moment equations. Again, if the condition $\Sigma M_A = 0$ is satisfied, $\mathbf{C} = \mathbf{0}$. In addition, for a point B (see Fig. 6-26c) on the x-axis on or off the body (except at point A), the equation $\Sigma M_B = 0$ can be satisfied once $\Sigma M_A = 0$, only if $\Sigma F_y = 0$. Thus, $\mathbf{R} = \Sigma F_x\mathbf{i}$. Finally, for any point C (see Fig. 6-26c) on or off the body that does not lie on the x-axis, the equation $\Sigma M_C = 0$ can be satisfied only if $\Sigma F_x = 0$. Thus, a second set of alternative scalar equilibrium equations for two-dimensional problems is

$$\Sigma M_A = 0 \qquad \Sigma M_B = 0 \qquad \Sigma M_C = 0 \tag{6-6}$$

where A, B, and C are any three points not on the same straight line.

6-3-1 The Two-Force Body (Two-Force Members)

Equilibrium of a body under the action of two forces occurs with sufficient frequency to warrant special attention. For example, consider the
link with negligible weight shown in Fig. 6-27a. Any forces exerted on
the link by the frictionless pins at A and B can be resolved into components along and perpendicular to the axis of the link as shown in Fig.
6-27b. From the equilibrium equations

$$\Sigma F_x = 0 \qquad A_x - B_x = 0 \qquad A_x = B_x$$
$$\Sigma F_y = 0 \qquad A_y - B_y = 0 \qquad A_y = B_y$$

Forces A_y and B_y, however, form a couple that must be zero when the
link is in equilibrium; therefore, $A_y = B_y = 0$. Thus, for two-force members, equilibrium requires that the forces be equal, opposite, and collinear as shown in Fig. 6-27c. The shape of the member, as shown in Fig.
6-27d, has no effect on this simple requirement. Also, the weights of the
members must be negligible.

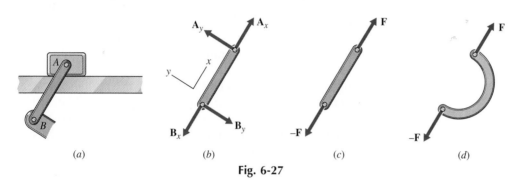

(a)　　　　(b)　　　　(c)　　　　(d)

Fig. 6-27

6-3-2 The Three-Force Body (Three-Force Members)

Equilibrium of a body under the action of three forces (see Fig. 6-28)
also represents a special situation. If a body is in equilibrium under the
action of three forces, then the lines of action of the three forces must
be concurrent (i.e., meet at a common point); otherwise, the nonconcurrent force would exert a moment about the point of concurrency of
the other two forces. A body subjected to three parallel forces represents a special case of the three-force body. For this case, the point of
concurrency is assumed to be at infinity.

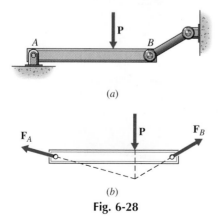

(a)

(b)

Fig. 6-28

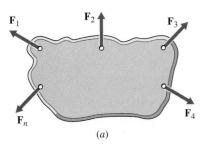

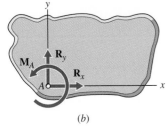

Fig. 6-29

6-3-3 Statically Indeterminate Reactions and Partial Constraints

Consider a body subjected to a system of coplanar forces F_1, F_2, F_3, F_4, $\cdots$, F_n, as shown in Fig. 6-29a. This system of forces can be replaced by an equivalent force–couple system at an arbitrary point A as shown in Fig. 6-29b. The resultant force has been represented by its rectangular components R_x and R_y and the couple by M_A. In order for the body to be in equilibrium, the supports must be capable of exerting an equal and opposite force–couple system on the body. As an example, consider the supports shown in Fig. 6-30a. The pin support at A, as shown in Fig. 6-30b, can exert forces in the x- and y-directions to prevent a translation of the body but no moment to prevent a rotation about an axis through A. The link at B, which exerts a force in the y-direction, produces the moment about point A required to prevent rotation. Thus, all motion of the body is prevented, and the body is in equilibrium under the action of the three forces A_x, A_y, and B_y. The forces exerted on the body by the supports are known as constraints (restrictions to movement). When the equations of equilibrium are sufficient to determine the unknown forces at the supports, as in Fig. 6-30, the body is said to be statically determinate with adequate (proper) constraints.

Three support reactions for a body subjected to a coplanar system of forces does not always guarantee that the body is statically determinate with adequate constraints. For example, consider the body and supports shown in Fig. 6-31a. The pin support at A (see Fig. 6-31b) can exert forces in the x- and y-directions to prevent a translation of the body, but since the line of action of force B_x passes through point A, it does not exert the moment required to prevent rotation of the body

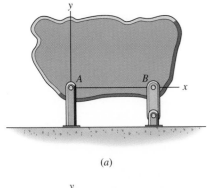

(a)

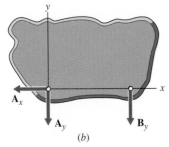

(b)

Fig. 6-30

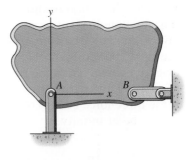

(a)

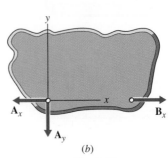

(b)

Fig. 6-31

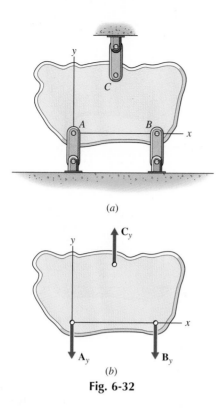

(a)

(b)

Fig. 6-32

about point A. Similarly, the three links shown in Fig. 6-32a can prevent rotation of the body about any point (see Fig. 6-32b) and translation of the body in the y-direction but not translation of the body in the
x-direction. The bodies in Figs. 6-31 and 6-32 are partially (improperly)
constrained, and the equations of equilibrium are not sufficient to determine all the unknown reactions. A body with an adequate number
of reactions is improperly constrained when the constraints are arranged in such a way that the support forces are either concurrent or
parallel. Partially constrained bodies can be in equilibrium for specific
systems of forces. For example, the reactions $\mathbf{R}_A$ and $\mathbf{R}_B$ for the beam
shown in Fig. 6-33a can be determined by using the equilibrium equa-

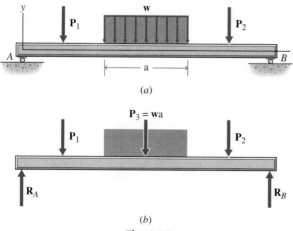

(a)

(b)

Fig. 6-33

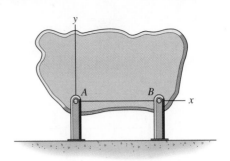

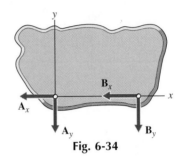

Fig. 6-34

tions $\Sigma F_y = 0$ and $\Sigma M_A = 0$. The beam is improperly constrained, however, since motion in the x-direction would occur if any of the applied loads had a small x-component.

Finally, if the link at B in Fig. 6-30a is replaced with a pin support, as shown in Fig. 6-34a, an additional reaction $\mathbf{B}_x$ (see Fig. 6-34b) is obtained that is not required to prevent movement of the body. Obviously, the three independent equations of equilibrium will not provide sufficient information to determine the four unknowns. Constrained bodies with extra supports are statically indeterminate since relations involving physical properties of the body, in addition to the equations of equilibrium, are required to determine some of the unknown reactions. Statically indeterminate problems are studied in Mechanics of Materials courses. The supports not required to maintain equilibrium of the body are called redundant supports. Typical examples of redundant supports for beams are shown in Fig. 6-35. The roller at support B of the cantilever beam shown in Fig. 6-35a can be removed without affecting equilibrium of the beam. Similarly, the roller support at either B or C could be removed from the beam of Fig. 6-35b without disturbing equilibrium of the beam.

6-3-4   Problem Solving

In Section 1-6, a procedure was outlined for solving engineering-type problems. The procedure consisted of three phases:

1. Problem definition and identification
2. Model development and simplification
3. Mathematical solution and interpretation of results

Application of this procedure to equilibrium-type problems yields the following steps.

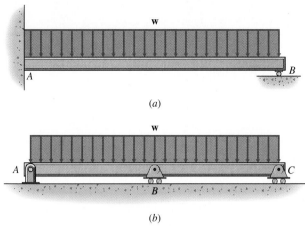

(a)

(b)

Fig. 6-35

Steps for Analyzing and Solving Problems

1. Read the problem carefully. Many student difficulties arise from failure to observe this preliminary step.
2. Identify the result requested.
3. Prepare a scaled sketch and tabulate the information provided.
4. Identify the equilibrium equations to be used to obtain the result.
5. Draw the appropriate free-body diagram. Carefully label all applied forces and support reactions. Establish a convenient set of coordinate axes. Use a right-handed system in case vector cross-products must be employed. Compare the number of unknowns on the free-body diagram with the number of independent equations of equilibrium. Draw additional diagrams if needed.
6. Apply the appropriate force and moment equations.
7. Report the answer with the appropriate number of significant figures and the appropriate units.
8. Study the answer and determine if it is reasonable. As a check, write some other equilibrium equations and see if they are satisfied by the solution.

Bodies subjected to coplanar force systems are not very complex, and a scalar solution is usually suitable for analysis. This results from the fact that moments can be expressed as scalars instead of vectors. For the more general case of rigid bodies subjected to three-dimensional force systems (discussed in the next section of this chapter), vector analysis methods are usually more appropriate.

The following examples illustrate the solution of coplanar force problems.

A pin-connected truss is loaded and supported as shown in Fig. 6-36a. The body W has a mass of 100 kg. Determine the components of the reactions at supports A and B.

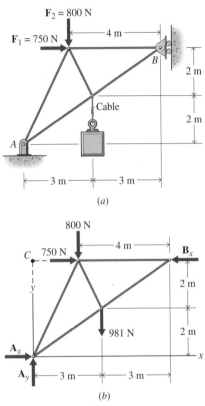

(a)

(b)

Fig. 6-36

SOLUTION

The tension T in the cable between the body W and the truss is

$$T = mg = 100(9.81) = 981 \text{ N}$$

A free-body diagram of the truss is shown in Fig. 6-36b. The action of the pin support at A is represented by force components $\mathbf{A}_x$ and $\mathbf{A}_y$. The action of the roller support at B is represented by the force $\mathbf{B}_x$, which acts perpendicular to the vertical surface at B. Since the truss is subjected to a general coplanar force system, three equilibrium equations are available to solve for the unknown magnitudes of the three forces $\mathbf{A}_x$, $\mathbf{A}_y$, and $\mathbf{B}_x$.
Determination of $\mathbf{B}_x$:

$$+\!\!\downarrow\!\Sigma M_A = 0 \qquad B_x(4) - 800(2) - 750(4) - 981(3) = 0$$
$$B_x = +1885.8 \text{ N} \qquad \mathbf{B}_x = 1886 \text{ N}\!\leftarrow \qquad \text{Ans.}$$

Determination of $\mathbf{A}_x$:

$$+\rightarrow\Sigma F_x = 0: \qquad A_x + 750 - B_x = A_x + 750 - 1885.8 = 0$$
$$A_x = +1135.8\,\text{N} \qquad \mathbf{A}_x = 1136\,\text{N}\rightarrow \qquad\qquad \text{Ans.}$$

Determination of $\mathbf{A}_y$:

$$+\uparrow\Sigma F_y = 0: \qquad A_y - 800 - 981 = 0$$
$$A_y = +1781\,\text{N} \qquad \mathbf{A}_y = 1781\,\text{N}\uparrow \qquad\qquad \text{Ans.}$$

The component $\mathbf{A}_x$ can also be determined directly by summing moments about point C.

$$+\downarrow\Sigma M_C = 0: \qquad A_x(4) - 800(2) - 981(3) = 0$$
$$A_x = +1135.8\,\text{N} \qquad \mathbf{A}_x = 1136\,\text{N}\rightarrow$$

SOLUTION BY VECTOR ANALYSIS

From Eqs. 6-3

$$\mathbf{R} = \Sigma F_x\mathbf{i} + \Sigma F_y\mathbf{j} = \mathbf{0} \qquad \mathbf{C} = \Sigma M_z\mathbf{k} = \mathbf{0}$$

The applied loads can be written in Cartesian vector form as

$$\mathbf{F}_1 = 750\mathbf{i}\,\text{N} \qquad \mathbf{F}_2 = -800\mathbf{j}\,\text{N} \qquad \mathbf{F}_3 = -981\mathbf{j}\,\text{N}$$

The reactions can be written as

$$\mathbf{A} = A_x\mathbf{i} + A_y\mathbf{j} \qquad \mathbf{B} = -B_x\mathbf{i}$$

Therefore, from Eqs. 6-3.

$$\mathbf{R} = \Sigma F_x\mathbf{i} + \Sigma F_y\mathbf{j}$$
$$= (A_x - B_x + 750)\mathbf{i} + (A_y - 800 - 981)\mathbf{j} = \mathbf{0}$$

Summing moments about point A to minimize the number of unknowns

$$\mathbf{C} = \Sigma M_A\mathbf{k}$$
$$= \mathbf{r}_1 \times \mathbf{F}_1 + \mathbf{r}_2 \times \mathbf{F}_2 + \mathbf{r}_3 \times \mathbf{F}_3 + \mathbf{r}_B \times \mathbf{B}$$
$$= [(2\mathbf{i} + 4\mathbf{j}) \times 750\mathbf{i}] + [(2\mathbf{i} + 4\mathbf{j}) \times (-800\mathbf{j})]$$
$$+ [(3\mathbf{i} + 2\mathbf{j}) \times (-981\mathbf{j})] + [(6\mathbf{i} + 4\mathbf{j}) \times (-B_x\mathbf{i})]$$
$$= -3000\mathbf{k} - 1600\mathbf{k} - 2943\mathbf{k} + 4B_x\mathbf{k} = (-7543 + 4B_x)\mathbf{k} = \mathbf{0}$$

Equating the coefficients of $\mathbf{i}$, $\mathbf{j}$, and $\mathbf{k}$ to zero gives

$$A_x - B_x + 750 = 0$$
$$A_y - 800 - 981 = 0$$
$$-7543 + 4B_x = 0$$

Solving simultaneously yields

$$A_x = 1136\,\text{N} \qquad A_y = 1781\,\text{N} \qquad B_x = 1886\,\text{n}$$

Therefore

$$\mathbf{A} = 1136\mathbf{i} + 1781\mathbf{j}\,\text{N} \qquad \mathbf{B} = -1886\mathbf{i}\,\text{N} \qquad\qquad \text{Ans.}$$

The results are obviously identical to those obtained using a scalar analysis. For two-dimensional problems, scalar methods are usually preferred.

A beam is loaded and supported as shown in Fig. 6-37a. Determine the components of the reactions at supports A and B.

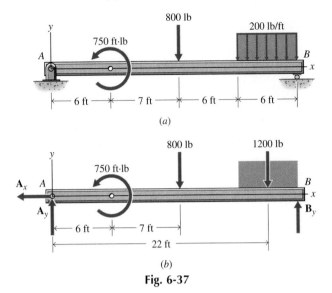

(a)

(b)

Fig. 6-37

SOLUTION

A free-body diagram of the beam is shown in Fig. 6-37b. The action of the pin support at A is represented by force components $\mathbf{A}_x$ and $\mathbf{A}_y$. The action of the roller support at B is represented by the force $\mathbf{B}_y$, which acts perpendicular to the horizontal surface at B. The 200 lb/ft distributed load can be represented on the free-body diagram by a fictitious resultant $\mathbf{R}$ with a line of action at a distance $\bar{x}$ from the left support. Thus,

$$R = A = 200(6) = 1200 \text{ lb} \qquad \bar{x} = 19 + \frac{1}{2}(6) = 22 \text{ ft}$$

The resultant $\mathbf{R}$ should be shown dotted on the free-body diagram to indicate that it can be used *only* for calculating external effects (reactions). For determinations involving internal surfaces (such as those presented later in Chapter 8), the actual distributed load must be used. The beam is subjected to a coplanar system of parallel forces in the y-direction; therefore, $\mathbf{A}_x = \mathbf{0}$. The two remaining equilibrium equations are available to solve for $\mathbf{A}_y$ and $\mathbf{B}_y$.
Determination of $\mathbf{B}_y$:

$$+\lfloor \Sigma M_A = 0: \qquad B_y(25) + 750 - 800(13) - 1200(22) = 0$$
$$B_y = +1442 \text{ lb} \qquad \mathbf{B}_y = 1442 \text{ lb}\uparrow \qquad \text{Ans.}$$

Determination of $\mathbf{A}_y$:

$$+\lfloor \Sigma M_B = 0: \qquad -A_y(25) + 750 + 800(12) + 1200(3) = 0$$
$$A_y = +558 \text{ lb} \qquad \mathbf{A}_y = 558 \text{ lb}\uparrow \qquad \text{Ans.}$$

Alternatively (or as a check):

$$+\uparrow \Sigma F_y = 0: \qquad A_y - 800 - 1200 + 1442 = 0$$
$$A_y = +558 \text{ lb} \qquad \mathbf{A}_y = 558 \text{ lb}\uparrow$$

A beam is loaded and supported as shown in Fig. 6-38a. Determine the components of the reactions at supports A and B.

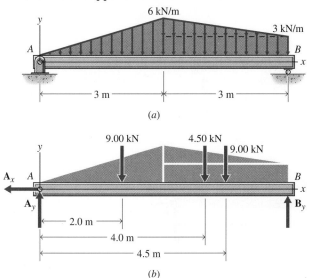

(a)

(b)

Fig. 6-38

SOLUTION

A free-body diagram of the beam is shown in Fig. 6-38b. The action of the pin support at A is represented by force components $\mathbf{A}_x$ and $\mathbf{A}_y$. The action of the roller support at B is represented by the force $\mathbf{B}_y$, which acts perpendicular to the horizontal surface at B. The distributed loads can be represented temporarily on the free-body diagram by fictitious resultants $\mathbf{R}_1$, $\mathbf{R}_2$, and $\mathbf{R}_3$ with lines of action at distances $\bar{x}_1$, $\bar{x}_2$, $\bar{x}_3$, respectively, from the left support. Thus,

$$R_1 = A_1 = \frac{1}{2}(6)(3) = 9.00 \text{ kN} \qquad \bar{x}_1 = \frac{2}{3}(3) = 2.0 \text{ m}$$

$$R_2 = A_2 = \frac{1}{2}(3)(3) = 4.50 \text{ kN} \qquad \bar{x}_2 = 3 + \frac{1}{3}(3) = 4.0 \text{ m}$$

$$R_3 = A_3 = 3(3) = 9.00 \text{ kN} \qquad \bar{x}_3 = 3 + \frac{1}{2}(3) = 4.5 \text{ m}$$

The beam is subjected to a coplanar system of parallel forces in the y-direction; therefore, $\mathbf{A}_x = \mathbf{0}$. The two remaining equilibrium equations are available to solve for $\mathbf{A}_y$ and $\mathbf{B}_y$.
Determination of $\mathbf{B}_y$:

$$+\,\zeta\Sigma M_A = 0: \quad B_y(6) - 9.00(2.0) - 4.50(4.0) - 9.00(4.5) = 0$$
$$B_y = +12.75 \text{ kN} \qquad \mathbf{B}_y = 12.75 \text{ kN}\uparrow \qquad \text{Ans.}$$

Determination of $\mathbf{A}_y$:

$$+\,\zeta\Sigma M_B = 0: \quad -A_y(6) + 9.00(4.0) + 4.5(2.0) + 9.00(1.5) = 0$$
$$A_y = +9.75 \text{ kN} \qquad \mathbf{A}_y = 9.75 \text{ kN}\uparrow \qquad \text{Ans.}$$

Alternatively: (or as a check)

$$+\uparrow\Sigma F_y = 0: \quad A_y - 9.00 - 4.50 - 9.00 + 12.75 = 0$$
$$A_y = +9.75 \text{ kN} \qquad \mathbf{A}_y = 9.75 \text{ kN}\uparrow$$

A pin-connected three-bar frame is loaded and supported as shown in Fig. 6-39a. Determine the reactions at supports A and B.

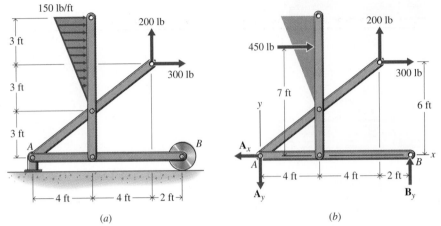

150 lb/ft

200 lb

3 ft

300 lb

3 ft

3 ft

A

B

4 ft | 4 ft | 2 ft

(a)

200 lb

450 lb

7 ft

300 lb

6 ft

A_x

y

A

B

x

4 ft | 4 ft | 2 ft

A_y

B_y

(b)

Fig. 6-39

SOLUTION

A free-body diagram of the frame is shown in Fig. 6-39b. The action of the pin support at A is represented by force components A_x and A_y. The action of the roller support at B is represented by the force B_y, which acts perpendicular to the horizontal surface at B. The distributed load can be represented temporarily on the free-body diagram by a fictitious resultant R, with a line of action at a distance $\bar{y}$ above support A. Thus,

$$R = A = \frac{1}{2}(150)(6) = 450 \text{ lb} \qquad \bar{y} = 3 + \frac{2}{3}(6) = 7.00 \text{ ft}$$

Since the frame is subjected to a general coplanar force system, three equilibrium equations are available to solve for A_x, A_y, and B_y.
Determination of B_y:

$$+\circlearrowleft \Sigma M_A = 0: \qquad B_y(10) - 450(7) + 200(8) - 300(6) = 0$$
$$B_y = +335 \text{ lb} \qquad \mathbf{B} = \mathbf{B_y} = 335 \text{ lb}\uparrow \qquad \text{Ans.}$$

Determination of A_x:

$$+\rightarrow \Sigma F_x = 0: \qquad -A_x + 450 + 300 = 0$$
$$A_x = +750 \text{ lb} \qquad \mathbf{A_x} = 750 \text{ lb}\leftarrow \qquad \text{Ans.}$$

Determination of A_y:

$$+\circlearrowleft \Sigma M_B = 0: \qquad A_y(10) - 450(7) - 200(2) - 300(6) = 0$$
$$A_y = +535 \text{ lb} \qquad \mathbf{A_y} = 535 \text{ lb}\downarrow$$

The reaction at support A is

$$A = \sqrt{(A_x)^2 + (A_y)^2} = \sqrt{(750)^2 + (535)^2} = 921 \text{ lb}$$
$$\theta = \tan^{-1} \frac{A_y}{A_x} = \tan^{-1} \frac{535}{750} = 35.5° \qquad \mathbf{A} = 921 \text{ lb} \nearrow 35.5° \qquad \text{Ans.}$$

A pin-connected two-bar frame is loaded and supported as shown in Fig. 6-40a. Determine the reactions at supports A and B.

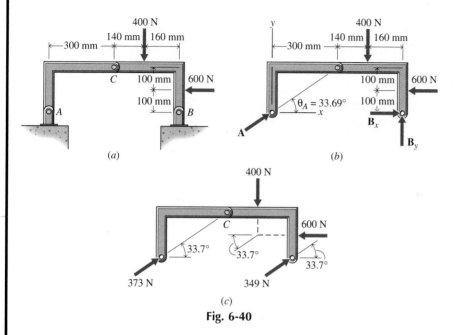

(a)

(b)

(c)

Fig. 6-40

SOLUTION

A free-body diagram of the frame is shown in Fig. 6-40b. The action of the pin support at A is represented by the single force $\mathbf{A}$ at a known angle (since bar AC is a two-force member) $\theta_A = \tan^{-1}(200/300) = 33.69°$. The action of the pin support at B is represented by force components $\mathbf{B}_x$ and $\mathbf{B}_y$. Since the frame is subjected to a general coplanar force system, three equilibrium equations are available to solve for the unknown magnitudes of forces $\mathbf{A}$, $\mathbf{B}_x$, and $\mathbf{B}_y$. Determination of $\mathbf{A}$:

$$+\circlearrowleft\Sigma M_B = 0: \quad -A\sin 33.69°(0.6) + 400(0.16) + 600(0.1) = 0$$
$$A = +372.6 \text{ N} \quad \mathbf{A} = 373 \text{ N}\angle 33.7° \qquad \text{Ans.}$$

Determination of $\mathbf{B}_x$:

$$+\rightarrow\Sigma F_x = 0: \quad A\cos 33.69° + B_x - 600 = 0$$
$$372.6\cos 33.69° + B_x - 600 = 0$$
$$B_x = +290 \text{ N} \quad \mathbf{B}_x = 290 \text{ N}\rightarrow$$

Determination of $\mathbf{B}_y$:

$$+\circlearrowleft\Sigma M_A = 0: \quad B_y(0.6) - 400(0.44) + 600(0.1) = 0$$
$$B_y = +193.3 \text{ N} \quad \mathbf{B}_y = 193.3 \text{ N}\uparrow$$

The reaction at support B is

$$B = \sqrt{(B_x)^2 + (B_y)^2} = \sqrt{(290)^2 + (193.3)^2} = 348.5 \text{ N}$$
$$\theta_B = \tan^{-1}\frac{B_y}{B_x} = \tan^{-1}\frac{193.3}{290} = 33.69° \quad \mathbf{B} = 349 \text{ N}\angle 33.7° \qquad \text{Ans.}$$

The results are shown in Fig. 6-40c.

A bar weighing 250 lb is supported by a post and cable as shown in Fig. 6-41a. Assume that all surfaces are smooth. Determine the tension in the cable and the forces on the bar at the contacting surfaces.

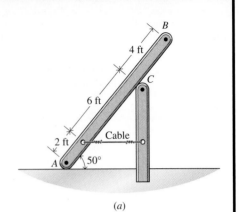

SOLUTION

A free-body diagram of the bar is shown in Fig. 6-41b. All surfaces are smooth; therefore, the reaction at A is a vertical force $\mathbf{A}$ and the reaction at C is a force $\mathbf{C}$ perpendicular to the bar. The cable exerts a tension $\mathbf{T}$ on the bar in the direction of the cable. Since the bar is subjected to a general coplanar force system, three equilibrium equations are available to solve for the unknown magnitudes of forces $\mathbf{A}$, $\mathbf{C}$, and $\mathbf{T}$.

Determination of $\mathbf{T}$:

The determination of the cable tension $\mathbf{T}$ can be simplified by taking moments about the point of concurrence (off the bar) of forces $\mathbf{A}$ and $\mathbf{C}$. Thus,

$$+\!\!\downarrow\!\Sigma M_D = 0: \quad T[(8/\sin 50°) - 2 \sin 50°] - 250(6 \cos 50°) = 0$$
$$T = +108.2 \text{ lb} \qquad \mathbf{T} = 108.2 \text{ lb} \rightarrow \qquad \text{Ans.}$$

Determination of $\mathbf{A}$:

$$+\!\!\downarrow\!\Sigma M_C = 0: \quad -A(8 \cos 50°) + T(6 \sin 50°) + 250(2 \cos 50°) = 0$$
$$-A(8 \cos 50°) + 108.2(6 \sin 50°) + 250(2 \cos 50°) = 0$$
$$A = +159.2 \text{ lb} \qquad \mathbf{A} = 159.2 \text{ lb}\!\uparrow \qquad \text{Ans.}$$

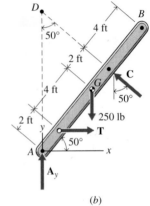

Determination of $\mathbf{C}$:

$$+\!\!\downarrow\!\Sigma M_A = 0: \quad -T(2 \sin 50°) - 250(6 \cos 50°) + C(8) = 0$$
$$-(108.2)(2 \sin 50°) - 250(6 \cos 50°) + C(8) = 0$$
$$C = +141.2 \text{ lb} \qquad \mathbf{C} = 141.2 \text{ lb} \searrow 40° \qquad \text{Ans.}$$

Fig. 6-41

A cylinder with a mass of 50 kg is supported on an inclined surface by a pin-connected two-bar frame as shown in Fig. 6-42a. Assume that all surfaces are smooth. Determine

a. The forces exerted on the cylinder by the contacting surfaces.
b. The reactions at supports A and C of the two-bar frame.

SOLUTION

a. A free-body diagram of the cylinder is shown in Fig. 6-42b. The actions of the inclined surface and frame are represented by forces N_1 and N_2, respectively. The action of the Earth (weight) is

$$W = mg = 50(9.81) = 490.5 \text{ N}$$

The three forces N_1, N_2, and W are a concurrent force system; therefore, only two equilibrium equations are available.
Determination of N_1:

$$+\nearrow \Sigma F_1 = 0: \quad N_1 - \frac{4}{5}(490.5) = 0$$

$$N_1 = 392.4 \text{ N} \qquad N_1 = 392 \text{ N} \angle 53.1° \quad \text{Ans.}$$

Determination of N_2:

$$+\nwarrow \Sigma F_2 = 0 \quad N_2 - \frac{3}{5}(490.5) = 0$$

$$N_2 = 294.3 \text{ N} \qquad N_2 = 294 \text{ N} \searrow 36.9° \quad \text{Ans.}$$

b. A free body diagram of the frame is shown in Fig. 6-42c. The action of the pin support at C is represented by the single force C at a known angle θ_C since bar BC is a two-force member:

$$\theta_C = \tan^{-1}\frac{600}{175} = 73.74°$$

The action of the pin support at A is represented by force components A_x and A_y. Since the frame is subjected to a general coplanar force system, three equilibrium equations are available to solve for the unknown magnitudes of the three forces C, A_x, and A_y.
Determination of C:

$$+\downarrow \Sigma M_A = 0: \quad C\sin 73.74°(0.4) - C\cos 73.74°(0.3) - 294.3(0.15) = 0$$
$$C = +147.15 \text{ N} \qquad C = 147.2 \text{ N} \searrow 73.7° \quad \text{Ans.}$$

Determination of A_x:

$$+\rightarrow \Sigma F_x = 0: \quad -A_x - C\cos\theta_C + \frac{4}{5}(294.3) = 0$$

$$-A_x - 147.15 \cos 73.74° + \frac{4}{5}(294.3) = 0$$

$$A_x = +194.2 \text{ N} \qquad A_x = 194.2 \text{ N} \leftarrow$$

Determination of A_y:

$$+\uparrow \Sigma F_y = 0: \quad A_y + C\sin\theta_C - \frac{3}{5}(294.3) = 0$$

$$A_y + 147.15 \sin 73.74° - \frac{3}{5}(294.3) = 0$$

$$A_y = +35.32 \text{ N} \qquad A_y = 35.3 \text{ N}\uparrow$$

The reaction at support A is

$$A = \sqrt{(A_x)^2 + (A_y)^2} = \sqrt{(194.2)^2 + (35.32)^2} = 197.4 \text{ N}$$
$$\theta_A = \tan^{-1}\frac{A_y}{A_x} = \tan^{-1}\frac{35.32}{194.2} = 10.31° \qquad A = 197.4 \text{ N} \searrow 10.31° \quad \text{Ans.}$$

The results are shown in Fig. 6-42d. Since the frame is a three-force body, the forces N_2, A, and C must be concurrent, as illustrated in the figure.

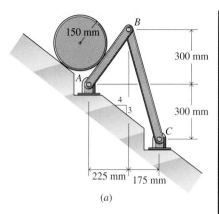

(a)

(b)

(c)

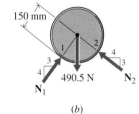

(d)

Fig. 6-42

PROBLEMS

6-27* A curved slender bar is loaded and supported as shown in Fig. P6-27. The bar has a uniform cross section and weighs 50 lb. Determine the reaction at support A.

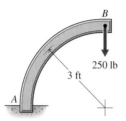

Fig. P6-27

6-28* A curved slender bar is loaded and supported as shown in Fig. P6-28. The bar has a uniform cross section and a mass of 15 kg. Determine the reactions at supports A and B.

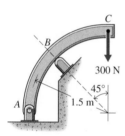

Fig. P6-28

6-29 A bar is loaded and supported as shown in Fig. P6-29. The bar has a uniform cross section and weighs 100 lb. Determine the reactions at supports A, B, and C.

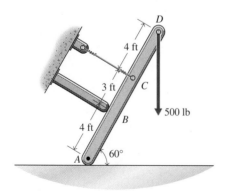

Fig. P6-29

6-30 A curved bar is loaded and supported as shown in Fig. P6-30. The bar has a uniform cross section and a mass of 75 kg. Determine the reactions at supports A and B.

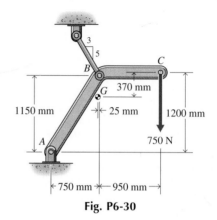

Fig. P6-30

6-31* A beam is loaded and supported as shown in Fig. P6-31. Determine the reactions at supports A and B. Neglect the weight of the beam.

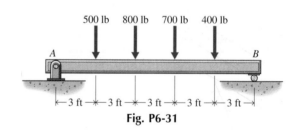

Fig. P6-31

6-32* A beam is loaded and supported as shown in Fig. P6-32. Determine the reaction at support A. Neglect the weight of the beam.

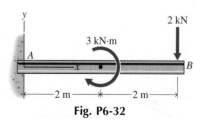

Fig. P6-32

6-33 A beam is loaded and supported as shown in Fig P6-33. Determine the reactions at supports A and B. Neglect the weight of the beam.

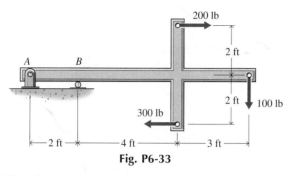

Fig. P6-33

6-34 A pipe strut BC is loaded and supported as shown in Fig. P6-34. The strut has a uniform cross section and a mass of 50 kg. Determine the reactions at supports B and C.

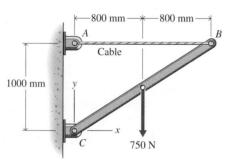

Fig. P6-34

6-35* A beam is loaded and supported as shown in Fig. P6-35. The beam has a uniform cross section and weighs 300 lb. Determine the reactions at supports A and B.

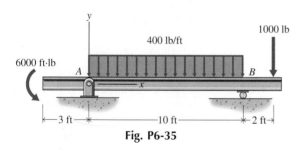

Fig. P6-35

6-36* A beam is loaded and supported as shown in Fig. P6-36. The beam has a uniform cross section and a mass of 200 kg. Determine the reaction at support A.

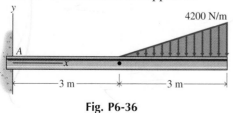

Fig. P6-36

6-37 A beam is loaded and supported as shown in Fig. P6-37. The beam has a uniform cross section and weighs 250 lb. Determine the reactions at supports A and B.

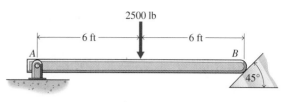

Fig. P6-37

6-38 A beam is loaded and supported as shown in Fig. P6-38. The beam has a uniform cross section and a mass of 120 kg. Determine the reactions at supports A and B.

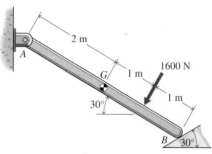

Fig. P6-38

6-39* The ladder shown in Fig. P6-39 weighs 60 lb. The ladder rests on a rough surface at A and against a smooth surface at B. Determine the reactions at supports A and B.

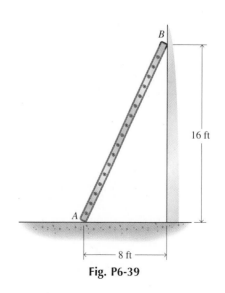

Fig. P6-39

241

6-40* The mass of the cylinder shown in Fig. P6-40 is 100 kg. Determine the reactions at contact points *A* and *B*. All surfaces are smooth.

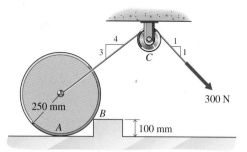

Fig. P6-40

6-41 A 75-lb load is supported by an angle bracket, pulley, and cable as shown in Fig. P6-41. Determine

a. The force exerted on the bracket by the pin at *C*.
b. The reactions at supports *A* and *B* of the bracket.

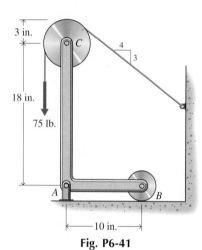

Fig. P6-41

6-42 A cylinder is supported by a bracket as shown in Fig. P6-42. The mass of the cylinder is 50 kg. If all surfaces are smooth, determine

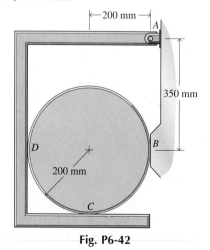

Fig. P6-42

a. The reaction at support *A* of the bracket.
b. The forces exerted on the cylinder at contact points *B*, *C*, and *D*.

6-43* A cylinder is supported by a bar as shown in Fig. P6-43. The weight of the cylinder is 100 lb and the weight of the bar is 20 lb. If all surfaces are smooth, determine the reactions at supports *A* and *B* of the bar.

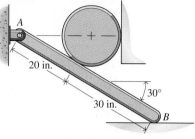

Fig. P6-43

6-44* A bar of uniform cross section rests against a post as shown in Fig. P6-44. The mass of the bar is 150 kg; the mass of the post is 80 kg. If all surfaces are smooth, determine

a. The forces exerted on the bar at contact points *A* and *C*.
b. The reaction at support *D* of the post.

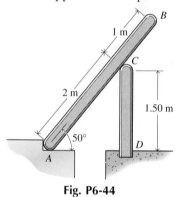

Fig. P6-44

6-45 A cylinder is supported by a bar and cable as shown in Fig. P6-45. The weight of the cylinder is 150 lb and the

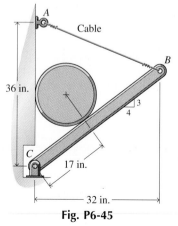

Fig. P6-45

weight of the bar is 20 lb. If all surfaces are smooth, determine the reaction at support C of the bar and the tension **T** in the cable.

6-46 A mass of 200 kg is supported by a bar, pulley, and cable as shown in Fig. P6-46. Determine

a. The force exerted on the bar by the pin at D.
b. The reaction at support A of the bar and the force exerted by link BC.

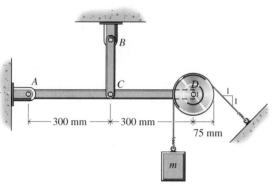

Fig. P6-46

6-47* A rope and pulley system is used to support a body W as shown in Fig. P6-47. Each pulley is free to rotate, and the rope is continuous over the pulleys. Determine the tension **T** in the rope required to hold body W in equilibrium if the weight of body W is 400 lb. Assume that all rope segments are vertical.

Fig. P6-47

6-48* A rope and pulley system is used to support a body W as shown in Fig. P6-48. Each pulley is free to rotate. One rope is continuous over pulleys A and B; the other is continuous over pulley C. Determine the tension **T** in the rope over pulleys A and B required to hold body W in equilibrium if the mass of body W is 175 kg.

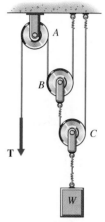

Fig. P6-48

6-49 Pulleys A and B of the chain hoist shown in Fig. P6-49 are connected and rotate as a unit. The chain is continuous, and each of the pulleys contain slots that prevent the chain from slipping. Determine the force **F** required to hold a 1000-lb block W in equilibrium if the radii of pulleys A and B are 3.5 and 4.0 in., respectively.

Fig. P6-49

243

6-50 Pulleys 1 and 2 of the rope and pulley system shown in Fig. P6-50 are connected and rotate as a unit. The radii of the pulleys are 100 mm and 300 mm, respectively. Rope A is fastened to pulley 1 at point A'. Rope B is fastened to pulley 2 at point B'. Rope C is continuous over pulleys 3 and 4. Determine the tension $\mathbf{T}$ in rope C required to hold body W in equilibrium if the mass of body W is 225 kg.

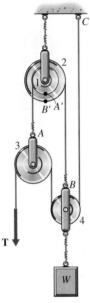

Fig. P6-50

6-51* Three pipes are supported in a pipe rack as shown in Fig. P6-51. Each of the pipes weighs 100 lb. Determine the reactions at supports A and B.

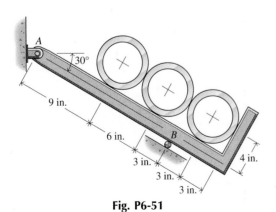

Fig. P6-51

6-52* A beam is loaded and supported as shown in Fig. P6-52. Determine the reactions at supports A and B when $m_1 = 75$ kg and $m_2 = 225$ kg.

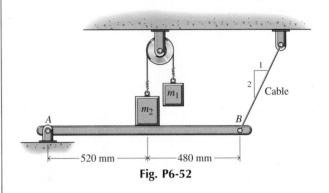

Fig. P6-52

6-53 Bar AD shown in Fig. P6-53 weighs 120 lb. Determine

a. The force exerted on the bar by link CE and the forces exerted on the bar at contact points B and D. All surfaces are smooth.

b. The reaction at support F of the post.

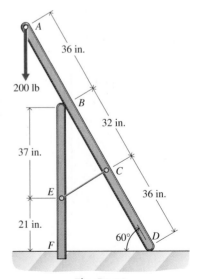

Fig. P6-53

6-54 A two-beam mechanism is loaded and supported as shown in Fig. P6-54. Determine

a. The force **F** required to hold the system in equilibrium.
b. The tension in cable *CD*.
c. The reaction at support *B*.
d. The reaction at support *A*.

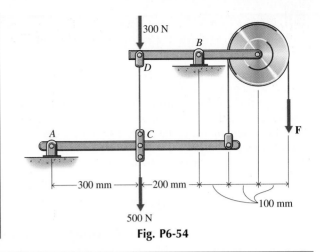

Fig. P6-54

6-4 EQUILIBRIUM IN THREE DIMENSIONS

A general, three-dimensional, system of forces $\mathbf{F}_1$, $\mathbf{F}_2$, $\cdots$, $\mathbf{F}_n$ and couples $\mathbf{C}_1$, $\mathbf{C}_2$, $\cdots$, $\mathbf{C}_n$ can be replaced by an equivalent system consisting of a system of noncoplanar, concurrent forces and a system of noncoplanar couples. The resultant **R** of the concurrent force system can be expressed as

$$\mathbf{R} = \Sigma F_x \mathbf{i} + \Sigma F_y \mathbf{j} + \Sigma F_z \mathbf{k} \qquad (4\text{-}30)$$

The resultant **C** of the noncoplanar system of couples can be expressed as

$$\mathbf{C} = \Sigma M_x \mathbf{i} + \Sigma M_y \mathbf{j} + \Sigma M_z \mathbf{k} \qquad (4\text{-}31)$$

The resultant force **R** and the resultant couple **C**, together, constitute the resultant of the general three-dimensional force system. Equations 4-30 and 4-31 indicate that the resultant of the force system may be a force **R**, a couple **C**, or both a force **R** and a couple **C**. Thus, a rigid body subjected to a general three-dimensional system of forces will be in equilibirum if $\mathbf{R} = \mathbf{C} = \mathbf{0}$, which requires that

and
$$\Sigma F_x = 0 \qquad \Sigma F_y = 0 \qquad \Sigma F_z = 0 \qquad (6\text{-}2)$$

$$\Sigma M_x = 0 \qquad \Sigma M_y = 0 \qquad \Sigma M_z = 0$$

Thus, there are six independent scalar equations of equilibrium for a rigid body subjected to a general three-dimensional system of forces. The first three equations indicate that the x-, y-, and z-components of the resultant force **R** acting on a body in equilibrium is zero. The second three equations express the further equilibrium requirement that there are no couple components acting on the body about any of the coordinate axes or about axes parallel to the coordinate axes. These six equations are both the necessary conditions and the sufficient conditions for equilibrium of the body. The six scalar equations (Eq. 6-2) are independent since each can be satisfied independently of the others.

The following examples illustrate the solution of three-dimensional problems.

A homogeneous flat plate that weighs 500 lb is supported by a shaft AB and a cable C as shown in Fig. 6-43a. The bearing at A is a ball bearing and the bearing at B is a thrust bearing. The bearings are properly aligned; therefore, they transmit only force components. When the three forces shown in Fig. 6-43a are being applied to the plate, determine the reactions at bearings A and B and the tension in cable C. Use a scalar analysis for the solution but express the final results in Cartesian vector form.

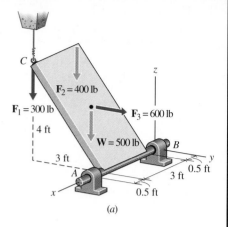

(a)

SOLUTION

A free-body diagram for the plate is shown in Fig. 6-43b. The bearing at A is a ball bearing; therefore, this reaction is represented by two force components A_y and A_z. The bearing at B is a thrust bearing; therefore, this reaction is represented by three force components B_x, B_y, and B_z. The force in the cable is represented by tension T_C. The plate is subjected to a general, three-dimensional system of forces; therefore, six equilibrium equations (Eqs. 6-2) are available for determining the six unknown force components indicated on the free-body diagram. Summing forces in the positive x-, y-, and z-directions and summing moments about the x-, y-, and z-axes in accordance with the right-hand rule yields

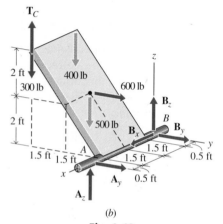

(b)

Fig. 6-43

$$\Sigma F_x = B_x = 0 \qquad\qquad B_x = 0 \qquad (a)$$

$$\Sigma F_y = A_y + B_y + F_3 = A_y + B_y + 600 = 0$$
$$A_y + B_y = -600 \text{ lb} \qquad (b)$$

$$\Sigma F_z = A_z + B_z + T_C - F_1 - F_2 - W$$
$$= A_z + B_z + T_C - 300 - 400 - 500 = 0$$
$$A_z + B_z + T_C = 1200 \text{ lb} \qquad (c)$$

$$\Sigma M_x = F_1(3) + F_2(3) - F_3(2) + W(1.5) - T_C(3)$$
$$= 300(3) + 400(3) - 600(2) + 500(1.5) - T_C(3) = 0$$
$$T_C = 550 \text{ lb} \qquad (d)$$

$$\Sigma M_y = F_1(3.5) + F_2(0.5) + W(2) - T_C(3.5) - A_z(4.0)$$
$$= 300(3.5) + 400(0.5) + 500(2) - T_C(3.5) - A_z(4.0) = 0$$
$$4.0A_z + 3.5T_C = 2250 \text{ ft} \cdot \text{lb} \quad (e)$$

$$\Sigma M_z = A_y(4) + F_3(2)$$
$$= A_y(4) + 600(2) = 0 \qquad\qquad A_y = -300 \text{ lb} \qquad (f)$$

From Eqs. b and f:

$$A_y + B_y = -300 + B_y = -600 \text{ lb}$$
$$B_y = -300 \text{ lb}$$

From Eqs. d and e:

$$4.0A_z + 3.5T_C = 4.0A_z + 3.5(550) = 2250 \text{ ft} \cdot \text{lb}$$
$$A_z = 81.3 \text{ lb}$$

From Eq. d:

$$A_z + B_z + T_C = 81.3 + B_z + 550 = 1200 \text{ lb}$$
$$B_z = 568.7 \text{ lb}$$

$$\mathbf{A} = -300\mathbf{j} + 81.3\mathbf{k} \text{ lb} \qquad\qquad \text{Ans.}$$
$$\mathbf{B} = -300\mathbf{j} + 569\mathbf{k} \text{ lb} \qquad\qquad \text{Ans.}$$
$$\mathbf{T}_C = 550\mathbf{k} \text{ lb} \qquad\qquad\qquad \text{Ans.}$$

A post and bracket is used to support a pulley as shown in Fig. 6-44a. A cable passing over the pulley transmits a 500-lb load as shown in the figure. Determine the reaction at support A of the post.

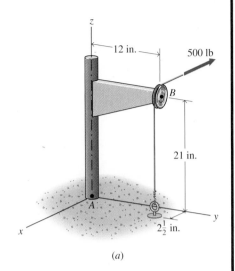

(a)

SOLUTION

A free-body diagram of the post and bracket with the pulley removed is shown in Fig. 6-44b. The applied load at B and the position vector from A to B can be written in Cartesian vector form as

$$\mathbf{B} = -500\mathbf{i} - 500\mathbf{k} \text{ lb} \qquad \mathbf{r}_{B/A} = 12\mathbf{j} + 21\mathbf{k} \text{ in.}$$

The fixed support at A can resist both a force and a couple. Therefore, the reaction at A can be written in Cartesian vector form as

$$\mathbf{A} = A_x\mathbf{i} + A_y\mathbf{j} + A_z\mathbf{k} \qquad \mathbf{C}_A = M_{Ax}\mathbf{i} + M_{Ay}\mathbf{j} + M_{Az}\mathbf{k}$$

The conditions for equilibrium are expressed by Eqs. 6-1 as

$$\mathbf{R} = \Sigma F_x\mathbf{i} + \Sigma F_y\mathbf{j} + \Sigma F_z\mathbf{k} = (A_x - 500)\mathbf{i} + A_y\mathbf{j} + (A_z - 500)\mathbf{k} = \mathbf{0} \qquad (a)$$

Summing moments about A to eliminate the maximum number of unknowns

$$\begin{aligned}
\mathbf{C} &= \Sigma M_x\mathbf{i} + \Sigma M_y\mathbf{j} + \Sigma M_z\mathbf{k} \\
&= M_{Ax}\mathbf{i} + M_{Ay}\mathbf{j} + M_{Az}\mathbf{k} + (\mathbf{r}_{B/A} \times \mathbf{B}) \\
&= M_{Ax}\mathbf{i} + M_{Ay}\mathbf{j} + M_{Az}\mathbf{k} + \begin{vmatrix} \mathbf{i} & \mathbf{j} & \mathbf{k} \\ 0 & 12 & 21 \\ -500 & 0 & -500 \end{vmatrix} \\
&= (M_{Ax} - 6000)\mathbf{i} + (M_{Ay} - 10{,}500)\mathbf{j} + (M_{Az} + 6000)\mathbf{k} = \mathbf{0} \qquad (b)
\end{aligned}$$

Equating the coefficients of $\mathbf{i}$, $\mathbf{j}$, and $\mathbf{k}$ to zero in Eqs. a and b and solving yields

$$A_x = 500 \text{ lb} \qquad A_y = 0 \text{ lb} \qquad A_z = 500 \text{ lb}$$
$$M_{Ax} = 6000 \text{ in.} \cdot \text{lb} \qquad M_{Ay} = 10{,}500 \text{ in.} \cdot \text{lb} \qquad M_{Az} = -6000 \text{ in.} \cdot \text{lb}$$

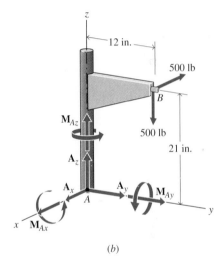

(b)

The reaction at support A can be expressed in Cartesian vector form as

$$\mathbf{A} = 500\mathbf{i} + 500\mathbf{k} \text{ lb} \qquad \mathbf{C}_A = 6000\mathbf{i} + 10{,}500\mathbf{j} - 6000\mathbf{k} \text{ in.} \cdot \text{lb} \qquad \text{Ans.}$$

The magnitude of the force at support A is

$$|\mathbf{A}| = \sqrt{A_x^2 + A_y^2 + A_z^2} = \sqrt{(500)^2 + (0)^2 + (500)^2} = 707 \text{ lb}$$

The magnitude of the moment at support A is

$$\begin{aligned}
|\mathbf{C}_A| &= \sqrt{M_{Ax}^2 + M_{Ay}^2 + M_{Az}^2} \\
&= \sqrt{(6000)^2 + (10{,}500)^2 + (-6000)^2} = 13{,}500 \text{ in.} \cdot \text{lb}
\end{aligned}$$

The direction angles associated with the couple $\mathbf{C}_A$ are obtained by using Eqs. 4-14. Thus

$$\theta_x = \cos^{-1}\frac{M_{Ax}}{|\mathbf{C}_A|} = \cos^{-1}\frac{6000}{13{,}500} = 63.6° \qquad \text{Ans.}$$

$$\theta_y = \cos^{-1}\frac{M_{Ay}}{|\mathbf{C}_A|} = \cos^{-1}\frac{10{,}500}{13{,}500} = 38.9° \qquad \text{Ans.}$$

$$\theta_z = \cos^{-1}\frac{M_{Az}}{|\mathbf{C}_A|} = \cos^{-1}\frac{-6000}{13{,}500} = 116.4° \qquad \text{Ans.}$$

The components of the force $\mathbf{A}$ and the couple $\mathbf{C}_A$ are shown in Fig. 6-44c.

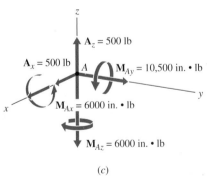

(c)

Fig. 6-44

EXAMPLE PROBLEM 6-15

The masses of cartons 1, 2, and 3, which rest on the platform shown in Fig. 6-45a, are 300 kg, 100 kg, and 200 kg, respectively. The mass of the platform is 500 kg. Determine the tensions in the three cables A, B, and C that support the platform.

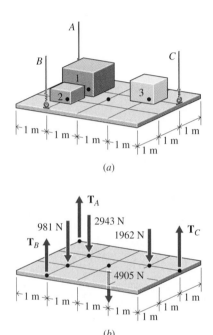

(a)

(b)

Fig. 6-45

SOLUTION

A free-body diagram for the platform is shown in Fig. 6-45b. The platform is subjected to a system of parallel forces in the z-direction; therefore, the equilibrium equations $\Sigma F_x = 0$, $\Sigma F_y = 0$, and $\Sigma M_z = 0$ are automatically satisfied. One force equation ($\Sigma F_z = 0$) and two moment equations ($\Sigma M_x = 0$ and $\Sigma M_y = 0$) are available to solve for the unknown cable tensions.

SOLUTION BY SCALAR ANALYSIS:

$$F_1 = m_1 g = 300(9.81) = 2943 \text{ N} \qquad F_3 = m_3 g = 200(9.81) = 1962 \text{ N}$$
$$F_2 = m_2 g = 100(9.81) = 981 \text{ N} \qquad W = m_p g = 500(9.81) = 4905 \text{ N}$$

$$+\uparrow \Sigma F_z = T_A + T_B + T_c - F_1 - F_2 - F_3 - W$$
$$= T_A + T_B + T_C - 2943 - 981 - 1962 - 4905 = 0$$

From which

$$T_A + T_B + T_C = 10{,}791 \text{ N} \qquad\qquad (a)$$

$$+\downarrow \Sigma M_x = T_A y_A + T_B y_B + T_C y_C - F_1 y_1 - F_2 y_2 - F_3 y_3 - W y_W$$
$$= T_A(0) + T_B(1) + T_C(4) - 2943(1) - 981(3) - 1962(1) - 4905(2) = 0$$

From which

$$T_B + 4T_C = 17{,}658 \text{ N} \cdot \text{m} \qquad\qquad (b)$$

$$+\downarrow \Sigma M_y = T_A y_A - T_B y_B - T_C y_C + F_1 y_1 + F_2 y_2 + F_3 y_3 - W y_W$$
$$= T_A(0) - T_B(3) - T_C(1) + 2943(1) + 981(1) + 1962(2) + 4905(1.5) = 0$$

From which

$$3T_B + T_C = 15{,}206 \text{ N} \cdot \text{m} \qquad\qquad (c)$$

Solving Eqs. a, b, and c simultaneously yields

$$\mathbf{T}_A = +3434 \text{ N} = 3430 \text{ N}\uparrow \qquad\qquad \text{Ans.}$$
$$\mathbf{T}_B = +3924 \text{ N} = 3920 \text{ N}\uparrow \qquad\qquad \text{Ans.}$$
$$\mathbf{T}_C = +3433 \text{ N} = 3430 \text{ N}\uparrow \qquad\qquad \text{Ans.}$$

SOLUTION BY VECTOR ANALYSIS:

The conditions for equilibrium are expressed by Eqs. 6-1 as

$$\mathbf{R} = \Sigma F_z \mathbf{k} = (T_A + T_B + T_C - F_1 - F_2 - F_3 - W)\mathbf{k}$$
$$= (T_A + T_B + T_C - 2943 - 981 - 1962 - 4905)\mathbf{k}$$
$$= (T_A + T_B + T_C - 10{,}971)\mathbf{k} = \mathbf{0} \qquad\qquad (d)$$

Summing moments about A:

$$\mathbf{C} = (\mathbf{r}_{B/A} \times \mathbf{T}_B) + (\mathbf{r}_{C/A} \times \mathbf{T}_C) + (\mathbf{r}_{1/A} \times \mathbf{F}_1) + (\mathbf{r}_{2/A} \times \mathbf{F}_2)$$
$$+ (\mathbf{r}_{3/A} \times \mathbf{F}_3) + (\mathbf{r}_{W/A} \times \mathbf{W}) = \mathbf{0}$$

$$(\mathbf{r}_{B/A} \times \mathbf{T}_B) = (3\mathbf{i} + 1\mathbf{j}) \times (T_B\mathbf{k}) = T_B\mathbf{i} - 3T_B\mathbf{j} \text{ N} \cdot \text{m}$$
$$(\mathbf{r}_{C/A} \times \mathbf{T}_C) = (1\mathbf{i} + 4\mathbf{j}) \times (T_C\mathbf{k}) = 4T_C\mathbf{i} - T_C\mathbf{j} \text{ N} \cdot \text{m}$$
$$(\mathbf{r}_{1/A} \times \mathbf{F}_1) = (1\mathbf{i} + 1\mathbf{j}) \times (-2943\mathbf{k}) = -2943\mathbf{i} + 2943\mathbf{j} \text{ N} \cdot \text{m}$$
$$(\mathbf{r}_{2/A} \times \mathbf{F}_2) = (1\mathbf{i} + 3\mathbf{j}) \times (-981\mathbf{k}) = -2943\mathbf{i} + 981\mathbf{j} \text{ N} \cdot \text{m}$$
$$(\mathbf{r}_{3/A} \times \mathbf{F}_3) = (2\mathbf{i} + 1\mathbf{j}) \times (-1962\mathbf{k}) = -1962\mathbf{i} + 3924\mathbf{j} \text{ N} \cdot \text{m}$$
$$(\mathbf{r}_{W/A} \times \mathbf{W}) = (1.5\mathbf{i} + 2\mathbf{j}) \times (-4905\mathbf{k}) = -9810\mathbf{i} + 7358\mathbf{j} \text{ N} \cdot \text{m}$$

Thus,

$$\mathbf{C} = \Sigma M_x\mathbf{i} + \Sigma M_y\mathbf{j} + \Sigma M_z\mathbf{k}$$
$$= (T_B + 4T_C - 17{,}658)\mathbf{i} + (-3T_B - T_C + 15{,}206)\mathbf{j} = \mathbf{0} \qquad\qquad (e)$$

From Eqs. d and e:

$$T_A + T_B + T_C = 10{,}791 \text{ N} \qquad \mathbf{T}_A = 3434\mathbf{k} \text{ N} = 3430 \text{ N}\uparrow \qquad \text{Ans.}$$
$$T_B + 4T_C = 17{,}658 \text{ N} \cdot \text{m} \qquad \mathbf{T}_B = 3924\mathbf{k} \text{ N} = 3920 \text{ N}\uparrow \qquad \text{Ans.}$$
$$3T_B + T_C = 15{,}206 \text{ N} \cdot \text{m} \qquad \mathbf{T}_C = 3433\mathbf{k} \text{ N} = 3430 \text{ N}\uparrow \qquad \text{Ans.}$$

The door shown in Fig. 6-46a has a mass of 25 kg and is supported in a horizontal position by two hinges and a bar. The hinges have been properly aligned; therefore, they exert only force reactions on the door. Assume that the hinge at B resists any force along the axis of the hinge pins. Determine the reactions at supports A, B, and D.

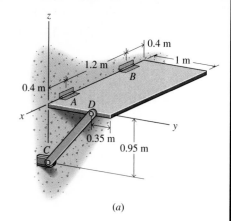

(a)

SOLUTION

A free-body diagram of the door is shown in Fig. 6-46b. The weight, the hinge reactions at A and B, and the bar reaction at D can be written as

$$\mathbf{W} = -mg\mathbf{k} = -25(9.81)\mathbf{k} = -245.3\mathbf{k} \text{ N}$$
$$\mathbf{A} = A_y\mathbf{j} + A_z\mathbf{k} \text{ N}$$
$$\mathbf{B} = B_x\mathbf{i} + B_y\mathbf{j} + B_z\mathbf{k} \text{ N}$$
$$\mathbf{D} = 0.5647D\mathbf{j} + 0.8253D\mathbf{k} \text{ N}$$

For the door to be in equilibrium, Eqs. 6-1 must be satisfied. Thus,

$$\mathbf{R} = \Sigma F_x\mathbf{i} + \Sigma F_y\mathbf{j} + \Sigma F_z\mathbf{k}$$
$$= (B_x)\mathbf{i} + (A_y + B_y + 0.5647D)\mathbf{j}$$
$$+ (A_z + B_z + 0.8253D - 245.3)\mathbf{k} = \mathbf{0} \qquad (a)$$

Summing moments about B to eliminate the maximum number of unknowns

$$\mathbf{C} = \Sigma C_x\mathbf{i} + \Sigma C_y\mathbf{j} + \Sigma C_z\mathbf{k}$$
$$= (\mathbf{r}_{A/B} \times \mathbf{A}) + (\mathbf{r}_{W/B} \times \mathbf{W}) + (\mathbf{r}_{D/B} \times \mathbf{D})$$

$$= \begin{vmatrix} \mathbf{i} & \mathbf{j} & \mathbf{k} \\ 1.2 & 0 & 0 \\ 0 & A_y & A_z \end{vmatrix} + \begin{vmatrix} \mathbf{i} & \mathbf{j} & \mathbf{k} \\ 0.6 & 0.5 & 0 \\ 0 & 0 & -245.3 \end{vmatrix} + \begin{vmatrix} \mathbf{i} & \mathbf{j} & \mathbf{k} \\ 1.6 & 0.65 & 0 \\ 0 & 0.5647D & 0.8253D \end{vmatrix}$$

$$\mathbf{C} = (-122.65 + 0.5364D)\mathbf{i} + (-1.2A_z + 147.18 - 1.3205D)\mathbf{j}$$
$$+ (1.2A_y + 0.9035D)\mathbf{k} = \mathbf{0} \qquad (b)$$

Equating the coefficients of **i**, **j**, and **k** to zero in Eqs. a and b and solving yields

$$
\begin{array}{ll}
B_x = 0 & B_x = 0 \\
A_y + B_y + 0.5647D = 0 & B_y = 43.03 \text{ N} \\
A_z + B_z + 0.8253D - 245.3 = 0 & B_z = 185.56 \text{ N} \\
-122.65 + 0.5364D = 0 & D = 228.65 \text{ N} \\
-1.2A_z + 147.18 - 1.3205D = 0 & A_z = -128.96 \text{ N} \\
1.2A_y + 0.9035D = 0 & A_y = -172.15 \text{ N}
\end{array}
$$

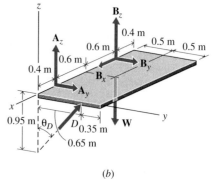

(b)

Fig. 6-46

The reactions at hinges A and B and the force exerted by the bar at D are

$$
\begin{array}{lll}
\mathbf{A} = -172.2\mathbf{j} - 129.0\mathbf{k} \text{ N} & |\mathbf{A}| = 215 \text{ N} & \text{Ans.} \\
\mathbf{B} = 43.0\mathbf{j} + 185.6\mathbf{k} \text{ N} & |\mathbf{B}| = 190.5 \text{ N} & \text{Ans.} \\
\mathbf{D} = 129.1\mathbf{j} + 188.7\mathbf{k} \text{ N} & |\mathbf{D}| = 229 \text{ N} & \text{Ans.}
\end{array}
$$

A scalar analysis can frequently be used in three-dimensional problems when a single unknown is the only quantity required. For example, the force exerted by the bar at support D can be determined by summing moments about the axis of the hinges. Thus,

$$+\big\lfloor\Sigma M_x = 0 \qquad D(0.95 \sin \theta_D) - W(0.50) = 0$$
$$D(0.95 \sin 34.38°) - 245.3(0.50) = 0$$
$$D = 228.6 = 229 \text{ N}$$

PROBLEMS

6-55* Determine the reaction at support *A* of the pipe system shown in Fig. P6-55 when the force applied to the pipe wrench is 50 lb.

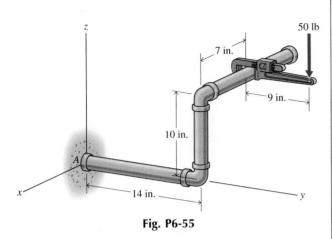

Fig. P6-55

6-56* A bar is supported by a ball-and-socket joint and two cables as shown in Fig. P6-56. Determine the reaction at support *A* (the ball-and-socket joint) and the tensions in the two cables.

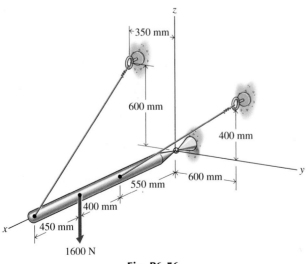

Fig. P6-56

6-57 The triangular plate of uniform thickness shown in Fig. P6-57 weighs 750 lb. Determine the tensions in the two cables supporting the plate and the reaction at the ball support.

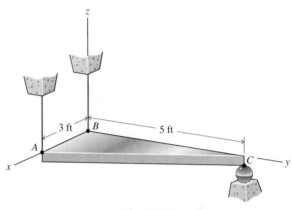

Fig. P6-57

6-58 The rectangular plate of uniform thickness shown in Fig. P6-58 has a mass of 500 kg. Determine the tensions in the three cables supporting the plate.

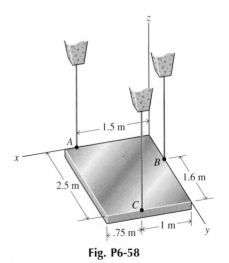

Fig. P6-58

6-59* A beam is supported by a ball-and-socket joint and two cables as shown in Fig. P6-59. Determine the reaction at support A (the ball-and-socket joint) and the tensions in the two cables.

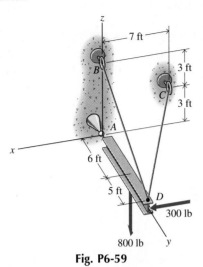

Fig. P6-59

6-60* A bar is supported by a ball-and-socket joint, link, and cable as shown in Fig. P6-60. Determine the reactions at supports A (ball-and-socket joint) and B (link) and the tension in the cable.

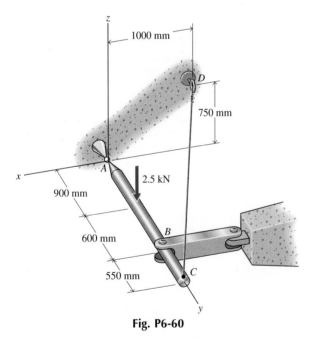

Fig. P6-60

6-61 The plate shown in Fig. P6-61 weighs 150 lb and is supported in a horizontal position by two hinges and a cable. The hinges have been properly aligned; therefore, they exert only force reactions on the plate. Assume that the hinge at B resists any force along the axis of the hinge

pins. Determine the reactions at supports A and B and the tension in the cable.

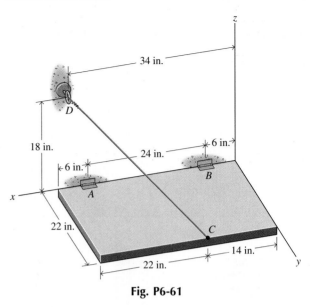

Fig. P6-61

6-62 The plate shown in Fig. P6-62 has a mass of 75 kg. The brackets at supports A and B exert only force reactions on the plate. Each of the brackets can resist a force along the axis of pins in one direction only. Determine the reactions at supports A and B and the tension in the cable.

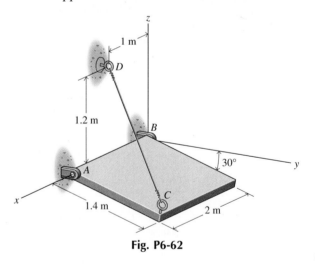

Fig. P6-62

6-63* A shaft is loaded through a pulley and a lever (see Fig. P6-63) that are fixed to the shaft. Friction between the belt and pulley prevents slipping of the belt. Determine the force $\mathbf{P}$ required for equilibrium and the reactions at supports A and B. The support at A is a ball bearing and the support at B is a thrust bearing. The bearings exert only force reactions on the shaft.

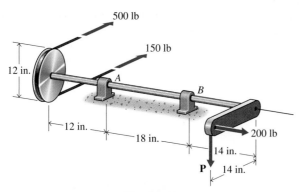

500 lb

150 lb

12 in.

A

B

12 in.

18 in.

200 lb

14 in.

P 14 in.

Fig. P6-63

6-64* The shaft with two levers shown in Fig. P6-64 is used to change the direction of a force. Determine the force **P** required for equilibrium and the reactions at supports A and B. The support at A is a ball bearing and the support at B is a thrust bearing. The bearings exert only force reactions on the shaft.

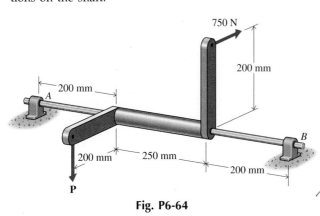

750 N

200 mm

A

200 mm

200 mm   250 mm

200 mm

B

P

Fig. P6-64

6-65 The shaft shown in Fig. P6-65 is part of a drive system in a factory. Friction between the belts and pulleys prevents slipping of the belts. Determine the torque **T** required for equilibrium and the reactions at supports A and

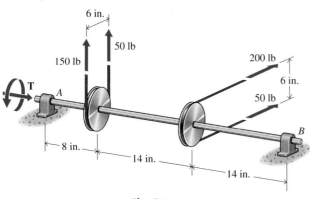

6 in.

50 lb

150 lb

200 lb

6 in.

50 lb

T

A

B

8 in.

14 in.

14 in.

Fig. P6-65

B. The support at A is a journal bearing and the support at B is a thrust bearing. The bearings exert only force reactions on the shaft.

6-66 The plate shown in Fig. P6-66 has a mass of 100 kg. The hinges at supports A and B exert only force reactions on the plate. Assume that the hinge at B resists any force along the axis of the hinge pins. Determine the reactions at supports A and B and the tension in the cable.

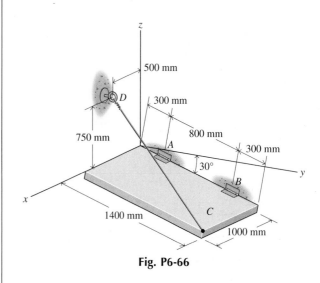

z

500 mm

D

300 mm

750 mm

800 mm

A

300 mm

30°

y

B

x

1400 mm

C

1000 mm

Fig. P6-66

6-67* The plate shown in Fig. P6-67 weighs 200 lb and is supported in a horizontal position by a hinge and a cable. Determine the reactions at the hinge and the tension in the cable.

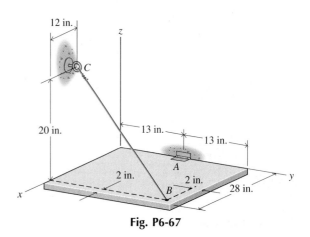

12 in.

z

C

20 in.

13 in.

13 in.

A

2 in.

2 in.

y

B

28 in.

x

Fig. P6-67

6-68* The block W shown in Fig. P6-68 has a mass of 250 kg. Bar AB rests against a smooth vertical wall at end B and is supported at end A with a ball-and-socket joint. The two cables are attached to a point on the bar midway between the ends. Determine the reactions at supports A and B and the tension in cable CD.

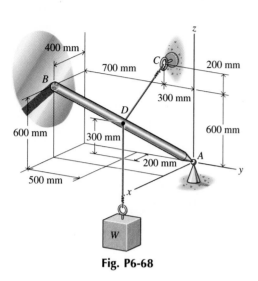

Fig. P6-68

6-70 The bent bar shown in Fig. P6-70 is supported with two brackets that exert only force reactions on the bar. The end of the bar at C rests against smooth horizontal and vertical surfaces. Determine the reactions at supports A, B, and C.

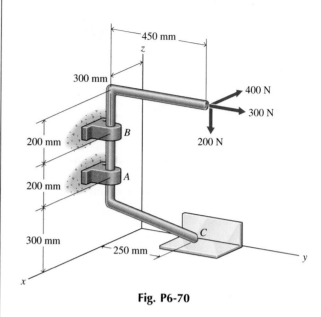

Fig. P6-70

6-69 The bent bar shown in Fig. P6-69 is supported with three brackets that exert only force reactions on the bar. Determine the reactions at supports A, B, and C.

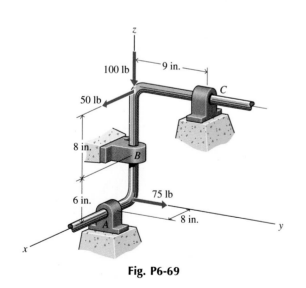

Fig. P6-69

6-71 Bar AB of Fig. P6-71 rests against a smooth vertical wall at end B and is supported at end A with a pin and bracket. Determine the reactions at supports A and B.

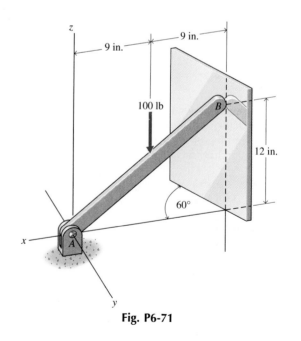

Fig. P6-71

SUMMARY

Any system of forces acting on a rigid body can be expressed in terms of a resultant force **R** and a resultant couple **C**. Therefore, for a rigid body to be in equilibrium, both the resultant force **R** and the resultant couple **C** must vanish. These two conditions can be expressed by the two vector equations

$$\mathbf{R} = \Sigma F_x \mathbf{i} + \Sigma F_y \mathbf{j} + \Sigma F_z \mathbf{k} = \mathbf{0}$$
$$\mathbf{C} = \Sigma M_x \mathbf{i} + \Sigma M_y \mathbf{j} + \Sigma M_z \mathbf{k} = \mathbf{0}$$

(6-1)

Equation 6-1 can be expressed in scalar form as

$$\Sigma F_x = 0 \qquad \Sigma F_y = 0 \qquad \Sigma F_z = 0$$
$$\Sigma M_x = 0 \qquad \Sigma M_y = 0 \qquad \Sigma M_z = 0$$

(6-2)

Equations 6-2 are the necessary conditions for equilibrium of a rigid body. They are also the sufficient conditions for equilibrium if all the forces acting on the body can be determined from these equations.

In order to study the force system acting on a body, it is necessary to identify all forces, both known and unknown, that act on the body. The best way to identify all forces acting on a body is to use the free-body diagram approach. Special care must be exercised when representing the actions of connections and supports on the free-body diagram.

The term "two-dimensional" is used to describe problems in which the forces involved are contained in a plane (say the xy-plane) and the axes of all couples are perpendicular to the plane containing the forces. For two-dimensional problems, the equations of equilibrium reduce to

$$\mathbf{R} = \Sigma F_x \mathbf{i} + \Sigma F_y \mathbf{j} = \mathbf{0}$$
$$\mathbf{C} = \Sigma M_z \mathbf{k} = \mathbf{0}$$

(6-3)

or in scalar form to

$$\Sigma F_x = 0 \qquad \Sigma F_y = 0 \qquad \Sigma M_A = 0$$

(6-4)

The third equation represents the sum of the moments of all forces about a z-axis through any point A on or off the body. Equations 6-4 are both the necessary and sufficient conditions for equilibrium of a body subjected to a two-dimensional system of forces. Alternative forms of Eqs. 6-4 are

$$\Sigma F_x = 0 \qquad \Sigma M_A = 0 \qquad \Sigma M_B = 0$$

(6-5)

where points A and B must have different x-coordinates, and

$$\Sigma M_A = 0 \qquad \Sigma M_B = 0 \qquad \Sigma M_C = 0$$

(6-6)

where A, B, and C are any three points not on the same straight line.

REVIEW PROBLEMS

6-72* Determine the reactions at supports A and B of the truss shown in Fig. P6-72.

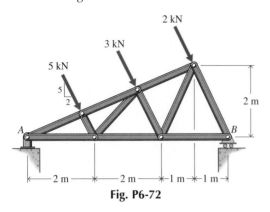

Fig. P6-72

6-73* Determine the force **P** required to pull the 250-lb roller over the step shown in Fig. P6-73.

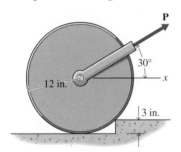

Fig. P6-73

6-74 Bar AB of Fig. P6-74 has a uniform cross section, a mass of 25 kg, and a length of 1 m. Determine the angle θ for equilibrium.

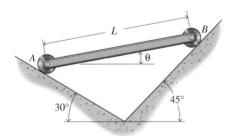

Fig. P6-74

6-75 The crane and boom, shown in Fig. P6-75, weigh 12,000 lb and 600 lb, respectively. When the boom is in the position shown, determine

a. The maximum load that can be lifted by the crane.
b. The tension in the cable used to raise and lower the boom when the load being lifted is 3600 lb.

c. The pin reaction at boom support A when the load being lifted is 3600 lb.

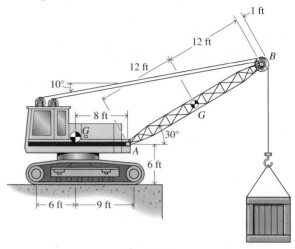

Fig. P6-75

6-76* A beam is loaded and supported as shown in Fig. P6-76. Determine the reactions at supports A and B. Neglect the weight of the beam.

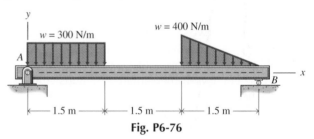

Fig. P6-76

6-77* A 500-lb homogeneous circular plate is supported by three cables as shown in Fig. P6-77. Determine the tensions in the three cables.

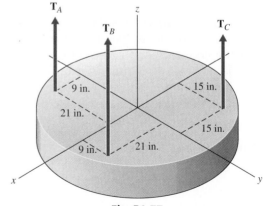

Fig. P6-77

6-78 A bracket is loaded and supported as shown in Fig. P6-78. Determine the reactions at supports A and B. Neglect the weight of the bracket.

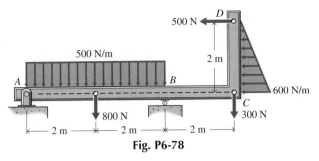

Fig. P6-78

6-79 Two beams are loaded and supported as shown in Fig. P6-79. Determine the reactions at supports A, B, and C. Neglect the weights of the beams.

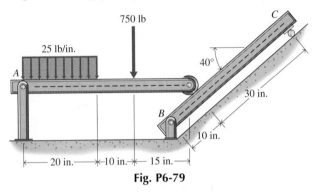

Fig. P6-79

6-80* Beam CD of Fig. P6-80 is supported at the left end C by a smooth pin and bracket and at the right end D by a continuous cable, which passes around a frictionless pulley. The lines of action of the force in the cable pass through point D. Determine the components of the reaction at support C and the force in the cable when a 5-kN load W is being supported by the beam.

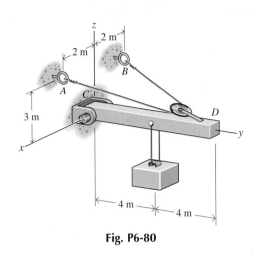

Fig. P6-80

6-81 The 570-lb block, shown in Fig. P6-81, is supported by a ball-and-socket joint at A, by a smooth pin at B, and by a cable at C. Determine the components of the reactions at supports A and B and the force in the cable at C.

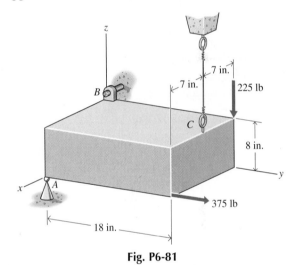

Fig. P6-81

Computer Problems

C6-82 The lever shown in Fig. P6-82 is formed in a quarter circular arc of radius 450 mm. Plot A and B, the magnitude of the pin force at A and the force on the smooth support B, as a function of θ ($0° \leq \theta \leq 85°$), the angle at which the support is located.

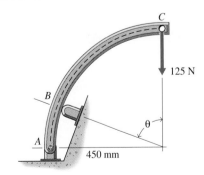

Fig. P6-82

C6-83 The crane and boom shown in Fig. P6-83 weigh 12,000 lb and 600 lb, respectively. The pulleys at D and E are small and the cables attached to them are essentially parallel.

a. Plot d, the location of the resultant force relative to point C, as a function of the boom angle θ ($0° \leq \theta \leq 80°$), when the crane is lifting a 3600-lb load.
b. Plot A, the magnitude of the reaction force on the pin at A, as a function of θ ($0° \leq \theta \leq 80°$) when the crane is lifting a 3600-lb load.
c. It is desired that the resultant force on the tread always be at least 1 ft behind C to ensure that the crane is never in danger of tipping over. Plot W_{max}, the maximum load that may be lifted, as a function of θ ($0° \leq \theta \leq 80°$).

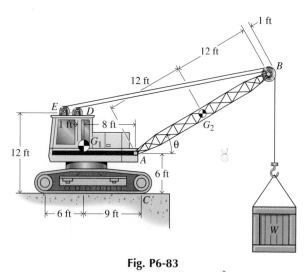

Fig. P6-83

C6-84 The tower crane shown in Fig. P6-84 is used to lift construction materials. The counterweight C weighs 31,000 N; the motor M weighs 4500 N; the weight of the boom AB is 36,000 N and can be considered as acting at

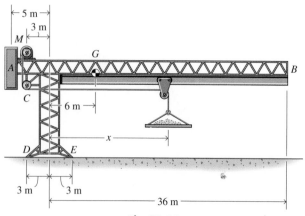

Fig. P6-84

point G; and the weight of the tower is 23,000 N and can be considered as acting at its midpoint.

a. If the tower is lifting a 9000-N load, calculate and plot the reaction forces on the feet at D and E as a function of the distance x at which the weight is being lifted.
b. It is desired that the reaction forces at the feet D and E always be greater than 4500 N to ensure that the tower is never in danger of tipping over. Plot W_{max}, the maximum load that can be lifted as a function of the distance x ($0 \leq x \leq 36$ m).

C6-85 The hydraulic cylinder BC is used to tip the box of the dump truck shown in Fig. P6-85. If the combined weight of the box and the load is 22,000 lb and acts through the center of gravity G

a. Plot C, the force in the hydraulic cylinder, as a function of the angle θ ($0 \leq \theta \leq 80°$).
b. Plot A, the magnitude of the reaction force on the pin at A, as a function of θ ($0° \leq \theta \leq 80°$).

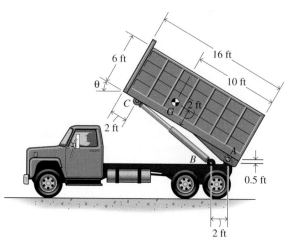

Fig. P6-85

C6-86 An oil rig is being raised into position by a hydraulic cylinder as shown in Fig. P6-86. If the rig has a mass of 50,000 kg and center of gravity at G

a. Plot B, the force in the hydraulic cylinder, as a function of the angle θ ($0 \leq \theta \leq 90°$).
b. Plot A, the magnitude of the reaction force on the pin at A, as a function of θ ($0° \leq \theta \leq 90°$).

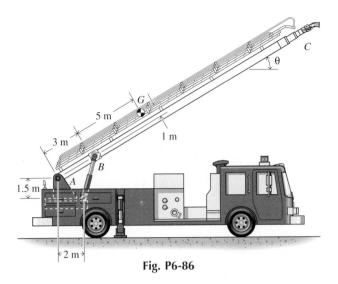

Fig. P6-86

geneous rectangular slab weighing 225 lb. Frictionless rollers (B and C) run in tracks at each side of the door as shown.

a. Plot T, the tension in the wire, as a function of d ($0 \leq d \leq 100$ in.).

b. Plot B and C, the forces on the frictionless rollers, as a function of d ($0 \leq d \leq 100$ in.).

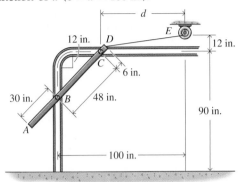

Fig. P6-89

C6-87 The wrecker truck of Fig. P6-87 has a weight of 15,000 lb and a center of gravity at G. The force exerted on the rear (drive) wheels by the ground consists of both a normal component B_y and a tangential component B_x while the force exerted on the front wheel consists of a normal force A_y only. Plot P, the maximum amount of pull the wrecker can exert as, a function of θ ($0° \leq \theta \leq 90°$) if B_x cannot exceed $0.8\ B_y$ (because of friction considerations) and the wrecker does not tip over backwards (the front wheels remain in contact with the ground).

C6-90 A hand winch is used to raise a 75-kg load as shown in Fig. P6-90. If the force P is always perpendicular to both the handle DE and the arm CD, plot A and B, the magnitudes of the bearing forces, as a function of the angle θ ($0 \leq \theta \leq 360°$).

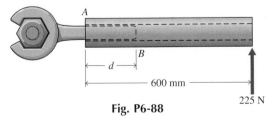

Fig. P6-87

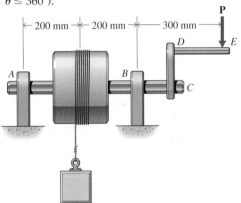

C6-88 A lightweight pipe is slipped over the handle of a wrench to give extra leverage (Fig. P6-88). If the inside of the pipe is smooth and fits slightly loosely about the handle of the wrench, plot the forces exerted on the wrench at A and B as a function of the overlap distance d ($0 \leq d \leq$ 200 mm).

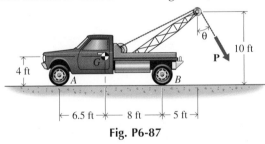

Fig. P6-88

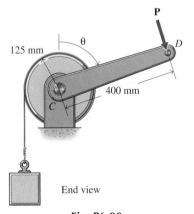

End view

Fig. P6-90

C6-89 The garage door ABCD shown in Fig. P6-89 is being raised by a cable DE. The one-piece door is a homo-

C6-91 An I-beam is supported by a ball-and-socket joint at A, by a rope BC, and by a horizontal force P as shown in Fig. P6-91. In addition, a 560-lb load is suspended from a movable support at D. If the uniform beam is 80 in. long and weighs 225 lb

a. Plot the force P required to keep the beam aligned with the y-axis as a function of the distance d ($0 \le d \le$ 80 in.).

b. On the same graph plot T_{BC}, the tension in the rope, and A, the magnitude of the force exerted on the ball-and-socket joint.

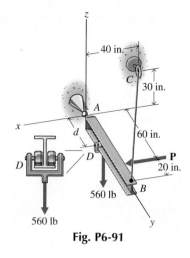

Fig. P6-91

TRUSSES, FRAMES, AND MACHINES

An extensive truss structure was used to support scaffolding for workers during renovation of the Statue of Liberty.

7-1   INTRODUCTION

In Chapter 6 the equations of equilibrium were used to determine the external support reactions on a rigid body. Determination of the support reactions, however, is only the first step in the analysis of engineering structures and machines. In this chapter the equations of equilibrium are used to determine the joint forces in structures composed of pin-connected members. Determination of these joint forces is a necessary first step in choosing the fasteners (type, size, material, etc.) used to hold the structure together. In Chapter 8 this analysis will be extended to show how to determine the resultant of the internal forces in a member. Determination of the internal forces is needed to design the members that make up the structure.

Like all forces (internal, external, applied, or reactive), the joint forces always occur in equal-magnitude, opposite-direction pairs. If not separated by means of a free-body diagram from the rest of the structure (or environment), the pair of forces need not be considered when writing the equations of equilibrium. Therefore, in order to determine the joint forces, the structure must be divided into two or more parts. Then the joint forces at the separation points become external forces on each free-body diagram and will enter the equations of equilibrium. Application of the equations of equilibrium to the various parts of a simple truss, frame, or machine allows the determination of all the forces acting at the connections.

Although there are many different types of engineering structures, this chapter will focus on two of the more important and common types:

Trusses, which are structures composed entirely of two force members. Trusses generally consist of triangular subelements and are constructed and supported so as to prevent any motion. Bridge supports such as shown in Fig. 7-1 are trusses. Their airy structure can carry a great load for a relatively small structural weight.

Frames, which are structures that always contain at least one member acted on by forces at three or more points. The table structure of Fig. 7-2 is a frame. Frames are also constructed and supported so as to prevent any motion. Framelike structures that are not fully constrained are often called **machines** or **mechanisms.**

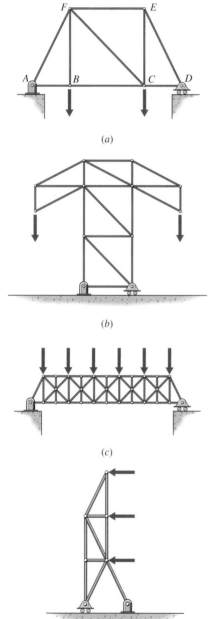

(a)

(b)

(c)

(d)

Fig. 7-1

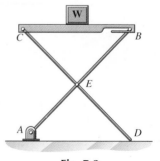

Fig. 7-2

7-2   PLANE TRUSSES

A truss is a structure composed of usually straight members joined together at their end points and loaded only at the joints (Fig. 7-1). The airy structure of a truss provides greater strength over large spans than would more solid types of structures. Trusses are commonly seen supporting the roofs of buildings as well as large railroad and highway bridges. Although not commonly seen, trusses also form the skeletal structure of many large buildings.

Planar trusses lie in a single plane, and all applied loads must lie in the same plane. Planar trusses are often used in pairs to support bridges, as shown in Fig. 7-3. All members of the truss *ABCDEF* lie in the same vertical plane. Loads on the floor of the bridge are carried by means of the floor construction to the joints *ABCD*. The loads thus transmitted to the joints act in the same vertical plane as the truss.

Space trusses are structures that are not contained in a single plane and/or are loaded out of the plane of the structure. Classic examples are the truss structures supporting large antennas and windmills. Space trusses will be discussed in Section 7-3.

There are four main assumptions made in the analysis of trusses. One result of the assumptions is that all members of the idealized structure are two-force members. Such structures are much easier to analyze than more general structures having the same number of members. Although the assumptions are idealizations of actual structures, real trusses behave according to the idealizations to a high degree of approximation. The resulting error is usually small enough to justify the assumptions.

Truss members are connected together at their ends only.

The first assumption means that the truss of Figs. 7-1*a* and 7-3 should be drawn as in Fig. 7-4. In actual practice, the main top and bottom chords frequently consist of members that span several joints rather than a series of shorter members between joints. The members of a truss are usually long and slender, however, and can support little lateral load or bending moment. Hence, the noncontinuous member assumption is usually acceptable. Because members are not continuous through a joint, no confusion should result from drawing trusses using the simple line representations of Fig. 7-1.

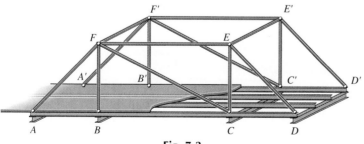

Fig. 7-3

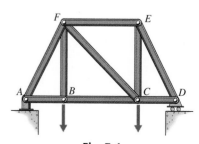

Fig. 7-4

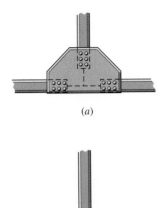

(a)

(b)

Fig. 7-5

Truss members are connected together by frictionless pins.

In real trusses, the members are usually bolted, welded, or riveted to a gusset plate, as shown in Fig. 7-5a, rather than connected by an idealized frictionless pin as shown in Fig. 7-5b. However, experience has shown the frictionless pin to be an acceptable idealization as long as the axes of the members all intersect at a single point.

The truss structure is loaded only at the joints.

As stated earlier, the members that make up a truss are usually long and slender. Like cables, such members can withstand large tensile (axial) load but cannot withstand moments or large lateral loads. Loads must either be applied directly to the joints as indicated in the diagrams of Fig. 7-1 or must be carried to the joints by a floor structure such as shown in Fig. 7-3.

The weights of the members may be neglected.

Frequently in the analysis of trusses, the weight of the members is neglected. While this may be acceptable for small trusses, it may not be acceptable for a large bridge truss. Again because the members can withstand little bending moment or lateral load, experience has shown that little error results from assuming that the load acts at the joints of the truss. Common practice is to assume that half of the weight of each member acts at the two joints that it connects.

The result of these four assumptions is that forces act only at the ends of the members. Also, because the pins are assumed frictionless, there is no moment applied to the ends of the members. Therefore, by the analysis of Section 6-3, each member is a two-force member supporting only an axial force, as shown in Fig. 7-6. In its simplest form, a truss (such as that shown in Fig. 7-7a) consists of a collection of two-force members held together by frictionless pins, as shown in Fig. 7-7b.

For general two-force members, the forces act along the line joining the points where the forces are applied. Since truss members are

Fig. 7-6

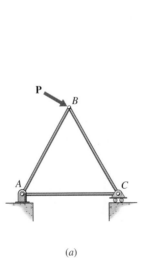

(a)

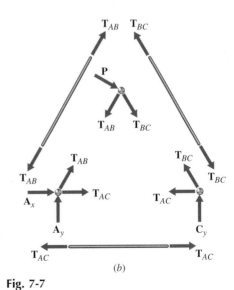

(b)

Fig. 7-7

usually straight, however, the forces will act along the axis of the member as shown in Figs. 7-6 and 7-7.[*] Forces that pull on the end of a member (as in Figs. 7-6 and 7-7) are called tensile and tend to elongate the member. Such forces are called tensile forces, and the member is said to be in tension. Forces that push on the end of the member tend to shorten the member. Such forces are called compressive forces, and the member is said to be in compression. It may be noted from Fig. 7-6 that when a joint exerts forces that pull on the end of a member, the member exerts forces that also pull on the joint.

It is important to distinguish which members of a truss are in compression and which are in tension. The long slender members that make up a truss are very strong in tension but tend to bend or buckle under large compressive loads. Truss members in compression either must be made thicker than the other truss members or must be braced to prevent buckling.

One end of large bridge trusses is usually allowed to "float" on a rocker or roller support as shown in Fig. 7-1. Aside from the mathematical requirement (in a planar equilibrium problem, only three support reactions can be determined), such a support is needed to allow for the expansion or contraction of the structure due to temperature variations.

The basic building block of all trusses is a triangle. In order to retain its shape and support the large loads applied to them, trusses must be rigid structures. The simplest structure that is rigid (independent of how it is supported) is a triangle. Of course, the word *rigid* does not mean that a truss will not deform at all under loading. It will undergo very small deformations, but will very nearly retain its original shape.

"Rigid" is often interpreted also to mean that the truss will retain its shape when removed from its supports or when one of the supports is free to slide. In this sense, the truss of Fig. 7-8 is rigid, whereas the truss of Fig. 7-9 is not. The truss of Fig. 7-9 is called a **compound truss,** and the lack of internal rigidity is made up for by an extra external support reaction.

The basic building block of all trusses is a triangle. Large trusses are constructed by attaching several triangles together. One method of construction starts with a basic triangular element such as triangle *ABC* of Fig. 7-10. Additional triangular elements are added one at a time by attaching one new joint (for example, *D*) to the truss, using two new members (for example, *BD* and *CD*). A truss that can be constructed in this fashion is called a **simple truss.** Although it might appear that all trusses composed of triangles are simple trusses, such is not the case. For example, neither of the trusses of Figs. 7-8 or 7-9 is a simple truss.

The importance of a simple truss is that it allows a simple way to check the rigidity and solvability of a truss. Clearly, since a simple truss is constructed solely of triangular elements, it is always rigid. Also, since each new joint brings two new members with it, a simple

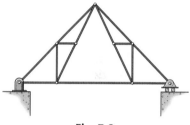

Fig. 7-8

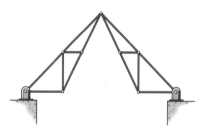

Fig. 7-9

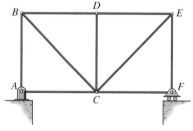

Fig. 7-10

---

[*]For curved two-force members, however, the line joining the ends is not the axis of the member. All the trusses considered in this chapter will contain only straight two-force members. For a brief discussion of the effect of curved two-force members, see Section 7-2-4.

relationship exists between the number of joints j and the number of members m in a simple plane truss:

$$m = 2j - 3 \qquad (7\text{-}1)$$

In the discussion of the method of joints that follows, this will be seen to be exactly the condition necessary to guarantee that the number of equations to be solved ($2j$) is the same as the number of unknowns to be solved for (m member forces and 3 support reactions).

Although Eq. 7-1 ensures that a simple plane truss is rigid and solvable, it is neither sufficient nor necessary to ensure that a nonsimple plane truss is rigid and solvable. For example, the nonsimple plane trusses of Figs. 7-8 and 7-9 are both rigid (at least while attached to their supports) and solvable, although one (Fig. 7-8) satisfies Eq. 7-1 and the other (Fig. 7-9) does not. A tempting generalization of Eq. 7-1 is

$$m = 2j - r \qquad (7\text{-}2)$$

where r is the number of support reactions. Both trusses of Figs. 7-8 and 7-9 satisfy Eq. 7-2, as do all simple trusses that have the customary three support reactions. However, constructions can be envisioned for which even Eq. 7-2 is not a proper test of the solvability of a truss.

7-2-1   Method of Joints

Consider the truss of Fig. 7-11*a*, whose free-body diagram is shown in Fig. 7-11*b*. Since the entire truss is a rigid body in equilibrium, each part must also be in equilibrium. The method of joints consists of taking the truss apart, drawing separate free-body diagrams of each part—each member and each pin—as in Fig. 7-12, and applying the equations of equilibrium to each part of the truss in turn.

The free-body diagrams of the members in Fig. 7-12 have only axial forces applied to their ends because of the assumptions about how a truss is constructed and loaded. The symbol T_{BC} is used to represent the unknown force in member BC. (No significance is attached to the order of the subscripts; that is, $T_{BC} = T_{CB}$.) Since the lines of action of the member forces are all known, the force in each member is completely specified by giving its magnitude and sense; that is, whether the force points away from the member as in Fig. 7-12 or toward the member. Thus the force (a vector) is represented by the scalar symbol T_{BC}. The sense of the force will be taken from the sign of T_{BC}; positive indicates the direction drawn on the free-body diagram, negative indicates the opposite direction.

Forces that point away from a member as in Fig. 7-12 tend to stretch the member and are called tensile. Forces that point toward a member tend to compress the member and are called compressive. Whether a member is in compression or tension is usually not known ahead of time. Although some people try to guess and draw some of the forces in tension and others in compression, it is not necessary to do so. In this book, the free-body diagrams will all be drawn as though all members are in tension. A negative value for a force in the solution will indicate that the member was really in compression. This can be reported either by saying that $T_{BC} = -2500$ lb or by saying that $T_{BC} = 2500$ lb (C). The latter is preferable since it does not depend on whether

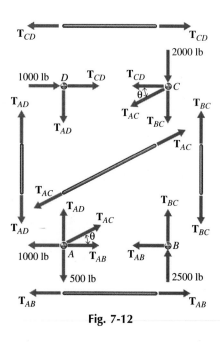

Fig. 7-11

Fig. 7-12

member BC was assumed to be in tension or in compression in the free-body diagram.

According to Newton's third law (of action and reaction), the force exerted on a member by a pin and the force exerted on a pin by a member are equal and opposite. Therefore the same symbol T_{AB} is used for the force exerted by the member AB on pin B and for the force exerted by pin B on member AB. Having drawn the free-body diagrams of the members as two-force members ensures that the members are in equilibrium. Then there is no further information to be gained from the free-body diagrams of the two-force members and they may be discarded for the remainder of the analysis. The analysis of the truss reduces to considering the equilibrium of the joints that make up the truss—hence the name *Method of Joints*.

Equilibrium of the joints that make up the truss is expressed by drawing a separate free-body diagram for each joint and writing the equilibrium equation

$$\Sigma \mathbf{F} = \mathbf{0} \tag{7-3}$$

for each joint. Since each joint consists of concurrent forces in a plane, moment equilibrium gives no useful information and Eq. 7-1 has only two independent components. Therefore, for a plane truss containing j pins, there will be a total of $2j$ independent scalar equations available. But according to Eq. 7-1, this is precisely the number of independent equations needed to solve for the m member forces and 3 support reactions of a simple truss.

Solution of the $2j$ equations is significantly simplified if a joint can be found on which only two unknown forces and one or more known forces act (for example joint D of Fig. 7-12). In this case, the two equations for this joint can be solved independently of the rest of the equations. If such a joint is not readily available, one can usually be created by solving the equilibrium equations for the entire truss first. Once two of the unknown forces have been determined, they can be treated as known forces on the free-body diagrams of the other joints. The joints are solved sequentially in this fashion until all forces are known.

As mentioned earlier, a negative value for a force (such as T_{DC}) indicates that the member is in compression rather than in tension. It is unnecessary to go back to the free-body diagram and change the direction of the arrow. In fact, doing so is likely only to cause confusion. The free-body diagrams should all be drawn consistently. A negative value for a symbol on one free body diagram translates to the same negative value for the same symbol on another free-body diagram.

Once all the forces have been determined, a summary should be made listing the magnitude of the force in each member and whether the member is in tension or compression (see the Example Problems).

Finally, it must be noted that the equations of equilibrium for the entire truss are contained in the equations of equilibrium of the joints (see Problem 7-31). That is, if all the joints are in equilibrium and all the members are in equilibrium, then the entire truss is also in equilibrium. A consequence of this is that the three support reactions can be determined along with the m member forces from the $2j$ equations of equilibrium of the joints. Overall equilibrium in this case may be used as a check of the solution. However, if overall equilibrium is used first to determine the support reactions and help start the method of joints, then three of the $2j$ joint equations of equilibrium will be redundant and may be used as a check of the solution.

Use the method of joints to find the force in each member of the truss of Fig. 7-11a.

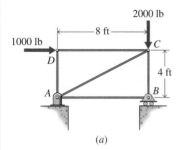 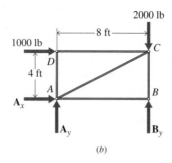

(a)   (b)

Fig. 7-11

SOLUTION

Truss $ABCD$ is a simple truss with $m = 5$ members and $j = 4$ joints. Therefore, the eight equations obtained from the equilibrium of the four joints can be solved for the three support reactions as well as the forces in all five members.

The first step is to draw a free-body diagram of the entire truss (Fig. 7-11b) and write the equilibrium equations

$$+\rightarrow\Sigma F = 1000 + A_x = 0$$
$$+\uparrow\Sigma F = -2000 + A_y + B_y = 0$$
$$\downarrow+\Sigma M_A = -(4)(1000) - (8)(2000) + 8B_y = 0$$

These equations can be solved to get

$$A_x = -1000 \text{ lb} \qquad B_y = 2500 \text{ lb} \qquad A_y = -500 \text{ lb}$$

Next, draw a free-body diagram of pin D and solve the equilibrium equations

$$+\rightarrow\Sigma F = 1000 + T_{CD} = 0$$
$$+\uparrow\Sigma F = -T_{AD} = 0$$

to get

$$T_{CD} = -1000 \text{ lb} \qquad T_{AD} = 0 \text{ lb}$$

Next draw a free-body diagram of pin C and solve the equilibrium equations

$$+\rightarrow\Sigma F = -T_{CD} - T_{AC} \cos \theta = 0$$
$$+\uparrow\Sigma F = -2000 - T_{BC} - T_{AC} \sin \theta = 0$$

where

$$\sin \theta = \frac{AD}{AC} = \frac{8}{\sqrt{4^2 + 8^2}} = 0.8944$$

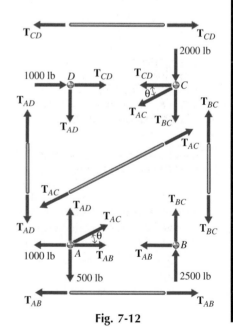

Fig. 7-12

and

$$\cos \theta = \frac{CD}{AC} = \frac{4}{\sqrt{4^2 + 8^2}} = 0.4472$$

But $T_{CD} = -1000$ lb, so

$$T_{AC} = -\frac{T_{CD}}{\cos \theta} = -\frac{-1000}{0.8944} = 1118 \text{ lb}$$

and

$$T_{BC} = -2000 - (1118)(0.4472) = -2500 \text{ lb}$$

Next, draw a free-body diagram of pin B and write the equilibrium equations

$$+\rightarrow \Sigma F = -T_{AB} = 0$$
$$+\uparrow \Sigma F = T_{BC} + 2500 = 0$$

The first of these equations can be solved to get

$$T_{AB} = 0 \text{ lb}$$

The second equation contains no unknowns, however, since the value of T_{BC} has already been found. The second equation can be used to check the consistency of the answers:

$$(-2500) + 2500 = 0 \qquad \text{(check)}$$

Finally, draw a free-body diagram of pin A and write

$$+\rightarrow \Sigma F = T_{AB} + T_{AC} \cos \theta - 1000 = 0$$
$$+\uparrow \Sigma F = T_{AD} + T_{AC} \sin \theta - 500 = 0$$

Again, there are no unknowns in these equations since the values of T_{AB}, T_{AC}, and T_{AD} have already been found. These two equations again reduce to a check of the consistency of the solution:

$$0 + (1118)(0.8944) - 1000 = -0.06$$
$$0 + (1118)(0.4472) - 500 = -0.03$$

The difference is less than the rounding performed on T_{AB}, T_{AC}, and T_{AD} and so the solution checks.

The desired answers then are

$$\begin{array}{ll} T_{AB} = T_{AD} = 0 \text{ lb} & \text{Ans.} \\ T_{AC} = 1118 \text{ lb } (T) & \text{Ans.} \\ T_{BC} = 2500 \text{ lb } (C) & \text{Ans.} \\ T_{CD} = 1000 \text{ lb } (C) & \text{Ans.} \end{array}$$

(The fact that T_{AB} and T_{AD} both came out zero is a peculiarity of the loading and does not mean that members AB and AD should be eliminated from the truss. For a slightly different loading situation, the forces in these members will not be zero. Even for the given loading condition, members AB and AD are necessary to ensure the rigidity of the truss. Without member AB, for example, the truss would collapse if the roller support at B were disturbed slightly to the right or left.)

The truss shown in Fig. 7-13a supports one side of a bridge; an identical truss supports the other side. Floor beams carry vehicle loads to the truss joints. A 2000 kg car is stopped on the bridge. Calculate the force in each member of the truss using the method of joints.

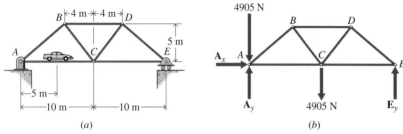

(a)   (b)

Fig. 7-13

SOLUTION

The truss of Fig. 7-13a is a simple truss with $m = 7$ members and $j = 5$ joints. Therefore, the ten equations obtained from the equilibrium of the five joints can be solved for the three support reactions as well as the forces in all seven members.

The first step is to divide the weight up between the joints of the truss. Half of the car's weight $(1000)(9.81) = 9{,}810$ Newtons (N) is carried by the truss shown and the other half is carried by the truss on the other side of the bridge. Since the car is midway between joints A and C, 4905 N will be applied to joint A and 4905 N will be applied to joint C.

The next step is to draw a free-body diagram of the entire truss (Fig. 7-13b) and write the equilibrium equations:

$$+\rightarrow \Sigma F = A_x = 0$$
$$+\uparrow \Sigma F = A_y - 4905 - 4905 + E_y = 0$$
$$\downarrow + \Sigma M_A = 20E_y - (10)(4905) = 0$$

These equations can be solved immediately to get

$$A_x = 0 \text{ N} \qquad E_y = 2453 \text{ N} \qquad A_y = 7357 \text{ N}$$

Now the free-body diagram (Fig. 7-14a) of pin A is drawn and the equilibrium equations written:

$$+\rightarrow \Sigma F = A_x + T_{AC} + T_{AB} \cos \theta = 0$$
$$+\uparrow \Sigma F = A_y - 4905 + T_{AB} \sin \theta = 0$$

where $A_x = 0$ N, $A_y = 7357$ N, and $\theta = \tan^{-1}(5/6) = 39.81°$. This gives

$$T_{AB} = -3830 \text{ N} \qquad T_{AC} = 2942 \text{ N}$$

Since one of the three forces applied at pin B is now known, the free-body diagram of pin B (Fig. 7-14b) is drawn next. The equilibrium equations for this pin are:

$$+\rightarrow \Sigma F = -T_{AB} \cos \theta + T_{BD} + T_{BC} \cos \phi = 0$$
$$+\uparrow \Sigma F = -T_{AB} \sin \theta - T_{BC} \sin \phi = 0$$

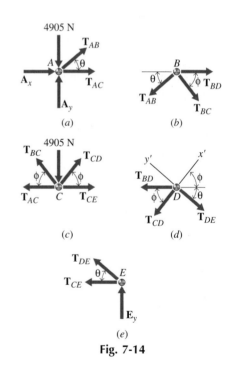

(a)

(b)

(c)

(d)

(e)

Fig. 7-14

where $T_{AB} = -3830$ N and $\phi = \tan^{-1}(5/4) = 51.34°$. They are solved to get

$$T_{BC} = 3140 \text{ N} \qquad T_{BD} = -4904 \text{ N}$$

At this point, pin C, pin D, or pin E could be solved next since each of these pins has only two forces whose values have not yet been found applied to them. For this example the free-body diagram of pin D (Fig. 7-14d) will be considered next. Writing the standard horizontal and vertical components of the equilibrium equations gives

$$+\rightarrow\Sigma F = -(-4904) - T_{CD}\cos\phi + T_{DE}\cos\theta = 0$$
$$+\uparrow\Sigma F = -T_{CD}\sin\phi - T_{DE}\sin\theta = 0$$

Both of these equations contain both of the unknown forces T_{CD} and T_{DE}. Although the solution of this pair of equations is not particularly difficult, the calculations can be simplified if the equilibrium equations are written in terms of components that are along and perpendicular to member CD. This gives

$$\Sigma F_{x'} = -(-4904)\cos\phi - T_{CD} + T_{DE}\cos(\theta + \phi) = 0$$
$$\Sigma F_{y'} = (-4904)\sin\phi - T_{DE}\sin(\theta + \phi) = 0$$

The second of these equations can be solved immediately to get

$$T_{DE} = -3830 \text{ N}$$

Then

$$T_{CD} = 3140 \text{ N}$$

Moving to pin C, the free-body diagram (Fig. 7-14c) is drawn and the equilibrium equations written

$$+\rightarrow\Sigma F = -(2942) - (3140)\cos\phi + (3140)\cos\phi + T_{CE} = 0$$
$$+\uparrow\Sigma F = (3140)\sin\phi - (4905) + (3140)\sin\phi = 0$$

The first of these equations gives

$$T_{CE} = 2942 \text{ N}$$

Since the values of all the forces in the second equation have already been found, this equation reduces to a check of the consistency of the results:

$$(3140)\sin(51.34) - (4905) + (3140)\sin(51.34) = -1.16$$

(The small number -1.16 is due to rounding all the intermediate answers to four significant figures. Keeping more accuracy in the intermediate values would reduce the residual, and so the solution checks.)

Finally, draw the free-body diagram of pin E (Fig. 7-14e) and write the equilibrium equations

$$+\rightarrow\Sigma F = -T_{CE} - T_{DE}\cos\theta = 0$$
$$+\uparrow\Sigma F = T_{DE}\sin\theta + E_y = 0$$

Again, there are no unknowns left to be solved for. These equations are used simply as a check

$$-(2942) - (-3830)\cos(39.81) = -0.10$$
$$(-3830)\sin(39.81) + (2453) = 0.87$$

and again the solution checks.

The required answers are

$$\begin{array}{ll} T_{AB} = T_{DE} = 3830 \text{ N } (C) & \text{Ans.} \\ T_{AC} = T_{CE} = 2942 \text{ N } (T) & \text{Ans.} \\ T_{BC} = T_{CD} = 3140 \text{ N } (T) & \text{Ans.} \\ T_{BD} = 4904 \text{ N } (C) & \text{Ans.} \end{array}$$

The truss shown (Fig. 7-15a) supports one end of a 40-ft wide, 24-ft high outdoor movie screen, which weighs 7000 lb. An identical truss supports the other end of the screen. A 20-mph wind blowing directly at the screen creates a wind loading of 1.042 lb/ft² on the screen. Calculate the maximum tensile and compressive forces in the members of the truss and indicate in which members they occur.

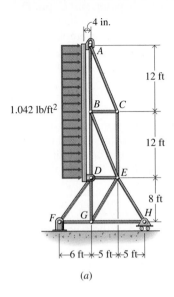

(a)

SOLUTION

Truss $ABCDEFGH$ is a simple truss with $m = 13$ members and $j = 8$ joints. Therefore, the 16 equations obtained from the equilibrium of the eight joints can be solved for the three support reactions as well as the forces in all 13 members.

The first step is to draw a free-body diagram of the entire truss (Fig. 7-15b) to find the support reactions. The distributed wind loading will be replaced with its equivalent force of

$$(1.042 \text{ lb/ft}^2)(960 \text{ ft}^2) = 1000 \text{ lb}$$

applied at the center of the screen. Only half of the weight and distributed force are carried by the truss at each end and are shown on the free-body diagram of Fig. 7-15b. The equilibrium equations are

$$+\rightarrow \Sigma F = 500 + F_x = 0$$
$$+\uparrow \Sigma F = F_y + H - 3500 = 0$$
$$\downarrow + \Sigma M_F = -(6 - 4/12)(3500) - (20)(500) + 16H = 0$$

Solution of these equations gives the support reactions as

$$F_x = -500 \text{ lb} \qquad F_y = 1635 \text{ lb} \qquad H = 1865 \text{ lb}$$

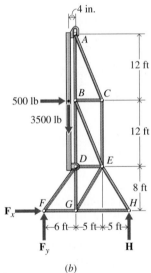

(b)

The next step is to draw a free-body diagram of the screen (Fig. 7-15c) to calculate the forces it exerts at the joints of the truss. Again the distributed wind loading is replaced with its equivalent force applied at the center of the screen. The equations of equilibrium are

$$+\rightarrow \Sigma F = 500 - A - D_x = 0$$
$$+\uparrow \Sigma F = D_y - 3500 = 0$$
$$\downarrow + \Sigma M_D = (4/12)(3500) + 24A - (12)(500) = 0$$

Solution of these equations gives

$$A = 201.4 \text{ lb} \qquad D_x = 298.6 \text{ lb} \qquad D_y = 3500 \text{ lb}$$

Now the joints will be solved in order starting from the top. First draw the free-body diagram of pin A (Fig. 7-16a) and write the equilibrium equations

$$+\rightarrow \Sigma F = A + T_{AC} \sin \theta = 0$$
$$+\uparrow \Sigma F = -T_{AB} - T_{AC} \cos \theta = 0$$

where $A = 201.4$ lb and $\theta = \tan^{-1} 5/12 = 22.62°$. These are solved to get

$$T_{AB} = 483.3 \text{ lb} \qquad \text{and} \qquad T_{AC} = -523.6 \text{ lb}$$

At this point, the free-body diagram of pin B involves three new unknowns and cannot yet be solved. Instead, draw the free-body diagram of pin C (Fig. 7-16c) and solve the equilibrium equations

$$+\rightarrow \Sigma F = -T_{BC} - (-523.6) \sin \theta = 0 \quad T_{BC} = 201.4 \text{ lb}$$
$$+\uparrow \Sigma F = (-523.6) \cos \theta - T_{CE} = 0 \quad T_{CE} = -483.3 \text{ lb}$$

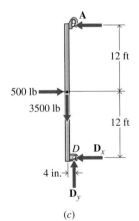

(c)

Fig. 7-15

Now draw the free-body diagram of pin B (Fig. 7-16b) and solve the equilibrium equations

$$+\rightarrow\Sigma F = 201.4 + T_{BE}\sin\theta = 0 \qquad T_{BD} = 966.6\text{ lb}$$
$$+\uparrow\Sigma F = 483.3 - T_{BD} - T_{BE}\cos\theta = 0 \quad T_{BE} = -523.6\text{ lb}$$

At this point, the free-body diagrams of pin D (Fig. 7-16d) and pin E (Fig. 7-16e) each contain three unknowns and are not directly solvable. Instead, draw the free-body diagram of pin F (Fig. 7-16f) and solve the equilibrium equations

$$+\rightarrow\Sigma F = T_{DF}(3/5) + T_{FG} + (-500) = 0 \quad T_{DF} = -2044\text{ lb}$$
$$+\uparrow\Sigma F = T_{DF}(4/5) + 1635 = 0 \qquad T_{FG} = 1726\text{ lb}$$

Now the equilibrium equations for pin D (Fig. 7-16d)

$$+\rightarrow\Sigma F = 298.6 - (-2044)(3/5) + T_{DE} = 0$$
$$+\uparrow\Sigma F = -3500 + 966.6 - (-2044)(4/5) - T_{DG} = 0$$

contain only two member forces that have not yet been found, and these equations can be solved to get

$$T_{DE} = -1525\text{ lb} \qquad\text{and}\qquad T_{DG} = -898.2\text{ lb}$$

Next, draw the free-body diagram of pin E (Fig. 7-16e) and write the equilibrium equations

$$+\rightarrow\Sigma F = -(-1525) - (-523.6)\sin\theta - T_{EG}\cos\phi + T_{EH}\cos\phi = 0$$
$$+\uparrow\Sigma F = (-483.3) + (-523.6)\cos\theta - T_{EG}\sin\phi - T_{EH}\sin\phi = 0$$

where $\phi = \tan^{-1}(8/5) = 58.0°$. These are solved to get

$$T_{EG} = 1059\text{ lb} \qquad\text{and}\qquad T_{EH} = -2199\text{ lb}$$

Then draw the free-body diagram of pin G (Fig. 7-16g) and write the equilibrium equations

$$+\rightarrow\Sigma F = T_{GH} + 1059\cos\phi - 1726 = 0$$
$$+\uparrow\Sigma F = (-898.2) + 1059\sin\phi = 0$$

The first equation gives

$$T_{GH} = 1165\text{ lb}$$

There are no unknowns left in the second equation, however, and it is used to check the consistency of the results

$$(-898.2) + (1059)\sin(58.0) = -0.12$$

The remainder is due to the roundoff of values above.

Finally, draw the free-body diagram of pin H (Fig. 7-16h) and write the equilibrium equations

$$+\rightarrow\Sigma F = -T_{GH} - T_{EH}\cos\phi = 0$$
$$+\uparrow\Sigma F = T_{EH}\sin\phi + H = 0$$

There are no unknowns left in these equations, and they reduce to a check of the results:

$$-(1165) - (-2199)\cos(58.0) = 0.29 \qquad\text{(check)}$$
$$(-2199)\sin(58.0) + 1865 = 0.14 \qquad\text{(check)}$$

A search through the member forces found above gives that the maximum tensile force occurs in member FG whereas the maximum compressive force occurs in member EH:

$$\text{max tensile} = T_{FG} = 1726\text{ lb }(T) \qquad\qquad\text{Ans.}$$
$$\text{max compressive} = T_{EH} = 2200\text{ lb }(C) \qquad\text{Ans.}$$

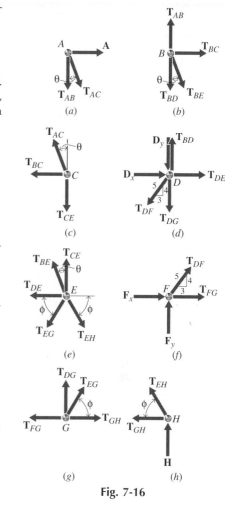

Fig. 7-16

PROBLEMS

7-1 through 7-12 Using the method of joints, determine the force in each member of the truss shown. State whether each member is in tension or compression.

7-1* $a = 20$ ft; $P = 2000$ lb

7-2* $a = 5$ m; $P = 9$ kN

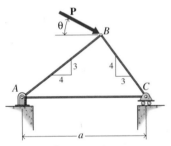

Fig. P7-1, P7-2

7-3 $a = 25$ ft; $P = 2500$ lb; $\theta = 30°$

7-4 $a = 6.25$ m; $P = 1$ kN; $\theta = 75°$

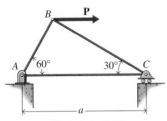

Fig. P7-3, P7-4

7-5* $a = 5$ ft; $P = 600$ lb

7-6 $a = 3$ m; $P = 2.5$ kN

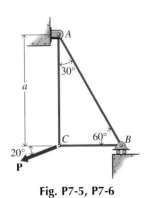

Fig. P7-5, P7-6

7-7 $a = 6$ ft; $P = 1500$ lb

7-8* $a = 2$ m; $P = 5$ kN

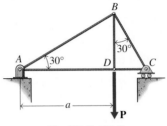

Fig. P7-7, P7-8

7-9 $a = 8$ ft; $P_1 = 800$ lb; $P_2 = 600$ lb

7-10* $a = 3$ m; $P_1 = 4$ kN; $P_2 = 3$ kN

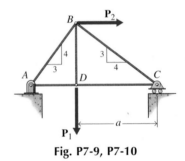

Fig. P7-9, P7-10

7-11* $a = 12$ ft; $P_1 = 600$ lb; $P_2 = 800$ lb

7-12 $a = 4$ m; $P_1 = 3$ kN; $P_2 = 4$ kN

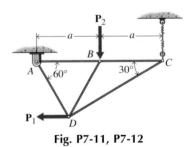

Fig. P7-11, P7-12

7-13 Assume that each member of the truss of Problem 7-9 is made of 4×6 in. lumber having a weight per length of 10 lb/ft. Assume that the weight of each member can be represented as a vertical force, half of which is applied at each end of the member, and determine the force in each member of the truss. Compare the results with the results of Problem 7-9.

7-14* Assume that each member of the truss of Problem 7-12 is made of steel having a mass per length of 6 kg/m. Assume that the weight of each member can be represented as a vertical force, half of which is applied at each end of the member, and determine the force in each member of the truss. Compare the results with the results of Problem 7-12.

7-15* Determine the force in each member of the pair of trusses that are loaded as shown in Fig. P7-15.

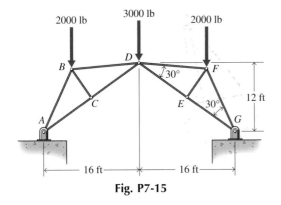

Fig. P7-15

7-16 Determine the force in each member of the truss loaded as shown in Fig. P7-16. All members are 3 m long.

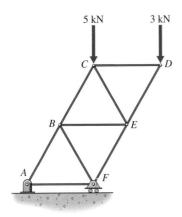

Fig. P7-16

7-17* A 4000-lb crate is attached by light, inextensible cables to the truss of Fig. P7-17. Determine the force in each member of the truss.

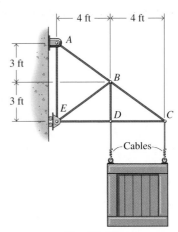

Fig. P7-17

7-18 The cables supporting the crate of Fig. P7-17 are rearranged as shown in Fig. P7-18. If the crate has a mass of 1800 kg, determine the forces in each member of the truss. Compare the ratios of the forces in members *BC* and *CD* to the weight of the crate with the same ratios for Problem 7-17.

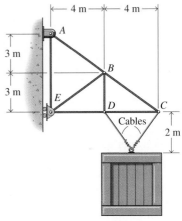

Fig. P7-18

7-19* Determine the forces in members *CG* and *FG* of the inverted Mansard truss of Fig. P7-19.

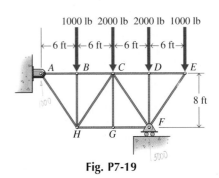

Fig. P7-19

275

7-20* Determine the forces in members *CG* and *FG* of the Warren bridge truss of Fig. P7-20.

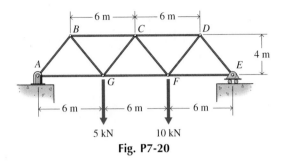

Fig. P7-20

7-21 Assume that each member of the truss of Problem 7-19 is made of timber having a weight per length of 6 lb/ft. Assume that the weight of each member can be represented as a vertical force, half of which is applied at each end of the member, and determine the force in members *CG* and *FG*. Compare the results with the results of Problem 7-19.

7-22* Assume that each member of the truss of Problem 7-20 is made of steel having a mass per length of 6 kg/m. Assume that the weight of each member can be represented as a vertical force, half of which is applied at each end of the member, and determine the force in members *CG* and *FG*. Compare the results with the results of Problem 7-20.

7-23* Snow on a roof supported by the Howe truss of Fig. P7-23 can be approximated as a distributed load of 20 lb/ft (measured along the roof). Treat the distributed load as you would the weight of the members; that is, replace the total load on each of the upper members as a vertical force, half applied to the joint at each end of the member. Determine the force in members *BC*, *BG*, and *CG*.

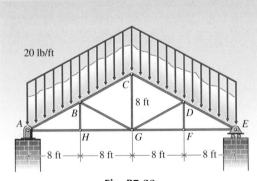

Fig. P7-23

7-24 Snow on a roof supported by the Pratt truss of Fig. P7-24 can be approximated as a distributed load of 250 N/m (measured along the roof). Treat the distributed load as you would the weight of the members; that is, replace the total load on each of the upper members as a vertical force, half applied to the joint at each end of the member. Determine the force in members *BC*, *CH*, and *CG*. Compare the ratios of the forces in *BC*, *CH*, and *CG* to the total load carried by the truss to the same ratios for *BC*, *CH*, and *CG* in Problem 7-23.

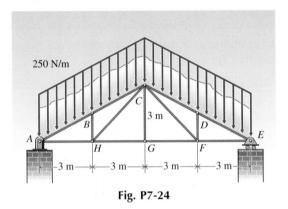

Fig. P7-24

7-25 Determine the forces in members *BC*, *CD*, and *CG* of the cathedral ceiling truss shown in Fig. P7-25.

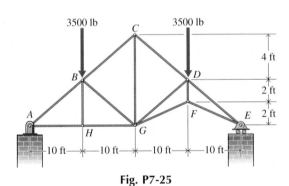

Fig. P7-25

7-26* Determine the forces in members *BC*, *CD*, and *CK* of the vaulted ceiling truss shown in Fig. P7-26.

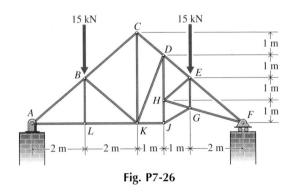

Fig. P7-26

7-27 The Gambrel truss shown in Fig. P7-27 supports one side of a bridge; an identical truss supports the other side. Floor beams carry vehicle loads to the truss joints. Calculate the forces in members BC, BG, and CG when a truck weighing 7500 lb is stopped in the middle of the bridge as shown. The center of gravity of the truck is midway between the front and rear wheels.

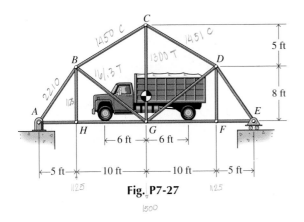

Fig. P7-27

7-28 The Gambrel truss shown in Fig. P7-28 supports one side of a bridge; an identical truss supports the other side. Floor beams carry vehicle loads to the truss joints. Calculate the forces in members BC, BG, and CG when a truck having a mass of 3500 kg is stopped in the middle of the bridge as shown. The center of gravity of the truck is 1 m in front of the rear wheels.

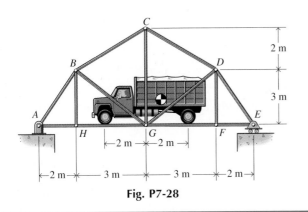

Fig. P7-28

7-29* Calculate the forces in members DE, DF, and EF of the scissors truss shown in Fig. P7-29.

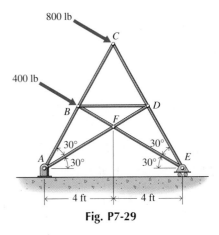

Fig. P7-29

7-30* The flat roof of a building is supported by a series of parallel plane trusses spaced 2 m apart (only one such truss is shown in Fig. P7-30). Calculate the forces in all the members of a typical truss when water collects to a depth of 0.2 m as shown. The density of water is 1000 kg/m^3.

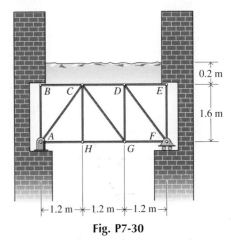

Fig. P7-30

7-31 For the simple three-bar truss of Fig. P7-31 show that the overall equilibrium of the truss is a consequence of the equilibrium of all of the pins; hence the equations of overall equilibrium give no new information. (*Hint:* Write the equations of equilibrium for each of the pins and eliminate the unknown member forces from these equations.)

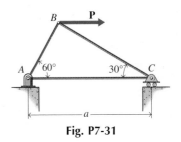

Fig. P7-31

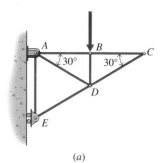

(a)

(b)

Fig. 7-17

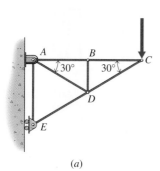

(a)

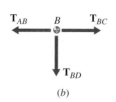

(b)

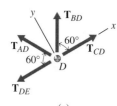

(c)

Fig. 7-18

7-2-2 Zero-force Members

It is often the case that certain members of a given truss carry no load. Zero-force members in a truss usually arise in one of two general ways. The first is:

> When only two members form a noncollinear truss joint and no external load or support reaction is applied to the joint, then the members must be zero-force members.

The truss of Fig. 7-17a is an example of this condition. The free-body diagram of pin C is drawn in Fig. 7-17b. The equations of equilibrium for this joint

$$+\rightarrow\Sigma F = -T_{BC} - T_{CD}\cos 30° = 0$$
$$+\uparrow\Sigma F = -T_{CD}\sin 30° = 0$$

are trivially solved to get

$$T_{CD} = 0 \quad\text{and}\quad T_{BC} = 0$$

That is, for this particular truss and for this particular loading, the two members BC and CD could be removed without affecting the solution or even (in this particular case) the stability of the truss.

The second way in which zero-force members normally arise in a truss is:

> When three members form a truss joint for which two of the members are collinear and the third forms an angle with the first two, then the noncollinear member is a zero-force member provided no external force or support reaction is applied to that joint. The two collinear members carry equal loads (either both tension or both compression).

Such a condition arises, for example, when the load of Fig. 7-17a is moved from pin B to pin C as in Fig. 7-18a. The free-body diagram of pin B is drawn in Fig. 7-18b. The equations of equilibrium for this joint are

$$+\rightarrow\Sigma F = -T_{AB} + T_{BC} = 0$$
$$+\uparrow\Sigma F = -T_{BD} = 0$$

Thus, since joint B is now unloaded, the force in member BD vanishes and the forces in members AB and BC are equal in magnitude—either both tension (both positive) or both compression (both negative).

Once it is known that BD is a zero-force member, the same reasoning can then be used to show that member AD carries no load. The free-body diagram of pin D is drawn in Fig. 7-18c. In order to simplify the calculations, coordinate axes are chosen along and normal to the collinear members CD and DE. The equations of equilibrium are then

$$+\nearrow\Sigma F_x = -T_{DE} - T_{AD}\cos 60° + T_{BD}\cos 60° + T_{CD} = 0$$
$$+\nwarrow\Sigma F_y = T_{AD}\sin 60° + T_{BD}\sin 60° = 0$$

But since $T_{BD} = 0$ (BD is already known to be a zero-force member), then

$$T_{AD} = 0 \quad\text{and}\quad T_{DE} = T_{CD}$$

Thus, for the loading of Fig. 7-18a, both members AD and BD are zero-force members.

These zero-force members cannot simply be removed from the truss and discarded, however. They are needed to guarantee the stabil-

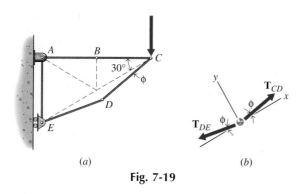

Fig. 7-19

ity of the truss. If members AD and BD were removed, there would be nothing to prevent some small disturbance from moving pin D slightly out of alignment as in Fig. 7-19a. Then the free-body diagram of pin D would look like Fig. 7-19b. Again choosing axes along and normal to the line CE gives the equilibrium equations

$$+\nearrow\Sigma F_x = -T_{DE}\cos\phi + T_{CD}\cos\phi = 0$$
$$+\nwarrow\Sigma F_y = T_{DE}\sin\phi + T_{CD}\sin\phi = 0$$

The first of these equations requires that $T_{CD} = T_{DE}$, whereas the second requires that $T_{CD} = -T_{DE}$. The only way both of these equations can be satisfied is if both forces equal zero. But equilibrium of pin C requires that T_{CD} not be zero. What has happened, of course, is that the truss is no longer in static equilibrium. Pin D will continue to buckle outward and the truss will collapse.

A seemingly trivial solution to the stability problem would be to replace the two members CD and DE with a single member CE and to replace the two members AB and BC with a single member AC. Though this solution would satisfy the statics part of the problem, it would not take care of the tendency for long slender members to buckle when subjected to large compressive loads. Therefore, long members such as member CE of Fig. 7-18 are usually replaced by a pair of shorter members and the midjoint braced if analysis of the truss indicates that the member is likely to be in compression for some expected loading. Long members such as member AC of Fig. 7-17 must also be replaced by a pair of shorter members and the midjoint braced if it is ever desired to load the truss at some point along the long member.

Thus one must not be too quick to discard truss members just because they carry no load for a given configuration. These members are often needed to carry part of the load when the applied loading changes, and they are almost always needed to guarantee the stability of the truss.

Although recognizing these and other special joint-loading conditions can simplify the analysis of a truss, such recognition is not required to solve the truss. If one does not recognize that a member is a zero-force member, drawing the free-body diagram and writing the equilibrium equations will immediately show that it is a zero-force member. Also, these shortcuts should be applied with care. If there is any doubt about whether or not a member is a zero-force member, the prudent choice is to draw the free-body diagram and solve for the member force.

EXAMPLE PROBLEM 7-4

Find all the zero force members in the simple Fink truss of Fig. 7-20a for the loading shown.

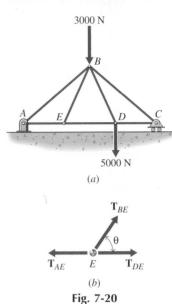

3000 N

5000 N

(a)

SOLUTION

Joint E connects three members, two of which (AE and DE) are collinear. The joint has no loads applied to it. Therefore the force in the noncollinear member (BE) must be zero. This is easily checked by drawing the free-body diagram for pin E (Fig. 7-20b) and writing the equation of equilibrium in the direction perpendicular to the collinear members AE and DE

$$+\uparrow \Sigma F = T_{BE} \sin \theta = 0$$

which gives immediately that

$$T_{BE} = 0$$

Although joint D looks similar to joint E, there is a load applied at this joint, and the foregoing argument does not apply (as is easily checked by drawing the free-body diagram and writing the equations of equilibrium for joint D).

The only zero-force member for the given loading condition is member BE. **Ans.**

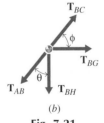

T_{BE}

θ

T_{AE} E T_{DE}

(b)

Fig. 7-20

EXAMPLE PROBLEM 7-5

Identify all the zero-force members in the scissors truss of Fig. 7-21a for the loading shown.

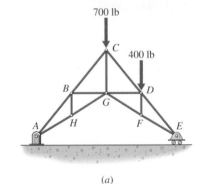

700 lb

400 lb

(a)

SOLUTION

Joint F connects three members, two of which (EF and FG) are collinear. The joint is unloaded and therefore the force in the noncollinear member (DF) must be zero. Similarly, at joint H members AH and GH are collinear and the joint is unloaded; hence the force in member BH is also zero.

Joint B connects two collinear members (AB and BC) and two noncollinear members (BG and BH). The joint is unloaded and the force in one of the noncollinear members (BH) is known to be zero. Therefore the force in the other noncollinear member (BG) must also be zero. Again, this can be easily checked by drawing the free-body diagram for pin B (Fig. 7-21b) and writing the equation of equilibrium in the direction perpendicular to the collinear members AB and BC

$$+\nwarrow \Sigma F = -T_{BH} \sin \theta - T_{BG} \sin \phi = 0$$

which gives

$$T_{BG} = 0$$

T_{BC}

ϕ

T_{BG}

θ

T_{AB} T_{BH}

(b)

Note that the argument applied to joint B does not apply to joint D since joint D has a load applied to it.

Therefore, the zero-force members for the given loading are BG, BH, and DF. **Ans.**

Fig. 7-21

PROBLEMS

7-32 through 7-43   Identify all of the zero-force members for the loadings shown in Figs. P7-32 through P7-43.

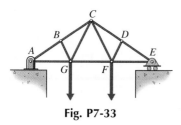

Fig. P7-32

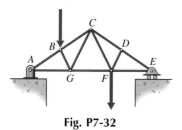

Fig. P7-33

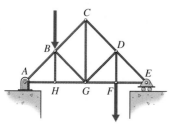

Fig. P7-34

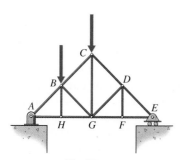

Fig. P7-35

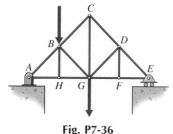

Fig. P7-36

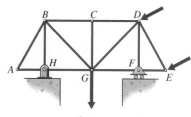

Fig. P7-37

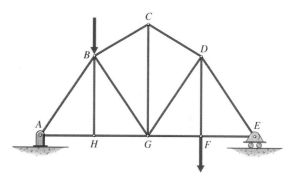

Fig. P7-38

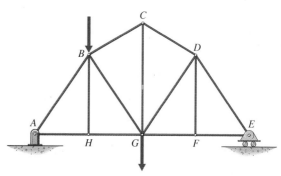

Fig. P7-39

281

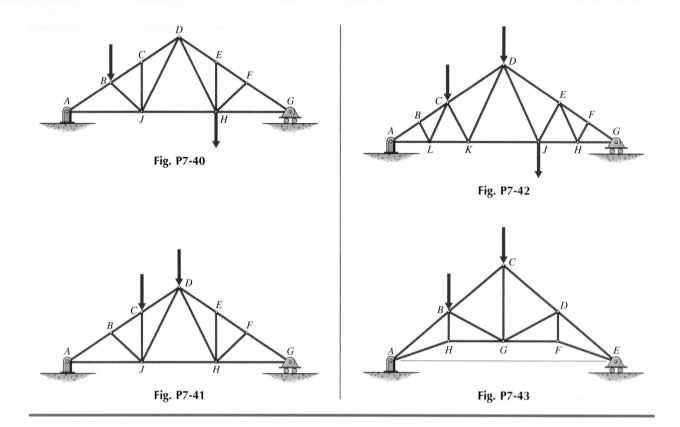

Fig. P7-40

Fig. P7-42

Fig. P7-41

Fig. P7-43

7-2-3 Method of Sections

As stated in the method of joints, if an entire truss is in equilibrium, then each and every part of the truss is also in equilibrium. That does not mean, however, that the truss must be broken up into its most elemental parts—individual members and pins. In the method of sections, the truss will be divided up into just two pieces. Each of these pieces is also a rigid body in equilibrium.

For example, the truss of Fig. 7-22a can be divided into two parts by passing an imaginary section *aa* through some of its members. Of course, the section must pass entirely through the truss so that complete free-body diagrams can be drawn for each of the two pieces. Since the whole truss is in equilibrium, the part of the truss to the left of section *aa* and the part of the truss to the right of section *aa* are both in equilibrium also.

The free-body diagrams of the two parts are drawn in Figs. 7-22b and 7-22c, and include the forces on each cut member that was exerted by the other part of the member, which was cut away. Since the members are all straight two-force members, the forces in these members must act along the members as shown. Forces in members that have not been cut are internal to the rigid bodies and are not shown on the free-body diagrams. Thus, in order to determine the force in member *CF*, the section must cut through that member.

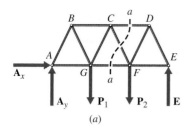
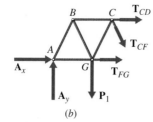
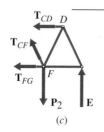

(a)   (b)   (c)

Fig. 7-22

As is the case with any rigid body in plane equilibrium, three independent equations of equilibrium can be written for each rigid body. The resulting six equations are sufficient to solve for the six unknowns—the forces in the three cut members and the three support reactions. As with the method of joints, the solution of the equations can be simplified if the support reactions are determined from overall equilibrium before the truss is sectioned. Then the equations for either part will yield the remaining three unknown forces. In this case, the equilibrium equations for the remaining portion of the truss give no new information; they merely repeat the other equilibrium equations (see Problem 7-75).

If a section cuts through four or more members whose forces are unknown, then the method of sections will not generate enough equations of equilibrium to solve for all the unknown forces. Though it still might be possible to obtain values for one or two of the forces (see Example Problem 7-7), it is usually best to use a section that cuts through no more than three members whose forces are unknown.

It will often happen that a section that cuts no more than three members and that passes through a given member of interest cannot be found. In such a case it may be necessary to draw a section through a nearby member and solve for the forces in it first. Then the method of joints can be used to find the forces in the members next to the cut section, or the truss can be further sectioned to find the force in the member of interest (see Example Problem 7-8).

One of the principle advantages of using the method of sections is that the force in a member near the center of a large truss usually can be determined without first obtaining the forces in the rest of the truss. As a result, the calculation of the force is independent of any errors in other internal forces previously calculated. To find the same force using the method of joints, however, would require that the force in a large number of other members be determined first. Any errors made in the determination of one member force will cause all subsequent forces to be in error as well.

Finally, the method of sections may be used as a spot check when the method of joints or a computer program is used to solve a large truss. Although it is unlikely that a computer will make an error in its computation, it is quite likely that the input data may be in error. Most often these errors occur when an operator incorrectly enters the coordinates of a joint, incorrectly specifies how the joints are connected, or incorrectly applies a load to the truss. In such cases, the method of sections can be used to independently check the forces in one or two interior members.

Use the method of sections to find the forces in members *EF*, *JK*, and *HJ* of the Baltimore truss drawn in Fig. 7-23*a*.

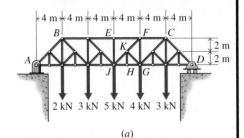

(*a*)

SOLUTION

First, draw the free-body diagram of the entire truss (Fig. 7-23*b*) to solve for the support reactions at *A* and *D*. Summing moments about *A* gives

$$\downarrow+\Sigma M_A = 24D - (4)(2) - (8)(3) - (12)(5) - (16)(4) - (20)(3) = 0$$

which can be solved for *D* to get

$$D = 9 \text{ kN}$$

Then force equilibrium gives

$$+\rightarrow\Sigma F = A_x = 0 \qquad\qquad A_x = 0 \text{ kN}$$
$$+\uparrow\Sigma F = A_y - 2 - 3 - 5 - 4 - 3 + D = 0 \quad A_y = 8 \text{ kN}$$

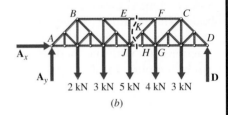

(*b*)

Next, pass a section through members *EF*, *JK*, and *HJ* as shown on Fig. 7-23*b* and draw the free-body diagrams of the resulting two parts (Figs. 7-23*c*, 7-23*d*). The force in member *JK* can be found by summing forces in the vertical direction for either free-body diagram. For example, from Fig. 7-23*c*

$$+\uparrow\Sigma F = 8 - 2 - 3 - 5 + T_{JK} \sin 45° = 0 \quad T_{JK} = 2.828 \text{ kN}$$

while from Fig. 7-23*d*

$$+\uparrow\Sigma F = 9 - 3 - 4 - T_{JK} \sin 45° = 0 \quad T_{JK} = 2.828 \text{ kN}$$

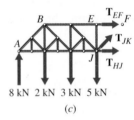

(*c*)

The force in member *HJ* can be found by summing moments about point *F* in Fig. 7-23*d*:

$$\downarrow+\Sigma M_F = (8)(9) - (4)(3) - 4T_{HJ} = 0$$

which gives

$$T_{HJ} = 15 \text{ kN}$$

Note that this is the same result as is obtained by summing moments about the point *F* in Fig. 7-23*c*:

$$\downarrow+\Sigma M_F = (12)(2) + (8)(3) + (4)(5) - (16)(8) + 4T_{HJ} = 0$$

which again gives

$$T_{HJ} = 15 \text{ kN}$$

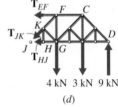

(*d*)

Fig. 7-23

Finally, the force in member *EF* can be found by summing forces in the horizontal direction or by summing moments about *J*. Choosing the latter method and using Fig. 7-23*d* gives

$$\downarrow+\Sigma M_J = (12)(9) + 4T_{EF} - (8)(3) - (4)(4) = 0$$

or

$$T_{EF} = -17 \text{ kN}$$

The desired answers are

$$T_{EF} = 17 \text{ kN } (C) \quad T_{HJ} = 15 \text{ kN } (T) \quad T_{JK} = 2.83 \text{ kN } (T) \qquad \text{Ans.}$$

(Since overall equilibrium was first used to find the support reactions, equilibrium of either Fig. 7-23*c* or Fig. 7-23*d* can be used to solve for the member forces. Usually the part with the fewest forces acting on it will result in the simplest equations of equilibrium.)

Use the method of sections to find the forces in members CD and FG of the truss in Fig. 7-24a.

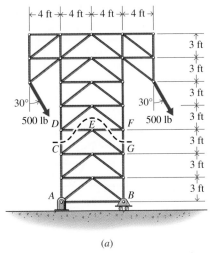

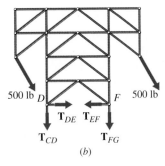

(a)   (b)

Fig. 7-24

SOLUTION

Cut a section through members CD, DE, EF, and FG as shown in Fig. 7-24a and draw the free-body diagram of the upper part of the truss (Fig. 7-24b). Summing moments about D

$$\left\lfloor +\Sigma M_D = (4)(500 \cos 30°) - (6)(500 \sin 30°) - 8T_{FG} \right.$$
$$- (12)(500 \cos 30°) - (6)(500 \sin 30°) = 0$$

gives
$$T_{FG} = -808.0 \text{ lb}$$

Then summing moments about F

$$\left\lfloor +\Sigma M_F = (12)(500 \cos 30°) - (6)(500 \sin 30°) + 8T_{CD} \right.$$
$$- (4)(500 \cos 30°) - (6)(500 \sin 30°) = 0$$

gives
$$T_{CD} = -58.01 \text{ lb}$$

The consistency of these answers can be checked using the vertical component of force equilibrium

$$+\uparrow\Sigma F = -2(500 \cos 30°) - (-808.0) - (-58.01) = -0.02$$

which is within the accuracy of the answers above.
The desired answers are

$$T_{CD} = 58.0 \text{ lb } (C) \qquad \text{Ans.}$$
$$T_{FG} = 808 \text{ lb } (C) \qquad \text{Ans.}$$

(Note that it was not necessary in this problem first to find the support reactions using overall equilibrium. Also note that neither T_{DE} nor T_{EF} can be found from this section. Either additional sections or the method of joints would be needed to find these forces if they needed to be found.)

Find the forces in members BC and BG in the Fink truss of Fig. 7-25a. All triangles are either equilateral or $30°–60°–90°$ right triangles and the loads are all perpendicular to side $ABCD$.

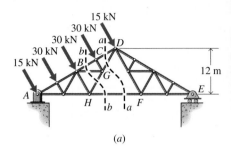

(a)

SOLUTION

First find the support reactions by drawing a free-body diagram of the entire truss (Fig. 7-25b) and writing the equilibrium equations

$$+\rightarrow\Sigma F = A_x + (15 + 30 + 30 + 30 + 15)(\sin 30°) = 0$$
$$+\uparrow\Sigma F = A_y - (15 + 30 + 30 + 30 + 15)(\cos 30°) + E = 0$$
$$\downarrow+\Sigma M_A = 41.57E - (6)(30) - (12)(30) - (18)(30) - (24)(15) = 0$$

which are solved to get

$$A_x = -60.00 \text{ kN} \qquad A_y = 69.28 \text{ kN} \qquad E = 34.64 \text{ kN}$$

A section cut through BC or BG would involve four unknown internal forces. The equilibrium equations cannot be solved until one or more of the forces are determined by some other means. A combination of the method of sections and the method of joints will be used to find the desired forces.

A section will be cut through the middle of the truss near to the members in which the forces are desired. First, cut a section aa (Fig. 7-25a) through members CD, DG, and FH and draw the free-body diagram of the right-hand part of the truss (Fig. 7-25c). Summing moments about point H

$$\downarrow+\Sigma M_H = (27.72)(34.64) - (13.86 \cos 30°)(15)$$
$$+ (13.86)(T_{CD} \sin 30°) = 0$$

gives

$$T_{CD} = -112.58 \text{ kN}$$

Next, draw a free-body diagram of pin C (Fig. 7-25d) and write the equilibrium equations along and perpendicular to members BC and CD:

$$\nearrow\Sigma F = T_{CD} - T_{BC} = 0$$
$$\nwarrow\Sigma F = -30 - T_{CG} = 0$$

which are immediately solved to get

$$T_{BC} = T_{CD} = -112.58 \text{ kN} \qquad \text{and} \qquad T_{CG} = -30.00 \text{ kN}$$

Finally, cut a section bb (Fig. 7-25a) through members BC, BG, GH, and FH and draw the free-body diagram of the left-hand part of the truss (Fig. 7-25e). Again summing moments about point H

$$\downarrow+\Sigma M_H = -(13.86)(69.28) + (12)(15) + (6)(30)$$
$$- (13.86 \sin 30°)T_{BC} - 6T_{BG} = 0$$

gives

$$T_{BG} = 29.99 \text{ kN}$$

Thus, the desired answers are

$$T_{BC} = 112.6 \text{ kN (C)} \qquad\qquad \text{Ans.}$$
$$T_{BG} = 30.0 \text{ kN (T)} \qquad\qquad \text{Ans.}$$

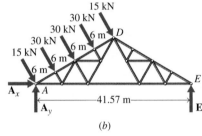

(b)

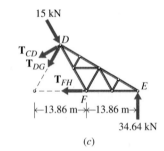

(c)

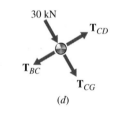

(d)

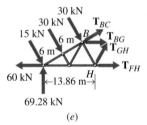

(e)

Fig. 7-25

PROBLEMS

Solve the following problems by the method of sections. Unless directed otherwise, neglect the weight of the members compared with the forces they support. Be sure to indicate whether the members are in tension or compression.

7-44* Find the forces in members *EJ* and *HJ* of the roof truss shown in Fig. P7-44.

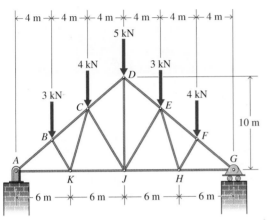

Fig. P7-44

7-45* Each truss member in Fig. P7-45 has length 5 ft. Find the forces in members *CD* and *EF*.

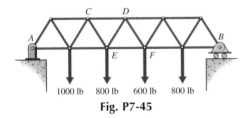

Fig. P7-45

7-46 Determine the forces in members *CD* and *EF* of the truss (Fig. P7-46), which serves to support the deck of a bridge.

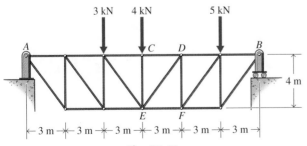

Fig. P7-46

7-47 The Howe roof truss of Fig. P7-47 supports the vertical loading shown. Determine the forces in members *CD* and *CK*.

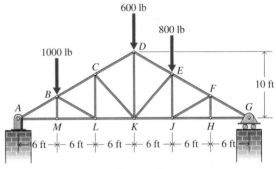

Fig. P7-47

7-48* Determine the forces in members *BC*, *BG*, and *GH* of the bridge truss shown in Fig. P7-48.

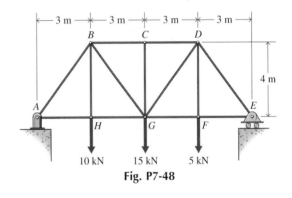

Fig. P7-48

7-49 Determine the forces in members *CD*, *DF*, and *EF* of the bridge truss shown in Fig. P7-49.

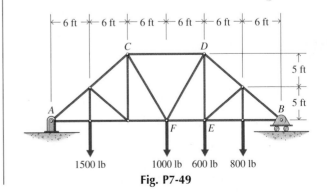

Fig. P7-49

287

7-50 Determine the forces in members CD, DE, and DF of the roof truss shown in Fig. P7-50. Triangle CDF is an equilateral triangle and joints E and G are at the midpoints of their respective sides.

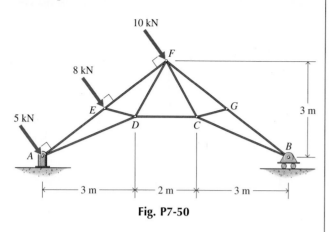

Fig. P7-50

7-51* The roof truss of Fig. P7-51 is composed of 30°–60°–90° right triangles and is loaded as shown. Determine the forces in members CD, CE, and EF.

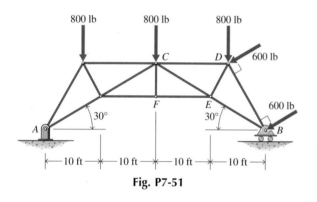

Fig. P7-51

7-52 The sign-board truss of Fig. P7-52 supports the load shown. Determine the forces in members CD, CE, and EF.

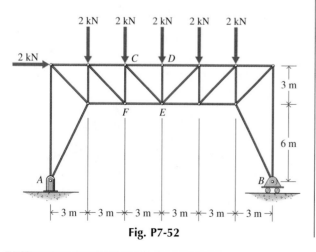

Fig. P7-52

7-53 The sign truss in Fig. P7-53 supports a sign that weighs 300 lb. The sign is connected to the truss at joints E, G, and H, and the connecting links are adjusted so that each joint carries one-third of the load. Determine the forces in members CD, CF, and FG.

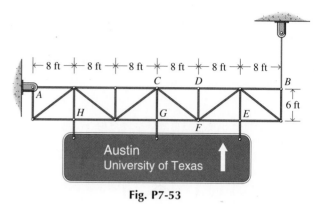

Fig. P7-53

7-54* A scissors truss (Fig. P7-54) is used to support a roof load. Determine the forces in members BC and BF.

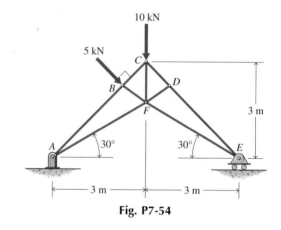

Fig. P7-54

7-55* For the compound truss of Fig. P7-55, find the forces in members CD and CE.

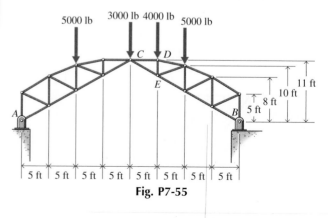

Fig. P7-55

288

7-56 For the compound truss of Fig. P7-56, find the forces in members *CD* and *DE*.

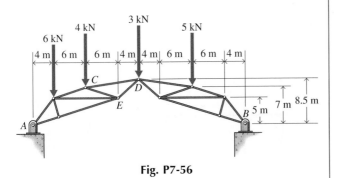

Fig. P7-56

7-57* Find the forces in members *BC* and *FE* of the stairs truss of Fig. P7-57.

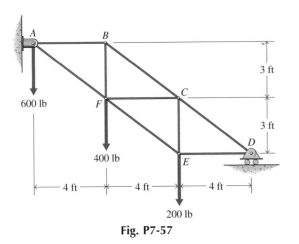

Fig. P7-57

7-58 Find the forces in members *CD* and *CG* of the stairs truss of Fig. P7-58.

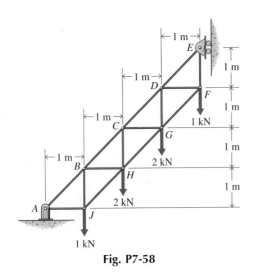

Fig. P7-58

7-59 Find the forces in members *DE*, *DJ*, and *JK* of the truss of Fig. P7-59.

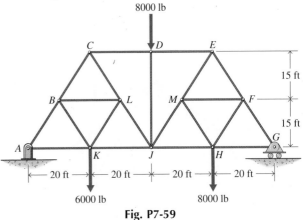

Fig. P7-59

7-60* Find the forces in members *CD*, *CE*, and *FG* of the Fink truss of Fig. P7-60.

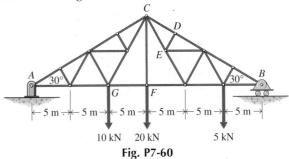

Fig. P7-60

7-61 Find the forces in members *CD*, *DH*, and *FH* of the Baltimore truss of Fig. P7-61.

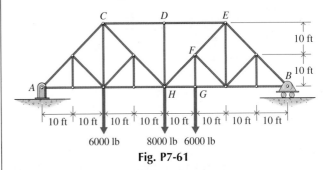

Fig. P7-61

7-62* The inverted bridge truss of Fig. P7-62 supports a roadway and vehicles giving the loading shown. Find the forces in members *CD*, *DG*, and *EG*.

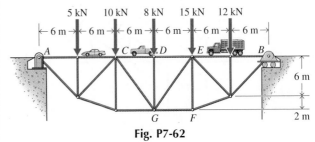

Fig. P7-62

7-63* Find the forces in members *EG*, *FG*, and *FH* in the truss of Fig. P7-63.

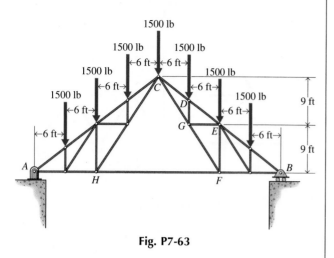

Fig. P7-63

7-64 Find the forces in members *AB* and *FG* of the truss shown in Fig. P7-64. (*Hint:* Use section *aa.*)

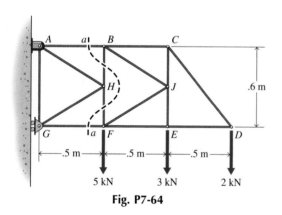

Fig. P7-64

7-65 Find the forces in members *CD* and *FG* of the K-truss shown in Fig. P7-65. (*Hint:* Use section *aa.*)

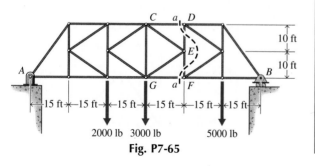

Fig. P7-65

7-66 The homogeneous 3-kN sign is attached by means of short links to the signboard truss of Fig. P7-66. Find the forces in members *CD*, *CE*, *EG*, and *FG*.

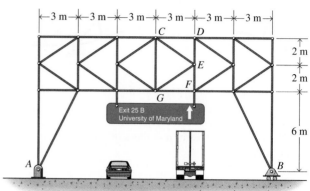

Fig. P7-66

7-67* Find the forces in members *CD*, *DG*, and *EG* of the transmission line truss shown in Fig. P7-67.

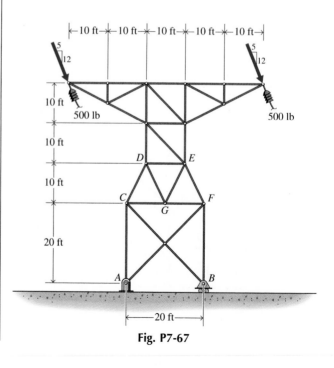

Fig. P7-67

7-68 Find the forces in members *CD, DF,* and *EF* of the transmission line truss shown in Fig. P7-68.

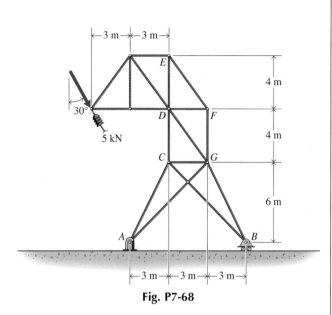

Fig. P7-68

7-69 Find the forces in members *CD, FG,* and *FH* of the truss shown in Fig. P7-69.

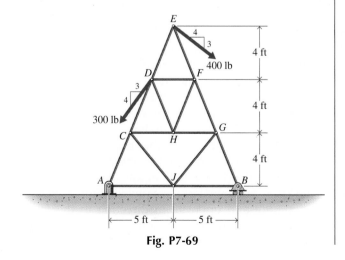

Fig. P7-69

7-70* Find the forces in members *CH, DF,* and *EF* of the truss shown in Fig. P7-70.

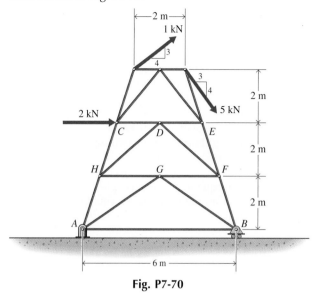

Fig. P7-70

7-71* Find the maximum load **P** that can be supported by the truss of Fig. P7-71 without producing a force of more than 2500 lbs in member *CD*.

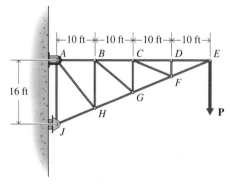

Fig. P7-71

7-72 Find the maximum load **P** that can be supported by the truss of Fig. P7-72 without producing a force of more than 3 kN in member *EJ*.

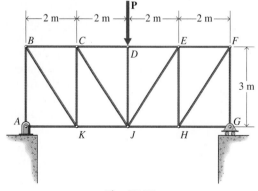

Fig. P7-72

7-73 The members that make up the truss of Fig. P7-71 cannot withstand tensile loads greater than 50,000 lb nor compressive loads greater than 35,000 lb. Find the maximum load **P** that can be supported.

7-74* The members that make up the truss of Fig. P7-72 cannot withstand tensile loads greater than 80 kN nor compressive loads greater than 50 kN. Find the maximum load **P** that can be supported.

7-75 Show that the overall equilibrium of a truss is a consequence of the equilibrium of the two separate parts generated by the method of sections. That is, section the bridge truss of Fig. P7-75 down the middle as indicated, and write the equilibrium equations for each piece. Eliminate the member forces from the resulting six equations and show that the result is equivalent to the equilibrium equations for the whole truss.

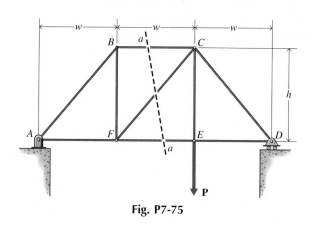

Fig. P7-75

7-2-4 Forces in Straight and Curved Two-Force Members

All the truss members used in this section have been straight. As a result, the resultants of the forces at the ends of the members have always acted along the axis of the members. When the forces are directed away from the ends of the member as in Fig. 7-26, the member is said to be in tension and the force **T** is referred to as the tension force *in the member* and not just the tension force *on the end of the member*. That this description of the force is correct is easily shown by considering the equilibrium of a portion of the truss member.

Suppose that the truss member of Fig. 7-26 is cut normal to its axis at section *aa*. In general, there will be some complex force distribution acting on the cut surface as shown on the free-body diagram of Fig. 7-27a. However, according to the discussion in Section 4-6, this system of forces can be replaced with an equivalent force-couple as shown on the free-body diagram of Fig. 7-27b. Then equilibrium of forces in the direction perpendicular to the axis of the member requires that **V**, the shear component of the equivalent force-couple, must be zero. Similarly, equilibrium of forces in the direction along the axis of the member requires that **P**, the axial component of the equivalent force-couple, must be equal in magnitude and opposite in direction to the force applied to the end of the member. Finally, moment equilibrium requires that **M**, the couple component of the equivalent force-couple,

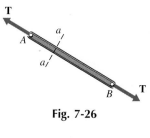

Fig. 7-26

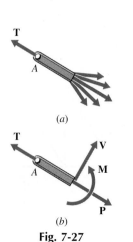

(a)

(b)

Fig. 7-27

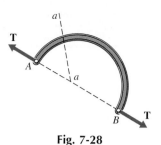

Fig. 7-28

must be zero. That is, *if the forces at the ends of a straight two-force member are pulling on the member, then the forces on any cut section of the member also represent an axial force pulling on the cut section of the member,* regardless of where the member is cut. Therefore, it is proper to talk about the force **T** as the tension *in the member.*

When a two-force member is curved, however, the forces at the ends of the member do not act along the axis of the member. Instead, the forces act along the line joining the points where the forces are applied as shown in Fig. 7-28. If the member is cut normal to its axis at section *aa*, there will again be a complex force distribution acting on the cut surface, as shown in Fig. 7-29*a*. Of course, this system of forces can still be replaced with an equivalent force-couple as shown on the free-body diagram of Fig. 7-29*b*. However, force equilibrium now requires that the resultant **R** of the axial **P** and shear **V** components of the equivalent force-couple be equal in magnitude and opposite in direction to the force **T** at the end of the member, as shown in Fig. 7-29*c*. Since the forces **R** and **T** are not collinear, moment equilibrium now requires that $M = Td \neq 0$.

It should be noted at this point that neither of the free-body diagrams shown in Fig. 7-29 represent two-force bodies even though they represent a part of a two-force body. The free-body diagram of Fig. 7-29*a* is not a two-force body since the forces are distributed over the entire cut surface and hence do not all act at a single point. The free-body diagram of Fig. 7-29*b* is not a two-force body since a couple acts on the body.

Therefore, *the design of straight two-force members need only consider axial forces,* but *curved two-force members must be designed to withstand shearing forces* **V** *and bending moments* **M** *as well as axial forces* **P**. To further complicate the problem, the strengths of the shear forces, bending moments, and axial forces depend on where the member is cut, as shown in Example Problem 7-9.

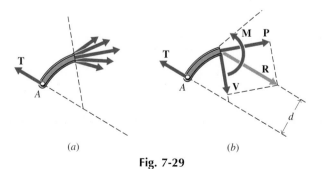

(a)   (b)

Fig. 7-29

EXAMPLE PROBLEM 7-9

An arch support shown in Fig. 7-30a consists of two quarter circle members. Determine the axial force P, shear force V, and bending moment M in member AB as a function of the angle θ.

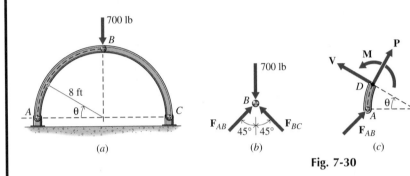

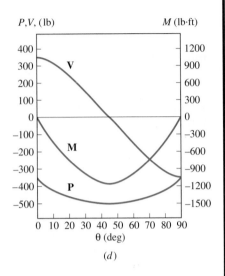

Fig. 7-30

SOLUTION

First draw the free-body diagram of pin B (Fig. 7-30b). Since members AB and BC are two-force members, the forces $\mathbf{F}_{AB}$ and $\mathbf{F}_{BC}$ act at a 45° angle to the vertical. Then the equilibrium equations

$$+\rightarrow \Sigma F = F_{AB} \sin 45° - F_{BC} \sin 45° = 0$$
$$+\uparrow \Sigma F = F_{AB} \cos 45° + F_{BC} \cos 45° - 700 = 0$$

give the forces in the members AB and BC

$$F_{AB} = F_{BC} = 495.0 \text{ lb}$$

Next, draw a free-body diagram of a section of member AB (Fig. 7-30c). Summing forces horizontally and vertically and summing moments about D gives

$$+\rightarrow \Sigma F = F_{AB} \sin 45° + P \sin \theta - V \cos \theta = 0$$
$$+\uparrow \Sigma F = F_{AB} \cos 45° + P \cos \theta + V \sin \theta = 0$$
$$\downarrow + \Sigma M_D = M + (8 \sin \theta)(F_{AB} \cos 45°)$$
$$- (8 - 8 \cos \theta)(F_{BC} \sin 45°) = 0$$

or

$$P = -350(\cos \theta + \sin \theta) \text{ lb} \qquad \text{Ans.}$$
$$V = 350(\cos \theta - \sin \theta) \text{ lb} \qquad \text{Ans.}$$
$$M = 2800(1 - \sin \theta - \cos \theta) \text{ ft} \cdot \text{lb} \qquad \text{Ans.}$$

These results are plotted in Fig. 7-30d.

7-3 SPACE TRUSSES

A truss whose joints do not all lie in a plane and/or that is loaded and supported out of its plane is called a **space truss.** As with a planar truss, the members of a space truss may be treated as two-force members provided that no member is continuous through a joint, the external loading is applied only at the joints, and the joints consist of frictionless ball and socket connections. As with plane trusses, when the weight of the members is to be included, half of the weight of each member is applied to the joints at each end of the member.

The three-dimensional equivalent of a triangle is a tetrahedron, as shown in Fig. 7-31a. Space trusses are constructed from tetrahedral subunits. By analogy with a simple plane truss, a **simple space truss** is formed by adding tetrahedron units to the truss as shown in Fig. 7-31b. Since now each new joint brings three new members with it, the relationship between the number of joints j and the number of members m in a simple space truss is given by

$$m = 3j - 6 \qquad (7\text{-}3)$$

Clearly, simple space trusses, like simple plane trusses, are always rigid.

As in the case of plane trusses, space trusses can be analyzed using either the method of joints or the method of sections. For the method of joints, the truss is divided up into all its elemental members and joints. Also as in the case of plane trusses, equilibrium of the members is guaranteed by the assumption that they are two-force members. Therefore, the members can again be discarded and only the free-body diagrams of the pins need be considered. Equilibrium of the pins is expressed by writing the force equilibrium equation

$$\Sigma \mathbf{F} = \mathbf{0} \qquad (7\text{-}4)$$

for each joint. Since each joint consists of a system of concurrent forces in three dimensions, moment equilibrium gives no useful information, but now Eq. 7-4 has three independent components. But according to Eq. 7-3, this is precisely the number of independent equations needed to solve for the m member forces and six support reactions of a simple space truss. Aside from the larger number of equations and unknowns, the solution procedure is identical to that of a planar truss.

The procedure for the method of sections is also essentially identical to that of a planar truss. The space truss is divided in two by passing a section completely through the truss and drawing free-body diagrams of each part. Application of the equations of equilibrium

$$\Sigma \mathbf{F} = \mathbf{0} \qquad \text{and} \qquad \Sigma \mathbf{M} = \mathbf{0} \qquad (7\text{-}5)$$

to these two parts yield a total of 12 equations—six for each part. These 12 equations are sufficient to determine the six support reactions and six internal member forces. However, it is usually difficult to pass a section through a typical space truss that cuts no more than six members. For this reason and because of the complexity of solving the large numbers of equations, the method of sections is not often used for space trusses.

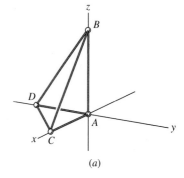

(a)

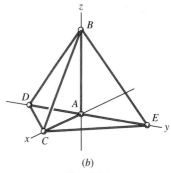

(b)

Fig. 7-31

The simple space truss of Fig. 7-32a is supported by a ball and socket joint at E and by short links at A, D, and C. Determine the forces in each of the members.

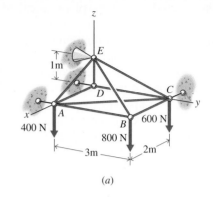

(a)

SOLUTION

As a first step, draw the free-body diagram of the entire truss (Fig. 7-32b) and solve for the support reactions. Moment equilibrium about point E gives

$$\Sigma M_E = (2\mathbf{i} - \mathbf{k}) \times (A\mathbf{j} - 400\mathbf{k}) + (2\mathbf{i} + 3\mathbf{j} - \mathbf{k}) \times (-800\mathbf{k})$$
$$+ (3\mathbf{j} - \mathbf{k}) \times (C\mathbf{i} - 600\mathbf{k}) + (-\mathbf{k}) \times (D\mathbf{j})$$
$$= 2A\mathbf{k} + 800\mathbf{j} + A\mathbf{i} + 1600\mathbf{j} - 2400\mathbf{i} - 3C\mathbf{k}$$
$$- 1800\mathbf{i} - C\mathbf{j} + D\mathbf{i}$$
$$= \mathbf{0}$$

The x-, y-, and z-components of this equation give

$$\begin{array}{lll} \mathbf{j}: & C = 2400 & C = 2400 \text{ N} \\ \mathbf{k}: & 2A - 3C = 0 & A = 3600 \text{ N} \\ \mathbf{i}: & A + D = 4200 & D = 600 \text{ N} \end{array}$$

Force equilibrium in the x-, y-, and z-directions gives

$$\begin{array}{lll} \Sigma F_x = E_x + C = 0 & E_x = -2400 \text{ N} \\ \Sigma F_y = E_y + A + D = 0 & E_y = -4200 \text{ N} \\ \Sigma F_z = E_z - 400 - 600 - 800 = 0 & E_z = 1800 \text{ N} \end{array}$$

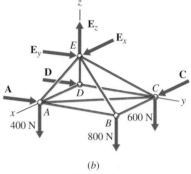

(b)

Fig. 7-32

The free-body diagram for pin A (Fig. 7-33a) still has too many unknowns to solve, and so pin B will be considered first. Forces are shown on the free-body diagram (Fig. 7-33b) with an arrow to symbolically show the direction and a value for the magnitude. The directions of the forces are along the members; therefore the force in member BE applied at pin B is

$$\mathbf{F}_{BE} = T_{BE} \frac{-2\mathbf{i} - 3\mathbf{j} + \mathbf{k}}{\sqrt{2^2 + 3^2 + 1^2}} = -0.5345 T_{BE}\mathbf{i} - 0.8018 T_{BE}\mathbf{j} + 0.2673 T_{BE}\mathbf{k}$$

Similarly, $\mathbf{F}_{AB}$, $\mathbf{F}_{BC}$, and the applied load $\mathbf{P}$ are

$$\mathbf{F}_{AB} = -T_{AB}\mathbf{j} \qquad \mathbf{F}_{BC} = -T_{BC}\mathbf{i} \qquad \mathbf{P} = -800\mathbf{k} \text{ N}$$

The x-, y-, and z-components of the equilibrium equation $\Sigma \mathbf{F} = \mathbf{0}$ for this joint then become

$$\begin{array}{llll} \mathbf{k}: & 0.2673 T_{BE} = 800 & T_{BE} = 2993 \text{ N} \\ \mathbf{i}: & -T_{BC} - 0.5345 T_{BE} = 0 & T_{BC} = -1600 \text{ N} \\ \mathbf{j}: & -T_{AB} - 0.8018 T_{BE} = 0 & T_{AB} = -2400 \text{ N} \end{array}$$

Now the member forces acting on pin A (Fig. 7-33a) are

$$\mathbf{F}_{AB} = T_{AB}\mathbf{j} = -2400\mathbf{j} \text{ N}$$
$$\mathbf{F}_{AC} = T_{AC} \frac{-2\mathbf{i} + 3\mathbf{j}}{\sqrt{2^2 + 3^2}} = -0.5547 T_{AC}\mathbf{i} + 0.8321 T_{AC}\mathbf{j}$$
$$\mathbf{F}_{AD} = -T_{AD}\mathbf{i}$$
$$\mathbf{F}_{AE} = T_{AE} \frac{-2\mathbf{i} + \mathbf{k}}{\sqrt{2^2 + 1^2}} = -0.8944 T_{AE}\mathbf{i} + 0.4472 T_{AE}\mathbf{k}$$

the support reaction and applied load are

$$\mathbf{A} = 3600\mathbf{j} \text{ N} \qquad \mathbf{P} = -400\mathbf{k} \text{ N}$$

Putting these into the force equilibrium equation gives

$$
\begin{array}{lll}
\mathbf{j}: & 0.8321T_{AC} = -1200 & T_{AC} = -1442.2 \text{ N} \\
\mathbf{k}: & 0.4472T_{AE} = 400 & T_{AE} = 894.5 \text{ N} \\
\mathbf{i}: & -0.5547T_{AC} - T_{AD} - 0.8944T_{AE} = 0 & T_{AD} = 0 \text{ N}
\end{array}
$$

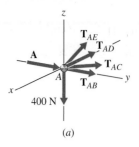

(a)

Next draw the free-body diagram of pin C (Fig. 7-33c). The force in member AC applied at pin C, $\mathbf{F}_{CA}$, is equal in magnitude but opposite in direction to the force in the same member applied at pin A, $\mathbf{F}_{AC}$, therefore

$$\mathbf{F}_{CA} = -\mathbf{F}_{AC} = 0.5547T_{AC}\mathbf{i} - 0.8321T_{AC}\mathbf{j} = -800\mathbf{i} + 1200\mathbf{j} \text{ N}$$

The rest of the member forces are again along the members,

$$\mathbf{F}_{CB} = T_{BC}\mathbf{i} = -1600\mathbf{i} \text{ N} \qquad \mathbf{F}_{CD} = -T_{CD}\mathbf{j}$$
$$\mathbf{F}_{CE} = T_{CE}\frac{-3\mathbf{j} + \mathbf{k}}{\sqrt{3^2 + 1^2}} = -0.9487T_{CE}\mathbf{j} + 0.3162T_{CE}\mathbf{k}$$

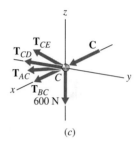

(b)

while the support reaction and applied load are

$$\mathbf{C} = C\mathbf{i} = 2400\mathbf{i} \text{ N} \qquad \mathbf{P} = -600\mathbf{k} \text{ N}$$

Substitution of these forces in the equilibrium equation yields

$$
\begin{array}{lc}
\mathbf{i}: & -1600 - 800 + 2400 = 0 \\
\mathbf{j}: & -T_{CD} - 0.9487T_{CE} = -1200 \\
\mathbf{k}: & 0.3162T_{CE} = 600
\end{array}
$$

The first equation has no unknowns and is just used as a check of the consistency of the answers. The remaining two equations give

$$T_{CD} = -600.2 \text{ N} \qquad T_{CE} = 1897.5 \text{ N}$$

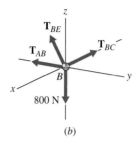

(c)

Next, consider pin D whose free-body diagram is drawn in Fig. 7-33d. The forces applied to this pin all act along the coordinate axes, and so force equilibrium is easily written

$$T_{AD} = 0 \text{ N} \qquad T_{CD} = -600 \text{ N} \qquad T_{DE} = 0 \text{ N}$$

The first two equations just repeat information already known. The third says that member DE is a zero-force member.

Finally, draw the free-body diagram of pin E (Fig. 7-33e). The member forces and support reaction acting on this pin are

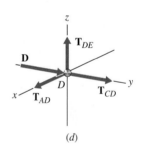

(d)

$$
\begin{array}{l}
\mathbf{F}_{EA} = -\mathbf{F}_{AE} = 0.8944T_{AE}\mathbf{i} - 0.4472T_{AE}\mathbf{k} = 800\mathbf{i} - 400\mathbf{k} \text{ N} \\
\mathbf{F}_{EB} = -\mathbf{F}_{BE} = 0.5345T_{BE}\mathbf{i} + 0.8018T_{BE}\mathbf{j} - 0.2673T_{BE}\mathbf{k} = 1600\mathbf{i} + 2400\mathbf{j} - 800\mathbf{k} \text{ N} \\
\mathbf{F}_{EC} = -\mathbf{F}_{CE} = 0.9487T_{CE}\mathbf{j} - 0.3162T_{CE}\mathbf{k} = 1800\mathbf{j} - 600\mathbf{k} \text{ N} \\
\mathbf{F}_{ED} = -T_{DE}\mathbf{k} = \mathbf{0} \text{ N} \\
\mathbf{E} = -2400\mathbf{i} - 4200\mathbf{j} + 1800\mathbf{k} \text{ N}
\end{array}
$$

There are no unknowns left in these forces and substitution into the force equilibrium equation just serves to check the consistency of the results

$$
\begin{array}{lc}
\mathbf{i}: & 800 + 1600 - 2400 = 0 \\
\mathbf{j}: & 2400 + 1800 - 4200 = 0 \\
\mathbf{k}: & -400 - 800 - 600 + 1800 = 0
\end{array}
$$

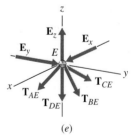

(e)

Fig. 7-33

The required answers are then

$$
\begin{array}{lll}
T_{AB} = 2400 \text{ N } (C) & T_{AC} = 1442 \text{ N } (C) & \text{Ans.} \\
T_{AD} = T_{DE} = 0 \text{ N} & T_{AE} = 894 \text{ N } (T) & \text{Ans.} \\
T_{BC} = 1600 \text{ N } (C) & T_{BE} = 2993 \text{ N } (T) & \text{Ans.} \\
T_{CD} = 600 \text{ N } (C) & T_{CE} = 1898 \text{ N } (T) & \text{Ans.}
\end{array}
$$

The space truss of Fig. 7-34 is supported by a ball and socket joint at A and by short links at B and C. A 125-lb force acts in the y-z plane at D. Determine the support reactions and forces in each of the members.

SOLUTION

The solution will be obtained using the method of joints. Since this truss is so simple, the support reactions will be found as part of the solution by joints rather than determined ahead of time using overall equilibrium.

The solution can start at pin D since there is a known force and no more than three unknown forces applied to this pin (Fig. 7-35d). The forces are shown on the free-body diagram with an arrow to symbolically show the direction and a value for the magnitude. The directions of the forces are along the members; hence the force in member CD applied at pin D is

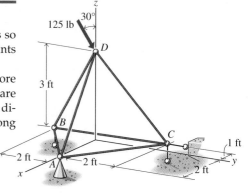

Fig. 7-34

$$\mathbf{F}_{DC} = T_{CD} \frac{-\mathbf{i} + 2\mathbf{j} - 3\mathbf{k}}{\sqrt{1^2 + 2^2 + 3^2}}$$
$$= -0.2673T_{CD}\mathbf{i} + 0.5345T_{CD}\mathbf{j} - 0.8018T_{CD}\mathbf{k}$$

Similarly for the other forces,

$$\mathbf{F}_{DB} = T_{BD} \frac{-\mathbf{i} - 2\mathbf{j} - 3\mathbf{k}}{\sqrt{1^2 + 2^2 + 3^2}}$$
$$= -0.2673T_{BD}\mathbf{i} - 0.5345T_{BD}\mathbf{j} - 0.8018T_{BD}\mathbf{k}$$
$$\mathbf{F}_{DA} = T_{AD} \frac{-2\mathbf{i} - 3\mathbf{k}}{\sqrt{2^2 + 3^2}} = 0.5547T_{AD}\mathbf{i} - 0.8321T_{AD}\mathbf{k}$$

Finally the applied load is

$$\mathbf{P} = 125.0 \sin 30\mathbf{j} - 125.0 \cos 30\mathbf{k} \text{ lb}$$
$$= 62.50\mathbf{j} - 108.25\mathbf{k} \text{ lb}$$

The x-, y-, and z-components of the equilibrium equation $\Sigma\mathbf{F} = \mathbf{0}$ for this joint then become

$$\mathbf{i}: \qquad -0.2673T_{CD} - 0.2673T_{BD} + 0.5547T_{AD} = 0$$
$$\mathbf{j}: \qquad 0.5345T_{CD} - 0.5345T_{BD} + 62.50 = 0$$
$$\mathbf{k}: \quad -0.8018T_{CD} - 0.8018T_{BD} - 0.8321T_{AD} - 108.25 = 0$$

respectively. These equations are easily solved to get

$$T_{AD} = -43.37 \text{ lb} \qquad T_{BD} = 13.466 \text{ lb} \qquad T_{CD} = -103.47 \text{ lb}$$

Next draw the free-body diagram of pin B (Fig. 7-35b). The force in member BD applied at pin B, $\mathbf{F}_{BD}$, is equal in magnitude but opposite in direction to the force in the same member applied at pin D, $\mathbf{F}_{DB}$; hence

$$\mathbf{F}_{BD} = 0.2673T_{BD}\mathbf{i} + 0.5345T_{BD}\mathbf{j} + 0.8018T_{BD}\mathbf{k}$$
$$= 3.599\mathbf{i} + 7.198\mathbf{j} + 10.797\mathbf{k} \text{ lb}$$

The other forces are again along the members; hence

$$\mathbf{F}_{BC} = T_{BC}\mathbf{j}$$
$$\mathbf{F}_{BA} = T_{AB} \frac{3\mathbf{i} + 2\mathbf{j}}{\sqrt{3^2 + 2^2}} = 0.8321T_{AB}\mathbf{i} + 0.5547T_{AB}\mathbf{j}$$
$$\mathbf{B} = B\mathbf{k}$$

Thus, the equilibrium equations for this joint are

$$\begin{array}{ll} \mathbf{i}: & 3.599 + 0.8321T_{AB} = 0 \\ \mathbf{j}: & 7.198 + T_{BC} + 0.5547T_{AB} = 0 \\ \mathbf{k}: & 10.797 + B = 0 \end{array}$$

which are solved to get

$$T_{BC} = -4.799\text{ lb} \qquad T_{AB} = -4.325\text{ lb} \qquad B = -10.797\text{ lb}$$

Next draw the free-body diagram of pin C (Fig. 7-35c). The forces applied at pin C are

$$\begin{aligned} \mathbf{F}_{CD} &= 0.2673T_{CD}\mathbf{i} - 0.5345T_{CD}\mathbf{j} + 0.8018T_{CD}\mathbf{k} \\ &= -27.66\mathbf{i} + 55.30\mathbf{j} - 82.96\mathbf{k}\text{ lb} \\ \mathbf{F}_{CB} &= -T_{BC}\mathbf{j} = 4.799\mathbf{j}\text{ lb} \\ \mathbf{F}_{CA} &= T_{AC}\frac{3\mathbf{i} - 2\mathbf{j}}{\sqrt{3^2 + 2^2}} = 0.8321T_{AC}\mathbf{i} - 0.5547T_{AC}\mathbf{j} \\ &= 0.8321T_{AC}\mathbf{i} - 0.5547T_{AC}\mathbf{j} \\ \mathbf{C} &= -C_y\mathbf{j} + C_z\mathbf{k} \end{aligned}$$

Putting these forces in the equilibrium equation produces

$$\begin{array}{ll} \mathbf{i}: & -27.66 + 0.8321T_{AC} = 0 \\ \mathbf{j}: & 55.30 + 4.799 - 0.5547T_{AC} - C_y = 0 \\ \mathbf{k}: & -82.96 + C_z = 0 \end{array}$$

which gives

$$C_y = 41.66\text{ lb} \qquad C_z = 82.96\text{ lb} \qquad T_{AC} = 33.24\text{ lb}$$

Finally, draw the free-body diagram of pin A (Fig. 7-35a). The forces acting at this joint are

$$\begin{aligned} \mathbf{F}_{AB} &= -0.8321T_{AB}\mathbf{i} - 0.5547T_{AB}\mathbf{j} \\ &= 3.599\mathbf{i} + 2.399\mathbf{j}\text{ lb} \\ \mathbf{F}_{AD} &= -0.5547T_{AD}\mathbf{i} + 0.8321T_{AD}\mathbf{k} \\ &= 24.06\mathbf{i} - 36.09\mathbf{k}\text{ lb} \\ \mathbf{F}_{AC} &= -0.8321T_{AC}\mathbf{i} + 0.5547T_{AC}\mathbf{j} \\ &= -27.66\mathbf{i} + 18.44\mathbf{j}\text{ lb} \\ \mathbf{A} &= -A_x\mathbf{i} + A_y\mathbf{j} + A_z\mathbf{k} \end{aligned}$$

Then the equilibrium equation becomes

$$\begin{array}{ll} \mathbf{i}: & 3.599 + 24.06 - 27.66 - A_x = 0 \\ \mathbf{j}: & 2.399 + 18.44 + A_y = 0 \\ \mathbf{k}: & -36.09 + A_z = 0 \end{array}$$

which gives

$$A_x = 0.00\text{ lb} \qquad A_y = -20.84\text{ lb} \qquad A_z = 36.09\text{ lb}$$

The desired answers are then

$$\begin{array}{ll} \mathbf{A} = -20.84\mathbf{j} + 36.1\mathbf{k}\text{ lb} & \text{Ans.} \\ \mathbf{B} = -10.80\mathbf{k}\text{ lb} & \text{Ans.} \\ \mathbf{C} = -41.7\mathbf{j} + 83.0\mathbf{k}\text{ lb} & \text{Ans.} \\ T_{AB} = 4.32\text{ lb }(C) \qquad T_{AC} = 33.2\text{ lb }(T) & \text{Ans.} \\ T_{AD} = 43.4\text{ lb }(C) \qquad T_{BC} = 4.80\text{ lb }(C) & \text{Ans.} \\ T_{BD} = 13.47\text{ lb }(T) \qquad T_{CD} = 103.5\text{ lb }(C) & \text{Ans.} \end{array}$$

(The consistency of these answers can be checked by considering the overall equilibrium of the entire truss.)

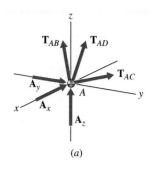

(a)

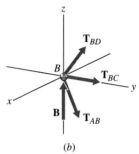

(b)

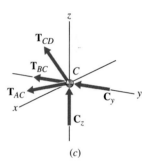

(c)

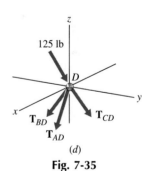

(d)

Fig. 7-35

PROBLEMS

7-76* The awning structure of Fig. P7-76 is supported by ball-and-socket joints at A and C and by a short link at B. The 750-N force is parallel to the z-axis. Determine the support reactions and the force in each member of this space truss.

7-78 A wire is stretched tight between two pylons, one of which is shown in Fig. P7-78. The 1.5-kN force is parallel to the x-y plane and makes an angle of 20° with the y-axis. The supports are equivalent to a ball-and-socket joint at B and short links at A and C. Determine the support reactions and the force in each member of this space truss.

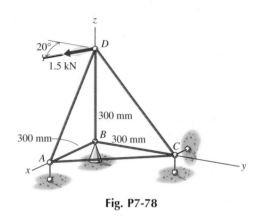

Fig. P7-78

7-79 A wire is stretched tight between two pylons, one of which is shown in Fig. P7-79. The 250-lb force is parallel to the x-y plane and makes an angle of 30° with the y-axis. The supports are equivalent to a ball-and-socket joint at B and short links at A and C. Determine the support reactions and the force in each member of this space truss.

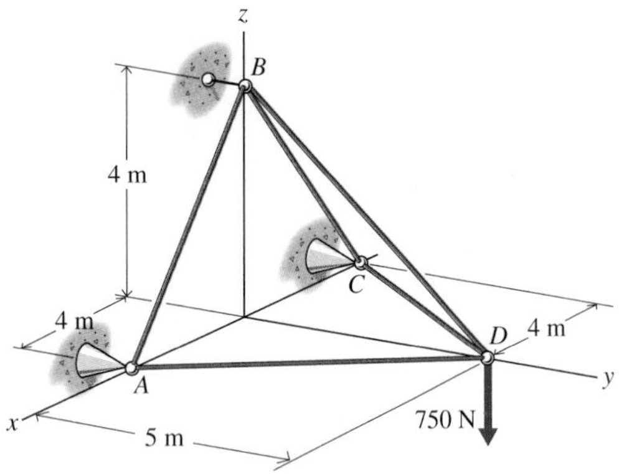

Fig. P7-76

7-77* The awning structure of Fig. P7-77 is supported by ball-and-socket joints at A and C and by a short link at B. The 50-lb force is parallel to the y-axis and the 150-lb force is parallel to the z-axis. Determine the support reactions and the force in each member of this space truss.

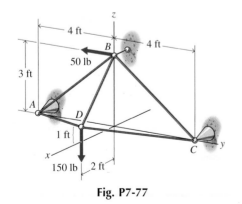

Fig. P7-77

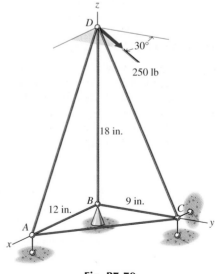

Fig. P7-79

7-80* A sign that weighs 650 N hangs from joints D and E of the space truss of Fig. P7-80. The weight is supported equally by the two joints. In addition, a wind blowing straight at the sign exerts a force of 200 N in the direction of the negative x-axis—also distributed evenly between the two joints. The truss supports are equivalent to a ball-and-socket joint at B and short links at A, C, and F. Determine the support reactions and the force in each member of the truss.

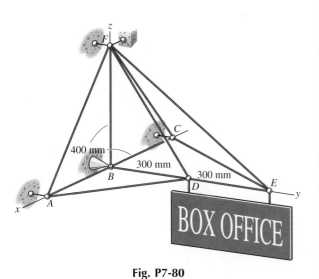

Fig. P7-80

7-81 A 120-lb sign hangs from joints D and E of the space truss of Fig. P7-81. The weight is supported equally by the two joints. In addition, a wind blowing straight at the sign exerts a force of 15 lb in the direction of the negative y-axis—also distributed evenly between the two joints. The truss supports are equivalent to a ball-and-socket joint at B and short links at A, C, and F. Determine the support reactions and the force in each member of the truss.

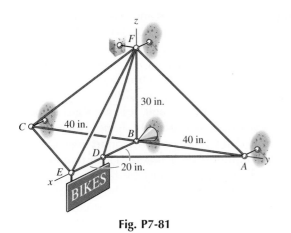

Fig. P7-81

7-82 A cable is attached to the tower of Fig. P7-82 and exerts a force of 2.5 kN in a horizontal plane and at an angle of 20° to the y-axis. The base of the tower is an equilateral triangle with sides 2-m long. The apex, joint G, is 3 m directly above the origin of the coordinate axes which is located at the centroid of triangle ABC. The triangle DEF is in a horizontal plane 1 m above the x-y plane. The supports are equivalent to a ball-and-socket joint at B and short links at A and C. Determine the support reactions and the force in members AB, EG, and FG of this space truss.

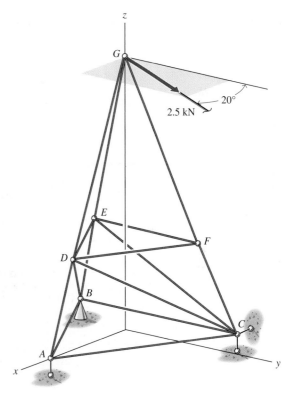

Fig. P7-82

7-83* A cable is attached to the tower of Fig. P7-83 and exerts a force of 450 lb in the *y-z* plane as shown. The base of the tower is an equilateral triangle with sides 3 ft long. The apex, joint *G*, is 5 ft directly above the origin of the coordinate axes which is located at the centroid of triangle *ABC*. The triangle *DEF* is in a horizontal plane 2 ft above the *x-y* plane. The supports are equivalent to a ball-and-socket joint at *B* and short links at *A* and *C*. Determine the support reactions and the force in members *BD*, *EG*, and *FG* of this space truss.

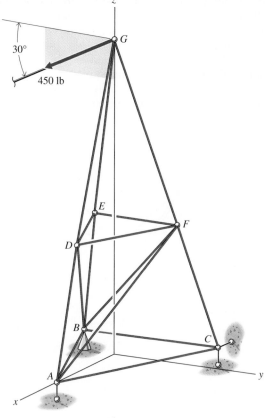

Fig. P7-83

7-4 FRAMES AND MACHINES

The first part of this chapter has dealt with very special types of structures, trusses, which are composed only of two-force members. Structures that contain other types of members are called frames or machines. Although frames and machines may also contain one or more two-force members, they always contain at least one member that is acted on by forces at more than two points or is acted on by both forces and moments.

The main distinction between frames and machines is that frames are rigid structures whereas machines are not. For example, the structure shown in Fig. 7-36a is a frame. Since it is a rigid body, three support reactions (Fig. 7-36b) are sufficient to fix it in place, and overall equilibrium is sufficient to determine the three support reactions.

The structure of Fig. 7-36c is a machine, although it is sometimes referred to as a nonrigid frame or a linkage. It is nonrigid is the sense that it depends on its supports to maintain its shape. The lack of internal rigidity is compensated for by an extra support reaction (Fig. 7-36d). In this case, overall equilibrium is not sufficient to determine all four support reactions. The structure must be taken apart and analyzed even if the only information desired is the support reactions.

In a more specific sense, the term *machine* is usually used to describe devices such as pliers, clamps, nutcrackers, and the like that are used to magnify the effect of forces. In each case, a force (input) is applied to the handle of the device and a much larger force (output) is applied by the device somewhere else. Like nonrigid frames, these machines must be taken apart and analyzed even if the only information desired is the relationship between the input and output forces.

As with the analysis of trusses, the method of solution for frames and machines consists of taking the structures apart, drawing free-body diagrams of each of the components, and writing the equations of equilibrium for each of the free-body diagrams. In the case of trusses, the direction of the force in all members was known and the method of joints reduced to solving a series of particle equilibrium problems. Since some of the members of frames and machines are not two-force members, however, the directions of the forces in these members is not known. The analysis of frames and machines will consist of solving for the equilibrium of a system of rigid bodies rather than a system of particles.

7-4-1 Frames

The method of analysis for frames can be demonstrated using the table of Fig. 7-2, which is reproduced here as Fig. 7-37a. None of the members that make up the table are two-force members, and so the structure is definitely not a truss. Although the table can be folded up by unhooking the top from the leg, in normal use the table is a stable, rigid structure. Therefore, the table is a frame.

The analysis will be started by first drawing the free-body diagram of the entire table (Fig. 7-37b), for which the equations of equilibrium

$$+\!\rightarrow\!\Sigma F_x = A_x = 0$$
$$+\!\uparrow\!\Sigma F_y = A_y + D_y - W = 0$$
$$\zeta + \Sigma M_A = 24D_y - 12W = 0$$

yield the support reactions

$$A_x = 0 \qquad A_y = \frac{W}{2} \qquad D_y = \frac{W}{2}$$

Next, the table is taken apart and the free-body diagrams of each of its parts drawn (Fig. 7-38). Since none of the members is a two-force member, the directions of the forces at the joints B, C, and E are not known—*they are not directed along the members!* Although the forces may be represented in terms of any convenient components, the free-body diagrams must take into account Newton's third law of action and reaction. That is, when drawing free-body diagrams, the forces exerted by one member on a second must be equal in magnitude and opposite in direction to the forces exerted by the second member on the first. For Fig. 7-38 this is effected by showing the components of the force exerted by member AB on member CD at joint E to have equal magnitude and opposite direction to the components of the force exerted by member CD on member AE at joint E and similarly for the other joints.

The floor at D can exert only an upward force on the leg and the force should be shown as such on the free-body diagrams. Similarly, the horizontal component of force exerted by the slot on the leg AB can act only to the left and should be shown as such. If the values of these forces come out to be negative, either the solution is in error or the table is not in equilibrium.

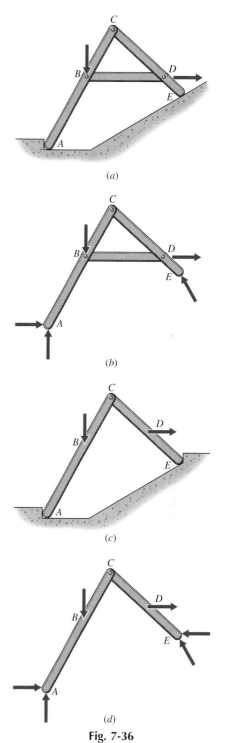

(a)

(b)

(c)

(d)

Fig. 7-36

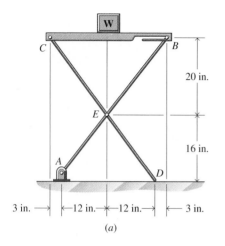

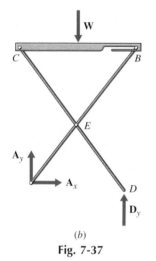

(a)

(b)

Fig. 7-37

The proper direction of the other force components are not all as clear. Though it is easy to guess that the vertical force components B_y and C_y act upward on the table top BC and downward on the legs, it is not as easy to decide whether to draw E_y acting upward or downward on leg AB. At this point it does not matter since the frictionless pin connections can support a force in either direction. As in the case of trusses, the direction that a force is shown on one member is unimportant as long as the force is represented consistent with Newton's third law on each part of the structure. If a force component is assumed in the wrong direction, its symbol will simply end up having a negative value. This can be accounted for in the report of final answers as shown in the Example Problems.

Although not all the members of a frame can be two-force members, it is possible and quite likely that one or more of the members will be two-force members. Take advantage of any such members and show that force as acting in its known direction. But, *be sure that all forces are not directed along the members.* Perhaps one of the most common mistakes that students make is to treat frames like trusses; that is, to draw all forces as acting along the members and trying to apply the method of joints.

Unlike the analysis of trusses, the free-body diagrams of the pins of a frame are usually not drawn and analyzed separately. In the case of the truss, equilibrium of the members was assured by the two-force member assumption, the direction of the force exerted by each member on a pin was known, and equilibrium of the pins contained all the useful information of the problem. None of these statements apply to the analysis of frames, however, and it is seldom useful to analyze the equilibrium of the pins separately.

In most cases, it does not matter to which member a pin is attached when the structure is taken apart. There are, however, a few special situations in which it does matter:

When a pin connects a support and two or more members, the pin must be assigned to one of the members. The support reactions are applied to the pin on this member.

When a pin connects two or more members and a load is applied to the pin, the pin must be assigned to one of the members. The load is applied to the pin on this member.

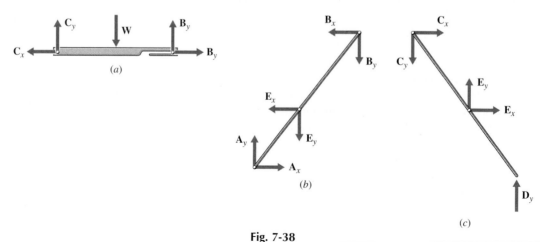

(a)

(b)

(c)

Fig. 7-38

Following these simple rules will avoid confusion as to where the loads and support reactions should be applied.

Special care is also warranted when one or more of the members that meet at a joint is a two-force member:

Pins should never be assigned to two-force members.

When all the members meeting at a joint are two-force members, the pin should be removed and analyzed separately as in the method of joints for a truss.

While these last two "rules" are not strictly necessary, following them will prevent confusion when dealing with two-force members in frames.

Finally, the equations of equilibrium are written for each part of the frame and are solved for the joint forces. There are three independent equations of equilibrium (two force and one moment) for each part, and hence for the table parts of Fig. 7-38 there will be nine equations to solve for the six remaining unknown forces ($\mathbf{B}_x$, $\mathbf{B}_y$, $\mathbf{C}_x$, $\mathbf{C}_y$, $\mathbf{E}_x$, $\mathbf{E}_y$). Prior solution of the overall equilibrium of the frame for the support reactions will have reduced three of these equations to a check of the consistency of the answers.

7-4-2 Machines

The method described for frames is also used to analyze machines and other nonrigid structures. In each case, the structure is taken apart, free-body diagrams are drawn for each part, and the equations of equilibrium are applied to each free-body diagram. For machines and nonrigid structures, however, the structure must be taken apart and analyzed even if the only information desired is the support reactions or the relationship between the external forces acting on it.

The method of analysis for machines can be demonstrated using the simple garlic press shown in Fig. 7-39a. Forces $\mathbf{H}_1$ and $\mathbf{H}_2$ applied to the handles (the *input forces*) are converted into forces $\mathbf{G}_1$ and $\mathbf{G}_2$ applied to the garlic clove (the *output forces*). Equilibrium of the entire press only gives that $H_1 = H_2$; it gives no information about the relationship between the input forces and the output forces.

In order to determine the relationship between the input forces and the output forces, the machine must be taken apart and free-body diagrams drawn for each of its parts as shown in Fig. 7-39b. Then the sum of moments about B gives

$$(a + b)H = bG$$

or

$$G = \frac{a + b}{b} H$$

The ratio of the output and input forces is called the *mechanical advantage* (M.A.) of the machine

$$\text{mechanical advantage} = \frac{\text{output force}}{\text{input force}}$$

For the garlic press, the mechanical advantage is just

$$\text{M.A.} = \frac{a + b}{b}$$

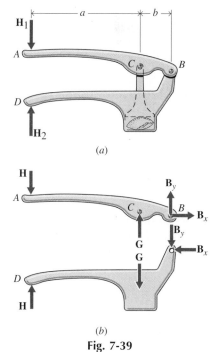

(a)

(b)

Fig. 7-39

A bag of potatoes is resting on the chair of Fig. 7-40a. The force exerted by the potatoes on the frame at one side of the chair is equivalent to horizontal and vertical forces of 24 N and 84 N, respectively, at E and a force of 28 N perpendicular to member BH at G (as shown in the free-body diagram of Fig. 7-40b). Find the forces acting on member BH.

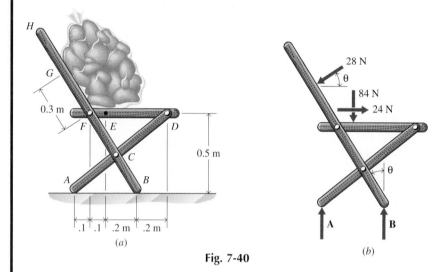

(a)

Fig. 7-40

(b)

SOLUTION

The equations of equilibrium for the entire chair are

$$+\rightarrow \Sigma F_x = 24 - 28 \cos \theta = 0$$
$$+\uparrow \Sigma F_y = A + B - 84 - 28 \sin \theta = 0$$
$$\downarrow + \Sigma M_B = (0.2)(84) - (0.5)(24) - 0.4\,A$$
$$+ \left(0.3 + \frac{0.5}{\cos \theta}\right)(28) = 0$$

where $\theta = \tan^{-1}(3/5)$. The first equation is satisfied identically. The remaining two equations give

$$A = 73.82 \text{ N} \qquad B = 24.58 \text{ N}$$

Next the chair is disassembled and free-body diagrams drawn for each part (Fig. 7-41). For member DF, the equilibrium equations can be written

$$+\rightarrow \Sigma F_x = D_x - F_x + 24 = 0$$
$$+\uparrow \Sigma F_y = F_y + D_y - 84 = 0$$
$$\downarrow + \Sigma M_D = (0.4)(84) - 0.5 F_y = 0$$

which gives

$$F_y = 67.2 \text{ N} \qquad D_y = 16.80 \text{ N} \qquad D_x = F_x - 24 \text{ N}$$

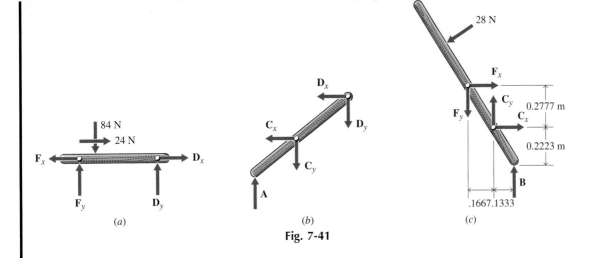

(a)　　　　　　　　(b)　　　　　　　　(c)

Fig. 7-41

Now the equations of equilibrium for member BH are

$$+\rightarrow\Sigma F_x = F_x + C_x - 28\cos\theta = 0$$
$$+\uparrow\Sigma F_y = 24.58 + C_y - 67.2 - 28\sin\theta = 0$$
$$\downarrow+\Sigma M_C = \left(0.3 + \frac{0.1667}{\sin\theta}\right)(28) + (0.1333)(24.58)$$
$$+ (0.1667)(67.2) - 0.2777\,F_x = 0$$

which have only three unknowns remaining and can be solved to get

$$F_x = 115.1\text{ N} \qquad C_x = -91.0\text{ N} \qquad C_y = 57.0\text{ N}$$

Then the forces acting on member BH are

$$\mathbf{B} = 24.58\mathbf{j}\text{ N} \qquad\qquad\qquad \text{Ans.}$$
$$\mathbf{C} = -91.0\mathbf{i} + 57.0\mathbf{j}\text{ N} \qquad\qquad \text{Ans.}$$
$$\mathbf{F} = 115.1\mathbf{i} + 67.2\mathbf{j}\text{ N} \qquad\qquad \text{Ans.}$$

plus the applied force of 28 N perpendicular to the bar at G. These forces are shown on the "report diagram" of Fig. 7-42.

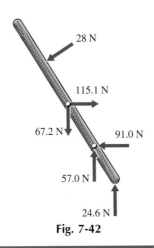

Fig. 7-42

The weight of books on a shelf bracket is equivalent to a vertical force of 75 lb as shown on Fig. 7-43a. In addition, a vertical load of 50 lb is suspended from the middle of the lower brace BC. Find all forces acting on all three members of this frame.

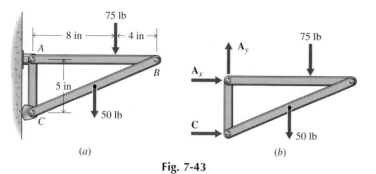

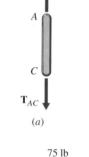

(a)

(a)

(b)

Fig. 7-43

SOLUTION

First draw the free-body diagram of the entire shelf bracket as in Fig. 7-43b. The equations of overall equilibrium are

$$\curvearrowleft + \Sigma M_A = 5C - (8)(75) - (6)(50) = 0$$
$$+\rightarrow \Sigma F_x = A_x + C = 0$$
$$+\uparrow \Sigma F_y = A_y - 75 - 50 = 0$$

which are solved to get the support reactions

$$A_x = -180 \text{ lb} \qquad A_y = 125 \text{ lb} \qquad C = 180 \text{ lb}$$

Next, dismember the bracket and draw separate free-body diagrams of each member (Fig. 7-44). Member AC is a two-force member, and so pin A will be assigned to member AB and pin C will be assigned to member BC. It does not matter to which member pin B is assigned since no two-force members are attached to joint B and joint B is neither loaded nor attached to a support. Just to be definite, pin B will be assigned to member AB as shown in Fig. 7-44a. Then the equations of equilibrium for member AB give

$$\curvearrowleft \Sigma M_B = (4)(75) + 12T_{AC} - (12)(125) = 0 \quad T_{AC} = 100 \text{ lb}$$
$$\curvearrowleft + \Sigma M_A = 12B_y - (8)(75) = 0 \qquad\qquad B_x = 180 \text{ lb}$$
$$+\rightarrow \Sigma F = (-180) + B_x = 0 \qquad\qquad\quad B_y = 50 \text{ lb}$$

It is easily verified that these values also satisfy the equations of equilibrium for the other free-body diagrams. These forces are all shown on the "report diagram" of Fig. 7-45.

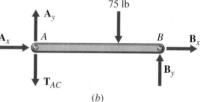

(b)

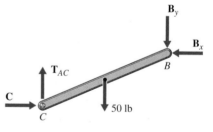

(c)

Fig. 7-44

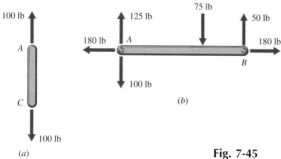

(a)

(b)

Fig. 7-45

(c)

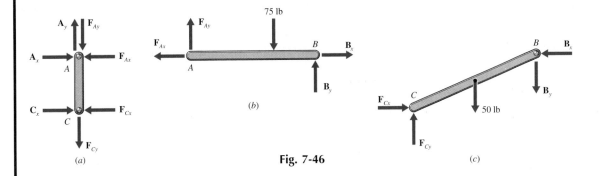

Fig. 7-46

(a)   (b)   (c)

ALTERNATIVE SOLUTION

Pins A and C can be assigned to the two-force member AC as shown on the free-body diagram of Fig. 7-46 if care is taken with the representation of the internal forces. The tension T_{AC} shown on Fig. 7-44a represents the force exerted on member AC by the pin C. Since the pin is now part of the member, the force T_{AC} is an internal force and is not shown on the free-body diagram. The symbols A_x and A_y have already been used to represent the components of the support reaction, so F_{Ax} and F_{Ay} will be used for the components of the forces of action and reaction at pin A. Similarly, F_{Cx} and F_{Cy} will be used for the components of the forces of action and reaction at pin C so as not to confuse these forces with the support reaction C.

As stated before, since no supports or two-force members are attached to pin B and no loads are applied at pin B, it does not matter to which member pin B is assigned. To illustrate this point, this time pin B will be assigned to member BC.

Now the equations of equilibrium for member AB give

$$\curvearrowleft + \Sigma M_A = 12B_y - (8)(75) = 0 \qquad B_y = 50 \text{ lb}$$
$$\curvearrowleft + \Sigma M_B = (4)(75) - 12F_{Ay} = 0 \qquad F_{Ay} = 25 \text{ lb}$$
$$+ \rightarrow \Sigma F = F_{Ax} + B_x = 0 \qquad B_x = -F_{Ax}$$

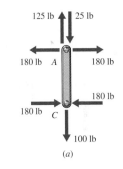

(a)

Next, equilibrium of member AC gives the equations

$$\curvearrowleft + \Sigma M_C = 5F_{Ax} - (5)(-180) = 0 \qquad F_{Ax} = -180 \text{ lb}$$
$$\curvearrowleft + \Sigma M_A = (5)(180) - 5F_{Cx} = 0 \qquad F_{Cx} = 180 \text{ lb}$$
$$+ \uparrow \Sigma F = 125 - 25 - F_{Cy} = 0 \qquad F_{Cy} = 100 \text{ lb}$$

Finally, returning to the free-body diagram of AB, the horizontal component of force equilibrium

$$+ \rightarrow \Sigma F = F_{Ax} + B_x = 0$$

gives

$$B_x = 180 \text{ lb}$$

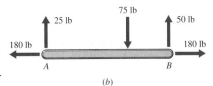

(b)

These forces are shown on the "report diagram" of Fig. 7-47. Although the forces at B are clearly the same on Fig. 7-45 and Fig. 7-47, the forces at A and C appear to be different on the two diagrams. However, the resultant forces on the two diagrams are actually the same on both diagrams; they are only expressed using different components.

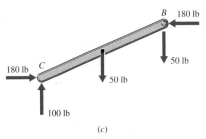

(c)

Fig. 7-47

PROBLEMS

7-84* In the linkage of Fig. P7-84, $a = 1.0$ m, $b = 0.5$ m, $\theta = 0°$, and $P = 300$ N. Determine all forces acting on member BCD.

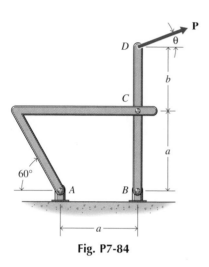

Fig. P7-84

7-85* In the linkage of Fig. P7-84, $a = 2.0$ ft, $b = 1.5$ ft, $\theta = 30°$, and $P = 40$ lb. Determine all forces acting on member BCD.

7-86 In the linkage of Fig. P7-86, $a = 50$ mm, $P_1 = 500$ N, and $P_2 = 250$ N. Determine all forces acting on member ABC.

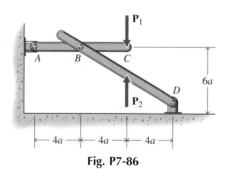

Fig. P7-86

7-87 In the linkage of Fig. P7-86, $a = 2.5$ in. and $P_1 = P_2 = 25$ lb. Determine all forces acting on member ABC.

7-88* The frame of Fig. P7-88 has a distributed load of $w = 200$ N/m applied to member CDE and a concentrated force $P = 200$ N applied to member ABC. If $a = 100$ mm, determine all forces acting on member ABC.

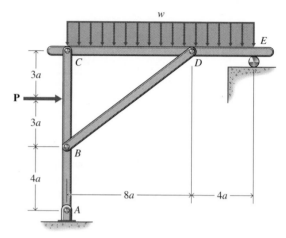

Fig. P7-88

7-89 The frame of Fig. P7-88 has a distributed load of $w = 30$ lb/ft applied to member CDE and a concentrated force $P = 75$ lb applied to member ABC. If $a = 6$ in., determine all forces acting on member ABC.

7-90 Determine all forces acting on member $ABCD$ of the frame of Fig. P7-90.

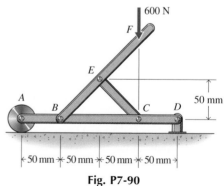

Fig. P7-90

310

7-91* The cord of Fig. P7-91 is wrapped around a frictionless pulley and supports a 40-lb weight. Determine all forces acting on member *EG*.

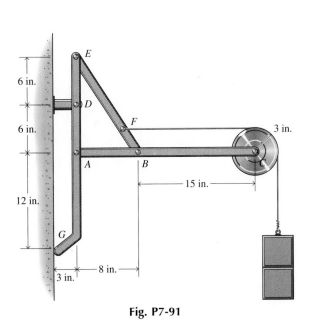

Fig. P7-91

7-92 Determine all forces acting on member *ABE* of the frame of Fig. P7-92.

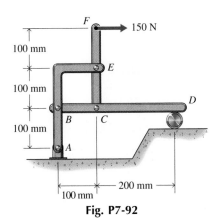

Fig. P7-92

7-93* Determine all forces acting on member *ABCD* of the frame of Fig. P7-93.

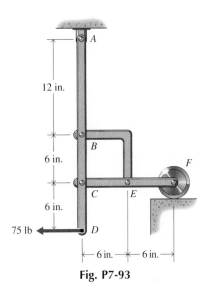

Fig. P7-93

7-94* The spring clamp of Fig. P7-94 is used to hold block *E* into the corner. The force in the spring is $F = k(\ell - \ell_0)$, where ℓ is the present length of the spring, $\ell_0 = 15$ mm is the unstretched length of the spring, and $k = 5000$ N/m is the spring constant. Determine all forces acting on member *ABC* of the spring clamp and the force exerted by the spring clamp on the block *E*.

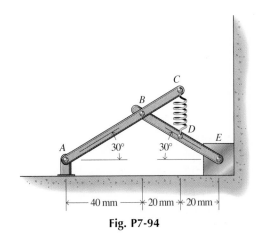

Fig. P7-94

7-95 The spring clamp of Fig. P7-95 is used to hold the block F against the floor. The force in the spring is $F = k(\ell - \ell_0)$, where ℓ is the present length of the spring, $\ell_0 = 3$ in. is the unstretched length of the spring, and $k = 240$ lb/ft is the spring constant. Determine all forces acting on member ABC of the spring clamp and the force exerted by the spring clamp on the block F.

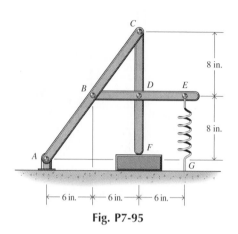

Fig. P7-95

7-96* A cord BD is used to keep the frame of Fig. P7-96 from collapsing under the distributed load w. Determine w when the tension in the cord is 600 N.

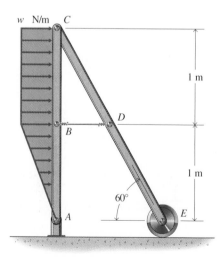

Fig. P7-96

7-97 The fold-down chair of Fig. P7-97 weighs 25 lb and has its center of gravity at G. Determine all forces acting on member ABC.

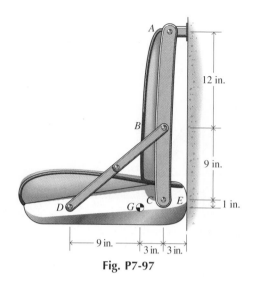

Fig. P7-97

7-98 The tower crane of Fig. P7-98 is rigidly attached to the building at F. A cable is attached at D and passes over small frictionless pulleys at A and E. The object suspended from C weighs 1500 N. Determine all forces acting on member $ABCD$.

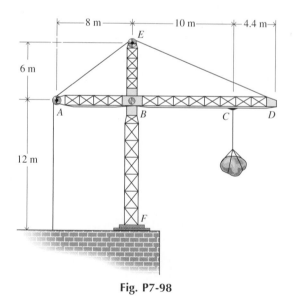

Fig. P7-98

7-99* The hoist pulley structure of Fig. P7-99 is rigidly attached to the wall at C. A load of sand hangs from the cable that passes around the 1-ft diameter, frictionless pulley at D. The weight of the sand can be treated as a triangular distributed load with a maximum of 70 lb/ft. Determine all forces acting on member ABC.

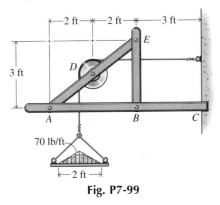

Fig. P7-99

7-100 The sand on the tray of Fig. P7-100 can be treated as a triangular distributed load with a maximum of 800 N/m. The wheel at C is frictionless. Determine all forces acting on member ABC.

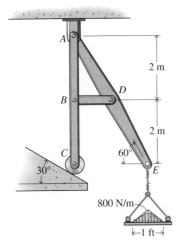

Fig. P7-100

7-101* In Fig. P7-101, a cable is attached to the structure at D, passes around a 1-ft diameter, frictionless pulley, and is then attached to a 250-lb weight W. Determine all forces acting on member $ABCDE$.

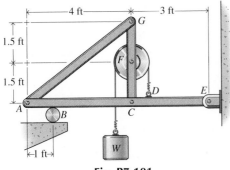

Fig. P7-101

7-102 In Fig. P7-102, a cable is attached to the structure at E, passes around the 0.8-m diameter, frictionless pulley at A, and then is attached to a 1000-N weight W. Determine all forces acting on member $ABCD$.

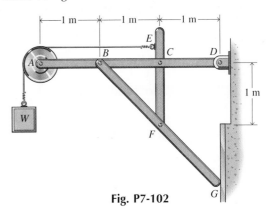

Fig. P7-102

7-103 In Fig. P7-103, a cable is attached to the structure at E, passes around the 2-ft diameter, frictionless pulley at C, and then is attached to a 200-lb weight W. A second cable is attached between A and F. Determine all forces acting on member $DEBFG$.

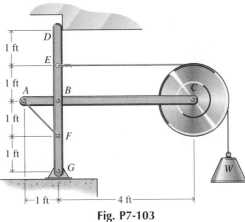

Fig. P7-103

7-104* A force $\mathbf{F}_2$ is applied to the cable that passes over the frictionless pulley at F in Fig. P7-104. Calculate the ratio of the force F_1 to the force F_2. If $F_2 = 400$ N and $a = 25$ mm, determine all forces acting on member ABC.

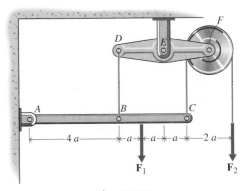

Fig. P7-104

313

7-105 Determine all forces acting on member *DEF* of the frame of Fig. P7-105.

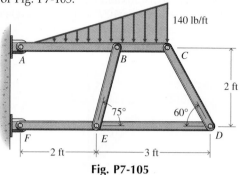

140 lb/ft

2 ft

2 ft

3 ft

75°

60°

Fig. P7-105

7-106 Determine all forces acting on member *ABC* of the frame of Fig. P7-106.

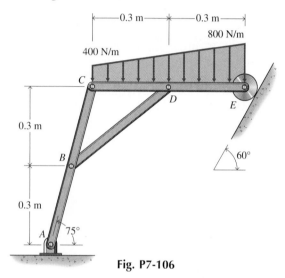

0.3 m — 0.3 m

800 N/m

400 N/m

0.3 m

0.3 m

0.3 m

60°

75°

Fig. P7-106

7-107* A force of 20 lb is required to pull the stopper *DE* in Fig. P7-107. Determine all forces acting on member *BCD*.

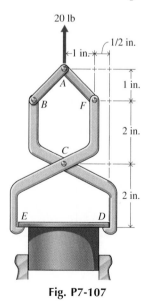

20 lb

1/2 in.

1 in.

1 in.

2 in.

2 in.

Fig. P7-107

7-108* Forces of 5 N are applied to the handles of the paper punch of Fig. P7-108. Determine the force exerted on the paper at *D* and the force exerted on the pin at *B* by handle *ABC*.

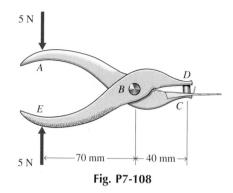

5 N

5 N

70 mm

40 mm

Fig. P7-108

7-109 Forces of 50 lb are applied to the handles of the bolt cutter of Fig. P7-109. Determine the force exerted on the bolt at *E* and all forces acting on the handle *ABC*.

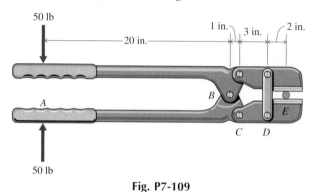

50 lb

50 lb

20 in.

1 in.

3 in.

2 in.

Fig. P7-109

7-110 Fig. P7-110 is a simplified sketch of the mechanism used to raise the bucket of a bulldozer. The bucket and its contents weigh 10 kN and have a center of gravity at *H*. Arm *ABCD* has a weight of 2 kN and a center of gravity at

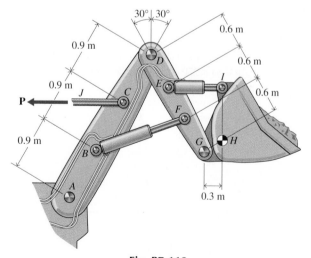

30° 30°

0.9 m

0.9 m

0.6 m

0.6 m

0.6 m

0.9 m

0.3 m

P

Fig. P7-110

B; arm *DEFG* has a weight of 1 kN and a center of gravity at *E*. The weight of the hydraulic cylinders can be ignored. Calculate the force in the horizontal cylinders *CJ* and *EI* and all forces acting on arm *DEFG* for the position shown.

7-111* The mechanism of Fig. P7-111 is designed to keep its load level while raising it. A pin on the rim of the 4-ft diameter pulley fits in a slot on arm *ABC*. Arms *ABC* and *DE* are each 4 ft long, and the package being lifted weighs 80 lb. The mechanism is raised by pulling on the rope that is wrapped around the pulley. Determine the force **P** applied to the rope and all forces acting on the arm *ABC* when the package has been lifted 4 ft as shown.

7-112 The jaws and bolts of the wood clamp in Fig. P7-112 are parallel. The bolts pass through swivel mounts so that no moments act on them. The clamp exerts forces of 300 N on each side of the board. Treat the forces on the boards as uniformly distributed over the contact areas and determine the forces in each of the bolts. Show on a sketch all forces acting on the upper jaw of the clamp.

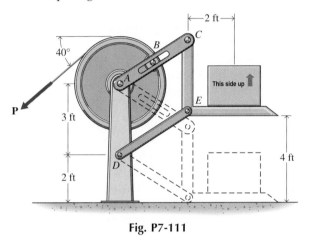

Fig. P7-111

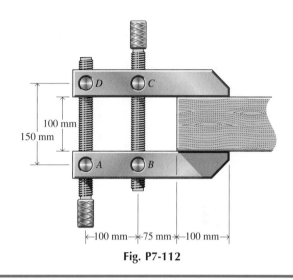

Fig. P7-112

SUMMARY

In previous chapters, equilibrium equations were used to determine rigid-body external support reactions. In this chapter, equilibrium equations are used to determine joint forces in structures comprised of pin-connected members. Like all forces (internal, external, applied, or reactive), the joint forces always occur in equal-magnitude, opposite-direction pairs. If not separated by means of a free-body diagram from the rest of the structure or environment, the pair of forces need not be considered when writing the equations of equilibrium. Therefore, in order to determine the joint forces, the structure must be divided into two or more parts.

Two broad categories of engineering structures were considered in this chapter; namely, trusses and frames. Trusses are rigid, fully constrained structures that are constructed using only pin-connected two-force members. Four main assumptions are made in the analysis of trusses: (1) Truss members are connected only at their ends; no member is continuous through a joint. (2) Members are connected by frictionless pins. (3) The truss structure is loaded only at the joints. (4) The weight of the members may be neglected. Because of these assumptions, truss members are modeled as two-force members with the forces acting at the ends of the member and directed along the axis of the member.

When the truss structure and the applied loads lie in the same plane, the structure is known as a planar truss. Space trusses are structures that are not contained in a single plane and/or are loaded out of the plane of the structure.

One method of analysis (the method of joints) for trusses is performed by drawing a free-body diagram for each pin (joint). Application of the vector equilibrium equation, $\Sigma \mathbf{F} = \mathbf{0}$, at each joint yields two algebraic equations, which can be solved for two unknowns. The pins are solved sequentially starting from a pin on which only two unknown forces and one or more known forces act. Once these forces are determined, their values can be applied to adjacent joints and treated as known quantities. This process is repeated until all unknown forces have been determined.

The method of joints is most often used when the forces in all of the members of a truss are to be determined.

A second method of analysis for trusses is the method of sections. When the method of sections is used, the truss is divided into two parts by passing an imaginary plane or curved section through the members of interest. Free-body diagrams may then be drawn for either or both parts of the truss. Since each part is a rigid body, three independent equations of equilibrium can be written for either part. Therefore, a section that cuts through no more than three members should be used.

It will often happen that a section that cuts no more than three members and that passes through a given member of interest cannot be found. In such a case it may be necessary to draw a section through a nearby member and solve for the forces in it first. Then the method of joints can be used to find the force in the member of interest.

One of the principle advantages of using the method of sections is that the force in a member near the center of a large truss usually can be determined without first obtaining the forces in the rest of the truss.

Structures that are not constructed entirely of two-force members are called frames or machines. Although frames and machines may also contain one or more two-force members, they always contain at least one member that is acted on by forces at more than two points or is acted on by both forces and moments. The main distinction between frames and machines is that frames are rigid structures, whereas machines are not.

As with the analysis of trusses, the method of solution for frames and machines consists of taking the structures apart, drawing free-body diagrams of each of the components, and writing the equations of equilibrium for each of the free-body diagrams. Since some of the members of frames and machines are not two-force members, however, the directions of the forces in these members is not known. The analysis of frames and machines consists of solving the equilibrium equations for a system of rigid bodies.

For machines and nonrigid structures, the structure must be taken apart and analyzed even if the only information desired is the support reactions or the relationship between the external forces (input and output forces) acting on it.

REVIEW PROBLEMS

7-113* Determine the force in each member of the truss shown in Fig. P7-113.

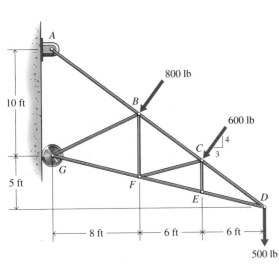

Fig. P7-113

7-114* Determine the force in each member of the truss shown in Fig. P7-114.

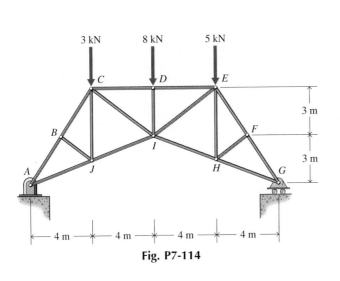

Fig. P7-114

7-115 Determine the forces in members *BC, CF, FG,* and *GE* of the truss shown in Fig. P7-115.

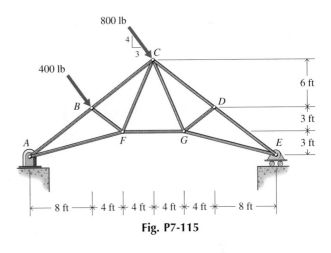

Fig. P7-115

7-116 Determine the forces in members *BC, BG, CG,* and *CF* of the truss shown in Fig. P7-116.

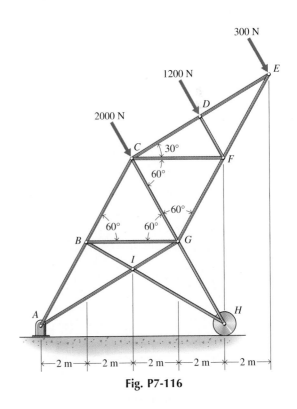

Fig. P7-116

7-117* Determine the force in each member of the space truss shown in Fig. P7-117. The support at *A* is a ball and socket. The supports at *D*, *E*, and *F* are links.

7-119 The weights of the bars in the structure shown in Fig. P7-119 are negligible. Determine

a. The force in the cable between pins B and E.
b. The reactions at supports A and F.

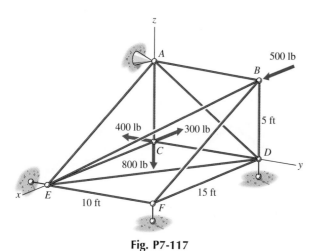

Fig. P7-117

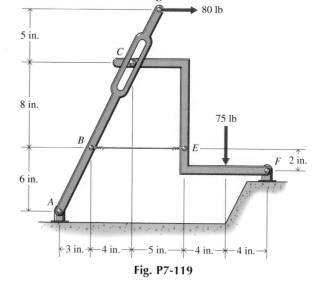

Fig. P7-119

7-118* Three bars are connected with smooth pins to form the frame shown in Fig. P7-118. The weights of the bars are negligible. Determine

a. The force exerted by the pin at D on member CDE.
b. The reactions at supports A and E.

7-120 Three bars are connected with smooth pins to form the frame shown in Fig. P7-120. The weights of the bars are negligible. Determine

a. The force exerted by the pin at *B* on member *ABC*.
b. The force exerted by the pin at *C* on member *ABC*.

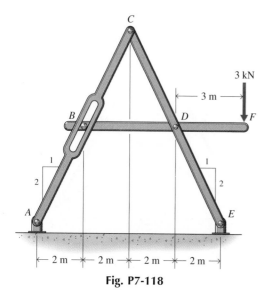

Fig. P7-118

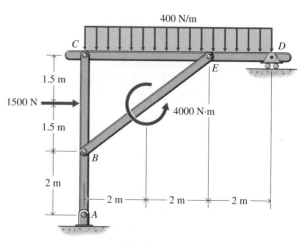

Fig. P7-120

7-121* A triangular plate is supported in a vertical plane by a bar and cable as shown in Fig. P7-121. The plate weighs 175 lb. Determine

a. The force in the cable between pins B and D.
b. The reaction at support A.
c. The force exerted by the pin at C on the plate.

7-123 A pin-connected system of levers and bars is used as a toggle for a press as shown in Fig. P7-123. Determine the force **F** exerted on the can at A when a force **P** = 100 lb is applied to the lever at G.

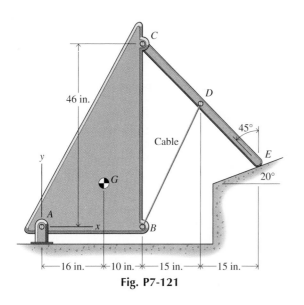

Fig. P7-121

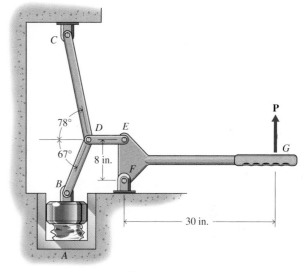

Fig. P7-123

7-122* A scissors jack for an automobile is shown in Fig. P7-122. The screw threads exert a force **F** on the blocks at joints A and B. Determine the force **P** exerted on the automobile if **F** = 800 N and (a) $\theta = 15°$, (b) $\theta = 30°$, and (c) $\theta = 45°$.

7-124 A pair of vise grip pliers is shown in Fig. P7-124. Determine the force **F** exerted on the block by the jaws of the pliers when a force **P** = 100 N is applied to the handles.

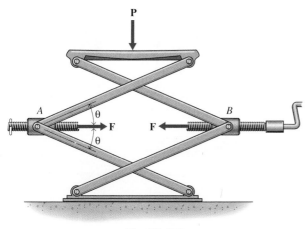

Fig. P7-122

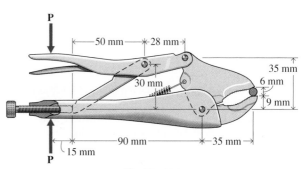

Fig. P7-124

Computer Problems

C7-125 An overhead crane consists of an I-beam supported by a simple truss as shown in Fig. P7-125. If the uniform I-beam weighs 400 lb, plot the force in members *AB*, *BC*, *EF*, and *FG* as a function of the position d ($0 \le d \le$ 8 ft).

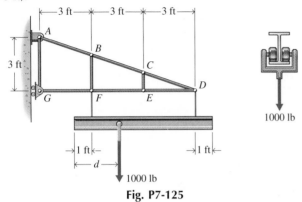

Fig. P7-125

C7-126 The simple truss shown in Fig. P7-126 supports one side of a bridge; an identical truss supports the other side. A 2000-kg car is stopped on the bridge at the location $c = 18$ m, and floor beams carry the vehicle loads to the joints. Treat the car as a point mass at the location c, and plot the force in members *AB*, *AE*, *BC*, and *BE* as a function of the bridge height b ($0.25a \le b \le a$, where $a = 12$ m).

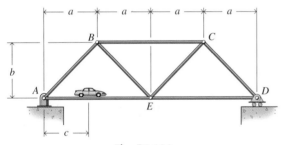

Fig. P7-126

C7-127 The simple truss shown in Fig. P7-126 supports one side of a bridge; an identical truss supports the other side. A 4500-lb car is stopped on the bridge at the location c, and floor beams carry the vehicle loads to the joints. The dimensions of the truss are $a = 40$ ft and $b = 20$ ft. Treat the car as a point mass at the location c, and plot the force in members *AB*, *AE*, *BC*, and *BE* as a function of the car's position c ($0 \le c \le 160$ ft).

C7-128 The Gambrel truss shown in Fig. P7-128 supports one side of a bridge; an identical truss supports the other side. A 3400-kg truck is stopped on the bridge at the location shown, and floor beams carry vehicle load to the truss joints. If the center of gravity of the truck is located 1.5 m in front of the rear wheels, plot the force in members *AB*, *BC*, *BG*, and *GH* as a function of the truck's location d ($0 \le d \le$ 20 m).

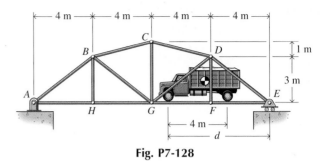

Fig. P7-128

C7-129 The mechanism shown in Fig. P7-129 is designed to keep its load level while raising it. A pin on the rim of the 4-ft diameter pulley fits in a slot on arm *ABC*. Arms *ABC* and *DE* are each 4 ft long and the package being lifted weighs 80 lb. The mechanism is raised by pulling on the rope, which is wrapped around the pulley.

a. Plot P, the force required to hold the platform as a function of the platform height h ($0 \le h \le 5.5$ ft).
b. Plot A, C, and E, the magnitudes of the pin reaction forces at A, C, and E as a function of h ($0 \le h \le 5.5$ ft).

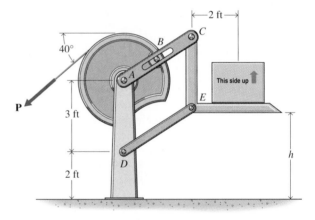

Fig. P7-129

C7-130 Forces of $P = 100$ N are being applied to the handles of the vice grip pliers shown in Fig. P7-130. Plot the force applied on the block by the jaws as a function of the distance d ($20 \le d \le 30$ mm).

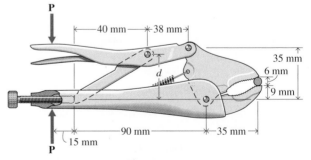

Fig. P7-130

C7-131 The door to an airplane hangar consists of two uniform sections, which are hinged at the middle as shown in Fig. P7-131. The door is raised by means of a cable, which is attached to a bar along the bottom edge of the door. Smooth rollers at the ends of the bar (C) run in a smooth vertical channel. If the door is 30 ft wide, 15 ft tall, and weighs 1620 lb:

a. Plot P, the force required to hold the door open, as a function of the door opening height h ($0.5 \le h \le 14.5$ ft).
b. Plot A and B, the hinge forces, as a function of the height h ($0.5 \le h \le 14.5$ ft).
c. What is the maximum height h if the force in the hinges is not to exceed 5000 lb?

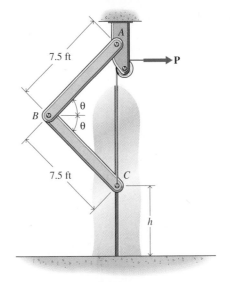

Fig. P7-131

8

INTERNAL FORCES IN STRUCTURAL MEMBERS

The gondola cars on an aerial tramway exert concentrated loads on the cable at discrete points.

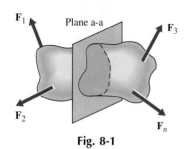

Fig. 8-1

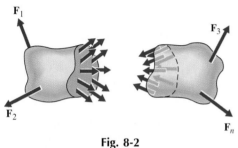

Fig. 8-2

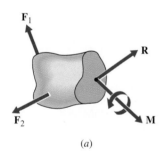

(a)

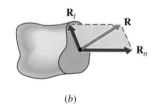

(b)

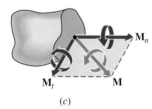

(c)

Fig. 8-3

8-1 INTRODUCTION

When a structural member or machine component (cable, bar, shaft, beam, or column) is subjected to a system of external loads (applied loads and support reactions), a system of internal resisting forces develops within the member to balance the external forces. Consider a body subjected to a system of balanced external forces $\mathbf{F}_1$, $\mathbf{F}_2$, $\mathbf{F}_3, \cdots, \mathbf{F}_n$, as shown in Fig. 8-1. These forces tend to either crush the body (compression) or pull it apart (tension). In either case, internal forces (resisting forces) develop within the body to resist the crushing or to hold the body together.

The resultant of the internal forces on a specific plane aa within a body can be determined by assuming that the plane separates the body into two parts as shown in Fig. 8-2. Since the body is in equilibrium, each part of the body must also be in equilibrium under the action of the internal forces that develop on the plane separating the body into two parts. Therefore, the resultant of the internal forces acting on plane aa can be determined by using either the left or right part of the body. In general, the distributed forces acting over the small elements of area dA that make up the cross section A are not uniformly distributed. The intensities of these internal forces (force per unit area) are called stresses. The problem of determining stress distributions in a specific body for a specific set of external loads is covered in textbooks dealing with the mechanics of deformable bodies.

A free-body diagram for the left part of the body is shown in Fig. 8-3a. The internal force distribution on plane aa has been replaced by a resultant force $\mathbf{R}$ at a point on plane aa and a resultant moment $\mathbf{M}$. The resultant force $\mathbf{R}$ can be resolved, as shown in Fig. 8-3b, into a component $\mathbf{R}_n$ (normal force) perpendicular to plane aa and a component $\mathbf{R}_t$ (shear force) tangent to plane aa. Similarly, the moment $\mathbf{M}$ can be resolved into a component $\mathbf{M}_n$ (torsional or twisting moment) about an axis perpendicular to plane aa and a component $\mathbf{M}_t$ (bending moment) about an axis tangent to plane aa, as shown in Fig. 8-3c. The components $\mathbf{R}_t$ and $\mathbf{M}_t$ can be resolved into rectangular components if an xyz-reference system is used.

The following procedure is suggested for determining the internal forces at a specific location in a member.

1. **Determine the Support Reactions.** Prepare a sketch of the body showing significant dimensions and all external loads (forces, bending moments, and torques) on the body in their exact locations. Support reactions should be determined before the plane of interest is exposed.

2. **Draw a Complete Free-Body Diagram.** Identify the plane of interest in the body. Prepare a free-body diagram for a portion of the body with the exposed plane of interest. Show all external loads on this part of the body and the force and moment resultants (or their components) on the exposed plane of interest.

3. **Apply the Equations of Equilibrium.** In the most general case, it will be possible to solve for six unknowns by using the equations:

$$\Sigma F_x = 0 \qquad \Sigma F_y = 0 \qquad \Sigma F_z = 0$$
$$\Sigma M_x = 0 \qquad \Sigma M_y = 0 \qquad \Sigma M_z = 0$$

8-2   AXIAL FORCE AND TORQUE IN BARS AND SHAFTS

In Chapter 7 axially loaded two-force truss members and multiforce frame and machine elements were discussed in detail. Application of the equations of equilibrium to the various parts of the truss, frame, or machine allowed the determination of all forces acting at the smooth pin connections. In many other types of engineering applications, knowledge of the maximum axial force, maximum shear force, maximum twisting moment, or maximum bending moment transmitted by any cross section through the member is required to establish the adequacy of the member for its intended use. In this section, the equations of equilibrium are used to establish the variation of internal axial force along the length of an axially loaded member and the variation of resisting torque transmitted by transverse cross sections of a shaft. Axial force and torque diagrams are introduced as a way to visualize the distributions for the full lengths of the members.

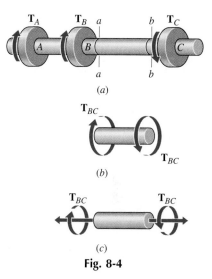

(a)

(b)

(c)

Fig. 8-4

An axial force diagram is a graph in which abscissas represent distances along the member and ordinates represent the internal axial forces at the corresponding cross sections. In plotting an axial force diagram, tensile forces are positive and compressive forces are negative. Example Problem 8-1 illustrates the computations required to construct an axial force diagram for a simple tension member subjected to four axial loads.

In a similar manner, a torque diagram is a graph in which abscissas represent distances along the member and ordinates represent the internal resisting torques at the corresponding cross sections. The sign convention used for torques is illustrated in Fig. 8-4. In the shaft shown in Fig. 8-4a, torque is applied to the shaft at gear C and is removed at gears A and B. Torques transmitted by cross sections aa and bb in the interval between gears B and C are shown in a pictorial fashion in Fig. 8-4b. A vector representation of these torques is shown in Fig. 8-4c. Positive torques point outward from the cross section when represented as a vector, according to the right-hand rule. Example Problem 8-2 illustrates the computations required to construct a torque diagram for a shaft subjected to four torques at different positions along the length of the shaft.

A steel bar with a rectangular cross section is used to transmit four axial loads as shown in Fig. 8-5a.

a. Determine the axial forces transmitted by cross sections in intervals AB, BC, and CD of the bar.

b. Draw an axial force diagram for the bar.

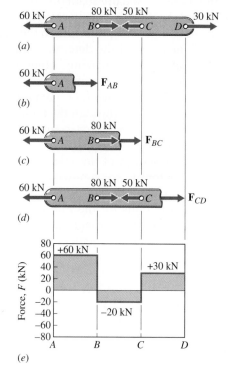

SOLUTION

a. The forces transmitted by cross sections in intervals AB, BC, and CD of the bar shown in Fig. 8-5a are obtained by using the three free-body diagrams shown in Figs. 8-5b, 8-5c, and 8-5d. Applying the force equilibrium equation $\Sigma F = 0$ along the axis of the bar yields

$$+\rightarrow\Sigma F = F_{AB} - 60 = 0 \qquad F_{AB} = +60 \text{ kN} \qquad \text{Ans.}$$
$$+\rightarrow\Sigma F = F_{BC} - 60 + 80 = 0 \qquad F_{BC} = -20 \text{ kN} \qquad \text{Ans.}$$
$$+\rightarrow\Sigma F = F_{CD} - 60 + 80 - 50 = 0 \qquad F_{CD} = +30 \text{ kN} \qquad \text{Ans.}$$

For all the above calculations, a free-body diagram of the part of the bar to the left of the imaginary cut has been used. A free-body diagram of the part of the bar to the right of the cut would have yielded identical results. In fact, for the determination of F_{CD}, the free-body diagram to the right of the cut would have been more efficient since only the unknown force F_{CD} and the 30-kN load would have appeared on the diagram.

b. The axial force diagram for the bar, constructed by using the results from part a, is shown in Fig. 8-5e. Note in the diagram that the abrupt changes in internal force are equal to the applied loads at pins A, B, C, and D. Thus, the axial force diagram could have been drawn directly below the sketch of the loaded bar of Fig. 8-5a, without the aid of the free-body diagrams shown in Figs. 8-5b, 8-5c, and 8-5d, by using the applied loads at pins A, B, C, and D.

A steel shaft is used to transmit torque from a motor to operating units in a factory. The torque is input at gear B (see Fig. 8-6a) and is removed at gears A, C, D, and E.

a. Determine the torques transmitted by cross sections in intervals AB, BC, CD, and DE of the shaft.
b. Draw a torque diagram for the shaft.

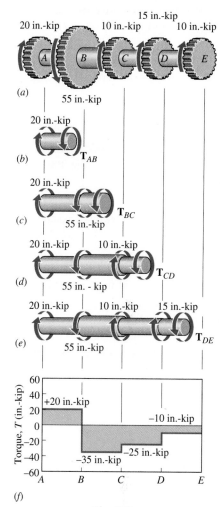

(a)

(b)

(c)

(d)

(e)

(f)

Fig. 8-6

SOLUTION

a. The torques transmitted by cross sections in intervals AB, BC, CD, and DE of the shaft shown in Fig. 8-6a are obtained by using the four free-body diagrams shown in Figs. 8-6b, 8-6c, 8-6d, and 8-6e. Applying the moment equilibrium equation $\Sigma M = 0$ about the axis of the shaft yields,

$$+\circlearrowleft\Sigma M = T_{AB} - 20 = 0 \qquad T_{AB} = +20 \text{ in.} \cdot \text{kip} \qquad \text{Ans.}$$
$$+\circlearrowleft\Sigma M = T_{BC} - 20 + 55 = 0 \qquad T_{BC} = -35 \text{ in.} \cdot \text{kip} \qquad \text{Ans.}$$
$$+\circlearrowleft\Sigma M = T_{CD} - 20 + 55 - 10 = 0 \qquad T_{CD} = -25 \text{ in.} \cdot \text{kip} \qquad \text{Ans.}$$
$$+\circlearrowleft\Sigma M = T_{DE} - 20 + 55 - 10 - 15 = 0 \qquad T_{DE} = -10 \text{ in.} \cdot \text{kip} \qquad \text{Ans.}$$

For all the above calculations, a free-body diagram of the part of the shaft to the left of the imaginary cut has been used. A free-body diagram of the part of the shaft to the right of the cut would have yielded identical results. In fact, for the determination of T_{CD} and T_{DE}, the free-body diagram to the right of the cut would have been more efficient since fewer torques would have appeared on the diagram.

b. The torque diagram for the shaft, constructed by using the results from part a, is shown in Fig. 8-6f. Note in the diagram that the abrupt changes in torque are equal to the applied torques at gears A, B, C, D, and E. Thus, the torque diagram could have been drawn directly below the sketch of the shaft of Fig. 8-6a, without the aid of the free-body diagrams shown in Figs. 8-6b, 8-6c, 8-6d, and 8-6e, by using the applied torques at gears A, B, C, D, and E.

327

PROBLEMS

8-1* A steel bar with a rectangular cross section is used to transmit four axial loads as shown in Fig. P8-1.

a. Determine the axial forces transmitted by cross sections in intervals *AB*, *BC*, and *CD* of the bar.
b. Draw an axial force diagram for the bar.

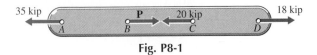

Fig. P8-1

8-2* A steel bar has three axial loads applied as shown in Fig. P8-2.

a. Determine the axial forces transmitted by cross sections in intervals *AB*, *BC*, and *CD* of the bar.
b. Draw an axial force diagram for the bar.

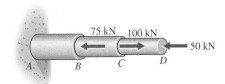

Fig. P8-2

8-3 A bar is loaded and supported as shown in Fig. P8-3.

a. Determine the maximum axial load transmitted by any transverse cross section of the bar.
b. Draw an axial force diagram for the bar.

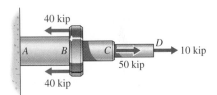

Fig. P8-3

8-4 A bar is loaded and supported as shown in Fig. P8-4.

a. Determine the maximum axial load transmitted by any transverse cross section of the bar.
b. Draw an axial force diagram for the bar.

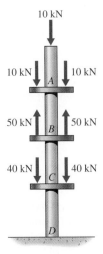

Fig. P8-4

8-5* For the steel shaft shown in Fig. P8-5,

a. Determine the torques transmitted by cross sections in intervals *AB*, *BC*, *CD*, and *DE* of the shaft.
b. Draw a torque diagram for the shaft.

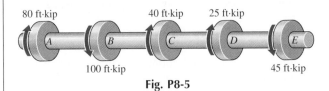

Fig. P8-5

8-6* The motor shown in Fig. P8-6 supplies a torque of 500 N · m to shaft *BCDE*. The torques removed at gears *C*, *D*, and *E* are 100 N · m, 150 N · m, and 250 N · m, respectively.

a. Determine the torques transmitted by cross sections in intervals *BC*, *CD*, and *DE* of the shaft.
b. Draw a torque diagram for the shaft.

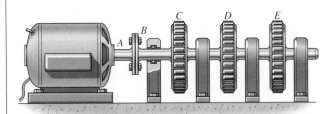

Fig. P8-6

8-7 For the steel shaft shown in Fig. P8-7,

a. Determine the maximum torque transmitted by any transverse cross section of the shaft.
b. Draw a torque diagram for the shaft.

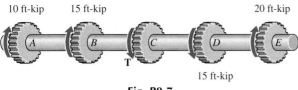

Fig. P8-7

8-8 For the steel shaft shown in Fig. P8-8,

a. Determine the maximum torque transmitted by any transverse cross section of the shaft.
b. Draw a torque diagram for the shaft.

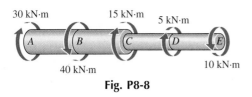

Fig. P8-8

8-9 Five 24-in.-diameter pulleys are keyed to a steel shaft as shown in Fig. P8-9. The pulleys carry belts that are used to drive machinery in a factory. Belt tensions for normal operating conditions are indicated on the figure.

a. Determine the maximum torque transmitted by any transverse cross section of the shaft.
b. Draw a torque diagram for the shaft.

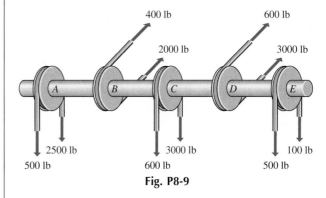

Fig. P8-9

8-10* For the steel shaft shown in Fig. P8-10,

a. Determine the maximum torque transmitted by any transverse cross section of the shaft.
b. Draw a torque diagram for the shaft.

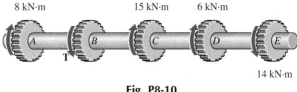

Fig. P8-10

8-3 AXIAL FORCE, SHEAR FORCE, AND BENDING MOMENTS IN MULTIFORCE MEMBERS

The frames and machines discussed in Chapter 7 all contained at least one multiforce member. Recall that the forces in these members were not directed along the axis of the member; therefore, a transverse cross section may be required to transmit axial forces, shear forces, and bending moments. In a general three-dimensional body, as discussed in Section 8-1, the internal forces on a specific plane are statically equivalent to a force–couple system, with the couple being dependent on the choice of location of the force. The point of reference to be used for all problems in this textbook is the centroid of the cross-sectional area. Many formulas derived later in Mechanics of Materials courses will require such a resolution of the internal forces.

For the case of the planar systems of external loads to be considered in this section, the internal force–couple system will consist of an axial force **P**, a shear force **V**, and a bending moment **M**. The procedure for determining these internal resisting forces and moments is illustrated in the following example.

Two bars and a cable are used to support a 500-lb load as shown in Fig. 8-7a. Determine the internal resisting forces and the moment transmitted by

a. Section *aa* in bar *BCD*.
b. Section *bb* in bar *EF*.

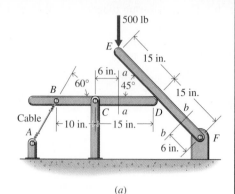

(a)

SOLUTION

The contact force F_D at the point of contact between bar *BCD* and bar *EDF* (see Fig. 8-7a) can be determined by drawing a free-body diagram of bar *EDF* and summing moments about pin *F*, as shown in Fig. 8-7b. Since the line of action of force F_D is perpendicular to the axis of bar *EF*,

$$+\!\downarrow\Sigma M_F = F_D(15) - 500(30)\cos 45° = 0 \qquad F_D = 707 \text{ lb}$$

a. Once the contact force F_D is known, the internal resisting forces and moment on section *aa* can be determined by using the free-body diagram shown in Fig. 8-7c for the portion of bar *BCD* to the right of section *aa*. Thus,

$$+\!\rightarrow\Sigma F_n = P - 707\cos 45° = 0 \qquad P = 500 \text{ lb} \qquad \text{Ans.}$$
$$+\!\uparrow\Sigma F_t = V - 707\sin 45° = 0 \qquad V = 500 \text{ lb} \qquad \text{Ans.}$$
$$+\!\downarrow\Sigma M_0 = M - 707(9)\sin 45° = 0 \qquad M = 4500 \text{ in.}\cdot\text{lb} \qquad \text{Ans.}$$

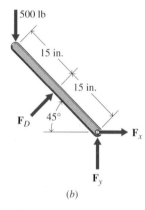

(b)

b. Similarly, the internal resisting forces and moment on section *bb* can be determined by using the free-body diagram shown in Fig. 8-7d for the portion of bar *EF* above section *bb*. Thus,

$$+\!\nwarrow\Sigma F_n = P - 500\cos 45° = 0 \qquad P = 354 \text{ lb} \qquad \text{Ans.}$$
$$+\!\swarrow\Sigma F_t = V - 707 + 500\sin 45° = 0 \qquad V = 354 \text{ lb} \qquad \text{Ans.}$$
$$+\!\downarrow\Sigma M_0 = M + 707(9) - 500(24)\cos 45° = 0 \qquad M = 2120 \text{ in.}\cdot\text{lb} \quad \text{Ans.}$$

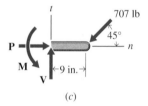

(c)

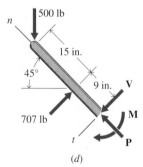

(d)

Fig. 8-7

PROBLEMS

8-11* A bracket mounted on a column carries three loads as shown in Fig. P8-11. Determine the internal resisting forces and moment transmitted by section *aa* in the column.

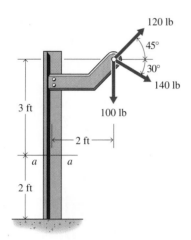

Fig. P8-11

8-12* A simple three-bar frame supports a 3-kN load as shown in Fig. P8-12. Determine the internal resisting forces and moment transmitted by section *aa* in bar *ABC*.

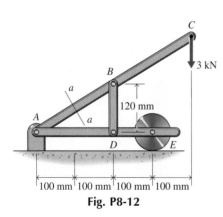

Fig. P8-12

8-13 An angle bracket is loaded and supported as shown in Fig. P8-13. Determine the internal resisting forces and moment transmitted by section *aa* in the bracket.

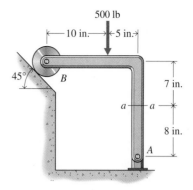

Fig. P8-13

8-14 An angle bracket is loaded and supported as shown in Fig. P8-14. Determine the internal resisting forces and moment transmitted by section *aa* in the bracket.

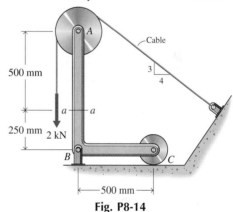

Fig. P8-14

8-15* A simple three-bar frame is loaded and supported as shown in Fig. P8-15. Determine the internal resisting forces and moment transmitted by

a. Section *aa* in bar *BEF*.
b. Section *bb* in bar *ABCD*.

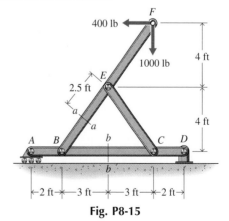

Fig. P8-15

331

8-16* The hook shown in Fig. P8-16 supports a 10-kN load. Determine the internal resisting forces and moment transmitted by

a. Section *aa*.
b. Section *bb*.
c. Section *cc*.

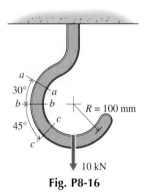

Fig. P8-16

8-17 A simple three-bar frame is loaded and supported as shown in Fig. P8-17. Determine the internal resisting forces and moment transmitted by

a. Section *aa* in bar *BDF*.
b. Section *bb* in bar *ABC*.

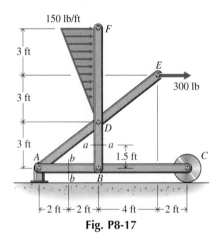

Fig. P8-17

8-18 A simple three-bar frame is loaded and supported as shown in Fig. P8-18. Determine the internal resisting forces and moment transmitted by

a. Section *aa* in bar *BD*.
b. Section *bb* in bar *ABC*.

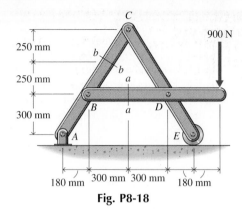

Fig. P8-18

8-19* A simple three-bar frame is loaded and supported as shown in Fig. P8-19. Determine the internal resisting forces and moment transmitted by

a. Section *aa* in bar *ABC*.
b. Section *bb* in bar *CDE*.

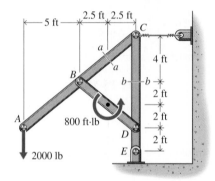

Fig. P8-19

8-20* A simple three-bar frame is loaded and supported as shown in Fig. P8-20. Determine the internal resisting forces and moment transmitted by

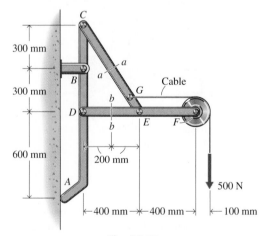

Fig. P8-20

332

a. Section *aa* in bar *CGE*.
b. Section *bb* in bar *DEF*.

8-21 A simple two-bar frame is loaded and supported as shown in Fig. P8-21. Determine the internal resisting forces and moment transmitted by section *aa* in bar *ABC*.

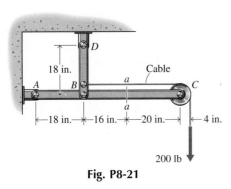

Fig. P8-21

8-22 A simple-three-bar frame is loaded and supported as shown in Fig. P8-22. Determine the internal resisting forces and moment transmitted by

a. Section *aa* in bar *DF*.
b. Section *bb* in bar *ABCD*.

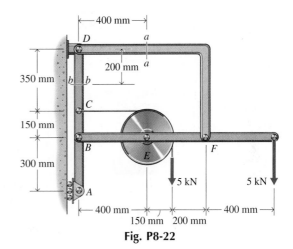

Fig. P8-22

8-4 SHEAR FORCES AND BENDING MOMENTS IN BEAMS

A structural member or machine component designed primarily to support forces acting perpendicular to the axis of the member is called a beam. The principal difference between beams and the axially loaded bars and the torsionally loaded shafts of Section 8-2 is in the direction of the applied loads. In general, the length of a beam is large in comparison to the two cross-sectional dimensions. The line connecting the centroids of the cross sections of a beam is usually referred to as the centroidal or longitudinal axis of the beam. A beam may be straight or curved depending on the shape of this centroidal axis. Many beams are used in a horizontal position in building construction, but vertical and inclined beams are also found in other applications. The deformation of a beam is primarily one of bending. Some beams are loaded in pure bending, while others are subjected to bending loads in combination with axial, shear, and torsional loads. Most of the multiforce members of Section 8-3 were beams of this type. Slender members subjected primarily to axial compressive loads are known as columns. Slender members subjected to axial compressive loads in combination with loads that produce bending are usually referred to as beam columns. In this section, only long slender beams with transverse loads in a single plane (called the plane of bending) will be considered. Such loads will cause only a shear force V_r and a bending moment M_r to be transmitted by an arbitrary cross section of the beam.

Before proceeding with an analysis of the shear force V_r and the bending moment M_r transmitted by a cross section of a beam, a few words will be said about beam supports, beam loadings, and the classification of beams.

333

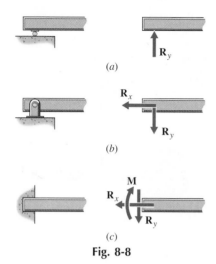

Fig. 8-8

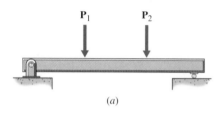

(a)

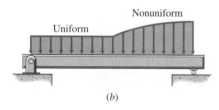

(b)

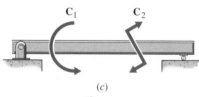

(c)

Fig. 8-9

The three types of beam supports commonly used in engineering practice include roller supports, pin supports, and fixed (or clamp) supports.

1. A roller support (see Fig. 8-8a) resists motion of the beam in a direction perpendicular to the axis of the beam. Hence, the reaction at a roller support for a horizontal beam is a vertical force. The beam is free to rotate at a roller support.

2. A pin support (see Fig. 8-8b) resists motion of the beam in any direction in the plane of loading. The reaction at a pin support for a horizontal beam is often represented by the horizontal and vertical components of the force. The beam is free to rotate at a pin support.

3. A fixed (or clamp) support (see Fig. 8-8c) prevents both rotation of the beam and motion of the beam in any direction in the plane of loading. The reaction at a fixed support can be represented by two force components and a moment.

Beams are classified according to the type of load they support. Beams may be subjected to concentrated loads, distributed loads, or couples (concentrated moments) that act alone or in any combination.

1. Loads that are applied to a very small portion of the length of a beam are called concentrated loads. A concentrated load (see Fig. 8-9a) can be idealized as a discrete force that acts at a specific point on the beam.

2. Loads that act over a finite length of the beam are called distributed loads. The distribution (see Fig. 8-9b) may be uniform or nonuniform. The weight of a beam is an example of a uniformly distributed load.

3. A concentrated moment (see Fig. 8-9c) is a couple produced by two equal and opposite forces applied to the beam at a particular section. The two ways used to represent the couple are shown on Fig. 8-9c.

Beams are also classified into types according to the kind of support used.

1. A beam supported by pins, rollers, or smooth surfaces at the ends (see Fig. 8-10a) is called a simply supported beam.

2. A simply supported beam that extends beyond its supports at either or both ends (see Fig. 8-10b) is called an overhanging beam.

3. A beam that is fixed at one end and free at the other end (see Fig. 8-10c) is called a cantilever beam.

4. A beam that is fixed at one end and simply supported at the other end (see Fig. 8-10d) is called a propped beam.

5. A beam with more than two simple supports (see Fig. 8-10e) is called a continuous beam.

6. A beam that is either fixed (no rotation) or restrained (limited rotation) at the ends (see Fig. 8-10f) is called a built-in beam.

Beams can also be separated into statically determinate beams and statically indeterminate beams. When the support reactions can be obtained from the equations of statics alone, the beam is statically determinate. If the forces applied to a beam are limited to a plane (say the

xy-plane), there are three equilibrium equations available for the determination of the support reactions. The equations are

$$\Sigma F_x = 0 \qquad \Sigma F_y = 0 \qquad \Sigma M_{zA} = 0$$

where A is any point in the plane of loading (the xy-plane). Thus, three reaction components, at most, can be determined. If all the applied forces and support reactions are perpendicular to the longitudinal axis of the beam, the equation $\Sigma F_x = 0$ is automatically satisfied. For such a beam to be statically determinate, only two unknown reactive forces can exist, since the number of equilibrium equations then available for the solution is reduced to two, namely

$$\Sigma F_y = 0 \qquad \text{and} \qquad \Sigma M_{zA} = 0$$

Examples of statically determinate beams are simple beams, overhanging beams, and cantilever beams.

The equations of equilibrium are not sufficient to determine the support reactions when the beam has more supports than are necessary to maintain equilibrium. Such beams are known as statically indeterminate beams, and the load-deformation properties of the beam, in addition to the equations of equilibrium, are required to determine the support reactions. Examples of statically indeterminate beams include the propped cantilever beam, the continuous beam, and the built-in beam. Only statically determinate beams will be treated in this textbook. Statically indeterminate beams will be treated in follow-on courses dealing with mechanics of materials.

Stresses and deflections in beams are functions of internal forces. The internal forces transmitted by a transverse cross section of a beam are those forces and moments required to resist the external forces and maintain equilibrium. Consider the beam shown in Fig. 8-11a, which is subjected to a uniformly distributed load w, two concentrated loads $\mathbf{P}_1$ and $\mathbf{P}_2$, and support reactions at the ends of the beam. The support reactions can be determined by using the equilibrium equations and a

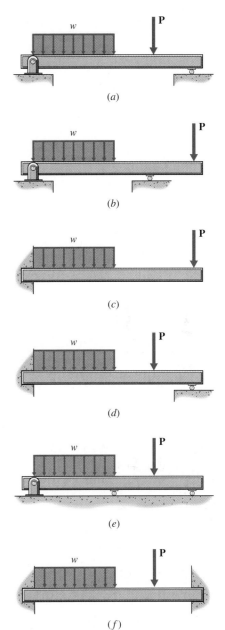

(a)

(b)

(c)

(d)

(e)

(f)

Fig. 8-10

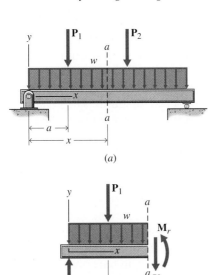

(a)

(b)

Fig. 8-11

free-body diagram of the entire beam. The internal forces transmitted by an arbitrary transverse cross section of the beam (say section aa) can be determined by passing a plane through the section so that the beam is separated into two parts. Since the entire beam is in equilibrium, each part of the beam separated by section aa must also be in equilibrium under the action of the internal forces.

When the equilibrium equation $\Sigma F_y = 0$ is applied to the free-body diagram of Fig. 8-11b, the result can be written as

$$R - wx - P_1 = V_r$$

or

$$V_a = V_r$$

where V_a is defined as the resultant of the external transverse forces acting on the part of a beam to either side of a section. This resultant force V_a is called the transverse shear or just the shear at the section. As seen from Fig. 8-11b, the shear V_a is equal in magnitude and opposite in sense to the resisting shear V_r. Since these shear forces are always equal in magnitude, they are frequently treated as though they were identical. For simplicity, the symbol V will be used henceforth to represent both the transverse shear V_a and the resisting shear V_r.

When the equilibrium equation $\Sigma M_{zA} = 0$ is applied to the free-body diagram of Fig. 8-11b, the result can be written as

$$Rx - \frac{wx^2}{2} - P_1(x - a) = M_r$$

or

$$M_a = M_r$$

where M_a is defined as the algebraic sum of the moments of the external forces, acting on the part of the beam to either side of the section, with respect to an axis at the section perpendicular to the plane of bending (the xy-plane). The moment M_a is called the bending moment or just the moment at the section. As seen from Fig. 8-11b, the bending moment M_a is equal in magnitude and opposite in sense to the resisting moment M_r. Since these moments are always equal in magnitude, they are frequently treated as though they were identical. For simplicity, the symbol M will be used henceforth to represent both the bending moment M_a and the resisting moment M_r.

The bending moment M_a and the transverse shear V_a are not normally shown on a free-body diagram. The usual procedure is to show each external force individually as indicated in Fig. 8-11a. The variation of V_r and M_r along the beam can be expressed by means of equations or shown graphically by means of shear-force and bending-moment diagrams to be developed later, in Section 8-5.

A sign convention is necessary for the correct interpretation of results obtained from equations or diagrams for shear and moment. The convention illustrated in Fig. 8-12 is widely used in the engineering profession. Note in Fig. 8-12b that a bending moment in a horizontal beam is positive at sections for which the top of the beam is in compression and the bottom is in tension. The signs of the terms in the preceding equations for V_r and M_r agree with this convention.

Since M and V vary with x, they are functions of x, and equations for M and V can be obtained from free-body diagrams of portions of the beam. The procedure is illustrated in Example Problem 8-4.

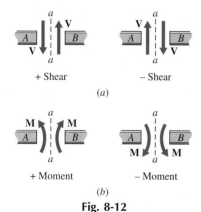

+ Shear   − Shear

(a)

+ Moment   − Moment

(b)

Fig. 8-12

A beam is loaded and supported as shown in Fig. 8-13a. Write equations for the shear V and bending moment M for any section of the beam under the distributed load for the interval $2 < x < 8$).

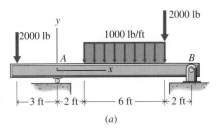

(a)

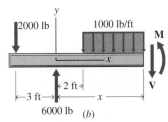

6000 lb   (b)

Fig. 8-13

SOLUTION

A free-body diagram of a portion of the beam from the left end to an arbitrary section under the distributed load is shown in Fig. 8-13b. Note that the resisting shear V and the resisting moment M are shown as positive values. In general, it is impossible to tell without calculation whether the shear and moment at a particular section are positive or negative. For this reason V and M should be represented as positive quantities on free-body diagrams so that the resulting equations will provide values with the correct algebraic sign.

The reaction at support A is determined by using a free-body diagram for the entire beam and summing moments about support B. Thus,

$$R_A(10) - 2000(13) - 1000(6)(5) - 2000(2) = 0$$

from which

$$R_A = 6000 \text{ lb}$$

From the definition of V or from the equilibrium equation $\Sigma F_y = 0$,

$$V = 6000 - 2000 - 1000(x - 2) = -1000x + 6000 \text{ lb} \qquad \text{Ans.}$$

From the definition of M, or from the equilibrium equation $\Sigma M_0 = 0$,

$$M = 6000x - 2000(x + 3) - 1000(x - 2)\left(\frac{x - 2}{2}\right)$$

$$= -500x^2 + 6000x - 8000 \text{ ft} \cdot \text{lb} \qquad \text{Ans.}$$

The equations for V and M in the other intervals can be determined in a similar manner.

PROBLEMS

In Problems 8-23–8-29, beams are loaded and supported as shown in the accompanying figures. Write equations for the shear V and the bending moment M (use the coordinate axes shown) for any section in the specified intervals of the beams.

8-23* In the interval $0 < x < 10$ of the beam shown in Fig. P8-23.

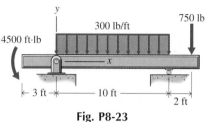

Fig. P8-23

8-24* In the interval $0 < x < 3$ of the beam shown in Fig. P8-24.

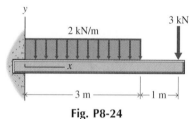

Fig. P8-24

8-25 In the interval $0 < x < 5$ of the beam shown in Fig. P8-25.

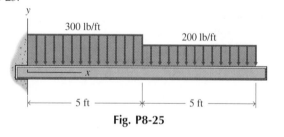

Fig. P8-25

8-26 In the interval $0 < x < 4$ of the beam shown in Fig. P8-26.

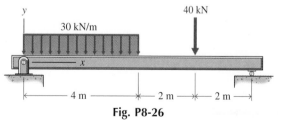

Fig. P8-26

8-27* In the interval $10 < x < 20$ of the beam shown in Fig. P8-27.

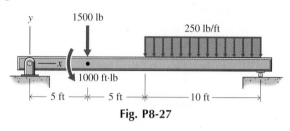

Fig. P8-27

8-28* In the interval $0 < x < 4$ of the beam shown in Fig. P8-28.

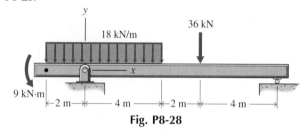

Fig. P8-28

8-29 In the interval $8 < x < 20$ of the beam shown in Fig. P8-29.

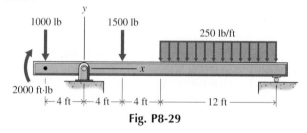

Fig. P8-29

8-30 A beam is loaded and supported as shown in Fig. P8-30. Using the coordinate axes shown, write equations for the shear V and bending moment M for any section of the beam

a. In the interval $0 < x < 3$.
b. In the interval $3 < x < 6$.

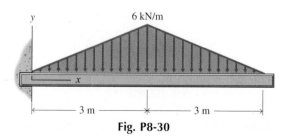

Fig. P8-30

8-31* A beam is loaded and supported as shown in Fig. P8-31. Using the coordinate axes shown, write equations for the shear V and bending moment M for any section of the beam

a. In the interval $0 < x < 12$.
b. In the interval $12 < x < 16$.
c. Use the results of parts a and b to determine the magnitudes and locations of the maximum shear V_{max} and the maximum bending moment M_{max} in the portion of the beam between the supports.

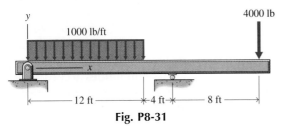

Fig. P8-31

8-32* A beam is loaded and supported as shown in Fig. P8-32. Using the coordinate axes shown, write equations for the shear V and bending moment M for any section of the beam

a. In the interval $0 < x < 2$.
b. In the interval $2 < x < 4$.
c. In the interval $4 < x < 8$.
d. Use the results of parts a, b, and c to determine the magnitudes and locations of the maximum shear V_{max} and the maximum bending moment M_{max} in the beam.

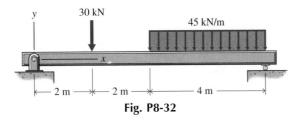

Fig. P8-32

8-33 A beam is loaded and supported as shown in Fig. P8-33. Using the coordinate axes shown, write equations for the shear V and bending moment M for any section of the beam

a. In the interval $0 < x < 6$.
b. In the interval $6 < x < 12$.
c. Use the results of parts a and b to determine the magnitudes and locations of the maximum shear V_{max} and the maximum bending moment M_{max} in the beam.

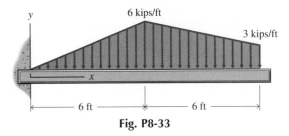

Fig. P8-33

8-34 A beam is loaded and supported as shown in Fig. P8-34. Using the coordinate axes shown, write equations for the shear V and bending moment M for any section of the beam

a. In the interval $0 < x < 2$.
b. In the interval $2 < x < 6$.
c. In the interval $6 < x < 9$.

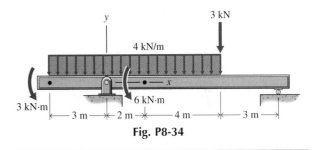

Fig. P8-34

8-5 SHEAR-FORCE AND BENDING-MOMENT DIAGRAMS

Shear-force and bending-moment diagrams are used to provide a graphical representation of the variation of shear force V and bending moment M over the length of a beam. On a shear diagram, the transverse shear force V is plotted as a function of position along the beam. On a moment diagram, the bending moment M is plotted as a function of position along the beam. From such diagrams, maximum values of shear force and maximum values of bending moment and their locations are easily established.

Shear-force and bending-moment diagrams can be drawn by calculating the values of shear and moment at various sections along the

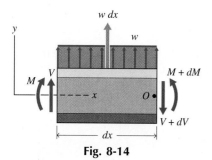

Fig. 8-14

beam and plotting enough points to obtain a smooth curve. Such a procedure is rather time consuming since no single elementary expression can be written for the shear V or the moment M, which applies to the entire length of the beam, unless the load is uniformly distributed or varies according to a known equation along the entire beam. Instead, it is necessary to divide the beam into intervals bounded by the abrupt changes in the loading.

Shear-force and bending-moment diagrams can also be established directly from a free-body diagram (load diagram) for the beam by using several simple mathematical relationships that exist between distributed loads and shears and between shears and moments. These relationships can be developed from a free-body diagram of an elemental length of a beam, as shown in Fig. 8-14. On this diagram, the upward direction is considered positive for the applied load w, and the shears and moments are shown as positive quantities according to the sign convention established in Section 8-4. Since the beam is in equilibrium, the element must also be in equilibrium. Applying the equilibrium equation $\Sigma F_y = 0$ to the element yields,

$$\Sigma F_y = V + w\,dx - (V + dV) = 0$$

from which

$$dV = w\,dx \qquad \text{or} \qquad \frac{dV}{dx} = w \qquad (8\text{-}1)$$

The preceding equation indicates that, at any section in the beam, the slope of the shear diagram is equal to the intensity of loading. When w is known as a function of x, the equation can be integrated between definite limits to yield:

$$\int_{x_1}^{x_2} w\,dx = \int_{V_1}^{V_2} dV = V_2 - V_1 \qquad (8\text{-}2)$$

Thus, the change in shear between sections at x_1 and x_2 is equal to the area under the load diagram between the two sections, provided there are no concentrated forces in the interval $x_1 < x < x_2$.

Applying the equilibrium equation $\Sigma M_O = 0$, to the element shown in Fig. 8-14 yields

$$\Sigma M_O = M + V\,dx + w\,dx\,\frac{dx}{2} - (M + dM) = 0$$

from which

$$dM = V\,dx + w\,\frac{(dx)^2}{2}$$

Dividing by dx and passing to the limit yields

$$\frac{dM}{dx} = V \qquad \text{or} \qquad dM = V\,dx \qquad (8\text{-}3)$$

The preceding equation indicates that at any section in the beam the slope of the moment diagram is equal to the shear. The equation can be integrated between definite limits to give

$$\int_{x_1}^{x_2} V\,dx = \int_{M_1}^{M_2} dM = M_2 - M_1 \qquad (8\text{-}4)$$

Thus, the change in moment between sections at x_1 and x_2 is equal to the area under the shear diagram between the two sections, provided there are no applied couples in the interval $x_1 < x < x_2$.

Note that the equations in this section were derived with the x-axis positive to the right, the applied loads positive upward, and the shear and moment signs as indicated in Fig. 8-12. If one or more of these assumptions are changed, the algebraic signs in the equation may need to be altered.

The relations just developed provide a second means for drawing shear and moment diagrams and for computing values of shear and moment at various sections along a beam. The method consists of drawing the shear diagram from the load diagram and the moment diagram from the shear diagram by means of Eqs. 8-1 through 8-4. This latter method, though it may not produce a precise curve, provides the shear and moment information usually required for designing beams and is less time consuming than the first method.

Complete shear and moment diagrams should indicate values of shear and moment at each section where the load changes abruptly and at sections where they are maximum or minimum (negative maximum values). Sections where the shear and moment are zero should also be located. When all loads and reactions are known, the shear and moment at the ends of the beam can be determined by inspection. Both shear and moment are zero at the free end of a beam unless a force or a couple or both are applied there; in this case, the shear is the same as the force and the moment the same as the couple. At a simply supported or pinned end, the shear must equal the end reaction and the moment must be zero. At a built-in or fixed end, the reactions are the shear and moment values.

Once a starting point for the shear diagram is established, the diagram can be sketched immediately below and aligned with the load diagram by using the definition of shear and the fact that the slope of the shear diagram can be obtained from the load diagram. When positive directions are chosen as upward and to the right, a positive distributed load, one acting upward, will result in a positive slope on the shear diagram, and a negative load will give a negative slope. A concentrated force will produce an abrupt change in shear. The change in shear between any two sections is given by the area under the load diagram between the two sections. The change of shear at a concentrated force is equal to the concentrated force.

The moment diagram is drawn immediately below and aligned with the shear diagram in the same manner. The slope at any point on the moment diagram is given by the shear at the corresponding point on the shear diagram, a positive shear representing a positive slope, where upward and to the right are positive, and a negative shear representing a negative slope. The change in moment between any two sections is given by the area under the shear diagram between the two sections. A couple applied to a beam will cause the moment to change abruptly by an amount equal to the moment of the couple.

Example Problems 8-5 and 8-6 illustrate the construction of shear and moment diagrams, directly from the load diagram, by using Eqs. 8-1 through 8-4.

A beam is loaded and supported as shown in Fig. 8-15*a*. Draw complete shear and moment diagrams for the beam.

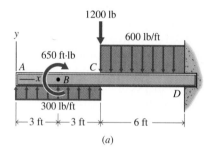

(a)

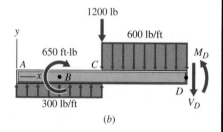

(b)

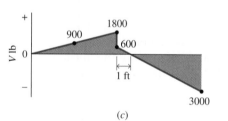

(c)

SOLUTION

A free-body diagram, or load diagram, for the beam is shown in Fig. 8-15*b*. The reactions at D are computed from the equations of equilibrium. It is not necessary to compute the reactions for a cantilever beam in order to draw shear and moment diagrams from the load diagram, but the reactions provide a convenient check. From the equilibrium equation $\Sigma F_y = 0$,

$$V_D = 300(6) - 1200 - 600(6) = -3000 \text{ lb}$$

From the equilibrium equation $\Sigma M_D = 0$,

$$M_D = 300(6)(9) + 650 - 1200(6) - 600(6)(3) = -1150 \text{ ft} \cdot \text{lb}$$

The shear and moment diagrams can be drawn directly from the load diagram without writing the shear and moment equations. The shear diagram, shown in Fig. 8-15*c*, is drawn below the load diagram. The shear at the left end of the beam is zero. The slope of the shear diagram is equal to the load w; therefore, the slope of the diagram between A and C is $+300$ lb/ft. The change in shear from A to C is equal to the area under the load diagram; thus,

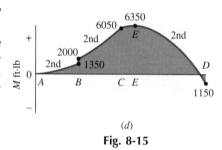

(d)

Fig. 8-15

$$V_C = V_A + \Delta V = 0 + 300(6) = 1800 \text{ lb}$$

A concentrated load **P** is applied at C; therefore, the shear changes abruptly by the magnitude of the load and in the direction of the load. Thus,

$$V_{C'} = V_C + P = 1800 - 1200 = 600 \text{ lb}$$

A uniform load w is applied to the beam between C' and D. Since the slope of the shear diagram is equal to the load w, the slope of the diagram between C' and D is -600 lb/ft. The change in shear from C' to D is equal to the area under the load diagram; thus,

$$V_D = V_{C'} + \Delta V = 600 + (-600)(6) = -3000 \text{ lb}$$

This shear is the same as the reaction at D, which provides a check.

The moment diagram is drawn below the shear diagram, as shown in Fig. 8-15d. The moment at the left end of the beam is zero. The slope of the moment diagram is equal to the shear V; therefore, the slope of the diagram increases uniformly from 0 at A to 900 lb at B, where a concentrated moment **C** is applied. The change in moment from A to B is equal to the area under the shear diagram; thus,

$$M_B = M_A + \Delta M = 0 + \frac{1}{2}(900)(3) = 1350 \text{ ft} \cdot \text{lb}$$

In the interval AB of the beam,

$$\frac{dM}{dx} = V = 300x \qquad (a)$$

because of the positive (upward) 300 lb/ft distributed load on the beam. Thus,

$$M = 150x^2 + C_1 = 150x^2 \qquad (b)$$

since $C_1 = 0$ because of the boundary condition $M = 0$ at $x = 0$. Equations a and b show that a constant distributed load distribution produces a linear shear distribution and a second-order moment distribution in the interval AB of the beam. The degree (2nd) of the moment equation between A and B is indicated on the moment diagram. At B, the moment changes abruptly by an amount equal to the moment of the couple ($+650$ ft $\cdot$ lb) applied to the beam at B. This moment does not appear on the shear diagram, so some procedure should be established to recall its presence on the beam when the moment diagram is being drawn by using shear diagram information. Thus,

$$M_{B'} = M_B + C = 1350 + 650 = 2000 \text{ ft} \cdot \text{lb}$$

From B' to C, the slope of the moment diagram increases uniformly from 900 lb at B' to 1800 lb at C. Again, the change in moment from B' to C is equal to the area under the shear diagram; thus,

$$M_C = M_{B'} + \Delta M = 2000 + \frac{1}{2}(900 + 1800)(3) = 6050 \text{ ft} \cdot \text{lb}$$

The slope of the moment diagram changes abruptly at C from $+1800$ lb to $+600$ lb. It then decreases uniformly from $+600$ lb at C to -3000 lb at D. The slope of the moment diagram is zero at point E, which is 1 ft to the right of point C. From the areas under the shear diagram, moments at E and D are

$$M_E = M_C + \Delta M = 6050 + \frac{1}{2}(600)(1) = 6350 \text{ ft} \cdot \text{lb}$$

$$M_D = M_E + \Delta M = 6350 + \frac{1}{2}(-3000)(5) = -1150 \text{ ft} \cdot \text{lb}$$

This moment is the same as the reaction at D, which provides a check.

A beam is loaded and supported as shown in Fig. 8-16a. Draw complete shear and moment diagrams for the beam.

SOLUTION

A free-body diagram, or load diagram, for the beam is shown in Fig. 8-16b. The reactions at B, C, and E are computed by using the equations of equilibrium for the entire beam. Since the pin at D will not transmit a moment, applying the equilibrium equation $\Sigma M_D = 0$ to the right portion of the beam yields $R_E = 50$ kN. Once R_E is known, the equilibrium equation $\Sigma M_B = 0$ yields $R_C = -35$ kN. Finally, the equilibrium equation $\Sigma F_y = 0$ yields $R_B = 215$ kN.

Once reactions R_B, R_C, and R_E are known, the shear and moment diagrams can be drawn directly from the load diagram as shown in Fig. 8-16c. The shear at the left end of the beam is zero. The slope of the shear diagram is equal to the load w; therefore, the slope of the diagram between A and B is -80 kN/m. The change in shear from A to B is equal to the area under the load diagram; thus,

$$V_B = V_A + \Delta V = 0 + (-80)(1.5) = -120 \text{ kN}$$

The shear changes abruptly at support B. Thus,

$$V_{B'} = V_B + R_B = -120 + 215 = +95 \text{ kN}$$

The slope of the shear diagram between B and C is -40 kN/m. The change in shear from B' to C is equal to the area under the load diagram; thus,

$$V_C = V_{B'} + \Delta V = 95 + (-40)(2) = +15 \text{ kN}$$

The shear changes abruptly at support C. Thus,

$$V_{C'} = V_C + R_C = 15 + (-35) = -20 \text{ kN}$$

The beam is not loaded between C and E; therefore, the slope of the shear diagram is zero and

$$V_E = V_{C'} + \Delta V = -20 + (0)(2.5) = -20 \text{ kN}$$

The shear changes abruptly at support E. Thus,

$$V_{E'} = V_E + R_E = -20 + 50 = 30 \text{ kN}$$

The slope of the shear diagram decreases uniformly from 0 at E to -40 kN/m at F. Thus, from the area under the load diagram,

$$V_F = V_{E'} + \Delta V = 30 + \frac{1}{2}(-40)(1.5) = 0$$

The shear at the right end of the beam must be zero since it is a free end. If the shear at F is not zero, it indicates that an error has occurred.

The moment diagram, shown in Fig. 8-16d, is drawn below the shear diagram. The moment at the left end of the beam is zero. The slope of the moment diagram is equal to the shear V; therefore, the slope of the moment diagram decreases uniformly from 0 at A to -120 kN at support B. The change in mo-

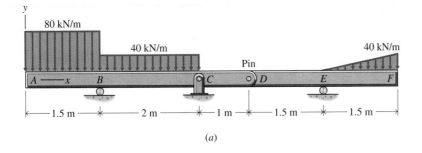

(a)

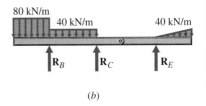

(b)

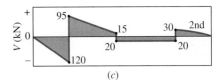

(c)

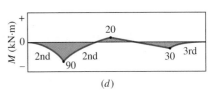

(d)

Fig. 8-16

ment from A to B is equal to the area under the shear diagram; thus,

$$M_B = M_A + \Delta M = 0 + \frac{1}{2}(-120)(1.5) = -90 \text{ kN} \cdot \text{m}$$

At support B the slope of the moment diagram changes abruptly from -120 kN to $+95$ kN; it then decreases uniformly from $+95$ kN at support B to $+15$ kN at support C. From the area under the shear diagram, the moment at C is

$$M_C = M_B + \Delta M = -90 + \frac{1}{2}(95 + 15)(2) = +20 \text{ kN} \cdot \text{m}$$

At support C, the slope of the moment diagram changes abruptly from $+15$ kN to -20 kN; it then remains constant at -20 kN between supports C and E. From the area under the shear diagram, the moment at E is

$$M_E = M_C + \Delta M = +20 + (-20)(2.5) = -30 \text{ kN} \cdot \text{m}$$

At support E, the slope of the moment diagram changes abruptly from -20 kN to $+30$ kN. It then decreases parabolically from $+30$ kN at support E to 0 at end F of the beam. From the area under the shear diagram, the moment at F is

$$M_F = M_E + \Delta M = -30 + \frac{2}{3}(30)(1.5) = 0$$

The moment at the right end of the beam must be zero since it is a free end. If the moment at F were not zero, it would indicate that an error has occurred. Also,

$$M_D = M_C + \Delta M = 20 - 20(1) = 0$$

The moment at hinge D must be zero. If M_D were not zero, it would indicate that an error has occurred.

Calculus would suggest that the locations of maximum shear and maximum moment are where $dV/dx = w = 0$ and $dM/dx = V = 0$, respectively. However, the functions $w(x)$ and $V(x)$ usually are not smooth, continuous functions and the maximum values of V and M often occur at points where concentrated loads or couples are applied to the beam rather than where $w(x) = 0$ or where $V(x) = 0$. Both possibilities need to be checked.

Sections where the bending moment is zero, called points of inflection or contraflexure, can be located by equating the expression for M to zero. The fiber stress is zero at such sections, and if a beam must be spliced, the splice should be located at or near a point of inflection if there is one.

PROBLEMS

In Problems 8-35–8-50, beams are loaded and supported as shown in the accompanying figures. Draw complete shear and moment diagrams for the beams.

8-35* The beam shown in Fig. P8-35.

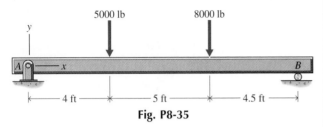

Fig. P8-35

8-36* The beam shown in Fig. P8-36.

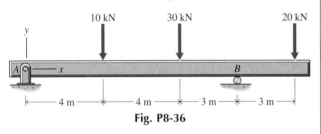

Fig. P8-36

8-37 The beam shown in Fig. P8-37.

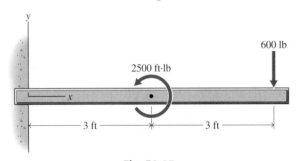

Fig. P8-37

8-38 The beam shown in Fig. P8-38.

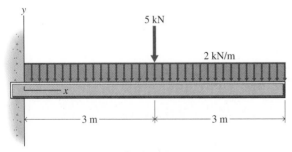

Fig. P8-38

8-39* The beam shown in Fig. P8-39.

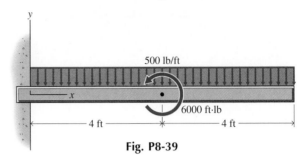

Fig. P8-39

8-40* The beam shown in Fig. P8-40.

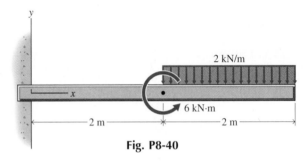

Fig. P8-40

8-41 The beam shown in Fig. P8-41.

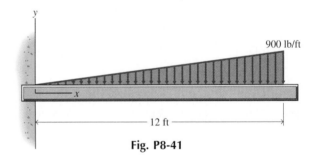

Fig. P8-41

8-42 The beam shown in Fig. P8-42.

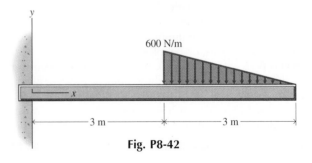

Fig. P8-42

8-43* The beam shown in Fig. P8-43.

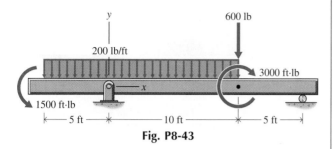

Fig. P8-43

8-44* The beam shown in Fig. P8-44.

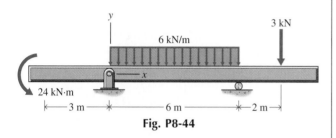

Fig. P8-44

8-45 The beam shown in Fig. P8-45.

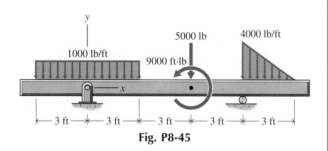

Fig. P8-45

8-46 The beam shown in Fig. P8-46.

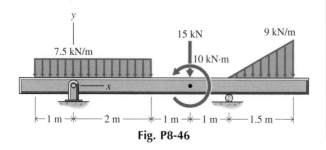

Fig. P8-46

8-47* The beam shown in Fig. P8-47.

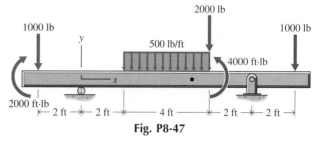

Fig. P8-47

8-48* The beam shown in Fig. P8-48.

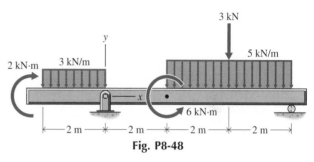

Fig. P8-48

8-49 The beam shown in Fig. P8-49.

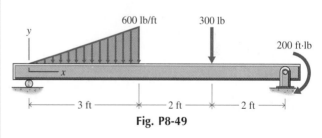

Fig. P8-49

8-50 The beam shown in Fig. P8-50.

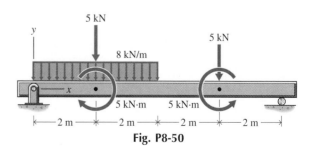

Fig. P8-50

Moment diagrams are shown in the figures accompanying Problems 8-51–8-56. Draw load and shear diagrams for the beams.

8-51 The beam shown in Fig. P8-51.

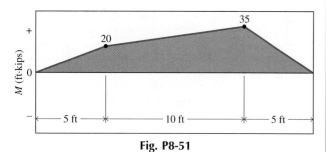

Fig. P8-51

8-52 The beam shown in Fig. P8-52.

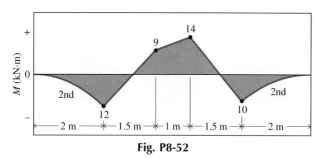

Fig. P8-52

8-53 The beam shown in Fig. P8-53.

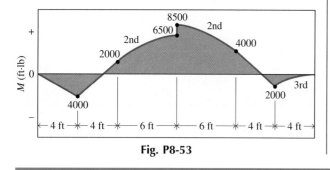

Fig. P8-53

8-54 The beam shown in Fig. P8-54.

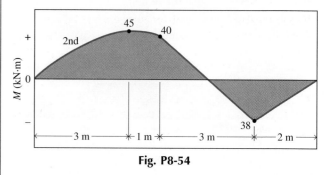

Fig. P8-54

8-55 The beam shown in Fig. P8-55.

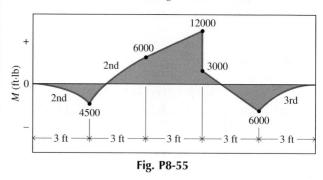

Fig. P8-55

8-56 The beam shown in Fig. P8-56.

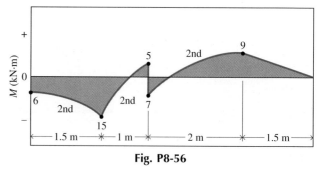

Fig. P8-56

8-6   FLEXIBLE CABLES

Flexible cables are used for suspension bridges and aerial tramways, for power transmission and telephone lines, for guy wires on radio and television towers, and for many other engineering applications. A cable is said to be perfectly flexible when it offers no resistance to bending. Actual cables are not perfectly flexible; however, the resistance they offer to bending is generally so small that any bending effects can be neglected in the analysis of the cable without introducing serious error. Once it is assumed that the cable offers no resistance to bending, the resultant internal force on any cross section must act along a tangent to the cable at that cross section.

In previous applications, cables were assumed to be straight two-force members capable of transmitting only axial tensile forces. When transverse loads are applied to a cable, it cannot remain straight but sags. Sag is defined as the difference in elevation between the lowest point on the cable and a support. When the supports are not at the same elevation, the sag measured from one support will be different from the sag measured from the other support. The span of a cable is defined as the horizontal distance between supports.

Flexible cables may be subjected to a series of distinct concentrated loads, or they may be subjected to loads that are uniformly distributed over the horizontal span of the cable or uniformly distributed over the length of the cable. The weights of cars and their contents on an aerial tramway is an example of a cable subjected to a series of concentrated loads. The weight of a suspension-bridge roadway is an example of a load that is uniformly distributed along the horizontal span of the cable. The weight of a power transmission cable of constant cross section is an example of a load that is uniformly distributed along the length of the cable.

In the following discussion of cables it will be assumed that the cables are perfectly flexible and inextensible. Relationships between the length, span, and sag of the cable, the tension in the cable, and the loads applied to the cable will be determined from equilibrium considerations.

8-6-1   Cables Subjected to Concentrated Loads

A cable subjected to concentrated loads P_1, P_2, and P_3 at discrete points along its length is depicted in Fig. 8-17a. The cable is anchored to rigid walls at ends A and B with pins. If the loads are much larger than the weight of the cable, the weight of the cable can be neglected in the analysis and the segments of the cable can be considered as straight two-force bars.

Assume for the following discussion that loads P_1, P_2, and P_3 together with distances x_1, x_2, x_3, and span a are known. The distances y_1, y_2, and y_3 are unknowns to be determined. A free-body diagram of the cable is shown in Fig. 8-17b. Since the distances y_1, y_2, and y_3 are unknown, the slopes of the cable segments at ends A and B are not known; therefore, the reactions at A and B are represented by two components each. Since four unknowns are involved, the three equations of equilibrium for this free-body diagram (Fig. 8-17b) are not sufficient to determine the reactions at A and B. The information that can be determined is as follows:

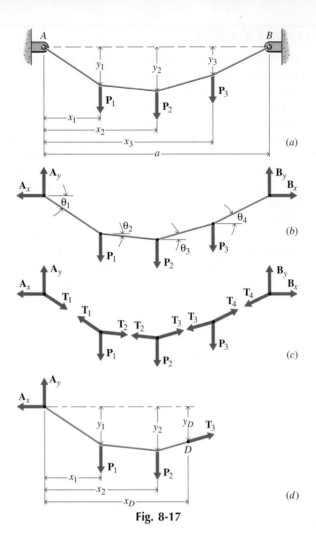

Fig. 8-17

From the equilibrium equation $\Sigma M_A = 0$:

$$+\lcircarrowdown B_y a - P_1 x_1 - P_2 x_2 - P_3 x_3 = 0$$

$$B_y = \frac{1}{a}(P_1 x_1 + P_2 x_2 + P_3 x_3)$$

From the equilibrium equation $\Sigma F_y = 0$:

$$+\uparrow A_y + B_y - P_1 - P_2 - P_3 = 0$$
$$A_y = P_1 + P_2 + P_3 - B_y$$
$$= P_1 + P_2 + P_3 - \frac{1}{a}(P_1 x_1 + P_2 x_2 + P_3 x_3)$$

From the equilibrium equation $\Sigma F_x = 0$:

$$+\rightarrow B_x - A_x = 0$$
$$B_x = A_x$$

Previously in the discussion of frames and trusses, additional equations were obtained by considering equilibrium of a portion of the structure. For a cable, if the weight of the cable is neglected, the internal forces in the different segments of the cable can be represented as shown in Fig. 8-17c. From this series of free-body diagrams and the equilibrium equation $\Sigma F_x = 0$, it is observed that

$$A_x = T_1 \cos \theta_1 = T_2 \cos \theta_2 = T_3 \cos \theta_3 = T_4 \cos \theta_4 = B_x$$

This equation indicates that the horizontal component of the tensile force at any point in the cable is constant and equal to the horizontal component of the pin reaction at the supports. The maximum tension T will occur in the segment with the largest angle of inclination θ since $\cos \theta$ will be minimum in this segment. Such a segment must be adjacent to one of the two supports. The equilibrium equations $\Sigma F_x = 0$ and $\Sigma F_y = 0$ applied to the pins at supports A and B yield

$$A_x = T_1 \cos \theta_1 \qquad B_x = T_4 \cos \theta_4$$
$$A_y = T_1 \sin \theta_1 \qquad B_y = T_4 \sin \theta_4$$

from which

$$T_1 = \sqrt{A_x^2 + A_y^2} \qquad T_4 = \sqrt{B_x^2 + B_y^2}$$
$$\theta_1 = \tan^{-1} \frac{A_y}{A_x} \qquad \theta_4 = \tan^{-1} \frac{B_y}{B_x} \qquad (a)$$

If either the maximum tension or the maximum slope is specified for a given problem, Eqs. a can be used to determine the unknown horizontal components of the support reactions. Once either A_x or B_x is known, all the remaining unknowns T_1, T_2, T_3, T_4, y_1, y_2, and y_3 can be determined by using the free-body diagrams shown in Fig. 8-17c. The procedure is illustrated in Example Problem 8-7.

The unknown horizontal components of the support reactions can also be determined if the vertical distance from a support to any point along the cable is known. For example, consider the free-body diagram shown in Fig. 8-17d, where it is assumed that the distance y_D is known. The equilibrium equation $\Sigma M_D = 0$ yields

$$A_y x_D - P_1(x_D - x_1) - P_2(x_D - x_2) - A_x y_D = 0$$

from which

$$A_x = \frac{1}{y_D}[A_y x_D - P_1(x_D - x_1) - P_2(x_D - x_2)]$$

Finally, problems of this type can be solved by specifying the length of the cable. In this case, the length of each segment of the cable is written in terms of the vertical distances y_1, y_2, and y_3 and the horizontal distances x_1, x_2, x_3, and span a. The required additional equation is then obtained by equating the sum of the individual segment lengths to the total length L:

$$L = \sqrt{x_1^2 + y_1^2} + \sqrt{(x_2 - x_1)^2 + (y_2 - y_1)^2}$$
$$+ \sqrt{(x_3 - x_2)^2 + (y_2 - y_3)^2} + \sqrt{(a - x_3)^2 + y_3^2}$$

Since the terms for the segment lengths in this equation involve square roots of the unknown vertical distances, any solution is extremely tedious and time consuming if performed by hand calculation. Computer solution is recommended for this formulation of cable problems.

A cable supports concentrated loads of 500 lb and 200 lb as shown in Fig. 8-18a. If the maximum tension in the cable is 1000 lb, determine

a. The support reactions A_x, A_y, D_x, and D_y.
b. The tensions T_1, T_2, and T_3 in the three segments of the cable.
c. The vertical distances y_B and y_C from the level of support A.
d. The length L of the cable.

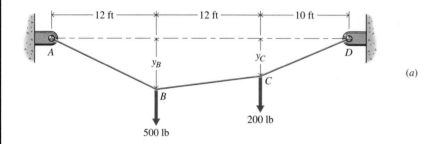

(a)

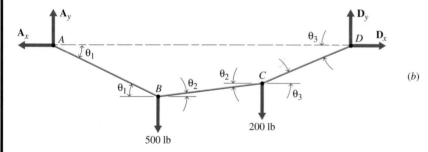

(b)

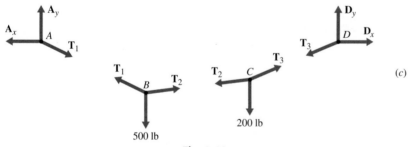

(c)

Fig. 8-18

SOLUTION

a. A free-body diagram for the cable is shown in Fig. 8-18b. From the equilibrium equation $\Sigma M_D = 0$,

$$A_y = \frac{1}{34}[500(22) + 200(10)] = 382.4 = 382 \text{ lb} \qquad \text{Ans.}$$

The maximum tension of 1000 lb will occur in interval AB of the cable. Thus, from the free-body diagram at point A of the cable (Fig. 8-18c),

$$A_x = \sqrt{T_1^2 - A_y^2} = \sqrt{1000^2 - 382.4^2} = 924.0 = 924 \text{ lb} \qquad \text{Ans.}$$

Also,

$$\theta_1 = \tan^{-1}\frac{A_y}{A_x} = \tan^{-1}\frac{382.4}{924.0} = 22.48°$$

Return now to the free-body diagram for the entire cable shown in Fig. 8-18b. From the equilibrium equation $\Sigma F_x = 0$,

$$D_x = A_x = 924.0 = 924 \text{ lb} \qquad \text{Ans.}$$

From the equilibrium equation $\Sigma F_y = 0$,

$$D_y = 500 + 200 - A_y = 700 - 382.4 = 317.6 = 318 \text{ lb} \qquad \text{Ans.}$$

b. Free-body diagrams at points B and C (Fig. 8-18c) can be used to determine the tensions T_2 and T_3. From the equilibrium equations $\Sigma F_x = 0$ and $\Sigma F_y = 0$,

At point B:

$$T_{2x} = T_2 \cos \theta_2 = T_1 \cos \theta_1 = 1000 \cos 22.48° = A_x = 924.0 \text{ lb}$$
$$T_{2y} = T_2 \sin \theta_2 = 500 - T_1 \sin \theta_1 = 500 - 1000 \sin 22.48° = 117.64 \text{ lb}$$
$$T_2 = \sqrt{T_{2x}^2 + T_{2y}^2} = \sqrt{924.0^2 + 117.64^2} = 931.4 = 931 \text{ lb} \qquad \text{Ans.}$$
$$\theta_2 = \tan^{-1}\frac{T_{2y}}{T_{2x}} = \tan^{-1}\frac{117.64}{924.0} = 7.256°$$

At point C:

$$T_{3x} = T_3 \cos \theta_3 = T_2 \cos \theta_2 = 931.4 \cos 7.256° = A_x = 924.0 \text{ lb}$$
$$T_{3y} = T_3 \sin \theta_3 = 200 + T_2 \sin \theta_2 = 200 + 931.4 \sin 7.256° = 317.6 \text{ lb}$$
$$T_3 = \sqrt{T_{3x}^2 + T_{3y}^2} = \sqrt{924.0^2 + 317.6^2} = 977.1 = 977 \text{ lb} \qquad \text{Ans.}$$
$$\theta_3 = \tan^{-1}\frac{T_{3y}}{T_{3x}} = \tan^{-1}\frac{317.6}{924.0} = 18.969°$$

As a check at point D:

$$T_3 = \sqrt{D_x^2 + D_y^2} = \sqrt{924.0^2 + 317.6^2} = 977.1 = 977 \text{ lb} \qquad \text{Ans.}$$

c. Once the angles are known, the vertical distances y_B and y_C are

$$y_B = 12 \tan \theta_1 = 12 \tan 22.48° = 4.966 = 4.97 \text{ ft} \qquad \text{Ans.}$$
$$y_C = 10 \tan \theta_3 = 10 \tan 18.969° = 3.439 = 3.44 \text{ ft} \qquad \text{Ans.}$$

d. The cable length L is obtained from the segment lengths as

$$L = \sqrt{12^2 + 4.966^2} + \sqrt{12^2 + (4.966 - 3.439)^2} + \sqrt{10^2 + 3.439^2}$$
$$= 35.66 = 35.7 \text{ ft} \qquad \text{Ans.}$$

PROBLEMS

8-57* A cable supports two vertical loads as shown in Fig. P8-57. If the maximum tension in the cable is 2000 lb, determine

a. The horizontal and vertical components of the reactions at supports A and D.
b. The vertical distances y_B and y_C.
c. The length L of the cable.

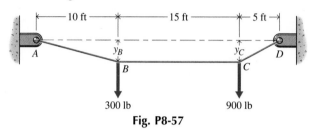

300 lb   900 lb

Fig. P8-57

8-58* A cable supports two vertical loads as shown in Fig. P8-58. The sag at point B of the cable is 2 m. Determine

a. The horizontal and vertical components of the reactions at supports A and D.
b. The tensions in the three segments of the cable.
c. The length L of the cable.

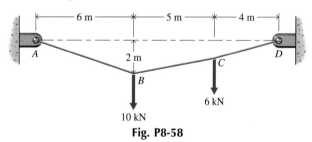

6 kN

10 kN

Fig. P8-58

8-59 The cable supports three vertical loads as shown in Fig. P8-59. The sag at point C of the cable is 4 ft. Determine

a. The horizontal and vertical components of the reactions at supports A and E.
b. The tensions in the four segments of the cable.
c. The length L of the cable.

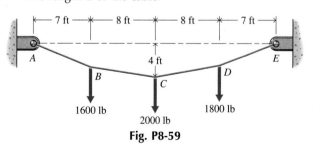

1600 lb   1800 lb

2000 lb

Fig. P8-59

8-60 A cable supports three vertical loads as shown in Fig. P8-60. If the maximum tension in the cable is 50 kN, determine

a. The horizontal and vertical components of the reactions at supports A and D.
b. The vertical distances y_B, y_C, and y_D.
c. The length L of the cable.

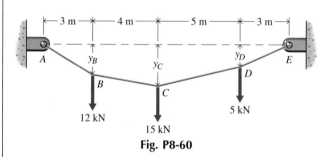

12 kN   5 kN

15 kN

Fig. P8-60

8-61* Solve Problem 8-57 if the maximum tension in the cable is 1500 lb.

8-62* Solve Problem 8-60 if the maximum tension in the cable is 35 kN.

8-63 A cable supports two vertical loads as shown in Fig. P8-63. If the maximum tension in the cable is 1000 lb, determine

a. The horizontal and vertical components of the reactions at supports A and D.
b. The vertical distances y_B and y_C.
c. The length L of the cable.

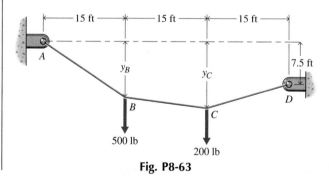

500 lb   200 lb

Fig. P8-63

8-64 A cable is loaded and supported as shown in Fig. P8-64. Determine

a. The horizontal and vertical components of the reactions at supports A and D.
b. The tensions in the three segments of the cable.
c. The vertical distance y_C.
d. The length L of the cable.

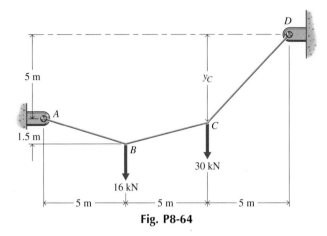

Fig. P8-64

8-65* A cable is loaded and supported as shown in Fig. P8-65. The vertical distance y_C is 4 ft. Determine

a. The horizontal and vertical components of the reactions at supports A and D.
b. The tensions in the three segments of the cable.
c. The vertical distance y_B.
d. The length L of the cable.

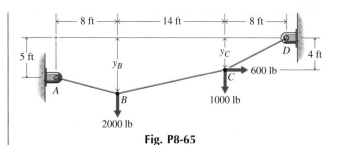

Fig. P8-65

8-66* A cable is loaded and supported as shown in Fig. P8-66. If the maximum tension in the cable is 40 kN, determine

a. The horizontal and vertical components of the reactions at supports A and D.
b. The vertical distances y_B and y_C.
c. The length L of the cable.

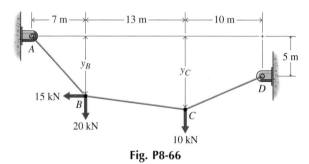

Fig. P8-66

8-67 Solve Problem 8-65 if the direction of the 600 lb load at point C is reversed.

8-68 Solve Problem 8-66 if the direction of the 15 kN load at point B is reversed.

8-6-2   Cables with Loads Uniformly Distributed Along the Horizontal

A cable carrying a load **W** distributed uniformly over a distance a is illustrated in Fig. 8-19a. Such a cable could be analyzed by using the procedures discussed in the previous section; however, the process would be time consuming and tedious due to the large number of loads involved. Such a cable can also be analyzed by representing the large number of discrete loads as a uniformly distributed load $w(x) = W/a$ along the horizontal. Little error is introduced by this simplification if the weight of the cable is small in comparison to the weight being supported by the cable. The loads on the cables of a suspension bridge closely approximate this type of loading since the weights of the cables are usually small in comparison to the weight of the roadway.

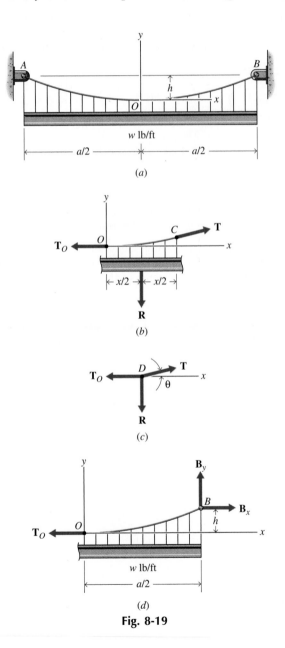

Fig. 8-19

The weight of the roadway is uniformly distributed along the length of the roadway.

Equations that express relationships between the length L, span a, and sag h of a cable with supports at the same height, the tension T in the cable, and the distributed load $w(x)$ applied to the cable can be developed by using equilibrium considerations and the free-body diagram of a portion of the cable shown in Fig. 8-19b. In this diagram, the lowest point on the cable (point O) has been taken as the origin of the xy-coordinate system.

The segment of cable shown in Fig. 8-19b is subjected to three forces; the tension $\mathbf{T}_O$ in the cable at point O, the tension $\mathbf{T}$ in the cable at an arbitrary point C, located a distance x from the origin, and the resultant $\mathbf{R}$ ($R = wx$) of the distributed load whose line of action is located a distance $x/2$ from point O. Since segment OC of the cable is in equilibrium under the action of these three forces, the three forces must be concurrent at point D, as shown in Fig. 8-19c. The two equilibrium equations $\Sigma F_x = 0$ and $\Sigma F_y = 0$ for this concurrent force system yield

$$+\rightarrow\Sigma F_x = T \cos\theta - T_O = 0 \qquad T_x = T\cos\theta = T_O \qquad (a)$$
$$+\uparrow\Sigma F_y = T \sin\theta - R = 0 \qquad T_y = T\sin\theta = R = wx \qquad (b)$$

Equation a shows that the horizontal component T_x of the tension T at any point in the cable is constant and equal to the tension T_O at the lowest point on the cable. Solving Eqs. a and b for T and θ yields

$$T = \sqrt{T_O^2 + w^2 x^2} \qquad (8\text{-}5)$$

$$\theta = \tan^{-1}\frac{wx}{T_O} \qquad (8\text{-}6)$$

Equation 8-5 shows that the tension is minimum at the lowest point in the cable (where $x = 0$) and maximum at the supports (where $x = a/2$). Thus,

$$T_{max} = \sqrt{T_O^2 + \frac{w^2 a^2}{4}} \qquad (8\text{-}7)$$

The shape of the curve can be determined by using Eq. 8-6, which gives the slope. Thus,

$$\frac{dy}{dx} = \tan\theta = \frac{wx}{T_O} \qquad (c)$$

Integrating Eq. c yields

$$y = \frac{wx^2}{2T_O} + C$$

The constant C is determined by using the boundary conditions resulting from the choice of axes. For the xy-coordinate system being used, y is equal to zero when x is equal to zero; therefore, C is equal to zero, and the equation of the loaded cable is

$$y = \frac{wx^2}{2T_O} = kx^2 \qquad (8\text{-}8)$$

Equation 8-8 indicates that the shape of the loaded cable is a parabola with its vertex at the lowest point of the cable. By applying the boundary condition $y = h$ (the sag) when $x = a/2$, an expression for the tension T_O is obtained in terms of the applied load w, the span a, and the sag h. Thus,

$$T_O = \frac{wa^2}{8h} \tag{8-9}$$

If Eq. 8-9 is substituted into Eq. 8-7, the maximum tension in the cable T_{max} is obtained in terms of the applied load w, the span a, and the sag h. Thus,

$$T_{max} = \sqrt{\frac{w^2 a^4}{64h^2} + \frac{w^2 a^2}{4}}$$

$$= \frac{wa}{8h} \sqrt{a^2 + (4h)^2} \tag{8-10}$$

Equations 8-9 and 8-10 can also be determined by using the free-body diagram of the right half of the cable shown in Fig. 8-19d. From the equilibrium equation $\Sigma M_B = 0$:

$$+\lfloor M_B = \frac{wa}{2}\left(\frac{a}{4}\right) - T_O(h) = 0 \qquad T_O = \frac{wa^2}{8h} \tag{8-9}$$

From the equilibrium equation $\Sigma F_x = 0$:

$$+\rightarrow \Sigma F_x = B_x - T_O = 0 \qquad B_x = T_O = \frac{wa^2}{8h}$$

From the equilibrium equation $\Sigma F_y = 0$:

$$+\uparrow \Sigma F_y = B_y - \frac{wa}{2} = 0 \qquad B_y = \frac{wa}{2}$$

Thus,

$$T_{max} = B = \sqrt{B_x^2 + B_y^2} = \sqrt{\left(\frac{wa^2}{8h}\right)^2 + \left(\frac{wa}{2}\right)^2}$$

$$= \frac{wa}{8h} \sqrt{a^2 + (4h)^2} \tag{8-10}$$

For any curve, the length dL of an arc of the curve can be obtained from the equation

$$dL = \sqrt{(dx)^2 + (dy)^2} = \sqrt{1 + \left(\frac{dy}{dx}\right)^2}\, dx \tag{d}$$

Thus, for the cable, from Eqs. 8-8 and 8-9,

$$y = kx^2 = \left(\frac{w}{2T_O}\right)x^2 = \left(\frac{4h}{a^2}\right)x^2$$

$$\frac{dy}{dx} = 2kx = \left(\frac{8h}{a^2}\right)x$$

$$L = \int_{span} dL = 2\int_0^{a/2} \sqrt{1 + 4k^2 x^2}\, dx$$

$$= \left[x\sqrt{1 + 4k^2 x^2} + \frac{1}{2k}\ln\left(2kx + \sqrt{1 + 4k^2 x^2}\right)\right]_0^{a/2}$$

$$= \frac{a}{2}\sqrt{1 + k^2 a^2} + \frac{1}{2k}\ln\left(ka + \sqrt{1 + k^2 a^2}\right)$$

Finally, substituting $k = 4h/a^2$ yields

$$L = \frac{1}{2}\sqrt{a^2 + 16h^2} + \frac{a^2}{8h}\ln\frac{1}{a}(4h + \sqrt{a^2 + 16h^2}) \qquad (8\text{-}11)$$

When the supports have different elevations, as shown in Fig. 8-20, the location of the lowest point on the cable (the origin of the xy-coordinate system) is not known and must be determined. The equation of the loaded cable (Eq. 8-8) remains valid since the free-body diagram (Fig. 8-19b) used in its development would be identical for a cable with supports at different elevations as shown in Fig. 8-20. If the sags h_A and h_B of the cable with respect to supports A and B are known, Eq. 8-8 can be used to locate the origin of the coordinate system. Thus, from Eq. 8-8,

$$y = kx^2$$
$$k = \frac{y}{x^2} = \frac{h_B}{x_B^2} = \frac{h_A}{x_A^2} \qquad (e)$$

Since $x_A = x_B - a$, Eq. e can be written

$$k = \frac{y}{x^2} = \frac{h_B}{x_B^2} = \frac{h_A}{(x_B - a)^2}$$

which leads to the quadratic equation

$$(h_B - h_A)x_B^2 - 2ah_B x_B + h_B a^2 = 0$$

Solving for x_B yields

$$x_B = \frac{h_B - \sqrt{h_B h_A}}{h_B - h_A}a \qquad (8\text{-}12)$$

Also, since $x_A = x_B - a$,

$$x_A = \frac{h_A - \sqrt{h_B h_A}}{h_B - h_A}a \qquad (8\text{-}13)$$

The horizontal distance from support A to the origin O of the coordinate system is shown as distance d in Fig. 8-20. At support A, $x_A = -d$. Thus,

$$d = \frac{\sqrt{h_B h_A} - h_A}{h_B - h_A}a \qquad (8\text{-}14)$$

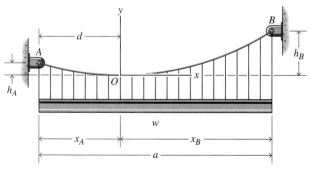

Fig. 8-20

The tension T_O in the cable can be found by using Eq. 8-8. Thus,

$$T_O = \frac{wx^2}{2y} = \frac{wx_A^2}{2h_A} = \frac{wx_B^2}{2h_B} \tag{8-15}$$

Equation 8-5 shows that the tension in the cable varies from a minimum value T_O at the lowest point in the cable to a maximum value at the highest (most distant) support. By substituting the magnitudes of x_A and x_B in Eq. 8-5 and using Eq. 8-15, we find that the tensions in the cable at the supports are

$$T_A = wd_A\left[1 + \frac{d_A^2}{4h_A^2}\right]^{1/2} \tag{8-16a}$$

$$T_B = wd_B\left[1 + \frac{d_B^2}{4h_B^2}\right]^{1/2} \tag{8-16b}$$

where $d_A = |x_A|$ and $d_B = |x_B|$.

The length of the curve between the lowest point O and a support can be determined by using Eq. d. Thus, for support B,

$$
\begin{aligned}
L_{OB} &= \int_0^{d_B} \sqrt{1 + 4k^2x^2}\, dx \\
&= \left[\frac{x}{2}\sqrt{1 + 4k^2x^2} + \frac{1}{4k}\ln\left[2kx + \sqrt{1 + 4k^2x^2}\right]\right]_0^{d_B} \\
&= \frac{d_B}{2}\sqrt{1 + 4k^2d_B^2} + \frac{1}{4k}\ln\left(2kd_B + \sqrt{1 + 4k^2d_B^2}\right)
\end{aligned}
$$

Finally, substituting $k = h_B/d_B^2$ yields

$$L_{OB} = \frac{1}{2}\sqrt{d_B^2 + 4h_B^2} + \frac{d_B^2}{4h_B}\ln\frac{1}{d_B}\left(2h_B + \sqrt{d_B^2 + 4h_B^2}\right) \tag{8-17}$$

Similarly,

$$L_{OA} = \frac{1}{2}\sqrt{d_A^2 + 4h_A^2} + \frac{d_A^2}{4h_A}\ln\frac{1}{d_A}\left(2h_A + \sqrt{d_A^2 + 4h_A^2}\right) \tag{8-18}$$

The total length L of the cable is the sum of the arc lengths L_{OA} and L_{OB} measured from the origin to each support.

EXAMPLE PROBLEM 8-8

A cable supports a uniformly distributed load of 300 lb/ft along the horizontal as shown in Fig. 8-21a. Determine

a. The minimum tension in the cable.
b. The tension in the cable at the supports and the angle it makes with the horizontal.
c. The length L of the cable.

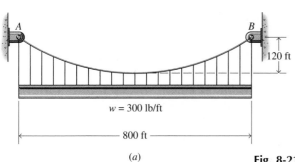

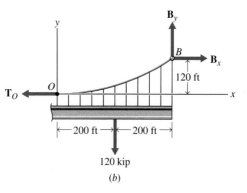

(a)

Fig. 8-21

(b)

SOLUTION

Parts a and b of the problem will be solved by using the free-body diagram shown in Fig. 8-21b. Since the supports are at the same elevation, the low point in the cable occurs at the middle of the span. Thus,

$$W = w\left(\frac{a}{2}\right) = 300(400) = 120{,}000 \text{ lb} = 120 \text{ kip}$$

a. From the equilibrium equation $\Sigma M_B = 0$:

$$+\downarrow M_B = W\left(\frac{x_B}{2}\right) - T_O(h_B) = 120(200) - T_O(120) = 0$$
$$T_O = 200 \text{ kip} \qquad\qquad \text{Ans.}$$

b. From the equilibrium equation $\Sigma F_x = 0$:

$$+\rightarrow\Sigma F_x = B_x - T_O = B_x - 200 = 0 \qquad B_x = 200 \text{ kip}$$

From the equilibrium equation $\Sigma F_y = 0$:

$$+\uparrow\Sigma F_y = B_y - W = B_y - 120 = 0 \qquad B_y = 120 \text{ kip}$$
$$T_A = T_B = T_{max} = \sqrt{B_x^2 + B_y^2}$$
$$= \sqrt{(200)^2 + (120)^2} = 233 \text{ kip} \qquad \text{Ans.}$$
$$\theta_x = \tan^{-1}\frac{B_y}{B_x} = \tan^{-1}\frac{120}{200} = 30.96 = 31.0° \qquad \text{Ans.}$$

c. The length of the cable is determined by using Eq. 8-11. Thus,

$$L = \frac{1}{2}\sqrt{a^2 + 16h^2} + \frac{a^2}{8h}\ln\frac{1}{a}(4h + \sqrt{a^2 + 16h^2})$$
$$= \frac{1}{2}\sqrt{(800)^2 + 16(120)^2}$$
$$+ \frac{(800)^2}{8(120)}\ln\frac{1}{800}\left[4(120) + \sqrt{(800)^2 + 16(120)^2}\right]$$
$$= 845.7 = 846 \text{ ft} \qquad\qquad \text{Ans.}$$

The cable shown in Fig. 8-22a supports a pipeline as it passes over a river. The support at the middle of the river is 25 m above the supports on the two sides. The lowest points on the cable are 45 m below the middle support. Determine

a. The minimum tension in the cable.
b. The maximum tension in the cable.
c. The length L of the cable.

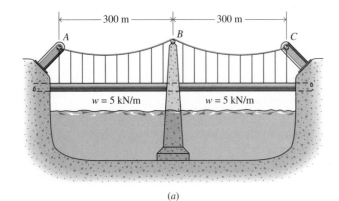

(a)

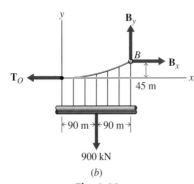

(b)

Fig. 8-22

SOLUTION

The lowest point in the cable (origin of the xy-coordinate system) can be located by using Eq. 8-14. Thus,

$$d = \frac{\sqrt{h_B h_A} - h_A}{h_B - h_A} a = \frac{\sqrt{45(20)} - 20}{45 - 20}(300) = 120.0 \text{ m}$$

Once the distance d is known, the free-body diagram shown in Fig. 8-22b can be used to solve parts a and b of the problem. The distance from the origin of coordinates to support B is

$$x_B = a - d = 300 - 120 = 180 \text{ m}$$
$$W = w(x_B) = 5(180) = 900 \text{ kN}$$

a. From the equilibrium equation $\Sigma M_B = 0$:

$$+\!\!\curvearrowleft M_B = W\!\left(\frac{x_B}{2}\right) - T_O(h_B) = 900(90) - T_O(45) = 0$$

$$T_O = T_{\min} = 1800 \text{ kN} \qquad\qquad \text{Ans.}$$

b. From the equilibrium equation $\Sigma F_x = 0$:

$$+\!\rightarrow\!\Sigma F_x = B_x - T_O = B_x - 1800 = 0 \qquad B_x = 1800 \text{ kN}$$

From the equilibrium equation $\Sigma F_y = 0$:

$$+\!\uparrow\!\Sigma F_y = B_y - W = B_y - 900 = 0 \qquad B_y = 900 \text{ kN}$$
$$T_A = T_B = T_{\max} = \sqrt{B_x^2 + B_y^2}$$
$$= \sqrt{(1800)^2 + (900)^2} = 2010 \text{ kN} \qquad \text{Ans.}$$

c. The length of the cable is determined by using Eqs. 8-17 and 8-18. Thus,

$$L_{OB} = \frac{1}{2}\sqrt{d_B^2 + 4h_B^2} + \frac{d_B^2}{4h_B}\ln\frac{1}{d_B}\left(2h_B + \sqrt{d_B^2 + 4h_B^2}\right)$$

$$= \frac{1}{2}\sqrt{(180)^2 + 4(45)^2}$$

$$+ \frac{(180)^2}{4(45)}\ln\frac{1}{180}\left[2(45) + \sqrt{(180)^2 + 4(45)^2}\right]$$

$$= 187.24 \text{ m}$$

Similarly,

$$L_{OA} = \frac{1}{2}\sqrt{d_A^2 + 4h_A^2} + \frac{d_A^2}{4h_A}\ln\frac{1}{d_A}\left(2h_A + \sqrt{d_A^2 + 4h_A^2}\right) \qquad (8\text{-}18)$$

$$= \frac{1}{2}\sqrt{(120)^2 + 4(20)^2}$$

$$+ \frac{(120)^2}{4(20)}\ln\frac{1}{120}\left[2(20) + \sqrt{(120)^2 + 4(20)^2}\right]$$

$$= 122.19 \text{ m}$$

Therefore,

$$L = 2(L_{OA} + L_{OB}) = 2(122.19 + 187.24) = 618.9 = 619 \text{ m} \qquad \text{Ans.}$$

PROBLEMS

8-69* A cable with supports at the same elevation has a span of 600 ft and a sag of 60 ft. The cable supports a uniformly distributed load of 750 lb/ft along the horizontal. Determine

a. The maximum tension in the cable.
b. The length of the cable.

8-70* A cable with supports at the same elevation is being designed to carry a uniformly distributed load of 1.50 kN/m along the horizontal. The sag of the cable at midspan is to be 4 m. If the maximum tension in the cable must be limited to 12 kN, determine

a. The maximum allowable span for the cable.
b. The required length for the cable.

8-71 The load carried by each cable of the suspension bridge shown in Fig. P8-71 is 2000 lb per foot of horizontal length. The span of the bridge is 1000 ft and the sag at midspan is 100 ft. Determine

a. The tension in the cable at the middle of the span.
b. The tension in the cable at the supports and the angle it makes with the horizontal.
c. The length of the cable.

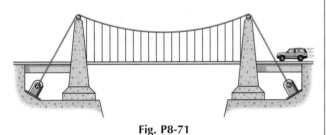

Fig. P8-71

8-72 The center span of the suspension bridge shown in Fig. P8-72 is 500 m. The sag at the center of the span is 50 m. The cables have been designed to resist a maximum tension of 5000 kN. Determine

a. The design load per horizontal meter of roadway.
b. The length of the cable in the center span.

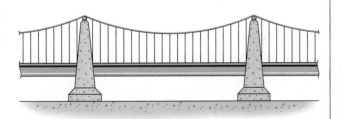

Fig. P8-72

8-73* A cable with supports at the same elevation carries a uniformly distributed load of 500 lb/ft along the horizontal. If the maximum tension in the cable is 100 kip and the sag at midspan is 40 ft, determine

a. The horizontal distance between supports.
b. The length of the cable.

8-74* A cable with supports at the same elevation has a span of 400 m. The cable supports a uniformly distributed load of 6 kN/m along the horizontal. The maximum tension in the cable is 5000 kN. Determine

a. The angle between the cable and the horizontal at a support.
b. The sag in the cable at the middle of the span.
c. The length of the cable.

8-75 A cable with supports at the same elevation has a span of 600 ft. The length of the cable is 630 ft. If the maximum tension in the cable resulting from a uniformly distributed load w along the horizontal is 360 kip, determine

a. The sag in the cable at the middle of the span.
b. The angle between the cable and the horizontal at a support.
c. The magnitude of the distributed load w.

8-76 The left support of the cable shown in Fig. P8-76 is located 10 m below the right support. The lowest point on the cable is 13 m below the right support. If the maximum tension in the cable resulting from a uniformly distributed load w along the horizontal is 400 kN, determine

a. The angle between the cable and the horizontal at the right support.
b. The magnitude of the distributed load w.
c. The length of the cable.

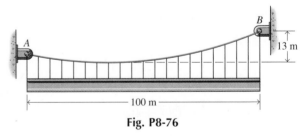

Fig. P8-76

8-77* A cable carries a uniformly distributed load of 500 lb per foot of horizontal length. The supports are 500 ft apart and the left support is 30 ft lower than the right support. The lowest point on the cable is 25 ft below the left support. Determine

a. The minimum tension in the cable.

364

b. The maximum tension in the cable.
c. The angle between the cable and the horizontal at the right support.
d. The length of the cable.

8-78* The right support of a cable is located 45 m below the left support. The horizontal distance between supports is 250 m. The lowest point on the cable is 15 m below the right support. If the maximum tension in the cable resulting from a uniformly distributed load w along the horizontal is 850 kN, determine

a. The angle between the cable and the horizontal at the right support.
b. The magnitude of the distributed load w.
c. The length of the cable.

8-79 The left support of the cable shown in Fig. P8-79 is located 40 ft below the right support. The cable is horizontal at the left support. If the maximum tension in the cable resulting from a uniformly distributed load w along the horizontal is 400 kip, determine

a. The angle between the cable and the horizontal at the right support.
b. The magnitude of the distributed load w.
c. The length of the cable.

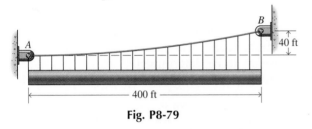

Fig. P8-79

a. The angle between the cable and the horizontal at the right support.
b. The magnitude of the distributed load w.
c. The length of the cable.

8-80 The cable shown in Fig. P8-80 carries a uniformly distributed load of 25 kN per meter of horizontal length. The left support of the cable is located 50 m below the right support. The angle between the cable and the horizontal at the left support is 10°. Determine

a. The tension in the cable at the left support.
b. The tension in the cable at the right support.
c. The angle between the cable and the horizontal at the right support.
d. The length of the cable.

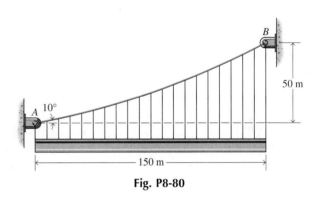

Fig. P8-80

8-6-3 Cables with Loads Uniformly Distributed Along Their Length

In the previous two sections, the weight of the cable was small in comparison to the loads being supported by the cable. For power transmission lines, telephone lines, and guy wires on radio and television towers, the weight of the cable is the only significant load being applied. In these applications, the weight is uniformly distributed along the length of the cable. When the supports are at the same elevation and the sag ratio is small (a taut cable), the curve assumed by the cable may be regarded with small error as being a parabola, since a uniformly distributed load along the cable does not differ significantly from the same load uniformly distributed along the horizontal. When the sag ratio is large ($h/a > 0.10$), the parabolic formulas of the previous section should not be used.

A uniform cable having a weight per unit length w over its length is illustrated in Fig. 8-23a. Equations that express relationships between the length L, span a, and sag h of the cable, the tension T in the cable, and the weight w of the cable can be developed by using equilibrium considerations and the free-body diagram of a portion of the cable shown in Fig. P8-23b. In this diagram, the lowest point on the cable (point O) has been taken as the origin of the xy-coordinate system.

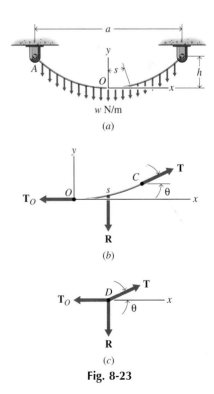

Fig. 8-23

365

The segment of cable shown in Fig. 8-23b is subjected to three forces; the tension $\mathbf{T}_O$ in the cable at point O, the tension $\mathbf{T}$ in the cable at an arbitrary point C, which is located a distance s from the origin, and the resultant $\mathbf{R}$ ($R = ws$) of the distributed load, whose line of action is located at an unknown distance x from point O. The weight acts at the centroid of the curve formed by the cable. The angle that $\mathbf{T}$ makes with the horizontal is denoted by θ. Since segment OC of the cable is in equilibrium under the action of these three forces, the three forces must be concurrent at point D as shown in Fig. 8-23c. The two equilibrium equations $\Sigma F_x = 0$ and $\Sigma F_y = 0$ for this concurrent force system yield:

$$+\rightarrow \Sigma F_x = T \cos \theta - T_O = 0 \qquad T_x = T \cos \theta = T_O \qquad (a)$$
$$+\uparrow \Sigma F_y = T \sin \theta - R = 0 \qquad T_y = T \sin \theta = R = ws \qquad (b)$$

Equation a shows that the horizontal component T_x of the tension T at any point in the cable is constant and equal to the tension T_O at the lowest point on the cable. Solving Eqs. a and b for T and θ yields:

$$T = [T_O^2 + w^2 s^2]^{1/2} \qquad (8\text{-}19)$$

$$\theta = \tan^{-1} \frac{ws}{T_O} \qquad (8\text{-}20)$$

Equation 8-19 shows that the tension is minimum at the lowest point in the cable (where $s = 0$) and maximum at the supports (where s is greatest). The distance s remains to be determined.

The shape of the curve can be determined by using Eq. 8-20, which gives the slope. Thus,

$$\frac{dy}{dx} = \tan \theta = \frac{ws}{T_O} \qquad (c)$$

Recall the relationship

$$(ds)^2 = (dx)^2 + (dy)^2$$

from which

$$\frac{dy}{dx} = \left[\left(\frac{ds}{dx} \right)^2 - 1 \right]^{1/2} \qquad (d)$$

Substituting Eq. d into Eq. c yields

$$\left[\left(\frac{ds}{dx} \right)^2 - 1 \right]^{1/2} = \frac{ws}{T_O}$$

from which

$$\frac{ds}{dx} = \left[1 + \left(\frac{ws}{T_O} \right)^2 \right]^{1/2}$$

Solving for dx yields

$$dx = \frac{(T_O/w)\, dx}{[(T_O/w)^2 + s^2]^{1/2}} \qquad (e)$$

Integrating Eq. e yields

$$x = \left(\frac{T_O}{w} \right) \ln \left\{ s + \left[\left(\frac{T_O}{w} \right)^2 + s^2 \right]^{1/2} \right\} + C$$

The integration constant C can be determined by substituting the boundary condition $s = 0$ when $x = 0$. Thus,

$$C = -\left(\frac{T_O}{w}\right) \ln \left(\frac{T_O}{w}\right)$$

therefore,

$$x = \left(\frac{T_O}{w}\right) \ln \frac{s + [(T_O/w)^2 + s^2]^{1/2}}{(T_O/w)} \tag{8-21}$$

Equation 8-21 can also be written in exponential form as

$$\left(\frac{T_O}{w}\right) e^{wx/T_O} = s + \left[\left(\frac{T_O}{w}\right)^2 + s^2\right]^{1/2}$$

Solving for s yields

$$s = \left(\frac{T_O}{w}\right)\left[\frac{e^{wx/T_O} - e^{-wx/T_O}}{2}\right] = \left(\frac{T_O}{w}\right) \sinh\left(\frac{wx}{T_O}\right) \tag{8-22}$$

Once the distance s is known in terms of x, Eq. 8-19 yields

$$T = T_O[1 + \sinh^2(wx/T_O)]^{1/2} = T_O \cosh(wx/T_O) \tag{8-23}$$

Similarly, Eq. 8-22 can be substituted into Eq. c to obtain

$$\frac{dy}{dx} = \sinh\left(\frac{wx}{T_O}\right)$$

which can be integrated to give

$$y = \left(\frac{T_O}{w}\right) \cosh\left(\frac{wx}{T_O}\right) + C$$

Substituting the boundary condition $y = 0$ at $x = 0$ yields

$$C = -\left(\frac{T_O}{w}\right)$$

Thus,

$$y = \left(\frac{T_O}{w}\right)\left[\cosh\left(\frac{wx}{T_O}\right) - 1\right] \tag{8-24}$$

Equation 8-24 is the Cartesian equation of a catenary.

The tension T at an arbitrary point along the cable can be expressed in terms of T_O and the y-coordinate of the point by solving Eq. 8-24 for $\cosh(wx/T_O)$ and substituting the result into Eq. 8-23. The final result is

$$T = T_O + wy \tag{8-25}$$

Equation 8-24 can also be used to determine T_O if both coordinates of a point on the cable are known. For example, the coordinates at a support for a cable with span a and sag h and supports at the same elevation are $x = a/2$ and $y = h$. Thus

$$h = \left(\frac{T_O}{w}\right)\left[\cosh\left(\frac{wa}{2T_O}\right) - 1\right] \tag{8-26}$$

Unfortunately, Eq. 8-26 cannot be solved directly for T_O; therefore, values for T_O must be obtained by using trial-and-error solutions, iterative procedures, numerical solutions, or graphs and tables.

A cable with supports at the same elevation has a span of 800 ft. The cable weighs 5 lb/ft and the sag at midspan is 200 ft. Determine

a. The tension in the cable at midspan.
b. The tension in the cable at a support.
c. The length of the cable.

SOLUTION

a. The tension in the cable at midspan can be determined by using Eq. 8-26. Thus,

$$h = \left(\frac{T_O}{w}\right)\left[\cosh\left(\frac{wa}{2T_O}\right) - 1\right]$$

$$200 = \frac{T_O}{5}\left[\cosh\frac{5(800)}{2T_O} - 1\right]$$

A trial-and-error solution for the above equation is shown in the following table. Equation 8-9 for a parabolic cable was used to obtain a first estimate. Thus $T_O = wa^2/8h = 5(800)^2/8(200) = 2000$ lb.

T_O	$\dfrac{T_O}{5}$	$\cosh\dfrac{2000}{T_O}$	$\dfrac{T_O}{5}\left[\cosh\dfrac{2000}{T_O} - 1\right]$	h	% Error
2000	400	1.54308	217.23	200	+8.62
2100	420	1.48885	205.32	200	+2.66
2200	440	1.44248	194.69	200	−2.66
2150	430	1.46478	199.86	200	−0.07

Thus,

$$T_O = 2150 \text{ lb} \qquad\qquad \text{Ans.}$$

The tension T_O in the cable at midspan can also be determined by using the Newton–Raphson Method presented in Appendix D. Equation 8-26 is first written in the form

$$f(T_O) = T_O\left[\cosh\left(\frac{2000}{T_O}\right) - 1\right] - 1000 = 0$$

Using the Newton–Raphson Method

$$(T_O)_{n+1} = (T_O)_n + \delta_n$$

$$\delta_n = -\frac{f(T_O)_n}{f'(T_O)_n}$$

$$f'(T_O) = \left[\cosh\left(\frac{2000}{T_O}\right) - 1\right] + T_O\left(-\frac{2000}{T_O^2}\right)\sinh\left(\frac{2000}{T_O}\right)$$

Using the initial value $(T_O)_0 = 2000$

n	$(T_O)_n$	$f(T_O)_n$	$f'(T_O)_n$	δ_n
0	2000	86.1614	-0.6321	136.3053
1	2136.31	6.6034	-0.5390	12.2505
2	2148.56	0.0440	-0.5317	0.0845
3	2148.64	0.0000		

Thus,

$$T_O = 2149 \text{ lb} \qquad \text{Ans.}$$

b. Once the tension at midspan is known, Eq. 8-25 can be used to determine the tension at a support. Thus,

$$T_B = T_{max} = T_O + wh = 2150 + 5(200) = 3150 \text{ lb} \qquad \text{Ans.}$$

c. The distance s along the cable from the low point to a support can be determined by using Eq. 8-22. Thus,

$$s = \left(\frac{T_O}{w}\right) \sinh \left(\frac{wx}{T_O}\right)$$

$$s_B = \frac{2150}{5} \sinh \frac{5(400)}{2150} = 460.2 \text{ ft}$$

Therefore

$$L = 2s_B = 2(460.2) = 920.4 = 920 \text{ ft} \qquad \text{Ans.}$$

The equivalent results obtained by using the parabolic equations are

$$T_B = \frac{1}{2}wa\left[1 + \left(\frac{a}{4h}\right)^2\right]^{1/2}$$

$$= \frac{1}{2}(5)(800)\left[1 + \left(\frac{800}{4(200)}\right)^2\right]^{1/2} = 2830 \text{ lb}$$

The length of the cable is determined by using Eq. 8-11. Thus,

$$L = \frac{1}{2}\sqrt{a^2 + 16h^2} + \frac{a^2}{8h} \ln \frac{1}{a}\left(4h + \sqrt{a^2 + 16h^2}\right)$$

$$= \frac{1}{2}\sqrt{(800)^2 + 16(200)^2}$$

$$+ \frac{(800)^2}{8(200)} \ln \frac{1}{800}\left[4(200) + \sqrt{(800)^2 + 16(200)^2}\right]$$

$$= 918.2 = 918 \text{ ft} \qquad \text{Ans.}$$

The results obtained by using the equations for parabolic and catenary cables are summarized in the following table.

Quantity	Catenary	Parabolic
T_O	2150	2000
T_{max}	3150	2830
L	920	918

PROBLEMS

8-81* A segment of an electric power transmission line with supports at the same elevation is shown in Fig. P8-81. A horizontal force of 3000 lb was applied to the line before it was attached to the towers. The weight of the line is 1.5 lb/ft. Determine

a. The maximum tension in the cable.
b. The sag at the middle of the span.
c. The length of the cable.

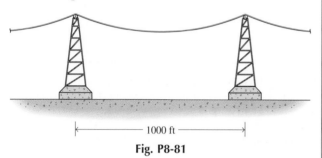

1000 ft

Fig. P8-81

8-82* A tugboat is used to tow a barge as shown in Fig. P8-82. The weight of the cable, after accounting for the buoyant force of the water on the submerged cable, is 120 N/m. The horizontal component of the force exerted on the barge by the cable is 40 kN. Determine

a. The maximum tension in the cable.
b. The sag in the cable at its lowest point.
c. The length of the cable.

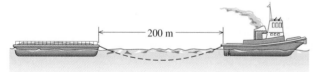

200 m

Fig. P8-82

8-83 A flexible cable with supports at the same elevation has a span of 400 ft. If the length of the cable is 500 ft, determine the sag midway between the supports.

8-84 A flexible cable with supports at the same elevation has a span of 300 m. If the sag midway between the supports is 80 m, determine the length of the cable.

8-85* A flexible cable with supports at the same elevation has a span of 900 ft. The cable weighs 0.75 lb/ft and is 1000 ft long. Determine

a. The sag midway between the supports.
b. The maximum tension in the cable.

8-86* A flexible cable with supports at the same elevation has a span of 250 m. The cable weighs 12 N/m. If the sag midway between the supports is 50 m, determine

a. The length of the cable.
b. The maximum tension in the cable.

8-87 A cable with supports at the same elevation has a span of 550 ft. The cable weighs 4 lb/ft and the sag at midspan is 100 ft. Determine

a. The tension in the cable at midspan.
b. The tension in the cable at a support.
c. The length of the cable.

8-88 A cable with supports at the same elevation has a span of 300 m. The cable weighs 75 N/m and the sag at midspan is 75 m. Determine

a. The tension in the cable at midspan.
b. The tension in the cable at a support.
c. The length of the cable.

8-89* The flexible cable shown in Fig. P8-89 weighs 5 lb/ft. The cable is horizontal at the left support. Determine

a. The tension in the cable at support A.
b. The tension in the cable at support B.
c. The length of the cable.

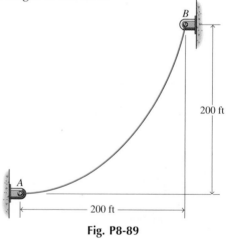

B

200 ft

A

200 ft

Fig. P8-89

8-90* The flexible cable used to tether the balloon shown in Fig. P8-90 weighs 7.0 N/m and is 100 m long. The balloon exerts a buoyant vertical force of 2000 N on the cable at end B. Determine

a. The force exerted by the wind on the balloon.
b. The height h of the balloon.
c. The horizontal distance d from the anchor A to the balloon.

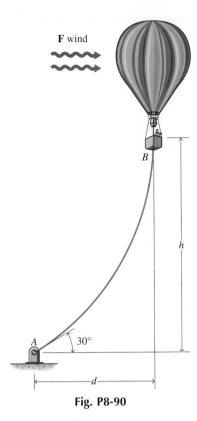

F wind

B

h

A

30°

d

Fig. P8-90

8-91 An electric power transmission cable weighing 2.5 lb/ft is supported by towers on two sides of a river valley as shown in Fig. P8-91. Determine

a. The tension in the cable at its lowest point.
b. The maximum tension in the cable.
c. The length of the cable.

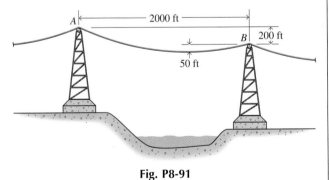

A |← 2000 ft →|

B | 200 ft

50 ft

Fig. P8-91

8-92 The cable shown in Fig. P8-92 weighs 30 N/m. If the sag of the cable is 200 m, determine

a. The tension in the cable at its lowest point.
b. The tension in the cable at support A.
c. The tension in the cable at support B.
d. The length of the cable.

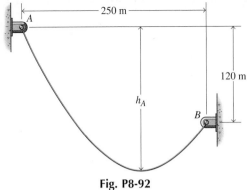

|← 250 m →|

A

120 m

h_A

B

Fig. P8-92

8-93 Three pairs of flexible cables, spaced at 120° intervals, are used to stabilize a 1000-ft television tower. The tower and one pair of cables is shown in Fig. P8-93. The weight of the cables is 3.0 lb/ft. The horizontal component of the force exerted by each cable is 6000 lb. At anchor A, the angle between cable AC and the horizontal must be 40°. Determine

a. The maximum tension in cable AC.
b. The horizontal distance d from the tower to anchor A.
c. The length of cable AC.

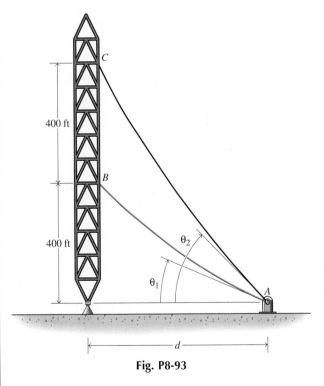

C

400 ft

B

400 ft

θ_2

θ_1

A

|← d →|

Fig. P8-93

SUMMARY

When a structural member or machine component is subjected to a system of external loads, a system of internal resisting forces develops within the member to balance the external forces. The resultant of the internal forces on a specific plane within a body can be determined by assuming that the plane separates the body into two parts. Since the body is in equilibrium, each part of the body must also be in equilibrium under the action of the internal forces that develop on the plane. The internal force distribution on the plane can be represented on a free-body diagram of either part of the body by a resultant force $\mathbf{R}$ at a point on the plane (usually the centroid) and a resultant couple $\mathbf{C}$. In most cases, it will be convenient to represent the resultant force $\mathbf{R}$ and the resultant couple $\mathbf{C}$ with rectangular components. Thus, with the x-axis normal to the plane at the point, the resultant force $\mathbf{R}$ is represented by a normal force $\mathbf{N}_x$ and two mutually perpendicular in-plane shearing forces $\mathbf{V}_y$ and $\mathbf{V}_z$. The resultant couple $\mathbf{C}$ is represented by a twisting moment (torque), $\mathbf{M}_x$ and two mutually perpendicular bending moments $\mathbf{M}_y$ and $\mathbf{M}_z$. Generally, it will be possible to solve for the six unknown force and moment components by using the equations

$$\Sigma F_x = 0 \qquad \Sigma F_y = 0 \qquad \Sigma F_z = 0$$
$$\Sigma M_x = 0 \qquad \Sigma M_y = 0 \qquad \Sigma M_z = 0$$

Axial force diagrams provide a graphical representation of the variation of internal axial force along the length of an axially loaded member. In a similar manner, torque diagrams provide a graphical representation of the variation in resisting torque transmitted by transverse cross sections of a shaft subjected to a series of twisting moments. Shear and bending moment diagrams provide similar graphical representations of the variation of shear force and bending moment along the length of a member. From such diagrams, the magnitudes and locations of maximum values of axial force, torque, shear force, or bending moment are easily established Shear-force and bending-moment diagrams can be established directly from a free-body diagram (load diagram) for a beam by using several simple mathematical relationships that exist between loads and shears and between shears and moments. The equations developed from a free-body diagram of an elemental length of a beam are

$$dV = w\,dx \qquad \text{or} \qquad \frac{dV}{dx} = w \qquad\qquad (8\text{-}1)$$

$$dM = V\,dx \qquad \text{or} \qquad \frac{dM}{dx} = V \qquad\qquad (8\text{-}3)$$

Flexible cables are used for suspension bridges, aerial tramways, power lines, and many other engineering applications. A cable is said to be perfectly flexible when it offers no resistance to bending. Actual cables are not perfectly flexible; however, the resistance they offer to bending is generally so small that any bending effects can be neglected in the analysis of the cable without introducing serious error. Once it is assumed that the cable offers no resistance to bending, the resultant internal force on any cross section must act along a tangent to the cable at that cross section. Typical types of loads applied to flexible cables include a series of distinct vertical concentrated loads, uniformly distributed vertical loads over the horizontal span of the cable, or uniformly distributed vertical loads (such as the weight of the cable) over the length of the cable. For these three types of loading, the horizontal component of the internal force on a cross section is constant for the length of the cable.

REVIEW PROBLEMS

8-94* A system of three bars supports a 750-N load as shown in Fig. P8-94. Determine the internal resisting force and moment transmitted by cross section *aa* in bar *BCE*.

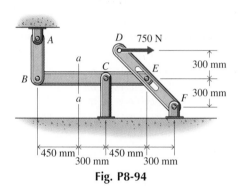

Fig. P8-94

8-95* A simple two-bar mechanism supports a 500-lb load as shown in Fig. P8-95. Determine the internal resisting forces and moment transmitted

a. By cross section *aa* in bar *AC*.
b. By the cross section at support *A*.

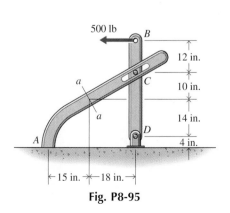

Fig. P8-95

8-96 Two bars, a pulley, and a cable are used to support a block as shown in Fig. P8-96. The two bars have negligible weight. The mass of the pulley is 50 kg and the mass of the block is 100 kg. Determine the internal resisting forces and moment transmitted by cross section *aa* in bar *AB*.

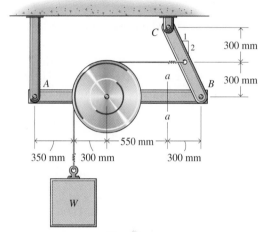

Fig. P8-96

8-97 An 800-lb automobile engine is supported by an engine hoist as shown in Fig. P8-97. Determine the internal resisting forces and moment transmitted by cross section *aa* in beam *ABC*.

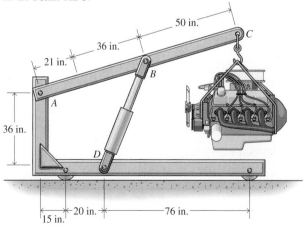

Fig. P8-97

8-98* Determine the internal resisting forces and moments transmitted by cross section *aa* in the bar shown in Fig. P8-98.

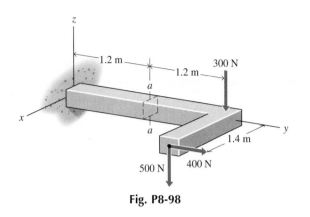

Fig. P8-98

8-99* A beam is loaded and supported as shown in Fig. P8-99.

a. Determine the shear force *V* and the bending moment *M* on a cross section 10 ft from the left end of the beam.
b. Draw complete shear and moment diagrams for the beam.

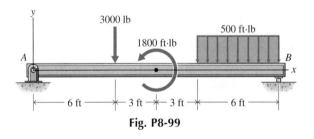

Fig. P8-99

8-100 A beam is loaded and supported as shown in Fig. P8-100.

a. Determine the shear force *V* and the bending moment *M* on a cross section 1.75 m from the right end of the beam.
b. Draw complete shear and moment diagrams for the beam.

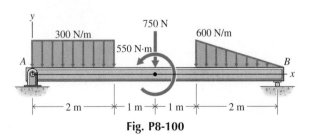

Fig. P8-100

8-101 A beam is loaded and supported as shown in Fig. P8-101.

a. Using the coordinate axes shown, write equations for the shear force *V* and bending moment *M* for any section of the beam in the interval $0 < x < 10$ ft.
b. Draw complete shear and moment diagrams for the beam.

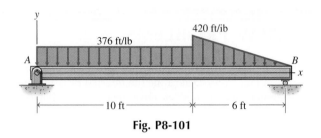

Fig. P8-101

8-102* The cable shown in Fig. P8-102 weighs 35 N/m. Determine the maximum tension in the cable and the sag h_A in the cable if

a. The length of the cable is 33 m.
b. The length of the cable is 45 m.
c. The length of the cable is 60 m.

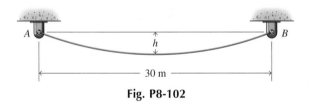

Fig. P8-102

8-103 A 500-lb block *W* is supported by a boom *CB* and a flexible cable *AB* as shown in Fig. P8-103. The boom weighs 250 lb and the cable weighs 2.5 lb/ft. Determine the maximum tension in the cable, the length *L* of the cable, and the sag *h* at midspan. Assume that the weight of the cable is uniformly distributed along its horizontal length.

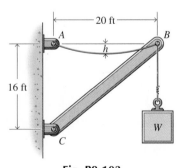

Fig. P8-103

C8-104 The hook shown in Fig. P8-104a supports a 10 kN load. Plot P, V, and M, the internal resisting forces and moment transmitted by a section of the beam, as a function of the angle θ ($0 \le \theta \le 150°$). (Use the directions shown in Fig. P8-104b for the positive directions on the graphs.)

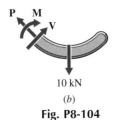

$R = 100$ mm

θ

10 kN

(a)

P **M**

V

10 kN

(b)

Fig. P8-104

C8-105 Forces of 100 N are being applied to the handles of the vice grip pliers shown in Fig. P8-105. Plot P, V, and M, the internal resisting forces and moment transmitted by section aa of the handle, as a function of the distance d ($20 \le d \le 30$ mm).

C8-106 A 2500 N load is supported by a roller on a beam as shown in Fig. P8-106.

a. Show that the maximum bending moment in the beam occurs at the roller.

b. Plot $|M|_{max}$, the maximum bending moment in the beam, as a function of the roller position b ($0 \le b \le 8$ m).

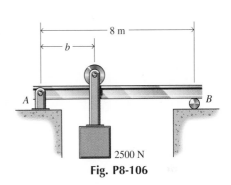

8 m

b

A

B

2500 N

Fig. P8-106

C8-107 A 3500 lb load is supported by a cart that rolls along a beam as shown in Fig. P8-107.

a. Show that the maximum bending moment in the beam occurs at the wheel that is closer to the middle of the beam.

b. Plot $|M|_{max}$, the maximum bending moment in the beam, as a function of the cart's position b ($1 \le b \le 19$ ft).

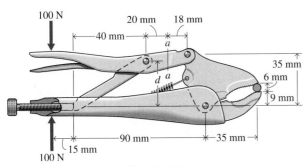

100 N

20 mm   18 mm

40 mm

a

35 mm

d a

6 mm

9 mm

90 mm   35 mm

15 mm

100 N

Fig. P8-105

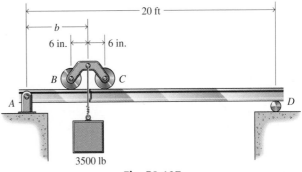

20 ft

b

6 in.   6 in.

B C

A D

3500 lb

Fig. P8-107

375

C8-108 An electric power transmission cable weighing 30 N/m is supported by towers on two sides of a river valley as shown in Fig. P8-108.

a. Plot h_B, the sag in the cable, as a function of the length of the cable L ($605 \leq L \leq 630$ m).
b. Plot T_{max}, the maximum tension in the cable, as a function of the length of the cable L ($605 \leq L \leq 630$ m).

C8-109 An electric power transmission cable weighing 2.5 lb/ft is supported by towers on two sides of a river valley as shown in Fig. P8-109. Design specifications require that the maximum tension in the cable be less than 12,000 lb and that the minimum clearance above the river be at least 75 ft. Determine the minimum allowable height of tower B.

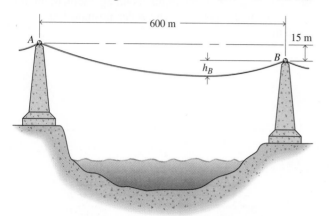

Fig. P8-108

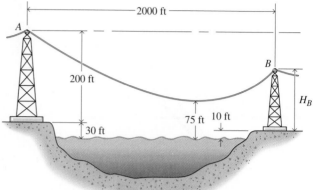

Fig. P8-109

9

FRICTION

Rock climbing is a sport that depends heavily on friction. Friction forces between the hands and the rope and between the shoes and the rock make climbing possible.

9-1   INTRODUCTION

Thus far in the book, the forces between two surfaces in contact have been either perfectly smooth or perfectly rough. Whenever tangential or friction forces were needed, they were assumed to supply whatever force was necessary to ensure equilibrium. This is the concept of a rough surface. However, no surface is perfectly rough any more than it is perfectly smooth. In the present chapter, it is desired to identify the origin and the nature of these tangential forces and to quantify the concept of roughness.

A perfectly smooth or frictionless surface that exerts only normal forces on bodies is a useful model for a large number of situations. However, frictional forces that act tangential to the surface are present in the contact between all real surfaces. Whether the friction forces are large or small depends on a number of things including the types of materials in contact.

Friction forces act to oppose the tendency of contacting surfaces to slip relative to one another and can be either good or bad. Without friction it would be impossible to walk or ride a bicycle or drive a car or pick up objects. Even in some machine applications such as brakes and belt drives, a design consideration is to maximize the friction. In many other machine applications, however, friction is undesirable. Friction causes energy loss and wears down sliding surfaces in contact. In these cases, a primary design consideration is to minimize friction.

There are two main types of friction that are commonly encountered in engineering practice: dry friction and fluid friction. As its name suggests, **dry friction** or **Coulomb friction** describes the tangential component of the contact force that exists when two dry surfaces slide or tend to slide relative to one another. Coulomb friction is the primary concern of this chapter and will be studied in considerable detail in Section 9-2.

Fluid friction describes the tangential component of the contact force that exists between adjacent layers in a fluid that are moving at different velocities relative to each other as in the thin layer of oil between bearing surfaces. The tangential forces developed between the adjacent fluid layers oppose the relative motion and are dependent primarily on the relative velocity between the two layers. Fluid friction is one of the primary concerns in the study of fluid mechanics and is more properly treated in a course in fluid mechanics.

9-2   CHARACTERISTICS OF COULOMB FRICTION

In order to investigate the behavior of frictional forces, consider a simple experiment consisting of a solid block of mass m resting on a rough horizontal surface and acted on by a horizontal force **P** (Fig. 9-1). Equilibrium of the block requires a force having both a normal component ($N = mg$) and a horizontal or friction component ($F = P$) acting on the contact surface. When the horizontal force **P** is zero, no horizontal component of force F will be required for equilibrium and friction will not exert a force on the surface. As the force **P** increases, the friction F also increases, as shown in the graph of Fig. 9-2. The friction force cannot increase indefinitely, however, and it eventually reaches its

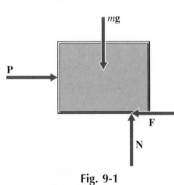

Fig. 9-1

maximum value F_{max}. The maximum value is also called the **limiting value of static friction.**

The condition when the friction force is at its maximum value is called the condition of **impending motion.** That is, if **P** increases beyond the point $P = F_{max}$, then friction can no longer supply the amount of force necessary for equilibrium. Therefore the block will no longer be in equilibrium, but will start moving in the direction of the force **P**. When the block starts moving, the friction force F normally decreases in magnitude by about 20 to 25 percent. From this point on, the block will slide with increasing speed while the friction force (the kinetic friction force) remains approximately constant (Fig. 9-2).

Repeating the experiment with a second block of mass $m_2 = 2m$ would produce similar results, but the limiting force at which the block starts to move would be observed to be twice as great. Repeating the experiment with two blocks of different sizes but the same mass and material would yield the same limiting force for both blocks. That is, the value of limiting friction is proportional to the normal force at the contact surface

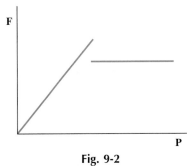

Fig. 9-2

$$F_{max} = \mu_s N \qquad (9\text{-}1)$$

The constant of proportionality μ_s is called the **coefficient of static friction,** and it depends on the types of material in contact. However, μ_s is observed to be relatively independent of both the normal force and the area of contact.

In order to understand how μ_s can be independent of the area of contact, one must consider where the friction forces come from. It is generally believed that dry friction results primarily from the roughness between two surfaces and to a lessor extent from attraction between the molecules of the two surfaces. Even two surfaces that are considered to be smooth have small irregularities, as the (idealized) enlargement of the contact surfaces of Fig. 9-3 shows. Therefore, contact between the block and the surface takes place only over a few very small areas of the common surface. The friction force **F** is then the resultant of the tangential component of the forces acting at each of these tiny contact points just as the normal force **N** is the resultant of the normal components of the forces acting at each of the contact points. (Normal and tangential here are relative to the overall contact plane and not the individual tiny contact points.) Increasing the number of contact points just means that the normal and frictional components at each point are proportionately smaller, but their sums **F** and **N** do not change. Therefore, μ_s will not change either.

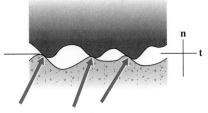

Fig. 9-3

Before going on it must be noted that since the normal force **N** is the resultant of a distributed force, it acts through the centroid of that force distribution. If the force distribution is uniform, **N** will act at the center of the surface. In general, however, **N** will not act at the center of the surface or through the center of the body. Since the actual distribution of forces is generally not known, the location of **N** is determined using moment equilibrium.

In many friction problems, it is easily recognized that a body is in no danger of tipping over. Since only force equilibrium is considered, it does not matter where the normal force is drawn on the free-body diagram. Even so, the student should not get into the habit of showing the force as always acting through the center of the body.

It must also be noted that friction is a resistive force. That is, friction always acts to opposite motion; it never acts to create motion. Equation 9-1 only tells how much friction is available $F_{avail} = F_{max} = \mu_s N$ to prevent motion. No matter how much frictional force is available on a surface, however, the *frictional force actually exerted is never greater than that required to satisfy the equations of equilibrium*

$$F \le \mu_s N \tag{9-2}$$

where *the equality holds only at the point of impending motion.*

Once the block starts to slip relative to the surface, the friction force will decrease to

$$F = \mu_k N \tag{9-3}$$

where μ_k is called the **coefficient of kinetic friction.** This coefficient is again independent of the normal force and is also independent of the speed of the relative motion—at least for low speeds. Of course, the presence of any oil or moisture on the surface can change the problem from one of dry friction, where the friction force is independent of the speed of the body, to one of fluid friction, where the friction force is a function of the speed. At higher speeds, the effect of lubrication by an intervening fluid film (such as oil, surface moisture, or even air) can become appreciable.

These results are summarized by

Coulomb's Laws of Dry Friction

The direction of the friction force on a surface is such as to oppose the tendency of the one surface to slide relative to the other. It is the relative motion or the impending relative motion of one body relative to another that is important.

The friction force is never greater than just sufficient to prevent motion.

For the static equilibrium case in which the two surfaces are stationary with respect to one another, the normal and tangential components of the contact force satisfy

$$F \le \mu_s N$$

where the equality holds for the case of **impending motion** in which the contacting surfaces are on the verge of sliding relative to each other.

For the case where two contacting surfaces are sliding over each other, the normal and tangential components of the contact force satisfy

$$F = \mu_k N$$

where $\mu_k < \mu_s$.

Of course, Coulomb's laws only apply when N is positive, that is, when the surfaces are being pressed together.

TABLE 9-1   COEFFICIENTS OF FRICTION FOR COMMON SURFACES

Materials	μ_s
Metal on metal	0.5
Wood on metal	0.5
Wood on wood	0.4
Leather on wood	0.4
Rubber on metal	0.5
Rubber on wood	0.5
Rubber on pavement	0.7

The values of μ_s and μ_k must be determined experimentally for each pair of contacting surfaces. Average values of μ_s for various types of materials are given in Table 9-1. Reported values for μ_s vary widely, however, depending on the exact nature of the contacting surfaces. Values for μ_k are generally 20 to 25 percent less than those reported for μ_s. Values for μ_k are not listed in Table 9-1 since the uncertainty in μ_s is much larger (by as much as 100 percent in some cases) than the difference between it and μ_k.

Because of the uncertainty in the values of μ_s and μ_k, Table 9-1 should be used to get only a rough estimate of the magnitude of the friction forces. If more accurate values are needed, experiments should be performed using the actual surfaces being studied.

Since the coefficients of friction are the ratio of two forces, they are dimensionless quantities and can be used with either SI or USCS units.

In many simple friction problems it is convenient to use the resultant of the friction and normal forces rather than their separate components. In the case of the block in Fig. 9-1 (which is redrawn in Fig. 9-4), this leaves only three forces acting on the block. Moment equilibrium is established simply by making the three forces concurrent and only force equilibrium need be considered. Since **N** and **F** are rectangular components of the resultant **R**, the magnitude and direction of the resultant are given by

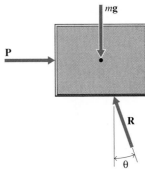

Fig. 9-4

$$R = \sqrt{N^2 + F^2} \qquad \text{and} \qquad \tan \theta = \frac{F}{N} \tag{9-4}$$

At the point of impending motion, Eq. 9-4 becomes

$$R = \sqrt{N^2 + F_{\max}^2} = \sqrt{N^2 + (\mu_s N)^2} = N\sqrt{1 + \mu_s^2} \tag{9-5a}$$

$$\tan \phi_s = \frac{F}{N} = \frac{\mu_s N}{N} = \mu_s \tag{9-5b}$$

where the angle ϕ_s, the angle between the resultant and the normal to the surface, is called the **angle of static friction.** For a given normal force **N**, if the friction force is less than the maximum $F < \mu_s N$, then the angle of the resultant will be less than the angle of static friction $\theta < \phi_s$. In no case can the angle of the resultant θ be greater than ϕ_s for a body in equilibrium. A similar relation is obtained in the case of kinetic friction,

$$\tan \phi_k = \mu_k$$

where ϕ_k is called the **angle of kinetic friction.**

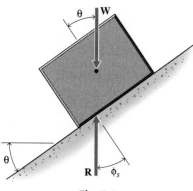

Fig. 9-5

When a block sits on an inclined surface and is acted on only by gravity, the resultant of the normal and friction force must be collinear, as shown in Fig. 9-5. But the angle between the resultant and the normal to the surface can never be greater than the angle of static friction ϕ_s. Therefore, the steepest inclination θ for which the block will be in equilibrium is equal to the angle of static friction. This angle is called the **angle of respose.**

There are three typical types of friction problems encountered in engineering analysis:

1. **Impending Motion Is not Assumed.**   This first case is just the type of equilibrium problem solved in the previous chapters of the book. The friction force F_{req} ($\neq \mu_s N$) and the normal force N are drawn on the free-body diagram and are determined using force equilibrium. The normal force should not be drawn as acting through the center of the body. Instead, its location is determined using moment equilibrium. The amount of friction required for equilibrium F_{req} and the location of the normal force are then checked against their maximum values. Three possibilities exist:

    a. If the amount of friction required for equilibrium F_{req} is smaller than or equal to the maximum amount of friction available $F_{avail} = F_{max} = \mu_s N$ and the location of the normal force is on the body, then the body is in equilibrium. In this case, the actual friction force supplied by the surface is

    $$F_{actual} = F_{req} < F_{avail} = \mu_s N$$

    That is, the surface supplies just enough friction force (resistance) to keep the body from moving.

    b. If the amount of friction required for equilibrium F_{req} is smaller than or equal to the maximum amount of friction available $F_{avail} = F_{max} = \mu_s N$ but the location of the normal force is not somewhere on the body, then the body is not in equilibrium and will tip over.

    c. If the amount of friction required for equilibrium F_{req} is greater than the maximum amount of friction available $F_{avail} = F_{max} = \mu_s N$, then the body is not in equilibrium and will slide. In this case, the actual friction force supplied by the surface is the kinetic friction force $F_{actual} = \mu_k N$.

2. **Impending Slipping Is Known to Occur at All Surfaces of Contact.** Since impending slipping is known to occur at all surfaces of contact, the magnitude of the friction forces can be shown as $\mu_s N$ on the free-body diagrams. The equations of equilibrium are then written and:

    a. If all the applied forces are given but μ_s is unknown, the equations of equilibrium can be solved for N and μ_s. This μ_s is the smallest coefficient for which the body will be in equilibrium.

    b. If the coefficient of friction is given but one of the applied forces is unknown, the equations of equilibrium can be solved for N and the unknown applied force.

3. **Impending Motion Is Known to Exist but the Type of Motion or Surface of Slip Is not Known.** Since it is not known whether the body tips or slips, the free-body diagrams must be drawn as in case 1 above. That is, the friction forces must not be shown as $\mu_s N$ on the free-body diagrams. At this point the three equations of equilibrium will contain more than three unknowns. Assumptions must be made about the type of motion that is about to occur until the number of equations equals the number of unknowns. The equations are then solved for the remaining unknowns and checked against the assumptions made about slipping or tipping. If F_{req} comes out greater than $F_{avail} = \mu_s N$ at some surface or if the location of the normal force is not on the body, then the assumptions must be changed and the problem solved again.

In both of the last two cases (2b and 3), the friction force is treated as if it is a known force. Care must be taken to be sure that it opposes the tendency of the other forces to cause motion. The result will not include a negative sign to indicate that the friction is in the wrong direction. If the direction of the friction force is drawn incorrectly, incorrect answers will result. The direction is easily determined by pretending for a moment that friction does not exist. Then apply the friction in such a direction as to opposite the motion that would occur in the absence of friction.

A 20-lb piece of electronic equipment is placed on a wooden skid that weighs 10 lb and that rests on a concrete floor (Fig. 9-6). Assume a coefficient of static friction of 0.45 and determine the minimum pushing force along the handle necessary to cause the skid to start sliding across the floor.

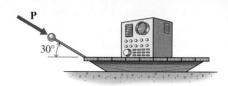

Fig. 9-6

SOLUTION

The free-body diagram of the skid is drawn in Fig. 9-7. Since motion is known to be impending (to the right), the friction force has magnitude $A_f = \mu_s A_n$ and is drawn pointing to the left so as to oppose the impending motion. (The symbol A_n was chosen for the normal force rather than N so as to avoid any confusion with the abbreviation for Newton.) The equilibrium equations are

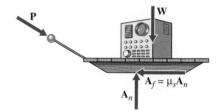

$$+\rightarrow \Sigma F_x = P \cos 30° - 0.45 A_n = 0$$
$$+\uparrow \Sigma F_y = A_n - P \sin 30° - 30 = 0$$

Fig. 9-7

These equations can be rewritten in the form

$$P \sin 30° = A_n - 30 \qquad (a)$$
$$P \cos 30° = 0.45 A_n \qquad (b)$$

Now dividing Eq. a by Eq. b gives

$$\tan 30° = \frac{A_n - 30}{0.45 A_n}$$

or

$$A_n = 40.53 \text{ lb}$$

This value can be substituted into either Eq. a or Eq. b to get

$$P = 21.06 \text{ lb} \qquad \text{Ans.}$$

the minimum force necessary to start the skid sliding.

In Example Problem 9-1, assume the equipment and skid weigh 100 N and 50 N, respectively, and a pulling force is applied to the handle. Determine the minimum force necessary to start the skid in motion now.

SOLUTION

The free-body diagram for this case is shown in Fig. 9-8. Note that now the direction of impending motion is to the left, and so the friction must act to the right. The equilibrium equations are now

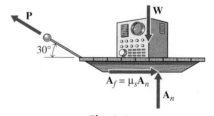

$$+\rightarrow \Sigma F_x = P \cos 30° - 0.45 A_n = 0$$
$$+\uparrow \Sigma F_y = A_n + P \sin 30° - 150 = 0$$

Solving as before gives

$$A_n = 119.1 \text{ N}$$
$$P = 61.9 \text{ N} \qquad \text{Ans.}$$

Fig. 9-8

(Note that this time the force P is only 41 percent of the total weight whereas in Example 9-1 the force required was 70 percent of the total weight.)

A 20-lb homogeneous box has tipped and is resting against a 40-lb homogeneous box (Fig. 9-9). The coefficient of friction between box A and the floor is 0.7; between box B and the floor, 0.4. Treat the contact surface between the two boxes as smooth and determine if the boxes are in equilibrium.

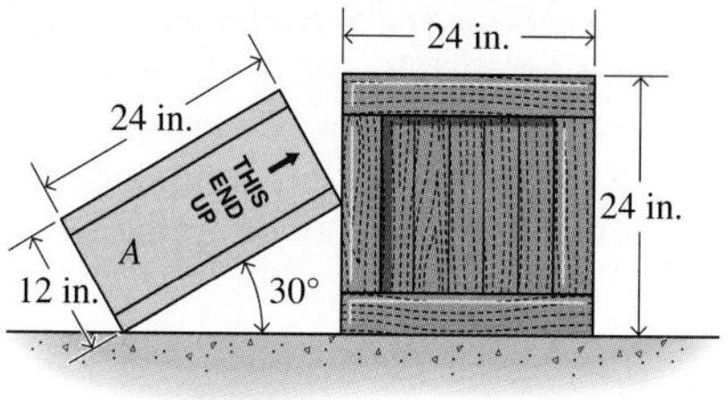

Fig. 9-9

SOLUTION

The free-body diagram of box A is drawn in Fig. 9-10a. The equilibrium equations

$$+\rightarrow \Sigma F_x = A_f - C = 0$$
$$+\uparrow \Sigma F_y = A_n - 20.0 = 0$$
$$\downarrow + \Sigma M_D = 12.00C - (7.392)(20) = 0$$

are solved to get

$$C = 12.320 \text{ lb} \qquad A_n = 20.000 \text{ lb} \qquad A_f = 12.320 \text{ lb}$$

The friction force available at this surface is

$$F_{max} = \mu_s A_n = (0.7)(20) = 14.00 \text{ lb}$$

Since the friction force required (12.32 lb) is less than the friction force available (14.00 lb), box A is in equilibrium.

The free-body diagram of box B is drawn in Fig. 9-10b. The equilibrium equations for this box

$$+\rightarrow \Sigma F_x = 12.320 - B_f = 0$$
$$+\uparrow \Sigma F_y = B_n - 40 = 0$$
$$\downarrow + \Sigma M_E = x_B B_n - (12.00)(12.230) - (12.00)(40.0) = 0$$

give

$$B_f = 12.320 \text{ lb} \qquad B_n = 40.00 \text{ lb} \qquad x_B = 15.669 \text{ in.}$$

The friction force available at this surface is

$$F_{max} = \mu_s B_n = (0.4)(40.00) = 16.00 \text{ lb}$$

Again the friction force available (16.00 lb) is greater than the friction force required (12.32 lb) and box B is also in equilibrium.

Thus, both boxes are in equilibrium. Ans.

(Note that while the normal force B_n does not act at the center of the crate, it does act on the bottom of the crate.)

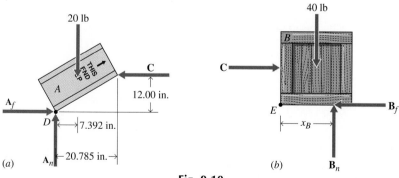

Fig. 9-10

The wheels of the refrigerator of Fig. 9-11 are stuck and will not turn. The refrigerator weighs 600 N. Assume a coefficient of friction between the wheels and the floor of 0.6 and determine the force necessary to cause the refrigerator to just start to move (impending motion). Also determine the maximum height h at which the force can be applied without causing the refrigerator to tip over.

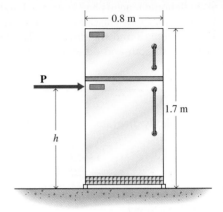

Fig. 9-11

SOLUTION

The free-body-diagram of the refrigerator is drawn in Fig. 9-12. The equilibrium equations

$$+\rightarrow \Sigma F_x = P - A_f - B_f = 0$$
$$+\uparrow \Sigma F_y = A_n + B_n - 600 = 0$$
$$\downdownarrows + \Sigma M_B = (0.4)(600) - hP - 0.8\, A_n = 0$$

give

$$A_n + B_n = 600$$
$$P = A_f + B_f = 0.6(A_n + B_n) = 360 \text{ N} \qquad\qquad \text{Ans.}$$

and

$$h = \frac{240 - 0.8A_n}{360}$$

When $h = 0$, $A_n = B_n = 300$ N, and the wheels share the load of the weight equally. As h increases, A_n gets smaller. However, the force at A cannot be negative, and so

$$h < \frac{240}{360} = 0.667 \text{ m} \qquad\qquad \text{Ans.}$$

Note that at the point of impending tipping, none of the weight is carried by the wheel at A; it has all been shifted to the wheel at the front corner of the refrigerator.

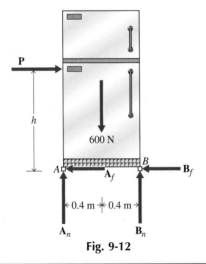

Fig. 9-12

The pickup of Fig. 9-13 is traveling at a constant speed of 50 mi/h and is carrying a 60-lb box in the back. The box projects 1 ft above the cab of the pickup. The wind resistance on the box can be approximated as a uniformly distributed force of 25 lb/ft on the exposed edge of the box. Calculate the minimum coefficient of friction required to keep the box from sliding on the bed of the pickup. Also determine whether or not the box will tip over.

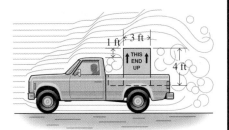

Fig. 9-13

SOLUTION

The free-body-diagram of the box is drawn in Fig. 9-14. The equilibrium equations for the box

$$+\rightarrow \Sigma F_x = (25)(1) - C_f = 0$$
$$+\uparrow \Sigma F_y = C_n - 60 = 0$$
$$\downarrow + \Sigma M_A = (1.5)(60) - (3.5)[(25)(1)] - x_C C_n = 0$$

are solved to get

$$C_f = 25.0 \text{ lb} \qquad C_n = 60.0 \text{ lb}$$

and

$$x_C = 0.042 \text{ ft} = 0.500 \text{ in.}$$

Thus the required coefficient of friction is

$$\mu_s = \frac{C_f}{C_n} = \frac{25.0}{60.0} = 0.417 \qquad\qquad \text{Ans.}$$

and since x_C is positive, the box will not tip.   Ans.

(*Note:* The friction and normal force must act on the box, and thus x_C must be a number between 0 and 3 ft. If the solution had given x_C to be negative, then no normal force on the bottom of the box could satisfy moment equilibrium and the box would tip over.)

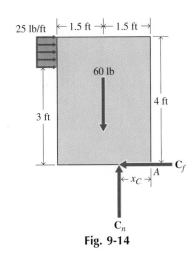

Fig. 9-14

PROBLEMS

9-1* A light, inextensible cord passes over a frictionless pulley (Fig. P9-1). One end of the rope is attached to a block; a force **P** is applied to the other end. Block A weighs 600 lb and block B weighs 1000 lb. The coefficient of friction between the blocks is 0.33, while the coefficient of friction between B and the floor is 0.25.

a. Determine if the system would be in equilibrium for $P = 400$ lb.
b. Determine the maximum **P** for which the system is in equilibrium.

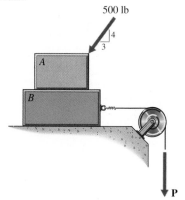

Fig. P9-1

9-2 A force of 500 N is applied to the end of the cord that is attached to block A of Fig. P9-2. Block A weighs 2000 N and block B weighs 4000 N. The coefficient of friction between the blocks is 0.3, while the coefficient of friction between B and the floor is 0.2.

a. Determine if the system would be in equilibrium for $P = 2500$ N.
b. Determine the maximum **P** for which the system is in equilibrium.

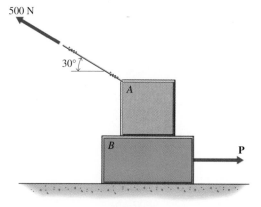

Fig. P9-2

9-3* The block in Fig. P9-3 weighs 500 lb and the coefficient of friction between the block and the floor is 0.2.

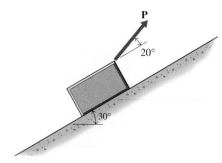

Fig. P9-3

a. Determine if the system would be in equilibrium for $P = 400$ lb.
b. Calculate the minimum **P** to prevent motion.
c. Determine the maximum **P** for which the system is in equilibrium.

9-4* The 25-kg block in Fig. P9-4 is held against the wall by the brake arm. The coefficient of friction between the wall and the block is 0.20; between the block and the brake arm, 0.50. Neglect the weight of the brake arm.

a. Determine if the system would be in equilibrium for $P = 230$ N.
b. Determine the minimum force **P** for which the system would be in equilibrium.
c. Determine the maximum force **P** for which the system would be in equilibrium.

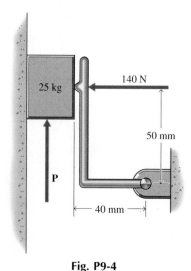

Fig. P9-4

9-5 Workers are pulling a 400-lb crate up an incline as shown in Fig. P9-5. The coefficient of friction between the crate and the surface is 0.2 and the rope on which the workers are pulling is horizontal.

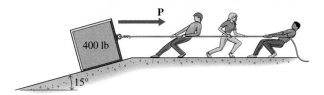

Fig. P9-5

a. Determine the force **P** that the workers must exert to start sliding the crate up the incline.
b. If one of the workers lets go of the rope for a moment, determine the minimum force the other worker must exert to keep the crate from sliding back down the incline.

9-6* The rods of Fig. P9-6 are lightweight and all pins are frictionless. The coefficient of friction between the 40-kg slider block and the floor is 0.40.

a. Assume that the force **P** is horizontal ($\theta = 0$) and determine the maximum force **P** for which motion does not occur.
b. Determine the angle θ that gives the absolute greatest force **P** for which motion does not occur.

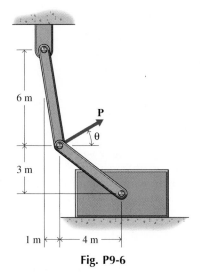

Fig. P9-6

9-7* The 200-lb crate of Fig. P9-7 is being moved by a rope that passes over a smooth pulley. The coefficient of friction between the crate and the floor is 0.30.

a. Assume that $h = 4$ ft and determine the force **P** necessary to produce impending motion.
b. Determine the value of h for which impending motion by slipping and by tipping would occur simultaneously.

9-8 The three blocks of Fig. P9-8 are connected by light, inextensible cords. The left block weighs 160 N and the center block weighs 300 N. The coefficient of friction between the 300-N block and the floor is 0.20; the two pulleys are frictionless.

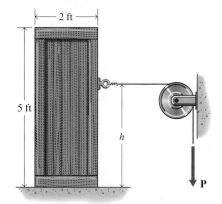

Fig. P9-7

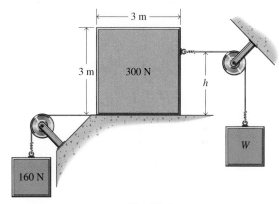

Fig. P9-8

a. Assume that $h = 1.8$ m and determine the minimum and maximum weight of the second block, W_{min} and W_{max}, such that motion does not occur.
b. Determine the value of h for which impending motion by slipping and by tipping would occur simultaneously.

9-9* A homogeneous, triangular block of weight W has height h and base width b (Fig. P9-9). Determine an expression for the coefficient of friction between the block and the surface for which impending motion by slipping and by tipping occur simultaneously.

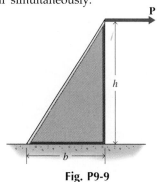

Fig. P9-9

9-10 Repeat Problem 9-9 for the case where the force **P** acts to the left.

389

9-11 The 200-lb block of Fig. P9-11 is sitting on a 30° inclined surface. The coefficient of friction between the block and the surface is 0.50. A light, inextensible cord is attached to the block, passes around a frictionless pulley, and is attached to a second block of mass M. Determine the minimum and maximum masses, M_{min} and M_{max}, such that the system is in equilibrium. Is impending motion by slipping or by tipping?

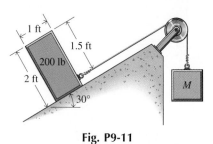

Fig. P9-11

9-12 A 20-kg triangular block sits on top of a 10-kg rectangular block (Fig. P9-12). The coefficient of friction is 0.40 at all surfaces. Determine the maximum horizontal force P for which motion will not occur.

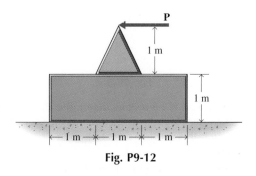

Fig. P9-12

9-13* The block A in Fig. P9-13 is pressed against the floor by a 65-lb force on the handle. The pin is frictionless and the weight of the handle can be neglected. The coefficient of friction is 0.2 at all surfaces. Determine the minimum weight W_A necessary to prevent slipping.

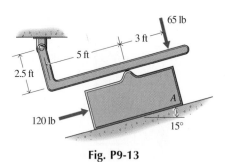

Fig. P9-13

9-14 A 30-kg box is sitting on an inclined surface as shown in Fig. P9-14. If the coefficient of friction between the box and the surface is 0.50 and the angle $\alpha = 60°$, deter-

mine the maximum force T for which the box will be in equilibrium. Is impending motion by tipping? Or by slipping?

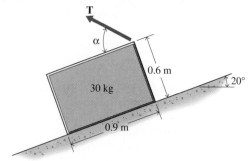

Fig. P9-14

9-15* A 120-lb girl is walking up a 48-lb uniform beam (Fig. P9-15). The coefficient of friction is 0.2 at all surfaces. Determine how far up the beam the girl can walk before the beam starts to slip.

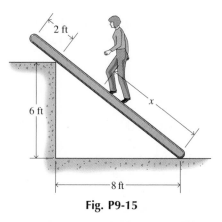

Fig. P9-15

9-16 A 75-kg man starts climbing a 5-m long ladder leaning against a wall (Fig. P9-16). The coefficient of friction is 0.25 at both surfaces. Neglect the weight of the ladder and

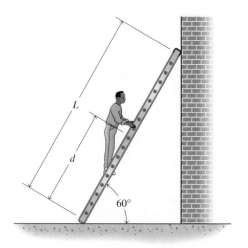

Fig. P9-16

determine how far up the ladder the man can climb before the ladder starts to slip.

9-17 In Problem 9-15 a piece of rubber is placed between the bottom end of the beam and the floor, increasing the friction there to 0.4. Determine how far up the beam the girl can now walk before the beam starts to slip.

9-18* In Problem 9-16 a piece of rubber is placed between the bottom end of the ladder and the floor, increasing the friction there to 0.4. Determine how far up the ladder the man can now climb before the ladder starts to slip.

9-19 A 100-lb uniform beam 16 ft long lies against a corner (Fig. P9-19). Determine the maximum force **P** for which the beam will be in equilibrium if the coefficient of friction μ_s is:

a.   0.6 at both surfaces.
b.   0.75 at the bottom surface and 0.4 at the corner surface.

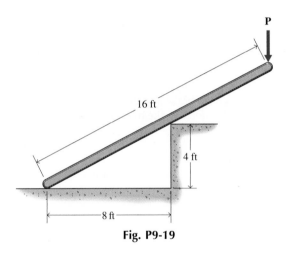

Fig. P9-19

9-20 A block of mass M rests on a 50-kg block, which in turn rests on an inclined plane (Fig. P9-20). The coefficient of friction between the blocks is 0.40; between the 50-kg block and the floor, 0.30. The pulley is frictionless and the weight of the cord can be neglected. Determine the minimum mass M_{min} necessary to prevent slipping.

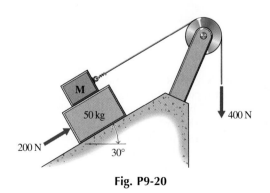

Fig. P9-20

9-21* In Fig. P9-21 box A weighs 10 lb and rests on an inclined surface, while box B weighs 20 lb and rests on a level surface. The coefficient of friction between box A and the surface is 0.45; between box B and the surface, 0.5. The pulleys are all frictionless. Determine the maximum weight of box C such that no motion will occur. Which motion would occur first?

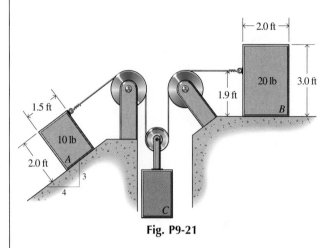

Fig. P9-21

9-22* A 50-kg uniform plank rests on rough supports at A and B (Fig. P9-22). The coefficient of friction is 0.60 at both surfaces. If a man weighing 800 newtons pulls on the rope with a force of $P = 400$ N, determine:

a.   The minimum and maximum angles θ_{min} and θ_{max} for which the system will be in equilibrium.
b.   The minimum coefficient of friction that must exist between the man's shoes and the ground for each of the cases in part a.

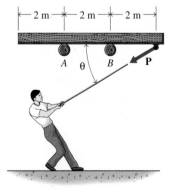

Fig. P9-22

391

9-23 The 45-lb plank rests against a wall in the corner (Fig. P9-23). The coefficient of friction between the floor and the plank is 0.4, while a roller at the top end eliminates friction there. Determine the minimum horizontal force **P** necessary to move the plank.

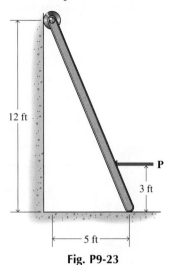

Fig. P9-23

9-24 A worker is lifting a 35-kg uniform beam with a force that is perpendicular to the beam (Fig. P9-24). The beam is 5 m long and the coefficient of friction between the beam and the ground is 0.2. Determine the height h at which the beam will begin to slip.

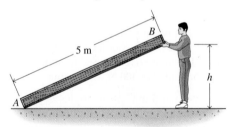

Fig. P9-24

9-25* A uniform plank 8 ft long is balanced on a corner (Fig. P9-25) using a horizontal force of 75 lb. If the plank weighs 45 lb, determine:

a. The angle θ for equilibrium.
b. The minimum coefficient of friction for which the plank is in equilibrium.

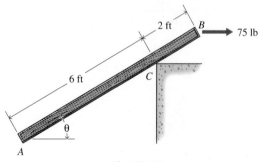

Fig. P9-25

9-26* A post is held up by a lightweight rope that passes around a frictionless pulley and is attached to a mass M (Fig. P9-26). Find the mass M in terms of μ_s, θ, and m_p (the mass of the post) for the case of impending slip at the floor.

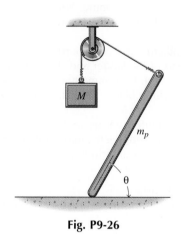

Fig. P9-26

9-27 A 25-lb weight is suspended from a lightweight rope wrapped around the inner cylinder of a drum (Fig. P9-27). A brake arm is pressed against the outer cylinder of the drum by a hydraulic cylinder. The coefficient of friction between the brake arm and the drum is 0.40. Determine the smallest force in the hydraulic cylinder necessary to prevent motion.

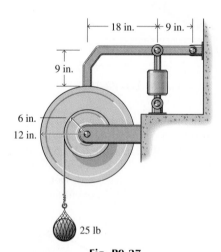

Fig. P9-27

9-28* A 250-N weight is suspended from a lightweight rope that is wrapped around the inner cylinder of a drum (Fig. P9-28). A T-bar brake mechanism rubs against the outer cylinder of the drum. The coefficient of friction between the T-bar and the drum is 0.25; the weight of the T-bar mechanism is 200 N; and the weight of the drum is 150 N. Determine the minimum force **P** necessary to prevent motion.

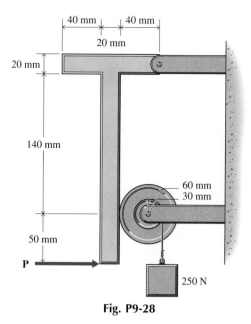

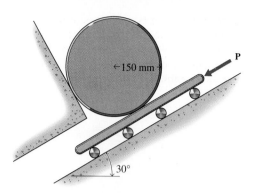

Fig. P9-30

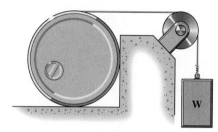

Fig. P9-31

Fig. P9-28

9-29 One jaw of an adjustable clamp is designed to slide along the frame to make the capacity adjustable (Fig. P9-29). The coefficient of friction between the jaw and the frame is 0.40. Determine the minimum value of the dimension x necessary to prevent the jaw of the clamp from slipping under load.

9-32 Three identical cylinders are stacked as shown in Fig. P9-32. The cylinders each weigh 100 N and are 200 mm in diameter, and the coefficient of friction is $\mu_s = 0.40$ at all surfaces. Determine the maximum force **P** that the cylinders can support without moving.

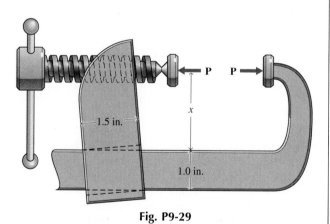

Fig. P9-29

Fig. P9-32

9-30 A 100-kg cylinder rests against a wall and a plate (Fig. P9-30). The coefficient of friction is 0.30 at both surfaces. The plate rests on rollers and the friction between the plate and the floor is negligible. Neglect the weight of the plate and determine the minimum force **P** necessary to move the plate.

9-31* A lightweight rope is wrapped around a 100-lb drum, passes over a frictionless pulley, and is attached to a weight W (Fig. P9-31). The coefficient of friction between the drum and the surfaces is 0.50. Determine the maximum amount of weight that can be supported by this arrangement.

9-33 A lightweight rope is wrapped around a 25-lb drum, which is 3 ft in diameter (Fig. P9-33). The coefficient of friction between the drum and the ground is 0.30. Determine whether or not the drum is in equilibrium.

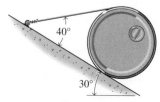

Fig. P9-33

9-34* A lightweight rope is wrapped around a drum as shown in Fig. P9-34. The coefficient of friction between the drum and the ground is 0.30. Determine the maximum angle θ such that the drum does not slip. Also determine the tension in the cable for this angle if the drum weighs 100 N.

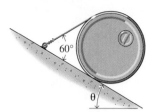

Fig. P9-34

9-35* When a drawer is pulled by only one of the handles, it tends to twist and rub as shown (highly exaggerated) in Fig. P9-35. The weight of the drawer and its contents is 2 lb and is uniformly distributed. The coefficient of friction between the sides of the drawer and the sides of the dresser is 0.6; between the bottom of the drawer and the side rails the drawer rides on $\mu_s = 0.10$. Determine the minimum amount of force necessary to pull the drawer out.

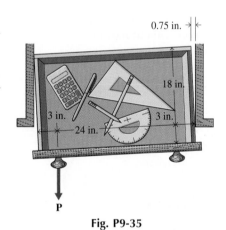

Fig. P9-35

9-36 An ill-fitting window is about 10 mm narrower than its frame (Fig. P9-36). The window weighs 40 N and the coefficient of friction between the window and the frame is 0.2. Determine the amount of force P that must be applied at the lower corner to keep the window from lowering.

9-37 For Problem 9-35 determine the minimum coefficient of friction such that no amount of force applied to a single handle will be able to pull the drawer out.

9-38* The broom shown in Fig. P9-38 weighs 8 N and is held up by the two cylinders, which are wedged between the broom handle and the side rails. The coefficient of fric-

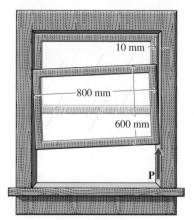

Fig. P9-36

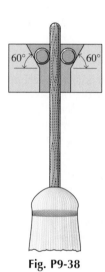

Fig. P9-38

tion between the broom and cylinders and between the cylinders and side rails is 0.30. The side rails are at an angle of $\theta = 60°$ to the horizontal. The weight of the cylinders may be neglected. Determine whether or not this system is in equilibrium. If it is in equilibrium, determine the force exerted on the broom handle by the rollers.

9-39 The simple mechanism of Fig. P9-38 is often used to hold the handles of brooms, mops, shovels, and other such tools. The weight of the tool causes the two otherwise free cylinders to become wedged into the corner between the handle and the rails. Although no amount of downward force will cause the handle to slip, the tool can be removed easily by lifting it upward and pulling it forward. Determine the minimum coefficient of friction necessary to make this device work.

9-40 Suppose that the coefficient of friction in Problem 9-38 is 0.20. Determine the minimum angle θ of the side rails such that this device will work.

9-41* A piece of paper weighing 0.01 lb is pinched between a free rolling cylinder and a fixed wall (Fig. P9-41). A force of 5 lb is applied to the paper to try to pull it out.

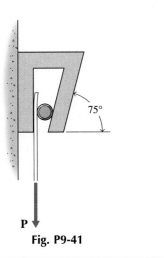

P

Fig. P9-41

The coefficient of friction is 0.20 at all surfaces. The cylinder rolls along a rail that makes an angle of 75° to the horizontal, and the weight of the cylinder can be neglected. Determine whether or not this system is in equilibrium. If it is in equilibrium, determine the force exerted on the paper by the roller.

9-42 The simple mechanism of Fig. P9-41 is often used to hold notes or signs on bulletin boards. The weight of the paper causes the otherwise free cylinder to become wedged into the corner between the paper and the rail. Although no amount of downward force will cause the paper to slip out, the paper can be removed easily by lifting it upward and pulling it forward. Determine the minimum coefficient of friction necessary to make this device work.

9-43 Suppose that the coefficient of friction in Problem 9-41 is 0.30. Determine the minimum angle θ of the side rails such that this device will work.

9-3 ANALYSIS OF SYSTEMS INVOLVING DRY FRICTION

Friction is often encountered in engineering applications in simple situations such as observed in the previous section. Just as often, however, friction is encountered in situations involving more complex applications. Some of these machine applications, which will be studied here, are (Fig. 9-15): wedges, screws, journal bearings, thrust bearings, and belts. As was the case with machine and non-rigid frames in Chapter 7, it will be necessary to consider equilibrium of the component parts of the machines—even when only external forces or reactions are desired.

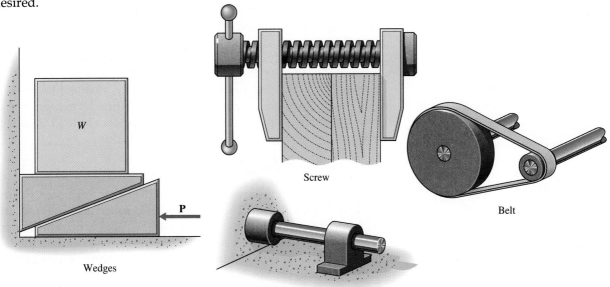

W

P

Wedges

Screw

Thrust Bearing

Journal Bearing

Belt

Fig. 9-15

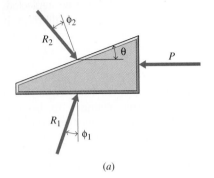

(a)

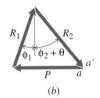

(b)

Fig. 9-16

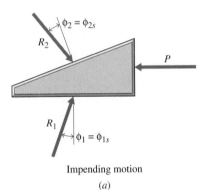

Impending motion

(a)

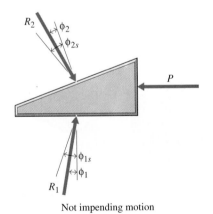

Not impending motion

(b)

Fig. 9-17

9-3-1 Wedges

A wedge is just a block that has two flat faces that make a small angle with each other. Wedges are often used in pairs as shown in Fig. 9-15 to raise heavy loads. Depending on the angle between the two surfaces of the wedge, the weight being lifted (the output force) can be many times that of the force **P** (the input force) applied to the wedge. Also, a properly designed wedge will stay in place and support the load even after the force **P** is removed.

Wedge problems can often be solved using a semigraphical approach. Wedges are almost always constrained against rotation so that only force equilibrium need be considered. Also, the number of forces acting on a wedge is usually small (the friction and normal forces are usually combined into a single resultant force as in Fig. 9-16a) so that force equilibrium can be expressed as a force polygon (Fig. 9-16b). The Law of Sines and the Law of Cosines can then be used to relate the forces and angles.

For the case of impending motion, the resultant of the normal and friction force is drawn at the angle of static friction (Fig. 9-17a) and the magnitude of the resultant or some other force determined. If motion is not impending, the resultant is drawn with whatever magnitude and at whatever angle ϕ_1 is required for equilibrium. This angle is then compared with the angle of static friction $\phi_1 \le \phi_{1s}$ (Fig. 9-17b) to determine whether or not equilibrium exists.

Like other machines, wedges are typically characterized by their **mechanical advantage** (M.A.) or the ratio of their output and input forces. In the case of wedges, the mechanical advantage is defined as the ratio

$$\text{M.A.} = \frac{\text{direct force}}{\text{wedge force}} \tag{9-6}$$

The numerator of Eq. 9-6 is the force that must be applied directly to some object to accomplish a desired task. In the case of the wedge in Fig. 9-15, this is just the weight of the object being raised. The denominator of Eq. 9-6 is the force that must be applied to the wedge to accomplish the same task. For the wedge of Fig. 9-15, this is P. Clearly, a well-designed wedge should have a mechanical advantage greater than one.

A wedge with a large mechanical advantage may not be the best overall design, however. A common design criterion for wedges is that the wedge remain in place after being forced under the load. A wedge that must be forcibly removed is called **self-locking.**

9-3-2 Square-Threaded Screws

Square-threaded screws are essentially wedges that have been wound around a cylindrical shaft. These simple devices can be found in nearly every facet of our lives. Screws are used as fasteners to hold machinery together. Screws are used in jacks to raise heavy loads and on the feet of heavy appliances such as refrigerators to level them. Screws are also used in vices and clamps to squeeze objects together. In each of these cases and many more like them, friction on the threads keeps the screws from turning and loosening.

For example, consider the simple C-clamp of Fig. 9-18. When a twisting moment **M** is applied to the screw, the clamp tightens and exerts an axial force **W** on whatever is held in the clamp. As the screw turns and tightens, however, a small segment of the screw's thread will travel around and up the groove in the frame. (Fig. 9-19). The distance that the screw moves in the axial direction during one revolution (from point A to A') is called the lead of the screw. For a single-threaded screw, the lead is the same as the distance between adjacent threads (Fig. 9-20). If a screw has two independent threads that wind around it, the lead would be twice the distance between the adjacent threads.

During each complete turn of the screw a small segment of the screw's thread will travel a distance $2\pi r$ around the shaft while advancing the distance L. It is as if the small segment of the screw thread is being pushed up a wedge or inclined plane of angle $\alpha = \tan^{-1}\left(\dfrac{L}{2\pi r}\right)$ (Fig. 9-21). In addition to the normal and friction forces on the thread, the free-body diagram for a typical segment of the screw includes a portion of the axial force dW and a force $d\mathbf{P}$ due to the twisting moment, $dM = r\, dP$. If the equilibrium equations for each little segment of the screw are added together, the resulting set of equations would be the same as the equilibrium equations for the free-body diagram shown in Fig. 9-22 in which $W = \int dW$ is the total axial force, $P = \int dP = \dfrac{\int dM}{r} = \dfrac{M}{r}$ is the total pushing force due to the moment **M**, and $\mathbf{F} = \int d\mathbf{F}$ and $\mathbf{N} = \int d\mathbf{N}$ are the total friction and normal forces, respectively.

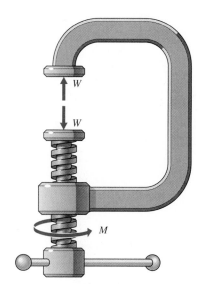

Fig. 9-18

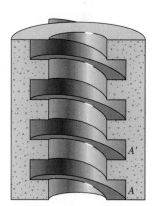

Fig. 9-19

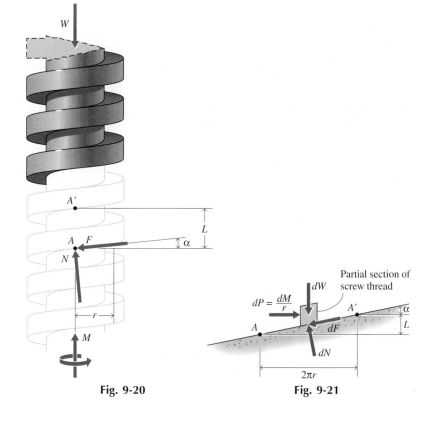

Fig. 9-20

Fig. 9-21

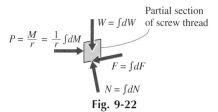

Fig. 9-22

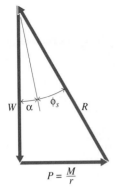

$P = \dfrac{M}{r}$

Fig. 9-23

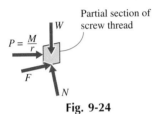

Partial section of
screw thread

$P = \dfrac{M}{r}$

Fig. 9-24

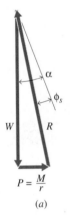

$P = \dfrac{M}{r}$

(a)

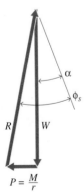

$P = \dfrac{M}{r}$

(b)

Fig. 9-25

If the twisting moment **M** is just sufficient to turn the screw, then **R**, the resultant of the friction and normal forces, will act at the angle of static friction ϕ_s to the normal and force equilibrium is neatly expressed by the force triangle (Fig. 9-23) from which

$$\tan(\alpha + \phi_s) = \frac{M/r}{W}$$

Therefore,

$$M = rW\tan(\alpha + \phi_s) \tag{9-7}$$

is the minimum torque **M** necessary to advance the screw against a load **W**.

When the twisting moment **M** is removed or reduced to near zero, the screw will tend to unwind and the friction force will change direction (Fig. 9-24). If the lead angle α is greater than the angle of static friction ϕ_s, then the screw will not be in equilibrium when the twisting moment is removed, but will require a twisting moment of

$$M = rW\tan(\alpha - \phi_s) \tag{9-8a}$$

to maintain equilibrium (Fig. 9-25a). However, if the lead angle α is less than the angle of static friction ϕ_s, then the screw will be in equilibrium even when the twisting moment is removed. This condition is called self-locking and is a design criterion in most screw designs. In this case a reverse moment of

$$M = rW\tan(\phi_s - \alpha) \tag{9-8b}$$

is required to remove the screw (Fig. 9-25b).

9-3-3 Journal Bearings

Journal bearings are used to support the sides of rotating shafts, as shown in Fig. 9-26. Journal bearings are usually lubricated to reduce friction, and the analysis of lubricated journal bearings is a topic for fluid mechanics. Some journal bearings, however, are not lubricated or are poorly lubricated. The analysis that follows can be used to estimate the amount of friction in these bearings.

Fig. 9-26

Suppose a torque **M** is applied to the shaft of Fig. 9-26 to cause it to rotate. Because of the lateral load **L** on the shaft, there will be contact between the shaft and bearing. As the shaft rotates, friction between the rotating shaft and the bearing causes the shaft to climb a small distance up the inner surface of the bearing. Since slip is occurring, the friction force is that of kinetic friction $F = \mu_k N$ and the resultant of the friction and normal forces acts at the angle of kinetic friction ϕ_k to the normal.

Applying force equilibrium to the free-body diagram of Fig. 9-27a gives that **R** must be a vertical force equal in magnitude to the load **L**. Then for moment equilibrium

$$M = Lr \sin \phi_k \qquad (9\text{-}9a)$$

where $r \sin \phi_k$ is the distance between the lines of action of **L** and **R** (Fig. 9-27b). Even for dry journal bearings, however, the coefficient of friction $\mu_k = \tan \phi_k$ is generally small so that $\sin \phi_k \cong \tan \phi_k = \mu_k$. Therefore, the torque required to rotate a shaft at constant speed is approximately

$$M \cong \mu_k Lr \qquad (9\text{-}9b)$$

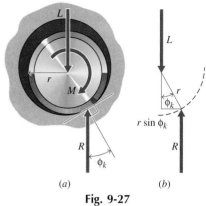

Fig. 9-27

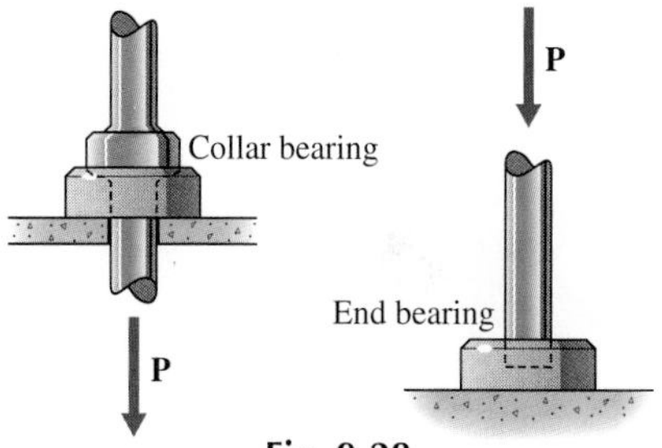

9-3-4 Thrust Bearings

Thrust bearings are used to support shafts acted on by axial loads. Thrust bearings include both the collar bearing and end bearing shown in Fig. 9-28. With collar bearings, friction forces act on the annular region of contact between the collar and the bearing. With end bearings, friction forces act on a circular area if the shaft is solid or on an annular region if the shaft is hollow. The analysis for clutch plates and disk brakes, which also involve friction on circular or annular regions, is identical to that for end bearings except that the kinetic coefficient of friction μ_k must be replaced with the static coefficient of friction μ_s since friction is only impending and not actually occurring.

The end bearing of Fig. 9-29 supports a hollow circular shaft that is subjected to an axial force **P**. If the contact pressure (normal force per unit area) between the shaft and the bearing is p, then the normal force on a small element of area dA is $dN = p\,dA$; the friction force on dA is $dF = \mu_k (p\,dA)$; and the moment of the friction force about the axis of the shaft is $dM = r\,dF = \mu_k pr\,dA$. Integrating these quantities over the contact area gives the total axial load carried by the bearing

$$P = \int p\,dA = \int_{R_1}^{R_2} p2\pi r\,dr \qquad (9\text{-}10a)$$

and the total moment

$$M = \int_A dM = \int_A \mu_k pr\,dA = \mu_k \int_{R_1}^{R_2} pr2\pi r\,dr \qquad (9\text{-}10b)$$

For new surfaces that are flat and well-supported, the pressure p is essentially constant over the contact area so that the total axial load carried by the bearing is

$$P = p\int_{R_1}^{R_2} 2\pi r\,dr = p(R_2^2 - R_1^2) \qquad (9\text{-}11)$$

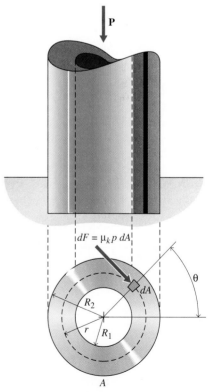

Fig. 9-28

Fig. 9-29

and the total moment is given by

$$M = \mu_k p \int_{R_1}^{R_2} 2\pi r^2 \, dr = \frac{2}{3}\mu_k \pi p(R_2^3 - R_1^3)$$

$$= \frac{2\mu_k P(R_2^3 - R_1^3)}{3(R_2^2 - R_1^2)} \quad (9\text{-}12a)$$

For a solid circular shaft of radius R, Eq. 9-12a simplifies to

$$M = \frac{2}{3}\mu_k PR \quad (9\text{-}12b)$$

As the shaft rotates in the bearing, however, the surfaces will gradually wear and the pressure will likely decrease with radial distance over the bearing surface. That is, during each rotation of the shaft, a small element of area travels a distance of $2\pi r$. Therefore, outer portions of the shaft, which travel farther, probably wear down faster than inner portions of the shaft, and contact will probably not be as strong there. Assuming that the pressure in a worn bearing decreases with radius according to $rp = C$, where C is a constant, gives the total axial force carried by the bearing as

$$P = \int p \, dA = \int_{R_1}^{R_2} \left(\frac{C}{r}\right)(2\pi r \, dr) = 2\pi C(R_2 - R_1) \quad (9\text{-}13)$$

and the total moment as

$$M = \mu_k C \int_{R_1}^{R_2} 2\pi r \, dr = \mu_k \pi C(R_2^2 - R_1^2)$$

$$= \frac{\mu_k P(R_2 + R_1)}{2} \quad (9\text{-}14a)$$

For a solid circular shaft of radius R, Eq. 9-14a simplifies to

$$M = \left(\frac{1}{2}\right)\mu_k PR \quad (9\text{-}14b)$$

which is just $\frac{3}{4}$ as much as for new surfaces.

9-3-5 Flat Belts and V-Belts

Many types of power machinery rely on belt drives to transfer power from one piece of equipment to another. Without friction, the belts would slip on their pulleys and no power transfer would be possible. Maximum torque is applied to the pulley when the belt is at the point of impending slip, and that is the case discussed below.

Although the analysis presented is for flat belts, it also applies to any shape belt as well as circular ropes as long as the only contact between the belt and the pulley is on the bottom surface of the belt. This section ends with a brief discussion of V-belts, which indicates the kind of modifications required when friction acts on the sides of the belt instead of the bottom.

Figure 9-30a shows a flat belt passing over a circular drum. The tensions in the belt on either side of the drum are $\mathbf{T}_1$ and $\mathbf{T}_2$ and the bearing reaction is $\mathbf{R}$. Friction in the bearing is neglected for this analysis, but a torque $\mathbf{M}$ is applied to the drum to keep it from rotating. If

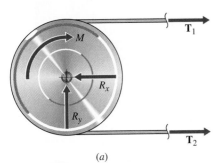

(a)

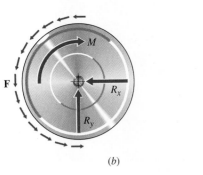

(b)

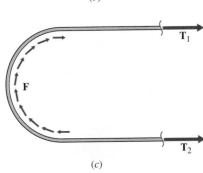

(c)

Fig. 9-30

there is no friction between the belt and the drum, the two tensions must be equal $T_1 = T_2$ and no torque is required for moment equilibrium $M = 0$. If there is friction between the belt and the drum, however, then the two tensions need not be equal and a torque $M = r$ $(T_2 - T_1)$ is needed to satisfy moment equilibrium. Assuming that $T_2 > T_1$, this means that friction must exert a counterclockwise moment on the drum (Fig. 9-30b) and the drum will exert an opposite frictional resistance on the belt (Fig. 9-30c). Because the friction force depends on the normal force and the normal force varies around the drum, care must be taken in adding up the total frictional resistance.

The free-body diagram (Fig. 9-31) of a small segment of the belt includes the friction force ΔF and the normal force ΔP. The tension in the belt increases from T on one side of the segment to $T + \Delta T$ on the other side. Equilibrium in the radial direction gives

$$\nwarrow\Sigma F_r = \Delta P - T \sin\left(\frac{\Delta\theta}{2}\right) - (T + \Delta T) \sin\left(\frac{\Delta\theta}{2}\right) = 0$$

or

$$\Delta P = 2T \sin\left(\frac{\Delta\theta}{2}\right) + \Delta T \sin\left(\frac{\Delta\theta}{2}\right) \qquad (a)$$

while equilibrium in the circumferential ($\theta-$) direction gives

$$\swarrow\Sigma F_\theta = (T + \Delta T) \cos\left(\frac{\Delta\theta}{2}\right) - T \cos\left(\frac{\Delta\theta}{2}\right) - \Delta F = 0$$

or

$$\Delta T \cos\left(\frac{\Delta\theta}{2}\right) = \Delta F \qquad (b)$$

In the limit as $\Delta\theta \to 0$ the normal force ΔP on the small segment of the belt must vanish according to Eq. a. But when the normal force vanishes ($\Delta P \to 0$), there can be no friction on the belt either ($\Delta F \to 0$). Therefore, the change in tension across the small segment of the belt must also vanish ($\Delta T \to 0$) in the limit as $\Delta\theta \to 0$ according to Eq. b.

Assuming that slip is impending gives $\Delta F = \mu_s \Delta P$ and Eqs. a and b can be combined to give

$$\Delta T \cos\left(\frac{\Delta\theta}{2}\right) = \mu_s 2T \sin\left(\frac{\Delta\theta}{2}\right) - \mu_s \Delta T \sin\left(\frac{\Delta\theta}{2}\right) \qquad (c)$$

which after dividing through by $\Delta\theta$ is

$$\frac{\Delta T}{\Delta\theta} \cos\left(\frac{\Delta\theta}{2}\right) = \mu_s T \frac{\sin\left(\frac{\Delta\theta}{2}\right)}{\Delta\theta/2} - \frac{\mu_s \Delta T}{2} \frac{\sin\left(\frac{\Delta\theta}{2}\right)}{\Delta\theta/2} \qquad (d)$$

Finally, taking the limits as $\Delta\theta \to 0$ and recalling that

$$\lim_{\Delta\theta\to 0} \frac{\Delta T}{\Delta\theta} \to \frac{dT}{d\theta} \qquad \lim_{x\to 0} \cos x \to x \qquad \lim_{x\to 0} \frac{\sin x}{x} \to 1$$

gives

$$\frac{dT}{d\theta} = \mu_s T \qquad (e)$$

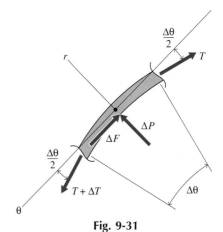

Fig. 9-31

(a)

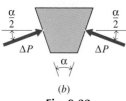

(b)

Fig. 9-32

Equation e can be rearranged in the form

$$\frac{dT}{T} = \mu_s \, d\theta \qquad (9\text{-}15)$$

which, since the coefficient of friction is a constant, can be immediately integrated from θ_1 where the tension is T_1 to θ_2 where the tension is T_2 to get

$$\ln\left(\frac{T_2}{T_1}\right) = \mu_s(\theta_2 - \theta_1) = \mu_s\beta \qquad (9\text{-}16a)$$

or

$$T_2 = T_1 e^{\mu_s\beta} \qquad (9\text{-}16b)$$

where $\beta = \theta_2 - \theta_1$ is the central angle of the drum for which the belt is in contact with the drum. The angle of wrap β must be measured in radians and must obviously be positive. Angles greater than 2π radians are possible and simply mean that the belt is wrapped more than one complete revolution around the drum.

It must be emphasized that Eq. 9-16 assumes impending slip at all points along the belt surface and therefore gives the maximum change in tension that the belt can have. Since the exponential function of a positive value is always greater than 1, Eq. 9-16 gives that T_2 (the tension in the belt on the side toward which slip tends to occur) will always be greater than T_1 (the tension in the belt on the side away from which slip tends to occur). Of course, if slip is not known to be impending, then Eq. 9-16 does not apply and T_2 may be larger or smaller than T_1.

V-belts as shown in Fig. 9-32a are handled similarly to the above. A view of the belt cross-section (Fig. 9-32b), however, shows that there are now two normal forces and there will also be two frictional forces (acting along the edges of the belt and pointing into the plane of the figure). Equilibrium in the circumferential ($\theta-$) direction now gives

$$\Delta T \cos\left(\frac{\Delta\theta}{2}\right) = 2\,\Delta F$$

while equilibrium in the radial direction gives

$$2\,\Delta P \sin\left(\frac{\alpha}{2}\right) = 2T \sin\left(\frac{\Delta\theta}{2}\right) + \Delta T \sin\left(\frac{\Delta\theta}{2}\right)$$

Continuing as above results finally in

$$T_2 = T_1 e^{(\mu_s)_{\text{enh}}\beta} \qquad (9\text{-}17)$$

in which $(\mu_s)_{\text{enh}} = \left[\dfrac{\mu_s}{\sin(\alpha/2)}\right] > \mu_s$ is an *enhanced coefficient of friction*. That is, V-belts always give a larger T_2 than flat belts for a given coefficient of friction μ_s and a given angle of wrap β.

Equations 9-16 and 9-17 can also be used when slipping is actually occurring by replacing the static coefficient of friction μ_s with the kinetic coefficient of friction μ_k.

A wedge is to be used to slide the 3000 N safe of Fig. 9-33a across the floor. Determine the minimum force **P** necessary if the coefficient of friction is 0.35 at all surfaces and the weight of the wedges may be neglected.

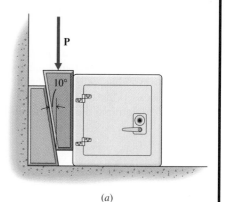

(a)

SOLUTION

SOLUTION 1. Using equilibrium equations

First draw the free-body diagram of the safe (Fig. 33b). Since motion is impending, the friction forces are determined by the normal forces $F = \mu_s N$ and must be drawn in the correct direction or the answer obtained will be incorrect. Clearly, the safe will tend to move to the right and the friction force $\mathbf{A}_f$ must act to the left to oppose the motion. However, the direction of the friction force $\mathbf{B}_f$ is not as easy to ascertain. Even though the safe is not moving up or down, it appears to be moving up relative to the wedge. Hence the friction force $\mathbf{B}_f$ must act downward on the safe to oppose this relative motion.

It may be easier to see the correct direction for this friction force on the free-body diagram of the wedge (Fig. 9-33c). The motion of the wedge is downward and the friction force $\mathbf{B}_f$ must act upward to oppose the motion. And if the safe exerts an upward frictional force on the wedge, then the wedge must exert an equal frictional force downward on the safe.

The equilibrium equations for the safe are

$$+\rightarrow\Sigma F_x = B_n - 0.35A_n = 0$$
$$+\uparrow\Sigma F_y = A_n - 0.35B_n - 3000 = 0$$

which are solved to get

$$A_n = 3419 \text{ N} \quad \text{and} \quad B_n = 1197 \text{ N}$$

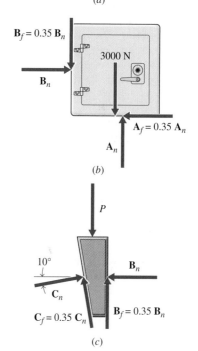

(b)

Equilibrium equations for the wedge are

$$+\rightarrow\Sigma F_x = C_n \cos 10° - 0.35C_n \sin 10° - 1197 = 0$$
$$+\uparrow\Sigma F_y = C_n \sin 10° + 0.35C_n \cos 10° + (0.35)(1197) - P = 0$$

which give

$$C_n = 1295 \text{ N} \quad \text{and} \quad P = 1090 \text{ N} \qquad \text{Ans.}$$

SOLUTION 2. Using the force equilibrium triangle

The free-body diagrams are drawn as above (Figs. 9-33b and 9-33c). Then the force equilibrium triangles are drawn for the safe (Fig. 9-33d) and for the wedge (Fig. 9-33e). In these diagrams, the normal and frictional forces have been combined into a single resultant force, which is drawn at the angle of static friction

$$\phi_s = \tan^{-1} 0.35 = 19.29°$$

since motion is impending.

Using the Law of Sines on the first force triangle (Fig. 9-33d)

$$\frac{B}{\sin (19.29°)} = \frac{3000}{\sin [90° - 2(19.29°)]}$$

gives immediately

$$B = 1268 \text{ N}$$

Then using the Law of Sines on the second force triangle (Fig. 9-33e)

$$\frac{P}{\sin [2(19.29°) + 10°]} = \frac{1268}{\sin (90° - 19.29° - 10°)}$$

gives

$$P = 1090 \text{ N} \qquad \text{Ans.}$$

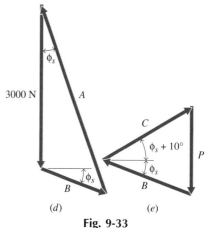

(d)   (e)

Fig. 9-33

A wedge is used to raise a 350-lb refrigerator (Fig. 9-34a). The coefficient of friction is 0.2 at all surfaces.

a. Determine the minimum force **P** needed to insert the wedge.
b. Determine if the system would still be in equilibrium if **P** = **0**.
c. If the system is not in equilibrium when **P** = **0**, determine the force necessary to keep the wedge in place, or if the system is in equilibrium when **P** = **0**, determine the force necessary to remove the wedge.

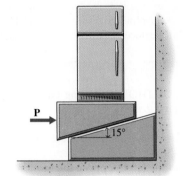

(a)

SOLUTION

a. First draw the free-body diagram of the upper wedge and the refrigerator (Fig. 9-34b) and its force equilibrium triangle (Fig. 9-34c). Since motion is impending, the friction force must be equal to the maximum available friction and must act to the left to oppose the motion. The normal and frictional forces are combined into a single resultant force acting at the angle of static friction

$$\phi_s = \tan^{-1} 0.2 = 11.31°$$

relative to the normal force or $15° + \phi_s$ relative to the vertical direction. The force triangle is just a right triangle so that

$$\tan (15° + 11.31°) = \frac{P}{350}$$

and

$$P = 173.1 \text{ lb} \qquad \text{Ans.}$$

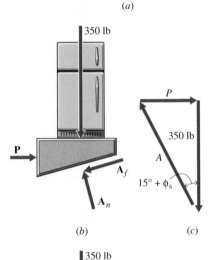

(b)   (c)

b. For **P** very small, the wedge will tend to move to the left and the friction will have to act to the right to oppose this motion (Fig. 9-34d). In the force equilibrium triangle (Fig. 9-34e), the resultant force is drawn at an angle ϕ relative to the normal force or $15° - \phi$ relative to the vertical direction. Since motion is not known to be impending, the angle ϕ is not necessarily equal to ϕ_s. In this case the angle ϕ is merely an unknown to be determined as part of the solution.

From the equilibrium force triangle, when $P = 0$

$$15 - \phi = \tan^{-1} \frac{P}{350} = 0$$

or $\phi = 15°$. But the angle of the resultant ϕ can never be greater than the angle of static friction $\phi_s = 11.31°$. Therefore, the wedge will not be in equilibrium if the force **P** is removed.   Ans.

c. Since the wedge will not stay in place by itself, a force **P** to the right will be required to hold the wedge in place. The minimum force necessary to hold the wedge in place is attained when $\phi = \phi_s = 11.31°$. The force equilibrium triangle (Fig. 9-34e) is again a right triangle so that

$$\tan (15° - 11.31°) = \frac{P}{350}$$

and   Ans.

$$P = 22.6 \text{ lb}$$

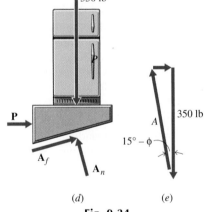

(d)   (e)

Fig. 9-34

In the C-clamp of Fig. 9-18, the screw has a mean radius of 3 mm and a single thread with a pitch of 2 mm. If the coefficient of friction is 0.2, determine:

a. The minimum twisting moment necessary to produce a clamping force of 600 N.
b. The minimum twisting moment necessary to release the clamp when the clamping force is 600 N.
c. The minimum coefficient of friction for which the clamp is self-locking.

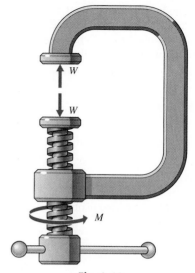

Fig. 9-18

SOLUTION

The screw has a single thread, and so the lead is equal to the pitch

$$\alpha = \tan^{-1}\left(\frac{L}{2\pi r}\right) = \tan^{-1}\left(\frac{2}{6\pi}\right) = 6.06°$$

The angle of static friction is

$$\phi_s = \tan^{-1} 0.2 = 11.31°$$

a. Since the twisting moment is just sufficient to tighten the screw, the free-body diagram and force triangle of Figs. 9-22 and 9-23 apply and

$$M = rW \tan (\alpha + \phi_s) = (0.003)(600) \tan (17.37°)$$
$$= 0.563 \text{ N} \cdot \text{m} \qquad \text{Ans.}$$

b. Now the twisting moment is just sufficient to loosen the screw, and the free-body diagram and force triangle of Figs. 9-24 and 9-25b apply and

$$M = rW \tan (\phi_s - \alpha) = (0.003)(600) \tan (5.25°)$$
$$= 0.1654 \text{ N} \cdot \text{m} \qquad \text{Ans.}$$

c. When $M = 0$, $\phi_s \geq \phi = \alpha$. The minimum coefficient of friction corresponds to $\phi = \phi_s = 6.06°$ and hence

$$\mu_s = \tan 6.06° = 0.106 \qquad \text{Ans.}$$

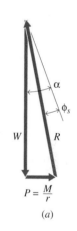

$$P = \frac{M}{r}$$

(a)

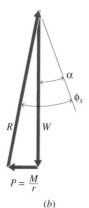

$$P = \frac{M}{r}$$

(b)

Fig. 9-25

$$P = \frac{M}{r} = \frac{1}{r}\int dM$$

$W = \int dW$

Partial section of screw thread

$F = \int dF$

$N = \int dN$

Fig. 9-22

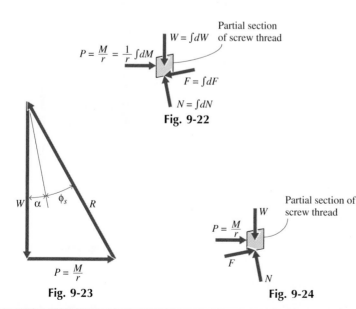

$$P = \frac{M}{r}$$

Fig. 9-23

$$P = \frac{M}{r}$$

W

F

N

Partial section of screw thread

Fig. 9-24

EXAMPLE PROBLEM 9-9

A one-inch-diameter shaft rotates inside a journal bearing. The coefficient of friction between the shaft and bearing is 0.12 and the shaft has a lateral load of 120 lb applied to it. Determine the torque required to rotate the shaft and the angle through which the shaft will climb up the bearing.

SOLUTION

The moment required is

$$M = \mu_k L r = (0.12)(120)(1) = 14.40 \text{ lb} \cdot \text{in.} \qquad \text{Ans.}$$

The angle that the shaft will climb is given by

$$\phi_k = \tan^{-1} \mu_k = \tan^{-1} 0.12 = 6.84° \qquad \text{Ans.}$$

(Note that for this angle, $\tan 6.84° = 0.1200$ while $\sin 6.84° = 0.1191$ and therefore the approximation $M = Lr \sin \phi_k \cong Lr \tan \phi_k = Lr \mu_k$ is within the accuracy of the knowledge of the coefficient of friction and other approximations of the problem.)

EXAMPLE PROBLEM 9-10

The pulley of Fig. 9-35a consists of a 100-mm-diameter wheel that fits loosely over a 15-mm-diameter axle. The axle is rigidly supported at its ends and does not turn. The static and dynamic coefficients of friction between the pulley and axle are 0.4 and 0.35, respectively, and the weight of the pulley may be neglected. Determine the force in the rope that is necessary:

a. To just start raising the 250-N load.
b. To raise the 250-N load at a constant rate.
c. To lower the 250-N load at a constant rate.

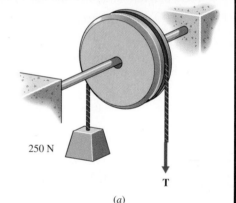

250 N

(a)

SOLUTION

a. The free-body diagram of the pulley is shown in Fig. 9-35b. The moment **M** is due to axle friction and for the case of impending motion is given by $M = \mu_s L r$, where L is the resultant contact force between the axle and pulley and r is the radius of the axle. The moment acts in a counterclockwise direction to oppose the impending motion. The equations of equilibrium give

$$+\rightarrow \Sigma F_x = A_x = 0 \qquad +\uparrow \Sigma F_y = A_y - T - 250 = 0$$
$$\zeta + \Sigma M_A = (50)(250) - 50T + (0.4) A_y (7.5) = 0$$

Solution of these equations gives

$$A_x = 0 \text{ N} \qquad A_y = 532 \text{ N} \qquad T = 281.9 \text{ N} \qquad \text{Ans.}$$

b. Now motion is occurring and the moment due to axle friction is given by $M = \mu_k L r$. Replacing the coefficient of static friction with the coefficient of kinetic friction and re-solving the above equations gives

$$A_x = 0 \text{ N} \qquad A_y = 528 \text{ N} \qquad T = 277.7 \text{ N} \qquad \text{Ans.}$$

c. Now a counterclockwise motion is occurring and the moment due to axle friction must act in a clockwise direction to oppose the motion. The magnitude of the moment is again given by $M = \mu_k L r$. Changing the sign on the third term in the moment equation above and re-solving the equations gives

$$A_x = 0 \text{ N} \qquad A_y = 475 \text{ N} \qquad T = 225.1 \text{ N} \qquad \text{Ans.}$$

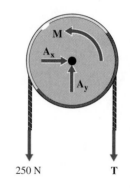

250 N   T

(b)

Fig. 9-35

EXAMPLE PROBLEM 9-11

A collar bearing supports an axial load of 15 lb. The inside and outside diameters of the bearing surface are 1 in. and 2 in., respectively. If the coefficient of kinetic friction is 0.15, compute the moment necessary to overcome friction for a new and for a worn-in bearing.

SOLUTION

For a new bearing the moment is

$$M = \frac{2\,\mu_k P(R_2^3 - R_1^3)}{3(R_2^2 - R_1^2)}$$

$$= \frac{2(0.15)(15)(2^3 - 1^3)}{3(2^2 - 1^2)}$$

$$= 3.50 \text{ lb} \cdot \text{in.} \qquad\qquad \text{Ans.}$$

For a worn-in bearing the moment would be

$$M = \frac{\mu_k P(R_2^2 - R_1^2)}{2(R_2 - R_1)}$$

$$= \frac{(0.15)(15)(2^2 - 1^2)}{2(2 - 1)}$$

$$= 3.38 \text{ lb} \cdot \text{in.} \qquad\qquad \text{Ans.}$$

EXAMPLE PROBLEM 9-12

The machine of Fig. 9-36 consists of a polishing disk that spins clockwise to shine a waxed floor. The coefficient of friction between the disk and the floor is 0.3 and the polishing unit weighs 175 N. Determine the forces that must be applied to the handle of the floor polisher to counteract the frictional moment of the polishing disk.

SOLUTION

Assuming that the pressure is distributed uniformly over the polishing disk, the frictional moment is given by

$$M = \frac{2}{3}\mu_k WR = \frac{2}{3}(0.3)(175)(250) = 8750 \text{ N} \cdot \text{mm}$$

and this must equal the moment of the couple applied to the handle

$$350\,P = 8750 \text{ N} \cdot \text{mm}$$

Therefore

$$P = 25.0 \text{ N} \qquad\qquad \text{Ans.}$$

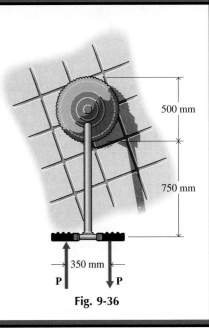

Fig. 9-36

EXAMPLE PROBLEM 9-13

An 80-lb child is sitting on a swing suspended by a rope that passes over a tree branch (Fig. 9-37). The coefficient of friction between the rope and the branch (which can be modeled as a flat belt over a drum) is 0.5, and the weight of the rope can be ignored. Determine the minimum force that must be applied to the other end of the rope to keep the child suspended.

SOLUTION

The rope is wrapped one-half turn or π-radians about the branch. The child is sitting on the side that motion is impending toward, so the tension in that side of the rope is designated T_2 and $T_1 = P$. Therefore,

$$80 \text{ lb} = P \, e^{0.5\pi}$$

or

$$P = 16.63 \text{ lb} \qquad \text{Ans.}$$

Fig. 9-37

EXAMPLE PROBLEM 9-14

A car is prevented from moving by pulling on a rope that is wrapped $n + \frac{1}{4}$ times around a tree (Fig. 9-38). The coefficient of friction between the rope and the tree is 0.35, and the force exerted by the car is 3600 N. If it is desired that the force exerted on the rope be no more than 125 N, determine n, the number of times the rope must be wrapped around the tree.

SOLUTION

The angle of twist necessary to hold the car is found from

$$3600 = 125 \, e^{0.35\beta}$$

Fig. 9-38

or

$$\beta = \frac{\ln \dfrac{3600}{125}}{0.35} = 9.60 \text{ radians}$$

which is 1.528 times around the tree. Any angle less than this will require a resisting force greater than 125 N, whereas any angle greater than this will require less force. Thus

$$n = 2 \qquad \text{Ans.}$$

will be sufficient.

Suppose the child of Example Problem 9-13 is holding the other end of the rope. Determine the minimum force he must exert on the rope to keep himself suspended.

SOLUTION

The free-body diagrams of the child and the tree limb are drawn in Fig. 9-39a and 9-39b, respectively. The rope is still wrapped one-half turn or π-radians about the branch so that

$$T_2 = T_1 \, e^{0.5\pi}$$

Vertical equilibrium of the child gives

$$+\uparrow \Sigma F_y = T_2 + T_1 - 80 = 0$$

Combining these two equations gives

$$80 - T_1 = T_1 \, (4.81)$$

or

$$T_1 = 13.77 \text{ lb} \qquad\qquad \text{Ans.}$$

T_2 T_1

80 lb

(a)

C

T_2 T_1

(b)

Fig. 9-39

PROBLEMS

Wedge Friction

9-44–9-61 A pair of wedges is used to move a crate of weight W. The coefficient of friction is the same at all surfaces and the weight of the wedges may be neglected. If the coefficient of static friction μ_s and the wedge angle θ are as given below, determine:

a. The force P necessary to insert the wedge.
b. If the system would be in equilibrium if the force P were removed.
c. The force P necessary to remove the wedge or to prevent the wedge from slipping out depending on the answer to part b.
d. The maximum angle θ for which the system would be in equilibrium if the force P were removed.

9-62–9-79 In Problems 9-44 through 9-61, the coefficient of friction between the contacting surfaces of the wedges is as given, but all other surfaces have been coated to reduce the coefficient of friction to 0.10. Determine:

a. The force P necessary to insert the wedge.
b. The force P necessary to prevent the wedge from slipping.
c. The maximum angle θ for which the system would be in equilibrium if the force P were removed.

9-44* & 9-62	$W = 3000$ N	$\theta = 15°$	$\mu_s = 0.25$
9-45* & 9-63*	$W = 2500$ lb	$\theta = 12°$	$\mu_s = 0.20$
9-46 & 9-64*	$W = 1200$ N	$\theta = 10°$	$\mu_s = 0.35$
9-47* & 9-65	$W = 4000$ lb	$\theta = 18°$	$\mu_s = 0.15$
9-48 & 9-66*	$W = 2400$ N	$\theta = 20°$	$\mu_s = 0.30$
9-49 & 9-67	$W = 5000$ lb	$\theta = 15°$	$\mu_s = 0.40$

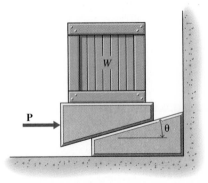

Fig. P9-44 through P9-49, P9-62 through P9-67

9-50* & 9-68	$W = 3000$ N	$\theta = 15°$	$\mu_s = 0.25$
9-51* & 9-69*	$W = 2500$ lb	$\theta = 12°$	$\mu_s = 0.20$
9-52 & 9-70	$W = 1200$ N	$\theta = 10°$	$\mu_s = 0.35$

9-53 & 9-71*	$W = 4000$ lb	$\theta = 18°$	$\mu_s = 0.15$
9-54* & 9-72*	$W = 2400$ N	$\theta = 20°$	$\mu_s = 0.30$
9-55 & 9-73	$W = 5000$ lb	$\theta = 15°$	$\mu_s = 0.40$

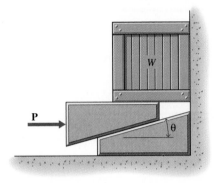

Fig. P9-50 through P9-55, P9-68 through P9-73

9-56 & 9-74*	$W = 3000$ N	$\theta = 15°$	$\mu_s = 0.25$
9-57* & 9-75*	$W = 2500$ lb	$\theta = 12°$	$\mu_s = 0.20$
9-58* & 9-76	$W = 1200$ N	$\theta = 10°$	$\mu_s = 0.35$
9-59 & 9-77*	$W = 4000$ lb	$\theta = 18°$	$\mu_s = 0.15$
9-60* & 9-78	$W = 2400$ N	$\theta = 20°$	$\mu_s = 0.30$
9-61 & 9-79	$W = 5000$ lb	$\theta = 15°$	$\mu_s = 0.40$

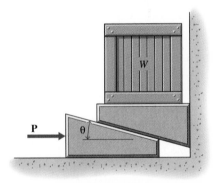

Fig. P9-56 through P9-61, P9-74 through P9-79

9-80–9-97 A pair of wedges is used to move a crate of weight W. The coefficient of friction between the contacting surfaces of the wedges is as given below while the coefficient of friction at all other surfaces is 0.10. The weight of the wedges may be neglected. Determine:

a. The force P necessary to insert the wedge for the given angle θ.
b. The maximum angle θ for which P is smaller than the force needed to push the wedge directly.
c. The minimum angle θ such that no value of P will cause the crate to move.

9-80*	$W = 3000$ N	$\theta = 15°$	$\mu_s = 0.25$
9-81	$W = 2500$ lb	$\theta = 12°$	$\mu_s = 0.20$
9-82	$W = 1200$ N	$\theta = 10°$	$\mu_s = 0.35$
9-83*	$W = 4000$ lb	$\theta = 18°$	$\mu_s = 0.15$
9-84*	$W = 2400$ N	$\theta = 20°$	$\mu_s = 0.30$
9-85	$W = 5000$ lb	$\theta = 15°$	$\mu_s = 0.40$

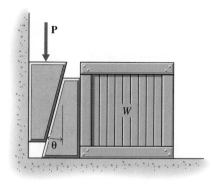

Fig. P9-80 through P9-85

9-86	$W = 3000$ N	$\theta = 15°$	$\mu_s = 0.25$
9-87*	$W = 2500$ lb	$\theta = 12°$	$\mu_s = 0.20$
9-88*	$W = 1200$ N	$\theta = 10°$	$\mu_s = 0.35$
9-89*	$W = 4000$ lb	$\theta = 18°$	$\mu_s = 0.15$
9-90	$W = 2400$ N	$\theta = 20°$	$\mu_s = 0.30$
9-91	$W = 5000$ lb	$\theta = 15°$	$\mu_s = 0.40$

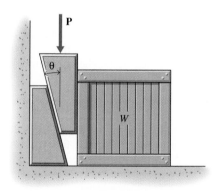

Fig. P9-86 through P9-91

9-92*	$W = 3000$ N	$\theta = 15°$	$\mu_s = 0.25$
9-93*	$W = 2500$ lb	$\theta = 12°$	$\mu_s = 0.20$
9-94	$W = 1200$ N	$\theta = 10°$	$\mu_s = 0.35$
9-95	$W = 4000$ lb	$\theta = 18°$	$\mu_s = 0.15$
9-96*	$W = 2400$ N	$\theta = 20°$	$\mu_s = 0.30$
9-97	$W = 5000$ lb	$\theta = 15°$	$\mu_s = 0.40$

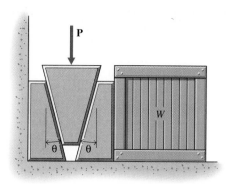

Fig. P9-92 through P9-97

9-98 A wedge is being forced under an 80-kg drum (Fig. P9-98). The coefficient of friction between the wedge and the drum is 0.10 while the coefficient of friction is 0.30 at all other surfaces. Assuming a wedge angle θ of 25° and that the weight of the wedge may be neglected, determine the minimum force **P** necessary to insert the wedge.

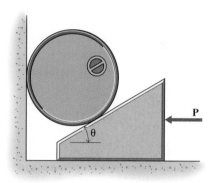

Fig. P9-98

9-99* A wedge rests between a 75-lb drum and a wall (Fig. P9-99). The coefficient of friction between the drum and the floor is 0.15 while the coefficient of friction is 0.50 at all other surfaces. Assuming a wedge angle θ of 40°, determine the minimum weight of the wedge that will cause motion.

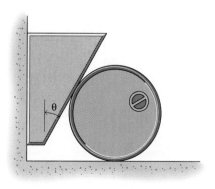

Fig. P9-99

9-100 The wedge of Problem 9-98 is to be designed so that slip occurs at all surfaces simultaneously. Determine the required wedge angle θ and the corresponding force **P**.

9-101 The wedge of Problem 9-99 is to be designed so that slip occurs at all surfaces simultaneously. Determine the required wedge angle θ and the corresponding wedge weight W.

9-102 The plunger of a door latch is held in place by a spring as shown in Fig. P9-102. Friction on the sides of the plunger may be ignored. If a force of 5 N is required to just start closing the door and the coefficient of friction between the plunger and the striker plate is 0.25, determine the force exerted on the plunger by the spring.

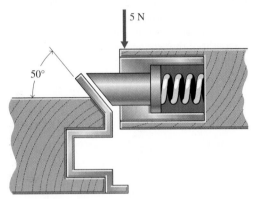

Fig. P9-102

Square-Threaded Screws Friction

9-103* The clamp of Fig. P9-103 is used to hold a cover in place. The screw of the clamp has a single thread with a mean radius of $\frac{1}{4}$ in. and a pitch of 0.1 in. The coefficient of friction is 0.40, and the required clamping force is 80 lb. Determine the twisting moment necessary to tighten the clamp and whether or not the clamp will stay in place if the twisting moment is removed.

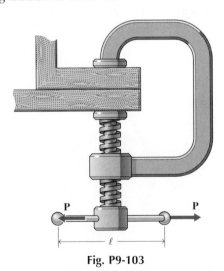

Fig. P9-103

9-104* In the old-fashioned printing press of Fig. P9-104, a square threaded screw is used to press the paper against the type bed. The single-threaded screw has a mean radius of 20 mm and a lead of 100 mm. The coefficient of friction is 0.15, and the clamping force necessary to guarantee clear printing on all parts of the paper is 400 N. Determine the twisting moment necessary to operate the press and whether or not a moment must be applied to release the press.

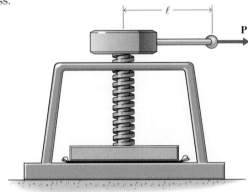

Fig. P9-104

9-105 The twisting moment of Problem 9-103 is applied by means of a force-couple applied to a rod through the head of the screw. Determine the minimum length rod necessary if the applied forces are not to exceed 8 lb and the clamping force does not exceed 150 lb.

9-106 The twisting moment of Problem 9-104 is applied by means of a single force on a long handle attached to the head of the screw. Determine the minimum length handle necessary such that the force required on the handle is 5 percent of the force applied by the press on the paper.

9-107* The screw jack of Fig. P9-107 uses a square-threaded screw having a coefficient of friction of 0.35 and a mean radius of 1 in. Determine the maximum lead such that:

a. The jack will stay in place when the moment is removed.

b. The moment necessary to lower the jack is 20 percent of the moment necessary to raise the jack.

Fig. P9-107

9-108 An upside-down screw jack is often used as a foot leveler on refrigerators and other heavy appliances (Fig. P9-108). Suppose the weight carried by a single foot screw is 800 N and the screw has a mean radius of 3 mm and a lead of 1 mm. The coefficient of friction is 0.3 between the screw and the refrigerator frame while friction between the screw head and the floor may be ignored. Determine the minimum moment necessary to raise and the minimum moment to lower the corner of the refrigerator.

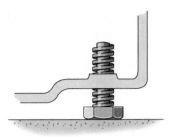

Fig. P9-108

9-109 The gear puller of Fig. P9-109 has a doubly threaded screw with a mean radius of $\frac{1}{4}$ in., a pitch of 0.1 in., and a coefficient of friction of 0.30. The gear is wedged onto its shaft and requires a force of 100 lb to remove it. Determine the moment that must be applied to the gear puller screw to remove the gear.

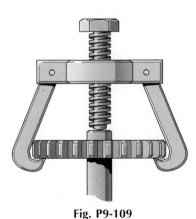

Fig. P9-109

9-110* The turnbuckle of Fig. P9-110 is used to secure a support cable on a sailboat. The doubly-threaded screws each have a mean radius of 4 mm and a pitch of 2 mm and are threaded in opposite directions. The coefficient of friction is 0.2 and the tension in the cables is 600 N. Determine the minimum moment necessary to tighten and the minimum moment necessary to loosen the turnbuckle.

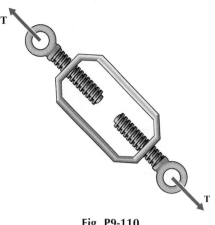

Fig. P9-110

Journal Bearing Friction

9-111* A grinding wheel weighing 2 lb is supported by a journal bearing at each end of the axle (Fig. P9-111). The coefficient of friction between the axle and bearing is $\mu_k = 0.10$ and the diameter of the axle is 0.75 in. Determine the torque required to rotate the wheel at a constant speed.

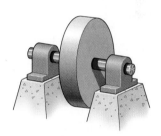

Fig. P9-111

9-112* A pulley consists of a 120-mm-diameter wheel that fits loosely over an 18-mm-diameter shaft (Fig. P9-112). If the coefficient of friction between the pulley and the shaft is $\mu_k = 0.17$ and the weight of the pulley is 5 N, determine the torque required to raise a 2.5-kg load at a constant speed.

Fig. P9-112

9-113* The pulley of Fig. P9-113 consists of a 4-in.-diameter wheel that fits loosely over a ½-in.-diameter shaft. A rope passes over the pulley and is attached to a 50-lb weight. The static and kinetic coefficients of friction between the pulley and the shaft are 0.35 and 0.25, respectively, and the weight of the pulley is 2 lb. The rope being pulled makes an angle $\theta = 90°$ with the horizontal. Determine the force necessary to:

a. Just start raising the weight.
b. Raise the weight at a constant speed.
c. Just start lowering the weight.
d. Lower the weight at a constant speed.

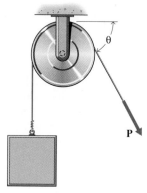

Fig. P9-113

9-114 Repeat Problem 9-113 for $\theta = 0°$ and a 110-mm diameter, 1-kg pulley on a 12-mm-diameter shaft raising/lowering a 225-N weight.

9-115 Repeat Problem 9-113 for $\theta = 30°$.

9-116 Repeat Problem 9-114 for $\theta = 60°$.

9-117 Television cable is being removed from a 36-in. diameter spool as shown in Fig. P9-117. The spool (including the wire) weighs 450 lb and fits loosely over a 2-in.-diameter shaft. If the static and kinetic coefficients of friction between the spool and the shaft are 0.40 and 0.25, respectively, determine:

a. The force necessary to just start the spool rotating.
b. The force necessary to remove cable at a constant rate.
c. The force exerted on the bearing by the spool when cable is being removed at a constant rate.

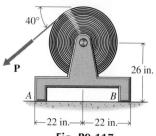

Fig. P9-117

d. The normal force on the footpads at A and B when cable is being removed at a constant rate.
e. The minimum coefficient of friction between the footpads and the floor to prevent slippage when cable is being removed at a constant rate.

9-118* The two-pan balance scale of Fig. P9-118 consists of a 400-mm-long bar that fits loosely over a 20-mm-diameter shaft. The center of mass of the bar coincides with the center of the shaft. The scale functions by placing an item to be weighed (for example, a mini-TV) in the right pan and then placing known weights in the left pan until the scale is in equilibrium. However, because of bearing friction, the system can be in equilibrium for more than a single value of weight in the left pan. If the coefficient of friction between the bar and the shaft is 0.3, determine the range of weights that can be placed in the left pan to balance an 80-N object placed in the right pan.

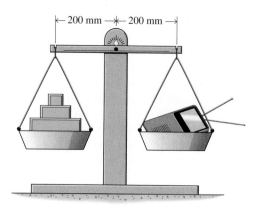

Fig. P9-118

9-119 The cable spool of Problem 9-117 is being pulled in a cart (Fig. P9-119). As the cart is pulled forward, the cable unwinds. Determine the force necessary to pull the cart at a constant speed:

a. If the cart wheels are frictionless.
b. If the two wheels of the cart fit loosely over a fixed axle and the coefficient of friction is 0.25 between the wheel and the axle.

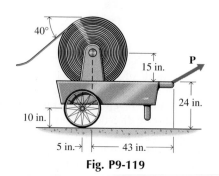

Fig. P9-119

414

9-120* An axial load of 750 N is applied to the 60-mm-diameter end bearing of Fig. P9-120. For a coefficient of friction of 0.2, determine the torque needed to rotate the shaft for:

a.   A new bearing surface.
b.   A worn-in bearing surface.

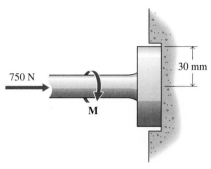

Fig. P9-120

9-121 An axial load of 160 lb is applied to the end bearing of Fig. P9-121. Contact between the end bearing and the seat takes place over an annular ring of radius 4 in. to 7 in. For a coefficient of friction of 0.2, determine the torque needed to rotate the shaft for:

a.   A new bearing surface.
b.   A worn-in bearing surface.

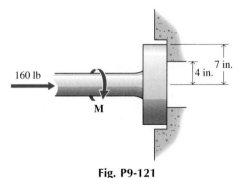

Fig. P9-121

9-122* Recalculate the minimum moment necessary to raise and the minimum moment to lower the foot leveler of Problem 9-108 when disk friction between the screw head and the floor is included. The coefficient of friction between the screw head and the floor is 0.4 and the area of contact between the screw head and the floor may be assumed to be a circle of radius 10 mm.

9-123* The electric auto polisher of Fig. P9-123 has an 8-in.-diameter polishing pad. The coefficient of friction is 0.3 between the pad and the auto surface. Determine the forces that must be applied to the handles to hold the polisher steady when it is pressed against the surface with a force of 20 lb.

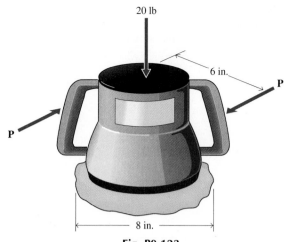

Fig. P9-123

9-124 An electric disk sander (Fig. P9-124) is pressed against a surface with a force of 50 N. The coefficient of friction between the sandpaper and the surface is 0.6 and the diameter of the sanding disk is 200 mm. Determine the torque that the motor must develop to overcome the developed friction.

Fig. P9-124

9-125* Recalculate the forces for Problem 9-123 for the case in which the polishing pad is attached to a flexible backing surface and the pressure distribution on the disk is as shown in Fig. P9-125.

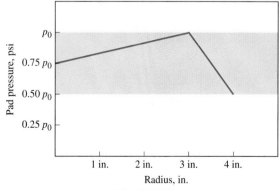

Fig. P9-125

415

9-126 Recalculate the torque for Problem 9-124 for the case in which the polishing pad is attached to a flexible backing surface and the pressure distribution on the disk is as shown in Fig. P9-126.

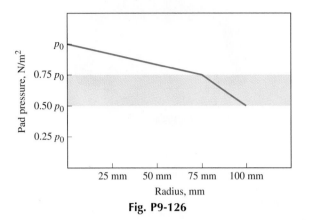

Fig. P9-126

9-127 Determine an expression for the torque **M** required to rotate a shaft carrying an axial load **P** and supported by a 60° conical end bearing as shown in Fig. P9-127. The coefficient of friction is μ.

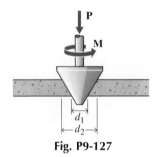

Fig. P9-127

Belt Friction

9-128* A rope attached to a 220-kg object passes over a fixed drum (Fig. P9-128). If the coefficient of friction between the rope and the drum is 0.30, determine:

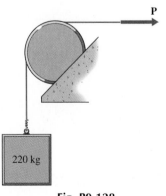

Fig. P9-128

a. The minimum force **P** that must be used to keep the block from falling.
b. The minimum force **P** that must be used to begin to raise the block.

9-129* A rope attached to a 35-lb crate passes around two fixed pegs (Fig. P9-129). The 45-lb crate is attached to a wall by a second cord. The coefficient of friction between the two crates is 0.25; between the crate and the floor, 0.25; and between the rope and the pegs, 0.20. Determine the minimum force **P**$_{min}$ that must be used to cause motion.

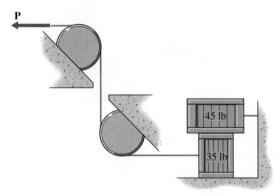

Fig. P9-129

9-130 A rope makes two complete turns around a circular post. If a pull of 200 N on one end of the rope will just support a force of 30 kN on the other end of the rope, determine the coefficient of friction between the post and the rope.

9-131* The belt drive of Fig. P9-131 consists of an 8-in.-diameter, 10-lb wheel that is driven by a torque of 3 lb-ft. The wheel is constrained to move in the frictionless vertical slot and is held taut against the belt by the spring. The coefficient of friction between the belt and the wheel is 0.30 while the two smaller pulleys are frictionless. Determine the minimum tension in the spring **T**$_{min}$ necessary to prevent the belt from slipping.

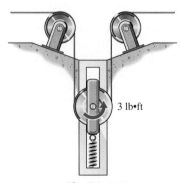

Fig. P9-131

416

9-132 The scaffolding of Fig. P9-132 is raised using an electric motor that sits on the scaffolding. Frictionless wheels at the ends of the scaffold restrict horizontal motion. The 250-mm-diameter pulley at the top is jammed and will not rotate. The coefficient of friction between the rope and the pulley is 0.25 and the weight of the scaffold, motor, and supplies is 2500 N. Determine the minimum torque that must be supplied by the motor:

a. To raise the scaffold at a constant rate.
b. To lower the scaffold at a constant rate.

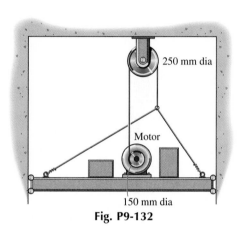

Fig. P9-132

9-133 The band brake of Fig. P9-133 is used to control the rotation of a drum. The coefficient of friction between the belt and the drum is 0.25 and the weight of the handle is 2 lb. If a force of 50 lb is applied to the end of the handle, determine the maximum torque for which no motion occurs if the torque is applied:

a. Clockwise.
b. Counterclockwise.

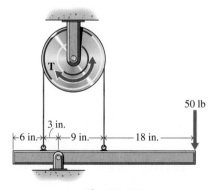

Fig. P9-133

9-134* The band brake of Fig. P9-134 is used to control the rotation of a drum. The coefficient of friction between the belt and the drum is 0.35 and the weight of the handle is 15 N. If a force of 200 N is applied to the end of the handle, determine the maximum torque for which no motion occurs if the torque is applied clockwise.

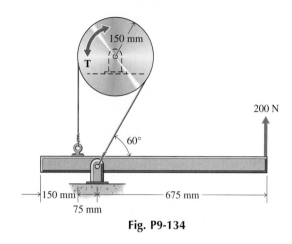

Fig. P9-134

9-135* Determine the minimum coefficient of friction between the belt and drum of Problem 9-133 for which the band brake is self-locking, that is, for which no force on the handle is necessary to restrain arbitrarily large torques on the drum.

9-136 Solve Problem 9-134 if the torque is applied counterclockwise.

9-137 The band wrench of Fig. P9-137 is used to unscrew an oil filter from a car. (The filter acts as if it were a wheel with a resisting torque of 40 lb·ft.) Determine the minimum coefficient of friction between the band and the filter that will prevent slippage. The weight of the handle can be neglected.

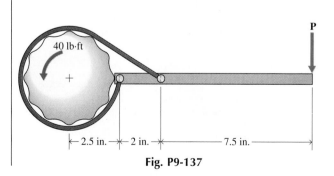

Fig. P9-137

9-138* A 100-N crate is sitting on a 200-N crate (Fig. P9-138). The rope that joins the crates passes around a frictionless pulley and a fixed drum. The coefficient of friction is 0.30 at all surfaces. Determine the minimum force **P** that must be applied to the 200-N crate to start motion.

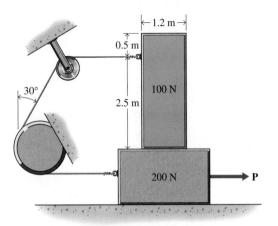

Fig. P9-138

9-139 A 2-ft-diameter drum is rigidly attached to a piece of machinery (Fig. P9-139). The coefficient of friction between the machine and the floor is 0.20 and the weight of the machine including the drum is 375 lb. A rope passes around the drum and is attached to two crates. The coefficient of friction between the crates is 0.30; between the crate and the floor, 0.40; and between the rope and the drum, 0.10. Determine the maximum force that can be applied to the lower crate without causing motion.

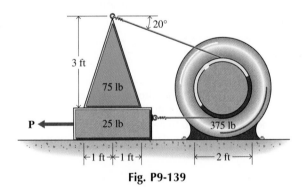

Fig. P9-139

9-140 The rope connecting the two blocks of Fig. P9-140 passes over a fixed drum. The coefficient of friction be-

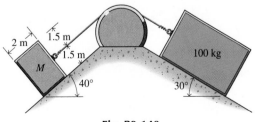

Fig. P9-140

tween the left block and the floor is 0.50; between the right block and the floor, 0.40; and between the rope and the drum, 0.30. Determine the minimum and maximum mass of the left block for which motion does not occur.

9-141* A uniform belt 36 in. long and weighing 1 lb hangs over a small fixed peg (Fig. P9-141). If the coefficient of friction between the belt and the peg is 0.40, determine the maximum distance between the two ends d for which the belt will not slip off the peg.

Fig. P9-141

9-142* A rope connecting two blocks passes around a frictionless pulley and a fixed drum (Fig. P9-142). The coefficient of friction between the rope and the drum and between the 10-kg block and the wall is 0.30; the pin supporting the 30-kg block is frictionless. Determine the maximum force **P** for which no motion occurs.

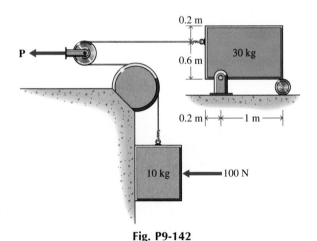

Fig. P9-142

9-143 A rope connecting two blocks is wrapped $\frac{3}{4}$-turn around a fixed peg (Fig. P9-143). The coefficient of friction between the peg and the rope is 0.15; between the 65-lb block and the floor, 0.40. Determine the minimum and maximum weight of block B for which no motion occurs.

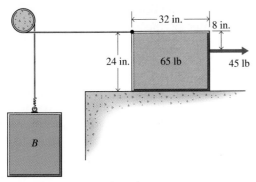

Fig. P9-143

9-144* A rope is attached to the top corner of a 450-N block, passes over a fixed drum, and is attached to another block. The coefficient of friction between the 450-N block and the floor is 0.30; between the rope and the drum, 0.60.

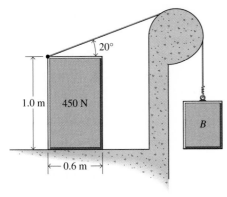

Fig. P9-144

Determine the maximum mass of block B for which no motion occurs.

9-145* Repeat Problem 9-1 for the case of a V-pulley with angle $\alpha = 30°$. The coefficient of friction between the pulley and the belt is 0.15 and the pulley is jammed and will not turn.

9-146 Repeat Problem 9-8 for the case of V-pulleys with angle $\alpha = 40°$. The coefficient of friction between the pulleys and the belts is 0.10 and the pulleys are jammed and will not turn.

9-147 Repeat Problem 9-11 for the case of a V-pulley with angle $\alpha = 25°$. The coefficient of friction between the pulley and the belt is 0.15 and the pulley is jammed and will not turn.

9-148* Repeat Problem 9-20 for the case of a V-pulley with angle $\alpha = 30°$. The coefficient of friction between the pulley and the belt is 0.10 and the pulley is jammed and will not turn.

9-149* Repeat Problem 9-133 for the case of a V-pulley with angle $\alpha = 40°$ and $\mu = 0.10$.

9-150 Repeat Problem 9-134 for the case of a V-pulley with angle $\alpha = 25°$.

9-151 Repeat Problem 9-139 for the case of a V-pulley with angle $\alpha = 30°$.

9-152* Repeat Problem 9-136 for the case of a V-pulley with angle $\alpha = 40°$.

9-4 ROLLING RESISTANCE

Rolling resistance refers to the forces that cause a rolling wheel to gradually slow down and stop. Rolling resistance, however, is not due to Coulomb friction and cannot be described with a coefficient of friction in the same sense as wedges and bearings. Therefore, it is worth examining the origin and nature of rolling resistance in a little more detail.

Several effects including air resistance and bearing friction combine to cause a rolling wheel to gradually slow down and stop. In what follows, it will be assumed that these sources of resistance have been minimized and that the primary source of resistance is the interaction between the wheel and the surface on which it is rolling.

Figure 9-40a shows a wheel that is being rolled at a constant speed across a flat horizontal surface by a force **P** at its center. The free-body diagram of the wheel is shown in Fig. 9-40b. The force **L** represents the weight of the wheel plus any vertical load that might be applied to its

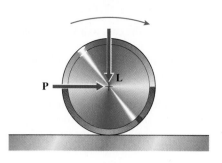

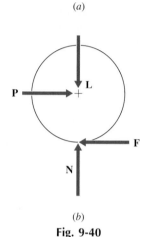

(a)

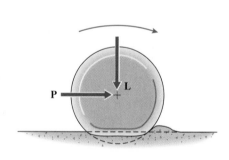

(b)

Fig. 9-40

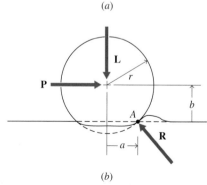

(a)

(b)

Fig. 9-41

axle. If the wheel and the surface are both perfectly rigid, the normal force will act through the center of the wheel. Summing moments about the center of the wheel then gives that $F = 0$. That is, no Coulomb friction acts on the bottom of the wheel. Furthermore, summing forces in the horizontal direction gives that $P = 0$ also. That is, no pushing force is required to keep the wheel rolling at constant speed.

The apparent contradiction with our common experience is caused by assuming the wheel and surface are perfectly rigid. All real materials deform to some extent. The softer the materials and/or the heavier the load, the more the deformation will be.

The real situation, then, is probably more like Fig. 9-41a, which shows a rubber wheel rolling across a rubber surface. The load L will cause both the surface and the wheel to be dented. In addition, since the wheel is being pushed to the right, the surface will probably stretch and be pushed up a little in front of the wheel. The free-body diagram for this situation is shown in Fig. 9-41b, in which **R** represents the resultant of all contact forces between the wheel and the surface.

Equilibrium of moments about the center of the wheel requires that **R** act through the center of the wheel as shown. Then, summing moments about A, the point through which **R** acts, gives

$$\curvearrowleft +\Sigma M_A = La - Pb = 0$$

In most cases of interest the deformation is small so that b is very nearly equal to the radius of the wheel r. Therefore, the force **P** required to keep the wheel rolling at a constant speed is given by

$$P = \frac{aL}{r} \tag{9-18}$$

The distance a is usually termed the coefficient of rolling resistance.* It is not a dimensionless coefficient as are μ_s and μ_k, but has a dimension of length. Like μ_s and μ_k, the coefficient of rolling resistance depends on the properties of the contacting surfaces. Since it arises due to deformation of the surface rather than from the roughness of the surface, however, it is not related to μ_s and μ_k. In fact, there is even considerable disagreement over whether or not a is even a constant.

Bearing these things in mind, Table 9-2 presents a list of rolling coefficients. The reported values vary by 50 percent or more, however, and no great accuracy should be expected from this procedure.

---

*Some people prefer to keep friction coefficients dimensionless and refer to the ratio a/r as the coefficient of rolling resistance.

TABLE 9-2   COEFFICIENTS OF ROLLING RESISTANCE

Materials	a (in.)	a (mm)
Hardened steel on hardened steel	0.0004	0.01
Mild steel on mild steel	0.015	0.38
Steel on wood	0.08	2
Steel on soft ground	5	130
Pneumatic tires on pavement	0.025	0.6
Pneumatic tires on dirt road	0.05	1.3

The two-wheel cart of Fig. 9-42a has 500-mm-diameter bicycle tires and a mass of 15 kg. The cart is carrying 95 kg of rocks on a dirt road. The wheels fit loosely over a 15-mm-diameter fixed axle. The coefficient of kinetic friction between the wheel and the axle is 0.25, and the center of mass of the cart and the rocks is 100 mm in front of the axle. Determine the force (both the horizontal and vertical components) that must be applied to the handle of the cart to pull it at a constant speed.

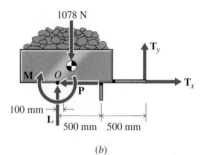

(a)

SOLUTION

Both the axle friction and the rolling friction depend on the normal load on the axle. Therefore, the first step is to draw a free-body diagram of the cart minus the wheels (Fig. 9-42b) to bring the axle load into the equations. The equilibrium equations give

$$+\rightarrow \Sigma F_x = T_x - P = 0$$
$$+\uparrow \Sigma F_y = L - 1078 + T_y = 0$$
$$\zeta + \Sigma M_0 = 1T_y - (0.1)(1078) - M = 0$$

where L is the total vertical load on the axle, P is the total horizontal force on the axle, and $M = 0.25L(0.015/2)$ is the total moment on the axle due to both wheels. Solution of these equations gives

$$T_x = P \qquad T_y = 109.6 \text{ N} \qquad L = 968 \text{ N}$$

Next draw a free-body diagram of one of the wheels (Fig. 9-42c). The forces L_f and P_f are the components of the vertical and horizontal forces that are carried by the wheel on the front side of the cart and $M_f = 0.25L_f(0.015/2)$ is the torque exerted on the front wheel by bearing friction. The ground reaction acts at point A, which is ahead of the axle by the distance $a = 1.3$ mm according to Table 9-2. Summing moments about point A gives

$$\zeta + \Sigma M_A = M_f + aL_f - rP_f = 0$$

or

$$P_f = \frac{aL_f}{r} + \frac{M_f}{r} = \frac{1.3L_f}{250} + \frac{0.25L_f(7.5)}{250}$$
$$= (0.00520 + 0.00750)L_f = 0.01270L_f$$

The first term in the foregoing equation represents the force necessary to overcome rolling resistance, and the second term represents the force necessary to overcome axle friction. Analysis of the back wheel similarly gives

$$P_b = 0.01270L_b$$

The total horizontal force is the sum of the horizontal forces on both wheels

$$P = P_f + P_b = 0.01270(L_f + L_b)$$

and the total vertical force is the sum of the vertical forces on both wheels

$$L = L_f + L_b = 968 \text{ N}$$

so finally,

$$T_x = P = 0.01270(968) = 12.30 \text{ N} \qquad \text{Ans.}$$

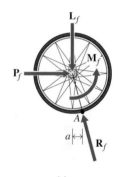

(c)

Fig. 9-42

PROBLEMS

9-153* Determine the horizontal force required to push a 2500-lb automobile on level pavement. The auto has four 23-in.-diameter tires. Neglect all friction except rolling resistance.

9-154* A 1200-kg automobile is observed to roll at a constant speed down a 1° incline. The auto has four 550-mm-diameter tires. Neglect all friction except rolling resistance and determine the coefficient of rolling friction between the tires and the surface.

9-155 Determine the angle of incline down which a 50-ton railroad car (Fig. P9-155) would roll at constant speed. The car has eight 20-in.-diameter steel wheels, which roll on steel rails. Neglect all friction except rolling resistance.

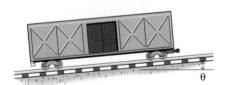

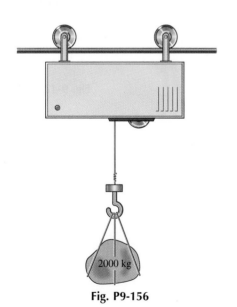

Fig. P9-155

9-156* An overhead crane has two 75-mm-diameter steel wheels that ride on steel rails (Fig. P9-156). Determine the horizontal force necessary to push the crane when it is carrying a load of mass 2000 kg. Neglect all friction except rolling resistance.

Fig. P9-156

9-157 A 600-lb safe rolls on four 2-in.-diameter steel wheels. Determine the horizontal force necessary to push the safe across a level wood floor. Neglect all friction except rolling resistance.

9-158 A horizontal force of 170 N is required to push a 180-kg refrigerator across a level linoleum floor. The refrigerator rolls on four 50-mm-diameter plastic wheels. Neglect all friction except rolling resistance and determine the coefficient of rolling friction between the plastic wheels and the linoleum floor.

9-159* A horizontal force of 200 lb is required to push an 800-lb piano across a level carpeted floor. The piano rolls on four 2-in.-diameter steel wheels. Neglect all friction except rolling resistance and determine the coefficient of rolling friction between the steel wheels and the carpeted floor.

9-160 A 35-kg child rides on a 10-kg bicycle with 500-mm-diameter wheels. Determine the horizontal force necessary to push the child and bicycle.

a. On a level sidewalk.
b. Through level sand.

Neglect all friction except rolling resistance.

9-161 The wheels of Problem 9-153 rotate on 1.5-in.-diameter shafts. If the coefficient of friction between the wheel and the shaft is 0.15, determine the force necessary to push the automobile and the ratio of the rolling friction to the axle friction.

9-162* The wheels of Problem 9-156 rotate on 5-mm-diameter axles. If the coefficient of friction between the wheel and the axle is 0.40, determine the force necessary to push the crane and the ratio of the rolling friction to the axle friction.

9-163* The wheels of Problem 9-157 rotate on $\frac{1}{4}$-in.-diameter axles. If the coefficient of friction between the wheel and the axle is 0.45, determine the force necessary to push the safe and the ratio of the rolling friction to the axle friction.

9-164 The wheels of Problem 9-160 rotate on 5-mm-diameter axles. If the coefficient of friction between the wheel and the axle is 0.20, determine the force necessary to push the bicycle and the ratio of the rolling friction to the axle friction.

SUMMARY

Tangential forces due to friction are always present at the interface between two contacting bodies, and these forces always act in a direction to oppose the tendency of the contacting surfaces to slip relative to one another. In some types of machine elements, such as bearings and skids, it is desired to minimize friction, while in other types, such as brakes and belt drives, it is desired to maximize friction.

Two main types of friction are commonly encountered in engineering practice: dry friction and fluid friction. Dry friction describes the force that opposes the slip of one solid surface on another. Dry friction depends on the normal force pressing the surfaces together and on the nature of the materials in contact. Fluid friction describes the force that opposes the slip of two layers of fluid moving at different speeds. Fluid friction depends on the relative speed of the moving fluid layers.

The strength of friction forces is limited. The condition when a friction force is at its maximum value is called impending motion. Beyond that point, friction can no longer supply the amount of force required for equilibrium. The value of limiting friction is proportional to the normal force N at the contact surface

$$F_{max} = \mu_s N \qquad (9\text{-}1)$$

The constant of proportionality μ_s is called the coefficient of static friction; it depends on the types of material in contact but is independent of both the normal force and the area of contact.

The frictional force actually exerted is never greater than that required to satisfy the equations of equilibrium

$$F \le \mu_s N \qquad (9\text{-}2)$$

The equality in Eq. 9-2 holds only at the point of impending motion. Once a body starts to slip relative to its supporting surface, the friction force will decrease slightly and

$$F = \mu_k N \qquad (9\text{-}3)$$

where μ_k is the coefficient of kinetic friction. This coefficient is usually about 20 to 25 percent smaller than μ_s and is again independent of the normal force. It is also independent of the speed of the relative motion— at least for low speeds. The presence of any moisture or oil on the surface, however, can change the problem from one of dry friction where the friction force is independent of the speed of the body to one of fluid friction where the friction force is a function of the speed.

Some components in engineering systems where frictional forces play an important role include wedges, screws, journal bearings, thrust bearings, and belts. The procedure followed in analyzing all these components is the same as that used previously in Chapter 7 for pin-connected machine and frames.

Rolling resistance is not due to Coulomb friction and cannot be described with the usual coefficients of friction. Rolling resistance arises because of the deformation of the wheel and/or the surface on which it is rolling.

The magnitude of the force **P** required to keep a wheel rolling at a constant speed is given by

$$P = a\frac{L}{r} \tag{9-18}$$

where r is the radius of the wheel and L is the load supported by the wheel. The coefficient of rolling resistance a is not a dimensionless coefficient as are μ_s and μ_k, but has a dimension of length. Although a depends on the properties of the contacting surfaces, it arises because of the deformation of the surface rather than the roughness of the surface. Therefore, the coefficient of rolling resistance is not directly related to μ_s and μ_k.

REVIEW PROBLEMS

9-165* A device for lifting rectangular objects such as bricks and concrete blocks is shown in Fig. P9-165. Determine the minimum coefficient of static friction between the contacting surfaces required to make the device work.

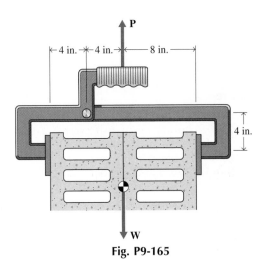

Fig. P9-165

9-166* The static coefficient of friction between the brake pad and the brake drum of Fig. P9-166 is 0.40. When a force

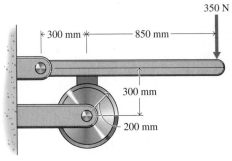

Fig. P9-166

of 350 N is being applied to the brake arm, determine the couple required to initiate rotation of the drum if the direction of rotation is (a) clockwise and (b) counterclockwise.

9-167 Blocks A and B of Fig. P9-167 weigh 50 lb and 500 lb, respectively. The static coefficients of friction are 0.25 for all contacting surfaces. Determine the force **P** required to initiate movement of the blocks.

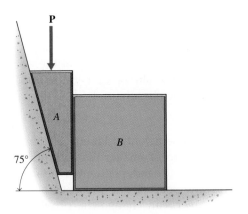

Fig. P9-167

9-168 A bumper jack, used to raise the wheel of an automobile in order to change a tire, is shown in Fig. P9-168. The mean diameter of the screw is 30 mm and the pitch of the screw is 6 mm. The coefficient of friction between the screw and the nut is 0.30. Determine the force that must be applied to the crank in order to:

a. Raise the automobile.
b. Lower the automobile.

Fig. P9-168

9-169* The device shown in Fig. P9-169 is used to raise boxes and crates between floors of a factory. The frame, which slides on the 4-in.-diameter vertical post, weighs 50 lb. If the coefficient of friction between the post and the frame is 0.10, determine the force **P** required to raise a 150-lb box.

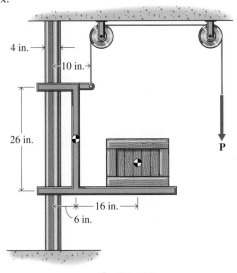

Fig. P9-169

9-170* A conveyor belt is driven with the 200-mm diameter multiple-pulley drive shown in Fig. P9-170. Couples

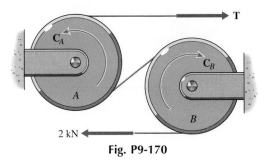

Fig. P9-170

C_A and C_B are applied to the system at pulleys A and B, respectively. The angle of contact between the belt and a pulley is 225° for each pulley and the coefficient of friction is 0.30. Determine the maximum force **T** that can be developed by the drive and the magnitudes of input couples C_A and C_B.

9-171 The electric motor shown in Fig. P9-171 weighs 30 lb and delivers 50 in. · lb of torque to pulley A of a furnace blower by means of a V-belt. The effective diameters of the 36° pulleys are 5 in. The coefficient of friction is 0.30. Determine the minimum distance a to prevent slipping of the belt if the rotation of the motor is clockwise.

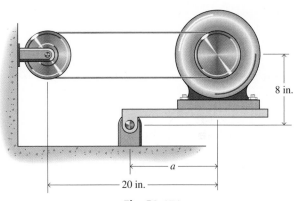

Fig. P9-171

9-172 The brake shown in Fig. P9-172 is used to control the motion of block B. If the mass of block B is 25 kg, and the kinetic coefficient of friction between the brake drum and brake pad is 0.30, determine the force **P** required for a constant velocity descent.

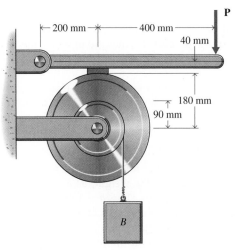

Fig. P9-172

425

9-173 A rotating shaft with a belt-type brake is shown in Fig. P9-173. The static coefficient of friction between the brake drum and the brake belt is 0.20. When a 75-lb force **P** is applied to the brake arm, rotation of the shaft is prevented. Determine the maximum torque that can be resisted by the brake:

a. If the shaft is tending to rotate counterclockwise.
b. If the shaft is tending to rotate clockwise.

9-174 Blocks A and B of Fig. P9-174 have 20- and 50-kg masses, respectively. The static coefficients of friction between blocks A and B and between block B and the horizontal surface are 0.25. The static coefficient of friction between the rope and the fixed pegs is 0.20. Determine the force **P** required to initiate motion of block B.

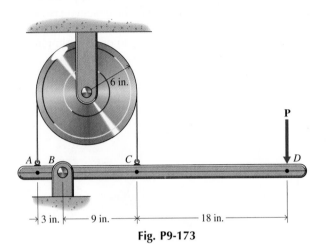

Fig. P9-173

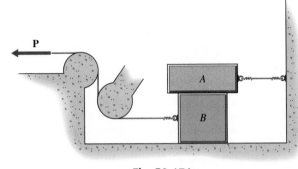

Fig. P9-174

Computer Problems

C9-175 A 20-lb weight is suspended from a lightweight rope, which is securely wrapped around the inner cylinder of a drum (Fig. P9-175). Rotation of the drum is controlled by a lightweight brake arm ABC, which is pressed against the outer cylinder of the drum by a hydraulic cylinder BD.

a. Plot T_{BD}, the minimum force in the cylinder that will hold the weight, as a function of μ_s, the coefficient of friction between the brake arm and the drum ($0.05 \le \mu_s \le 0.9$).
b. What happens to the system at $\mu_s = 0.75$?
c. What does the solution for $\mu_s > 0.75$ mean?
d. Repeat the problem for the case in which the weight is suspended from the other side of the drum.

C9-176 A 30-kg object is suspended from a lightweight rope, which is securely wrapped around a drum (Fig. P9-176). Rotation of the drum is controlled by the lightweight brake arm ABC. If the coefficient of friction between the brake arm and the drum is $\mu_s = 0.3$:

a. Plot **P**, the minimum force that must be applied to the brake arm to hold the weight, as a function of a, the location of the drum, ($10 \le a \le 1000$ mm).
b. What happens to the system at $a = 75$ mm?
c. What does the solution for $a < 75$ mm mean?
d. Repeat the problem for the case in which the object is suspended from the other side of the drum.

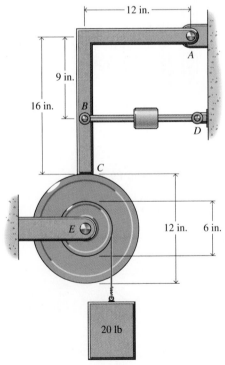

Fig. P9-175

426

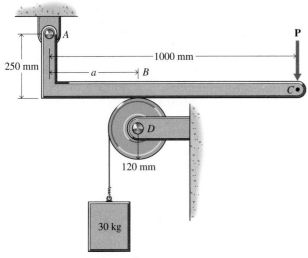

Fig. P9-176

C9-177 The simple mechanism of Fig. P9-177 is often used to hold the handles of brooms, mops, shovels, and other such tools. The weight of the tool causes the two otherwise free cylinders to become wedged into the corner between the handle and the rails. Although no amount of downward force will cause the handle to slip, the tool can be removed easily by lifting it upward and pulling it forward. The coefficient of friction is the same between the cylinder and the broom handle and between the cylinder and the side rail. If the broom shown weighs 1.5 lb and the weight of the two small cylinders may be neglected;

a. Plot μ_s, the minimum coefficient of the friction for which the system is in equilibrium, as a function of the rail angle θ ($5° \le \theta \le 75°$).

b. Plot A_n, the normal component of the force exerted on the broom handle by one of the cylinders, as a function of the angle θ ($5° \le \theta \le 75°$).

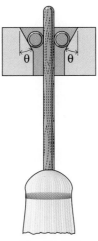

Fig. P9-177

C9-178 The simple mechanism of Fig. P9-178 is often used to hold notes or signs on bulletin boards. The weight of the paper causes the otherwise free cylinder to become wedged into the corner between the paper and the rail. Although no amount of downward force will cause the paper to slip out, the paper can be removed easily by lifting it upward and pulling it sideways. The coefficient of friction is the same between the cylinder and the paper and between the cylinder and the front rail. The small amount of friction between the back side of the paper and the back of the device may be neglected.

a. Plot μ_s, the minimum coefficient of friction for which the system is in equilibrium, as a function of the rail angle θ ($5° \le \theta \le 75°$).

b. If a downward force of 15 N is applied to the paper, plot A_n, the normal component of the force exerted on the paper by the cylinder, as a function of the angle θ ($5° \le \theta \le 75°$).

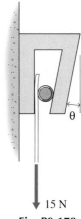

Fig. P9-178

C9-179 A 50-lb cylinder rests against a wall and a plate as shown in Fig. P9-179. The coefficient of friction is the same at both surfaces of contact with the cylinder.

a. Plot **P**, the maximum force that may be applied to the plate without moving the plate, as a function of the coefficient of friction μ_s ($0.05 \le \mu_s \le 0.8$).

b. On the same graph, plot $(A_f)_{actual}$ and $(B_f)_{actual}$, the actual amounts of friction that act at points A and B, and $(A_f)_{available}$ and $(B_f)_{available}$, the maximum amounts of friction available for equilibrium at points A and B.

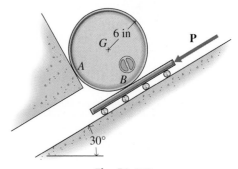

Fig. P9-179

427

C9-180 A 10-kg cylinder rests on a thin lightweight piece of cardboard as shown in Fig. P9-180. The coefficient of the friction is the same at all surfaces.

a. Plot **P**, the maximum force that may be applied to the cardboard without moving it, as a function of the coefficient of friction μ_s ($0.05 \le \mu_s \le 0.8$).
b. On the same graph, plot $(A_f)_{actual}$ and $(B_f)_{actual}$, the actual amounts of friction that act at points A and B, and $(A_f)_{available}$ and $(B_f)_{available}$, the maximum amounts of friction available for equilibrium at points A and B.
c. What happens to the system at $\mu_s \cong 0.364$?
d. What does the solution for $\mu_s > 0.364$ mean?

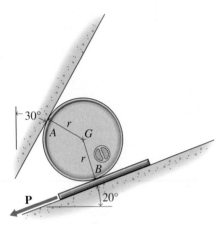

Fig. P9-180

C9-181 The hand brake of Fig. P9-181 is used to control the rotation of a drum. The coefficient of friction between the belt and the drum is $\mu_s = 0.15$, and the weight of the handle may be neglected. If a counterclockwise torque of 5 lb·ft is applied to the drum.

a. Plot **P**, the minimum force that must be applied to the handle to prevent motion, as a function of a, the location of the pivot point ($0.5 \le a \le 5.8$ in.).

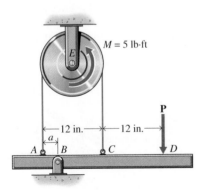

Fig. P9-181

b. Repeat the problem for a clockwise torque of 5 lb·ft. What happens to the system at $a \cong 4.6$ in.? What does the solution mean for $a > 4.6$ in.?

C9-182 The hand brake of Fig. P9-182 is used to control the rotation of a drum. The coefficient of friction between the belt and the drum is $\mu_s = 0.35$, and the weight of the handle may be neglected. If a clockwise torque of 475 N·m is applied to the drum.

a. Plot **P**, the minimum force that must be applied to the handle to prevent motion, as a function of a, the location of the pivot point ($5 \le a \le 500$ mm).
b. Repeat the problem for a counterclockwise torque of 475 N·m.

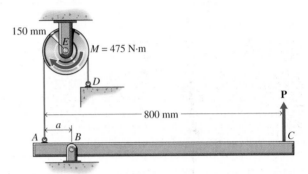

Fig. P9-182

C9-183 The band wrench of Fig. P9-183 is used to unscrew an oil filter from a car. (The filter acts as if it were a wheel with a resisting torque of 40 lb·ft.) The coefficient of friction between the band and the filter is $\mu_s = 0.15$, and the friction between the end of the wrench and the filter may be neglected.

a. Plot **P**, the minimum force that must be applied to the wrench to loosen the filter, as a function of a ($5 \le a \le 500$ mm).
b. Plot T_B, the tension on the handle at B, as a function of a ($5 \le a \le 500$ mm).
c. Plot A, the normal force exerted on the filter by the handle of the wrench, as a function of a ($5 \le a \le 500$ mm).

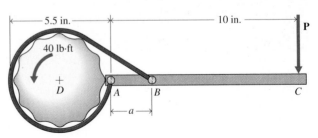

Fig. P9-183

SECOND MOMENTS OF AREA AND MOMENTS OF INERTIA

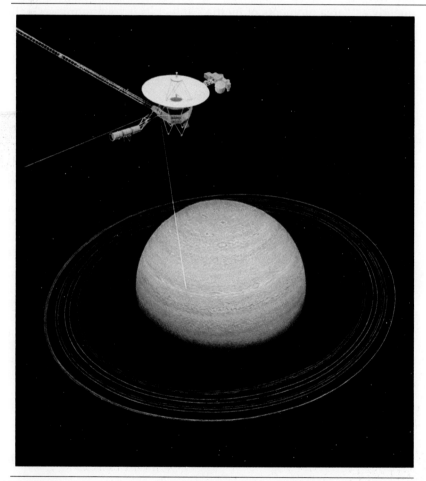

The thrusts required to change orientations of the Voyager 2 satellite during its encounter with Uranus depend on the moments of inertia of the satellite. Moments of inertia are modified by the size and placement of antennae.

10-1 INTRODUCTION

In Chapter 5 the centroid for an area was located by using an integral of the form $\int_A x \, dA$. In the analysis of stresses and deflections in beams and shafts, an expression of the form $\int_A x^2 \, dA$ is frequently encountered, in which dA represents an element of area and x represents the distance from the element to some axis in, or perpendicular to, the plane of the area. An expression of the form $\int_A x^2 \, dA$ is known as the second moment of the area. In the analysis of the angular motion of rigid bodies, an expression of the form $\int_m r^2 \, dm$ is encountered, in which dm represents an element of mass and r represents the distance from the element to some axis. Euler[1] gave the name "moment of inertia" to expressions of the form $\int_m r^2 \, dm$. Because of the similarity between the two types of integrals, both have become widely known as moments of inertia. In this text, the integrals involving areas will generally be referred to as "second moments of area" and less frequently as "area moments of inertia." Integrals involving masses will be referred to as "mass moments of inertia" or simply "moments of inertia."

Methods used to determine both second moments of area and mass moments of inertia will be discussed in later sections of this chapter.

10-2 SECOND MOMENT OF PLANE AREAS

The second moment of an area with respect to an axis will be denoted by the symbol I for an axis in the plane of the area and by the symbol J for an axis perpendicular to the plane of the area. The particular axis about which the second moment is taken will be denoted by subscripts. Thus, the second moments of the area A shown in Fig. 10-1 with respect to x- and y-axes in the plane of the area are

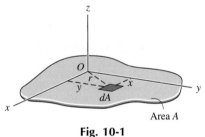

Fig. 10-1

$$I_x = \int_A y^2 \, dA \qquad \text{and} \qquad I_y = \int_A x^2 \, dA \qquad (10\text{-}1)$$

The quantities I_x and I_y are sometimes referred to as rectangular second moments of area A.

Similarly, the second moment of the area A shown in Fig. 10-1 with respect to a z-axis, which is perpendicular to the plane of the area at the origin O of the xy-coordinate system, is

$$J_z = \int_A r^2 \, dA = \int_A (x^2 + y^2) \, dA \qquad (10\text{-}2)$$

Thus,

$$J_z = \int_A x^2 \, dA + \int_A y^2 \, dA = I_y + I_x \qquad (10\text{-}3)$$

The quantity J_z is known as the polar second moment of the area A.

The second moment of an area can be visualized as the sum of a number of terms each consisting of an area multiplied by a distance squared. Thus, the units of a second moment are a length raised to the fourth power (L^4). Common units are mm^4 and $in.^4$. Also, the sign of

---

[1] Leonhard Euler (1707–83), a noted Swiss mathematician and physicist.

each term summed to obtain the second moment is positive since either a positive or negative distance squared is positive. Therefore, the second moment of an area is always positive.

10-2-1 Parallel-Axis Theorem for Second Moments of Area

When the second moment of an area has been determined with respect to a given axis, the second moment with respect to a parallel axis can be obtained by means of the parallel axis theorem (also known as the transfer formula). If one of the axes (say the x-axis) passes through the centroid of the area, as shown in Fig. 10-2, the second moment of the area about a parallel x'-axis is

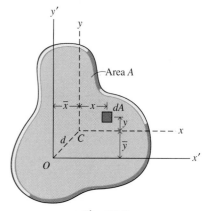

Fig. 10-2

$$I_{x'} = \int_A (y + \bar{y})^2 \, dA = \int_A y^2 \, dA + 2\bar{y} \int_A y \, dA + \bar{y}^2 \int_A dA$$

where $\bar{y}$ is the same for every element of area dA and has been taken outside the integral. The integral $\int y^2 \, dA$ is the second moment of the area I_x and the last integral is the total area A. Thus,

$$I_{x'} = I_x + 2\bar{y} \int_A y \, dA + \bar{y}^2 A \tag{a}$$

The integral $\int_A y \, dA$ is the first moment of the area with respect to the x-axis. Since the x-axis passes through the centroid C of the area, the first moment is zero and Eq. a becomes

$$I_{x'} = I_{xC} + \bar{y}^2 A \tag{10-4}$$

where I_{xC} is the second moment of the area with respect to the x-axis through the centroid and $|\bar{y}|$ is the distance between the x- and x'-axes. In a similar manner it can be shown that

$$J_{z'} = J_{zC} + (\bar{x}^2 + \bar{y}^2)A = J_{zC} + d^2 A \tag{10-5}$$

where J_{zC} is the polar second moment of the area with respect to the z-axis through the centroid and d is the distance between the z- and z'-axes.

The parallel-axis theorem states that the second moment of an area with respect to any axis in the plane of the area is equal to the second moment of the area with respect to a parallel axis through the centroid of the area added to the product of the area and the square of the distance between the two axes. The theorem also indicates that the second moment of an area with respect to an axis through the centroid of the area is less than that for any parallel axis since

$$I_{xC} = I_{x'} - \bar{y}^2 A \tag{10-6}$$

As a point of caution, note that the transfer formula is valid only for transfers to or from a centroidal axis. Thus,

$$\begin{aligned} I_{x''} = I_{xC} + \bar{y}_2^2 A &= (I_{x'} - \bar{y}_1^2 A) + \bar{y}_2^2 A \\ &= I_{x'} + (\bar{y}_2^2 - \bar{y}_1^2)A \neq I_{x'} + (\bar{y}_2 - \bar{y}_1)^2 A \end{aligned}$$

10-2-2 Second Moments of Areas by Integration

Rectangular and polar second moments of area were defined in Section 10-2. When the second moment of a plane area with respect to a line is

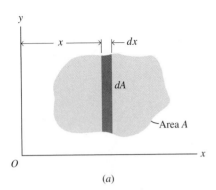

(a)

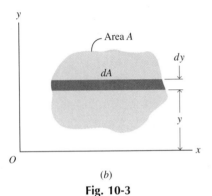

(b)

Fig. 10-3

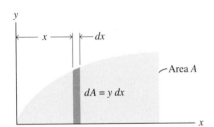

Fig. 10-4

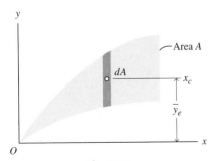

Fig. 10-5

determined by using Eqs. 10-1 or 10-2, it is possible to select the element of area dA in various ways and to express the area of the element in terms of either Cartesian or polar coordinates. In some cases, an element of area with dimensions $dA = dy\,dx$, as shown in Fig. 10-1, may be required. This type of element has the slight advantage that it can be used for calculating both I_x and I_y but it has the greater disadvantage of requiring double integration. Most problems can be solved with less work by choosing elements of the type shown in Figs. 10-3a and 10-3b. The following should be considered when selecting an element of area dA for a specific problem.

1. If all parts of the element of area are the same distance from the axis, the second moment can be determined directly by using Eqs. 10-1 or 10-2. Thus, the element shown in Fig. 10-1 can be used to determine either I_x or I_y directly, but a double integration is required. The element shown in Fig. 10-3a can be used to determine I_y directly since the dimension x is constant for the element. The element shown in Fig. 10-3a is not suitable for determining I_x directly since the y-dimension is not constant for the element. Similarly, the element shown in Fig. 10-3b is suitable for determining I_x directly but not I_y. A single integration would be required with elements of the type shown in Fig. 10-3.

2. If the second moment of the element of area with respect to the axis about which the second moment of the area is to be found is known, the second moment of the area can be found by summing the second moments of the individual elements that make up the area. For example, if the second moment dI_x for the rectangular area dA in Fig. 10-4 is known, the second moment I_x for the complete area A is simply $I_x = \int_A dI_x$.

3. If both the location of the centroid of the element and the second moment of the element about its centroidal axis parallel to the axis of interest for the complete area are known, the parallel-axis theorem can often be used to simplify the solution of a problem. For example, consider the area shown in Fig. 10-5. If both the distance $\bar{y}_e$ and the second moment dI_{xC} for the rectangular element dA are known, then by the parallel axis theorem $dI_x = dI_{xC} + \bar{y}_e^2\,dA$. The second moment for the complete area A is then simply $I_x = \int_A dI_x$.

From the previous discussion it is evident that either single or double integration may be required for the determination of second moments of area, depending on the element of area dA selected. When double integration is used, all parts of the element will be the same distance from the moment axis, and the second moment of the element can be written directly. Special care must be taken in establishing the limits for the two integrations to see that the correct area is included. If a strip element is selected, the second moment can usually be obtained by a single integration, but the element must be properly selected in order its second moment about the reference axis to be either known or readily calculated by using the parallel-axis theorem. The following example problems illustrate the procedure for determining the second moments of areas by integration.

Determine the second moment of area for the rectangle shown in Fig. 10-6a with respect to

a. The base of the rectangle.
b. An axis through the centroid parallel to the base.
c. An axis through the centroid normal to the area.

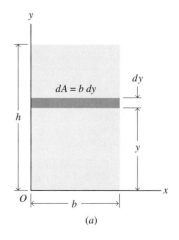

(a)

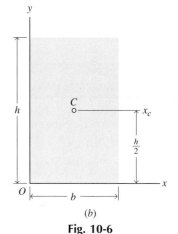

(b)

Fig. 10-6

SOLUTION

a. An element of area $dA = b \, dy$, as shown in Fig. 10-6a, will be used. Since all parts of the element are located a distance y from the x-axis, Eq. 10-1 can be used directly for the determination of the second moment I_x about the base of the rectangle. Thus

$$I_x = \int_A y^2 \, dA = \int_0^h y^2 b \, dy = \left[\frac{by^3}{3} \right]_0^h = \frac{bh^3}{3} \qquad \text{Ans.}$$

This result will be used frequently in later examples when elements of the type shown in Fig. 10-4 are used to determine second moments about the x-axis.

b. The parallel-axis theorem (Eq. 10-6) will be used to determine the second moment I_{xC} about an axis that passes through the centroid of the rectangle (see Fig. 10-6b) and is parallel to the base. Thus,

$$I_{xC} = I_x - \bar{y}^2 A = \frac{bh^3}{3} - \left(\frac{h}{2} \right)^2 (bh) = \frac{bh^3}{12} \qquad \text{Ans.}$$

This result will be used frequently in later examples when elements of the type shown in Fig. 10-5 are used to determine second moments about the x-axis.

c. The second moment I_{yC} for the rectangle can be determined in an identical manner. It can also be obtained from the preceding solution by interchanging b and h; that is

$$I_{yC} = \frac{hb^3}{12}$$

The polar second moment J_{zC} about a z-axis through the centroid of the rectangle is given by Eq. 10-3 as

$$J_{zC} = I_{xC} + I_{yC} = \frac{bh^3}{12} + \frac{hb^3}{12} = \frac{bh}{12} (h^2 + b^2) \qquad \text{Ans.}$$

EXAMPLE PROBLEM 10-2

Determine the second moment of area for the circle shown in Fig. 10-7 with respect to a diameter of the circle.

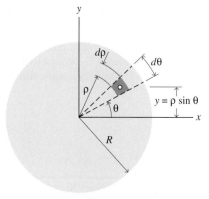

Fig. 10-7

SOLUTION

Polar coordinates are convenient for this problem. An element of area $dA = \rho \, d\theta \, d\rho$, as shown in Fig. 10-7, will be used. If the x-axis is selected as the diameter about which the second moment of area is to be determined, then $y = \rho \sin \theta$. Application of Eq. 10-1 yields

$$I_x = \int_A y^2 \, dA = \int_0^{2\pi} \int_0^R (\rho \sin \theta)^2 \, (\rho \, d\theta \, d\rho)$$

$$= \int_0^{2\pi} \int_0^R \rho^3 \sin^2 \theta \, d\rho \, d\theta = \frac{R^4}{4}\left[\frac{\theta}{2} - \frac{\sin 2\theta}{4}\right]_0^{2\pi} = \frac{\pi R^4}{4} \quad \text{Ans.}$$

EXAMPLE PROBLEM 10-3

Determine the polar second moment of area for the circle shown in Fig. 10-8 with respect to

a. An axis through the center of the circle and normal to the plane of the area.
b. An axis through the edge of the circle and normal to the plane of the area.

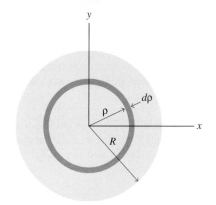

Fig. 10-8

SOLUTION

a. Polar coordinates are convenient for this problem. An element of area $dA = 2\pi\rho \, d\rho$, as shown in Fig. 10-8, will be used. Since all parts of the element are located a constant distance ρ from the center of the circle, Eq. 10-2 can be used directly for the determination of the polar second moment J_z about an axis through the center of the circle and normal to the plane of the area. Thus,

$$J_z = \int_A r^2 \, dA = \int_0^R \rho^2 \, (2\pi\rho \, d\rho) = \int_0^R 2\pi\rho^3 \, d\rho = \frac{\pi R^4}{2} \quad \text{Ans.}$$

This result could have been obtained from the solution of Example Problem 2-2 and use of Eq. 10-3. Thus,

$$J_z = I_x + I_y = \frac{\pi R^4}{4} + \frac{\pi R^4}{4} = \frac{\pi R^4}{2}$$

b. The parallel-axis theorem (Eq. 10-5) will be used to determine the polar second moment $J_{z'}$ about an axis that passes through the edge of the circle and is normal to the plane of the area. Thus,

$$J_{z'} = J_{zC} + d^2 A = \frac{\pi R^4}{2} + R^2(\pi R^2) = \frac{3\pi R^4}{2} \quad \text{Ans.}$$

434

Determine the second moment of area for the shaded region of Fig. 10-9 with respect to

a. The x-axis.
b. An axis through the origin of the xy-coordinate system and normal to the plane of the area.

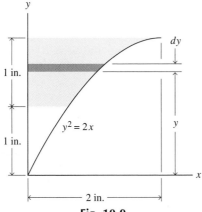

Fig. 10-9

SOLUTION

a. An element of area $dA = x \, dy = (y^2/2) \, dy$, as shown in Fig. 10-3b, will be used. Since all parts of the element are located a distance y from the x-axis, Eq. 10-1 can be used directly for the determination of the second moment I_x about the x-axis. Thus,

$$I_x = \int_A y^2 \, dA = \int_A y^2 \left(\frac{y^2}{2} \, dy\right) = \int_1^2 \frac{y^4}{2} \, dy = \left[\frac{y^5}{10}\right]_1^2 = 3.10 \text{ in.}^4 \quad \text{Ans.}$$

b. The same element of area can be used to obtain the second moment I_y if the result of Example Problem 10-1 is used as the known value for dI_y. Thus,

$$dI_y = \frac{bh^3}{3} = \frac{dy(x)^3}{3} = \frac{dy(y^2/2)^3}{3} = \frac{y^6}{24} \, dy$$

Summing all such elements yields

$$I_y = \int_A dI_y = \int_1^2 \frac{y^6}{24} \, dy = \left[\frac{y^7}{168}\right]_1^2 = \frac{127}{168} = 0.756 \text{ in.}^4$$

Once I_x and I_y are known for the area, the polar second moment for an axis through the origin of the xy-coordinate system and normal to the plane of the area is obtained by using Eq. 10-3. Thus,

$$J_z = I_x + I_y = 3.10 + 0.756 = 3.856 = 3.86 \text{ in.}^4 \qquad \text{Ans.}$$

PROBLEMS

10-1* Determine the second moment of area for the isosceles triangle shown in Fig. P10-1 with respect to

a. The base of the triangle (the x-axis).
b. An axis through the centroid parallel to the base.

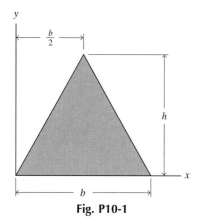

Fig. P10-1

10-2* Determine the second moment of area for the isosceles triangle shown in Fig. P10-1 with respect to an axis through the centroid of the triangle and parallel to the y-axis.

10-3 Determine the second moment of area for the half-ring shown in Fig. P10-3 with respect to

a. The y-axis.
b. An axis through the origin of the xy-coordinate system and normal to the plane of the area.

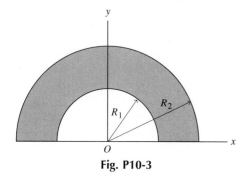

Fig. P10-3

10-4 Determine the second moment of area for the shaded region shown in Fig. P10-4 with respect to

a. The x-axis.
b. The y-axis.

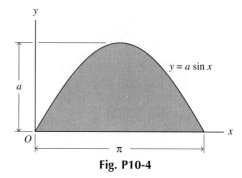

Fig. P10-4

10-5* Determine the second moment of area for the shaded region shown in Fig. P10-5 with respect to

a. The x-axis.
b. The y-axis.

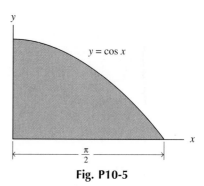

Fig. P10-5

10-6* Determine the second moment of area for the half-circle shown in Fig. P10-6 with respect to

a. The x-axis.
b. An axis through the centroid parallel to the x-axis.

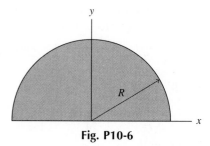

Fig. P10-6

10-7 Determine the polar second moment of area for the quarter-circle shown in Fig. P10-7 with respect to an axis that passes through the centroid and is normal to the plane of the area.

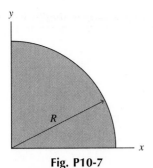

Fig. P10-7

10-8 Determine the second moment of area for the sector of a circle shown in Fig. P10-8 with respect to

a. The x-axis.
b. An axis through the origin of the xy-coordinate system and normal to the plane of the area.

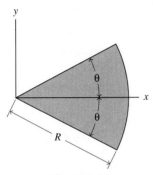

Fig. P10-8

10-9* Determine the second moment of area for the circle shown in Fig. P10-9 with respect to

a. The x-axis.
b. An axis through the origin of the xy-coordinate system and normal to the plane of the area.

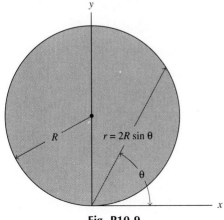

Fig. P10-9

10-10* Determine the second moment of area for the ellipse shown in Fig. P10-10 with respect to

a. The x-axis.
b. The y-axis.

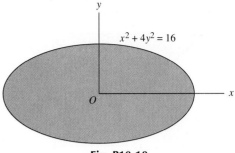

Fig. P10-10

10-11 Determine the second moment of area about the x-axis for the shaded region shown in Fig. P10-11.

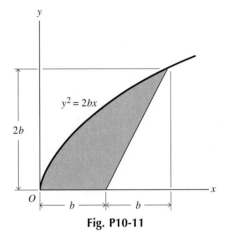

Fig. P10-11

10-12 Determine the second moment of area about the y-axis for the shaded region shown in Fig. P10-11.

10-13* Determine the second moment of area about the x-axis for the shaded region shown in Fig. P10-13.

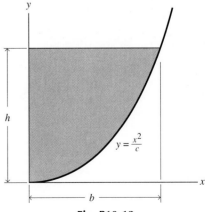

Fig. P10-13

437

10-14* Determine the second moment of area about the y-axis for the shaded region shown in Fig. P10-14.

10-15 Determine the second moment of area about the x-axis for the shaded region shown in Fig. P10-15.

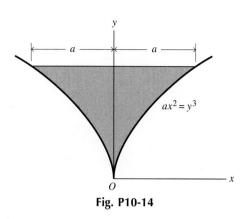

Fig. P10-14

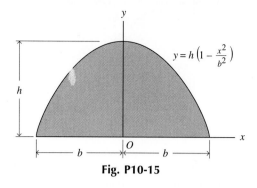

Fig. P10-15

10-16 Determine the second moment of area about the y-axis for the shaded region shown in Fig. P10-15.

10-2-3 Radius of Gyration of Areas

Since the second moment of an area has the dimensions of length to the fourth power, it can be expressed as the area A multiplied by a length k squared. Thus, from Eqs. 10-1 and 10-2

$$I_x = \int_A y^2 \, dA = A k_x^2 \qquad k_x = \sqrt{\frac{I_x}{A}}$$

$$I_y = \int_A x^2 \, dA = A k_y^2 \qquad k_y = \sqrt{\frac{I_y}{A}} \qquad (10\text{-}7)$$

$$I_z = \int_A r^2 \, dA = A k_z^2 \qquad k_z = \sqrt{\frac{I_z}{A}}$$

and from Eq. 10-3

$$k_z^2 = k_x^2 + k_y^2 \qquad (10\text{-}8)$$

The distance k is called the radius of gyration. The subscript denotes the axis about which the second moment of area is taken.

The parallel-axis theorem for second moments of area was discussed in Section 10-2-1. As indicated by Eq. a, a corresponding relation exists between the radii of gyration of the area with respect to two parallel axes, one of which passes through the centroid of the area. Thus,

$$k_{x'}^2 = k_{xC}^2 + \bar{y}^2 \qquad k_{y'}^2 = k_{yC}^2 + \bar{x}^2 \qquad (10\text{-}9)$$

Similarly for polar second moments of area and radii of gyration:

$$k_{z'}^2 = k_{zC}^2 + (\bar{x}^2 + \bar{y}^2) = k_{zC}^2 + d^2 \qquad (10\text{-}10)$$

Solving Eqs. 10-9 and 10-10 for rectangular and polar radii of gyration for arbitrary and centroidal x-, y-, and z-axes yields

$$k_{x'} = \sqrt{k_{xC}^2 + \bar{y}^2} \qquad k_{xC} = \sqrt{k_{x'}^2 - \bar{y}^2}$$

$$k_{y'} = \sqrt{k_{yC}^2 + \bar{x}^2} \qquad k_{yC} = \sqrt{k_{y'}^2 - \bar{x}^2} \qquad (10\text{-}11)$$

$$k_{z'} = \sqrt{k_{zC}^2 + d^2} \qquad k_{zC} = \sqrt{k_{z'}^2 - d^2}$$

The following example illustrates the concepts discussed in this section.

For the shaded area shown in Fig. 10-10a, determine

a. The radii of gyration k_x, k_y, and k_z.
b. The radius of gyration for an axis passing through the centroid and parallel to the y-axis.

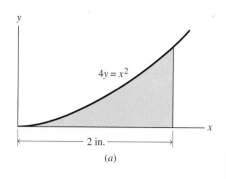

(a)

SOLUTION

a. The quantities required for the determination of k_x, k_y, and k_z are the area A and the second moments I_x and I_y. Since none of these quantities is readily available from known solutions, they will be determined by integration using the element of area shown in Fig. 10-10b. For area A:

$$A = \int_A y\, dx = \int_0^2 \frac{x^2}{4}\, dx = \left[\frac{x^3}{12}\right]_0^2 = \frac{2}{3} \text{ in.}^2$$

For the second moment I_x, the results of Example Problem 10-1 can be used to determine dI_x. Thus,

$$dI_x = \frac{bh^3}{3} = \frac{dx(y)^3}{3} = \frac{dx(x^2/4)^3}{3} = \frac{x^6}{192}\, dx$$

and

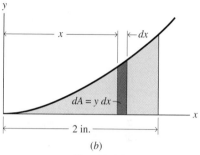

$dA = y\, dx$

(b)

Fig. 10-10

$$I_x = \int_A dI_x = \int_0^2 \frac{x^6}{192}\, dx = \left[\frac{x^7}{1344}\right]_0^2 = \frac{2}{21} \text{ in.}^4$$

For the second moment I_y:

$$I_y = \int_A x^2\, dA = \int_A x^2 y\, dx = \int_0^2 x^2\left(\frac{x^2}{4}\right) dx = \int_0^2 \frac{x^4}{4}\, dx = \left[\frac{x^5}{20}\right]_0^2 = \frac{8}{5} \text{ in.}^4$$

Once A, I_x, and I_y are known, the radii of gyration k_x and k_y are obtained from Eqs. 10-7. Thus,

$$k_x = \left[\frac{I_x}{A}\right]^{1/2} = \left[\frac{2/21}{2/3}\right]^{1/2} = 0.3780 = 0.378 \text{ in.} \qquad \text{Ans.}$$

$$k_y = \left[\frac{I_y}{A}\right]^{1/2} = \left[\frac{8/5}{2/3}\right]^{1/2} = 1.5492 = 1.549 \text{ in.} \qquad \text{Ans.}$$

The polar radius of gyration k_z obtained from k_x and k_y by using Eq. 10-8 is
$$k_z = \sqrt{k_x^2 + k_y^2} = \sqrt{(0.3780)^2 + (1.5492)^2} = 1.595 \text{ in.} \qquad \text{Ans.}$$

b. In order to determine the radius of gyration k_{yC}, the distance $\bar{x}$ between the y-axis and the centroid of the area must be determined. Thus,

$$A\bar{x} = \int_A x\, dA = \int_0^2 x\left(\frac{x^2}{4}\right) dx = \int_0^2 \frac{x^3}{4}\, dx = \left[\frac{x^4}{16}\right]_0^2 = 1.000 \text{ in.}^3$$

from which

$$\bar{x} = \frac{\int_A x\, dA}{A} = \frac{1.000}{2/3} = 1.500 \text{ in.}$$

The radius of gyration k_{yC} is then obtained by using Eqs. 10-11. Thus,
$$k_{yC} = \sqrt{k_y^2 - \bar{x}^2} = \sqrt{(1.5492)^2 - (1.500)^2} = 0.387 \text{ in.} \qquad \text{Ans.}$$

PROBLEMS

10-17* Determine the radii of gyration of the rectangular area shown in Fig. P10-17 with respect to

a. The x- and y-axes shown on the figure.
b. Horizontal and vertical centroidal axes.

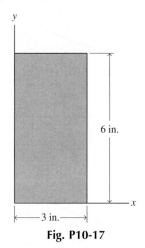

6 in.

3 in.

Fig. P10-17

10-18* Determine the radii of gyration of the triangular area shown in Fig. P10-18 with respect to

a. The x- and y-axes shown on the figure.
b. Horizontal and vertical centroidal axes.

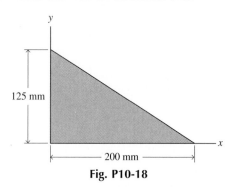

125 mm

200 mm

Fig. P10-18

10-19 Determine the radii of gyration of the shaded area shown in Fig. P10-19 with respect to

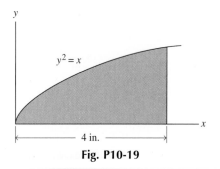

$y^2 = x$

4 in.

Fig. P10-19

a. The x- and y-axes shown on the figure.
b. Horizontal and vertical centroidal axes.

10-20 Determine the radii of gyration of the shaded area shown in Fig. P10-20 with respect to

a. The x- and y-axes shown on the figure.
b. Horizontal and vertical centroidal axes.

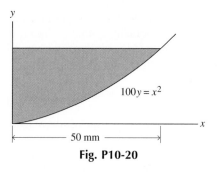

$100y = x^2$

50 mm

Fig. P10-20

10-21* Determine the polar radius of gyration of the shaded area shown in Fig. P10-21 with respect to an axis through the origin of the xy-coordinate system and normal to the plane of the area.

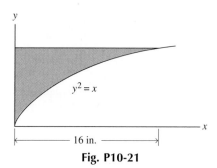

$y^2 = x$

16 in.

Fig. P10-21

10-22* Determine the radius of gyration of the shaded area shown in Fig. P10-22 with respect to an axis through the centroid of the area and normal to the plane of the area.

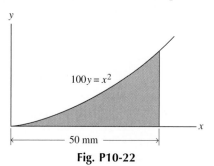

$100y = x^2$

50 mm

Fig. P10-22

10-23 Determine the radii of gyration of the circular area shown in Fig. P10-23 with respect to

a. Horizontal and vertical centroidal axes.
b. The x- and y-axes shown on the figure.

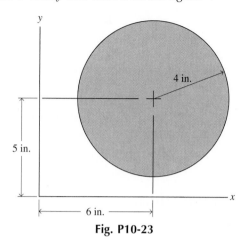

Fig. P10-23

10-24 Determine the radii of gyration of the area of the isosceles triangle shown in Fig. P10-24 with respect to

a. Horizontal and vertical centroidal axes.
b. The x- and y-axes shown on the figure.

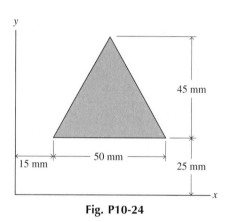

Fig. P10-24

10-25* Determine the polar radius of gyration of the circular area shown in Fig. P10-25 with respect to an axis through the origin of the xy-coordinate system and normal to the plane of the area.

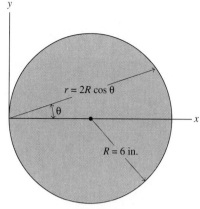

Fig. P10-25

10-26 Determine the polar radius of gyration of the shaded area shown in Fig. P10-26 with respect to an axis through the origin of the xy-coordinate system and normal to the plane of the area.

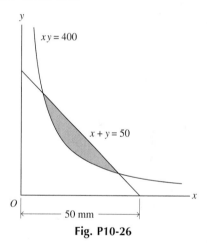

Fig. P10-26

10-2-4   Second Moments of Composite Areas

The second moments I_x, I_y, and I_z of an area A with respect any set of x-, y-, and z-coordinate axes were defined as

$$I_x = \int_A y^2 \, dA \qquad I_y = \int_A x^2 \, dA \qquad I_z = \int_A r^2 \, dA$$

Frequently in engineering practice, an irregular area A will be encountered that can be broken up into a series of simple areas $A_1, A_2, A_3, \ldots,$ A_n for which the integrals have been evaluated and tabulated. The second moment of the irregular area, the *composite area,* with respect to any axis is equal to the sum of the second moments of the separate parts of the area with respect to the specified axis. For example,

$$\begin{aligned}
I_x &= \int_A y^2 \, dA \\
&= \int_{A_1} y^2 \, dA_1 + \int_{A_2} y^2 \, dA_2 + \int_{A_3} y^2 \, dA_3 + \cdots + \int_{A_n} y^2 \, dA_n \\
&= I_{x1} + I_{x2} + I_{x3} + \cdots + I_{xn}
\end{aligned}$$

When an area such as a hole is removed from a larger area, its second moment must be subtracted from the second moment of the larger area to obtain the resulting second moment. Thus, for the case of a square plate with a hole

$$I_{\blacksquare} = I_{\square} + I_{\bullet}$$

Therefore,

$$I_{\square} = I_{\blacksquare} - I_{\bullet}$$

Table 10-1 contains a listing of the values of the integrals for frequently encountered shapes such as rectangles, triangles, circles, and semicircles. Tables listing second moments of area and other properties for the cross sections of common structural shapes are found in engineering handbooks and in data books prepared by industrial organizations such as the American Institute of Steel Construction. Properties of a few selected shapes are listed in Tables 10-2*a* and 10-2*b* for use in solving problems.

In some instances the second moments I_{xC}, I_{yC}, and I_{zC} of a composite shape with respect to centroidal x-, y-, and z-axes of the composite may be required. These quantities can be determined by first evaluating the second moments $I_{x'}$, $I_{y'}$, and $I_{z'}$ of the composite with respect to any convenient set of parallel x'-, y'-, and z'-axes and then transferring these second moments to the centroidal axes by using the parallel-axis theorem.

TABLE 10-1 SECOND MOMENTS OF PLANE AREAS

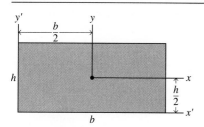

$$I_x = \frac{bh^3}{12}$$

$$I_{x'} = \frac{bh^3}{3}$$

$$A = bh$$

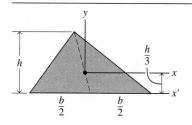

$$I_x = \frac{bh^3}{36}$$

$$I_{x'} = \frac{bh^3}{12}$$

$$A = \frac{1}{2}bh$$

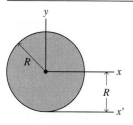

$$I_x = \frac{\pi R^4}{4}$$

$$I_{x'} = \frac{5\pi R^4}{4}$$

$$A = \pi R^2$$

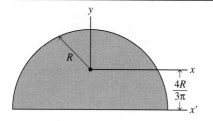

$$I_x = \frac{\pi R^4}{8} - \frac{8R^4}{9\pi}$$

$$I_y = \frac{\pi R^4}{8}$$

$$I_{x'} = \frac{\pi R^4}{8}$$

$$A = \frac{1}{2}\pi R^2$$

$$I_x = \frac{\pi R^4}{16} - \frac{4R^4}{9\pi}$$

$$I_{x'} = \frac{\pi R^4}{16}$$

$$A = \frac{1}{4}\pi R^2$$

$$I_x = \frac{R^4}{4}\left(\theta - \frac{1}{2}\sin 2\theta\right)$$

$$\bar{x} = \frac{2}{3}\frac{R\sin\theta}{\theta}$$

$$I_y = \frac{R^4}{4}\left(\theta + \frac{1}{2}\sin 2\theta\right)$$

$$A = \theta R^2$$

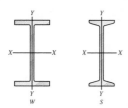

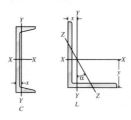

Shape	Area in.^{2}	Depth in.	Width in.	Flange Thickness in.	Web Thickness in.	Axis XX I_X in.^{4}	Axis XX k_X in.	Axis YY I_Y in.^{4}	Axis YY k_Y in.
W24 × 84	24.7	24.10	9.020	0.770	0.470	2370	9.79	94.4	1.95
W18 × 76	22.3	18.21	11.035	0.680	0.425	1330	7.73	152	2.61
W14 × 43	12.6	13.66	7.995	0.530	0.305	428	5.82	45.2	1.89
W12 × 30	8.79	12.34	6.520	0.440	0.260	238	5.21	20.3	1.52
W10 × 22	6.49	10.17	5.750	0.360	0.240	118	4.27	11.4	1.33
S18 × 70	20.6	18.00	6.251	0.691	0.711	926	6.71	24.1	1.08
S12 × 50	14.7	12.00	5.477	0.659	0.687	305	4.55	15.7	1.03
S10 × 35	10.3	10.00	4.944	0.491	0.594	147	3.78	8.36	0.901

Shape	Area in.^{2}	Depth in.	Width in.	Flange Thickness in.	Web Thickness in.	Axis XX I_X in.^{4}	Axis XX k_X in.	Axis YY I_Y in.^{4}	Axis YY k_Y in.	x in.
C15 × 50	14.7	15.00	3.716	0.650	0.716	404	5.24	11.0	0.867	0.798
C12 × 30	8.82	12.00	3.170	0.501	0.510	162	4.29	5.14	0.763	0.674
C10 × 30	8.82	10.00	3.033	0.436	0.673	103	3.42	3.94	0.669	0.649

Shape	Area in.^{2}	Axis XX I_X in.^{4}	Axis XX k_X in.	Axis XX y in.	Axis YY I_Y in.^{4}	Axis YY k_Y in.	Axis YY x in.
L8 × 8 × 1	15.00	89.0	2.44	2.37	89.0	2.44	2.37
L8 × 6 × 1	13.00	80.8	2.49	2.65	38.8	1.73	1.65
L8 × 4 × 1	11.00	69.6	2.52	3.05	11.6	1.03	1.05
L6 × 6 × 1	11.00	35.5	1.80	1.86	35.5	1.80	1.86
L6 × 4 × 3/4	6.94	24.5	1.88	2.08	8.68	1.12	1.08
L4 × 4 × 3/4	5.44	7.67	1.19	1.27	7.67	1.19	1.27
L4 × 3 × 1/2	3.25	5.05	1.25	1.33	2.42	0.864	0.827
L3 × 3 × 1/2	2.75	2.22	0.898	0.932	2.22	0.898	0.932

TABLE 10-2*B* PROPERTIES OF ROLLED-STEEL SHAPES (SI UNITS)

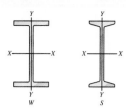
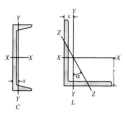

Shape	Area mm²	Depth mm	Width mm	Flange Thickness mm	Web Thickness mm	Axis XX I_X 10⁶mm⁴	Axis XX k_X mm	Axis YY I_Y 10⁶mm⁴	Axis YY k_Y mm
W610 × 125	15935	612	229	19.6	11.9	985	249	39.3	49.5
W457 × 113	14385	463	280	17.3	10.8	554	196	63.3	66.3
W356 × 64	8130	347	203	13.5	7.7	178	148	18.8	48.0
W305 × 45	5670	313	166	11.2	6.6	99.1	132	8.45	38.6
W254 × 33	4185	258	146	9.1	6.1	49.1	108	4.75	33.8
S457 × 104	13290	457	159	17.6	18.1	358	170	10.0	27.4
S305 × 74	9485	305	139	16.7	17.4	127	116	6.53	26.2
S254 × 52	6645	254	126	12.5	15.1	61.2	96.0	3.48	22.9

Shape	Area mm²	Depth mm	Width mm	Flange Thickness mm	Web Thickness mm	Axis XX I_X 10⁶mm⁴	Axis XX k_X mm	Axis YY I_Y 10⁶mm⁴	Axis YY k_Y mm	x mm
C381 × 74	9485	381	94.4	16.5	18.2	168	133	4.58	22.0	20.3
C305 × 45	5690	305	80.5	12.7	13.0	67.4	109	2.14	19.4	17.1
C254 × 45	5690	254	77.0	11.1	17.1	42.9	86.9	1.64	17.0	16.5

Shape	Area mm²	Axis XX I_X 10⁶mm⁴	Axis XX k_X mm	Axis XX y mm	Axis YY I_Y 10⁶mm⁴	Axis YY k_Y mm	Axis YY x mm
L203 × 203 × 25.4	9675	37.0	62.0	60.2	37.0	62.0	60.2
L203 × 152 × 25.4	8385	33.6	63.2	67.3	16.1	43.9	41.9
L203 × 102 × 25.4	7095	29.0	64.0	77.5	4.83	26.2	26.7
L152 × 152 × 25.4	7095	14.8	45.7	47.2	14.8	45.7	47.2
L152 × 102 × 19.1	4475	10.2	47.8	52.8	3.61	28.4	27.4
L102 × 102 × 19.1	3510	3.19	30.2	32.3	3.19	30.2	32.3
L102 × 76 × 12.7	2095	2.10	31.8	33.8	1.01	21.9	21.0
L76 × 76 × 12.7	1775	0.924	22.8	23.7	0.924	22.8	23.7

Determine the second moment of the shaded area shown in Fig. 10-11a with respect to

a. The x-axis.
b. The y-axis.
c. An axis through the origin O of the xy-coordinate system and normal to the plane of the area.

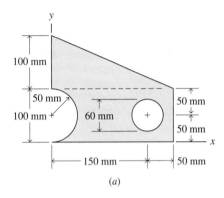

(a)

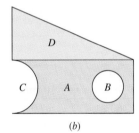

(b)

Fig. 10-11

SOLUTION

As shown in Fig. 10-11b, the shaded area can be divided into a 100×200-mm rectangle (A) with a 60-mm diameter circle (B) and a 100-mm diameter half circle (C) removed and a 100×200-mm triangle (D). The second moments for these four areas, with respect to the x- and y-axes, can be obtained by using information from Table 10-1, as follows.

a. For the rectangle (shape A):

$$I_{x1} = \frac{bh^3}{3} = \frac{200(100^3)}{3} = 66.667(10^6) \text{ mm}^4$$

For the circle (shape B):

$$I_{x2} = I_{xC} + \bar{y}^2 A = \frac{\pi R^4}{4} + \bar{y}^2 (\pi R^2)$$

$$= \frac{\pi (30^4)}{4} + (50^2)(\pi)(30^2) = 7.705(10^6) \text{ mm}^4$$

For the half circle (shape C):

$$I_{x3} = I_{xC} + \bar{y}^2 A = \frac{\pi R^4}{8} + \bar{y}^2 \left(\frac{\pi R^2}{2} \right)$$

$$= \frac{\pi (50^4)}{8} + (50)^2 \left[\frac{\pi (50)^2}{2} \right] = 12.272(10^6) \text{ mm}^4$$

For the triangle (shape D):

$$I_{x4} = I_{xC} + \bar{y}^2 A$$
$$= \frac{bh^3}{36} + \bar{y}^2 \left(\frac{bh}{2} \right)$$
$$= \frac{200(100^3)}{36} + \left(100 + \frac{100}{3} \right)^2 \left[\frac{200(100)}{2} \right] = 183.333(10^6) \text{ mm}^4$$

For the composite area:

$$I_x = I_{x1} - I_{x2} - I_{x3} + I_{x4}$$
$$= 66.667(10^6) - 7.705(10^6) - 12.272(10^6) + 183.333(10^6)$$
$$= 230.023(10^6) = 230(10^6) \text{ mm}^4 \qquad \text{Ans.}$$

b. For the rectangle (shape A):

$$I_{y1} = \frac{b^3 h}{3} = \frac{200^3(100)}{3} = 266.667(10^6) \text{ mm}^4$$

For the circle (shape B):

$$I_{y2} = I_{yC} + \bar{x}^2 A$$
$$= \frac{\pi R^4}{4} + \bar{x}^2 (\pi R^2)$$
$$= \frac{\pi (30^4)}{4} + (150^2)(\pi)(30^2) = 64.253(10^6) \text{ mm}^4$$

For the half circle (shape C):

$$I_{y3} = \frac{\pi R^4}{8} = \frac{\pi (50^4)}{8} = 2.454(10^6) \text{ mm}^4$$

For the triangle (shape D):

$$I_{y4} = \frac{bh^3}{12} = \frac{100(200^3)}{12} = 66.667(10^6) \text{ mm}^4$$

For the composite area:

$$I_y = I_{y1} - I_{y2} - I_{y3} + I_{y4}$$
$$= 266.667(10^6) - 64.253(10^6) - 2.454(10^6) + 66.667(10^6)$$
$$= 266.627(10^6) = 267(10^6) \text{ mm}^4 \qquad \text{Ans.}$$

c. For the composite area:

$$J_z = I_x + I_y$$
$$= 230.023(10^6) + 266.627(10^6)$$
$$= 496.650(10^6) = 497(10^6) \text{ mm}^4 \qquad \text{Ans.}$$

EXAMPLE PROBLEM 10-7

A column with the cross section shown in Fig. 10-12 is constructed from a W24 × 84 wide-flange section and a C12 × 30 channel. Determine the second moments and radii of gyration of the cross-sectional area with respect to horizontal and vertical axes through the centroid of the cross section.

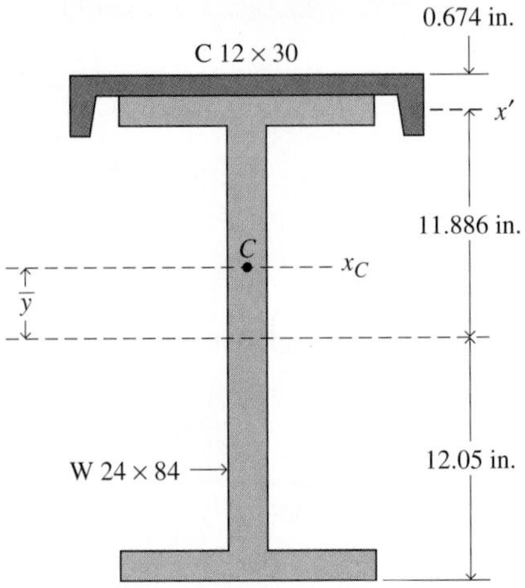

Fig. 10-12

SOLUTION

Properties and dimensions for the structural shapes can be obtained from Table 10-2*a*. In Fig. 10-12, the x-axis passes through the centroid of the wide-flange section and the x'-axis passes through the centroid of the channel. The centroidal x_C axis for the composite section can be located by using the principle of moments as applied to areas. The total area A_T for the composite section is

$$A_T = A_{WF} + A_{CH} = 24.7 + 8.82 = 33.52 \text{ in.}^2$$

The moment of the composite area about the x-axis is

$$M_x = A_{WF}\bar{y}_{WF} + A_{CH}\bar{y}_{CH} = 24.7(0) + 8.82(11.886) = 104.835 \text{ in.}^3$$

The distance $\bar{y}$ from the x-axis to the centroid of the composite section is

$$\bar{y} = \frac{M_x}{A_T} = \frac{104.835}{33.52} = 3.128 \text{ in.}$$

The second moment I_{xCWF} for the wide-flange section about the centroidal x_C-axis of the composite section is

$$I_{xCWF} = I_{xWF} + \bar{y}^2 A_{WF} = 2370 + (3.128)^2(24.7) = 2611.7 \text{ in.}^4$$

The second moment I_{xCCH} for the channel about the centroidal x_C-axis of the composite section is

$$\begin{aligned} I_{xCCH} &= I_{x'CH} + (\bar{y}_{CH} - \bar{y})^2 A_{CH} \\ &= 5.14 + (11.886 - 3.128)^2(8.82) = 681.7 \text{ in.}^4 \end{aligned}$$

For the composite area

$$\begin{aligned} I_{xC} &= I_{xCWF} + I_{xCCH} \\ &= 2611.7 + 681.7 = 3293.4 = 3290 \text{ in.}^4 \end{aligned} \qquad \text{Ans.}$$

The y-axis passes through the centroid of each of the areas; therefore, the second moment I_{yC} for the composite section is

$$I_{yC} = I_{yCWF} + I_{yCCH} = 94.4 + 162 = 256.4 = 256 \text{ in.}^4 \qquad \text{Ans.}$$

The radius of gyration about the x_C-axis for the composite section is

$$k_{xC} = \left(\frac{I_{xC}}{A_T}\right)^{1/2} = \left(\frac{3293.4}{33.52}\right)^{1/2} = 9.912 = 9.91 \text{ in.} \qquad \text{Ans.}$$

The radius of gyration about the y_C-axis for the composite section is

$$k_{yC} = \left(\frac{I_{yC}}{A_T}\right)^{1/2} = \left(\frac{256.4}{33.52}\right)^{1/2} = 2.766 = 2.77 \text{ in.} \qquad \text{Ans.}$$

PROBLEMS

10-27* Determine the second moments of the shaded area shown in Fig. P10-27 with respect to x- (horizontal) and y- (vertical) axes through the centroid of the area.

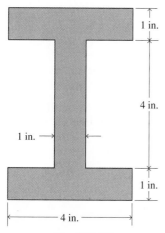

Fig. P10-27

10-28* Determine the second moments of the shaded area shown in Fig. P10-28 with respect to x- (horizontal) and y- (vertical) axes through the centroid of the area.

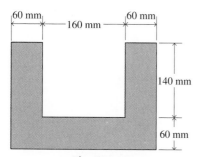

Fig. P10-28

10-29 Determine the second moment of the shaded area shown in Fig. P10-29 with respect to

a. The x-axis.
b. The y-axis.
c. An axis through the origin O of the xy-coordinate system and normal to the plane of the area.

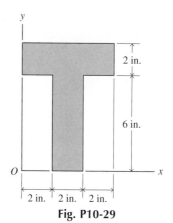

Fig. P10-29

10-30 Determine the second moment of the shaded area shown in Fig. P10-30 with respect to

a. The x-axis.
b. The y-axis.
c. An axis through the origin O of the xy-coordinate system and normal to the plane of the area.

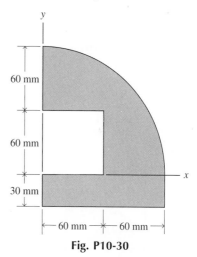

Fig. P10-30

10-31* Determine the second moments of the shaded area shown in Fig. P10-31 with respect to the x- and y-axes.

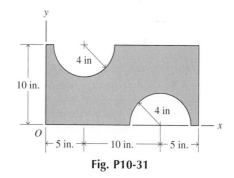

Fig. P10-31

449

10-32* Determine the second moments of the shaded area shown in Fig. P10-32 with respect to the *x*- and *y*-axes.

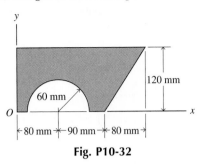

Fig. P10-32

10-33 Determine the second moments of the shaded area shown in Fig. P10-33 with respect to the *x*- and *y*-axes.

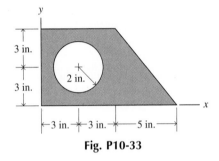

Fig. P10-33

10-34 Determine the second moments of the shaded area shown in Fig. P10-34 with respect to the *x*- and *y*-axes.

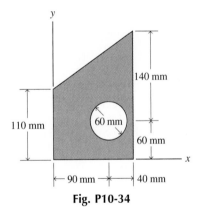

Fig. P10-34

10-35* Determine the second moments of the shaded area shown in Fig. P10-35 with respect to

a. The *x*- and *y*-axes shown on the figure.
b. The *x*- and *y*-axes through the centroid of the area.

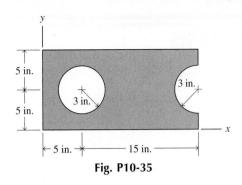

Fig. P10-35

10-36* Determine the second moments of the shaded area shown in Fig. P10-36 with respect to

a. The *x*- and *y*-axes shown on the figure.
b. The *x*- and *y*-axes through the centroid of the area.

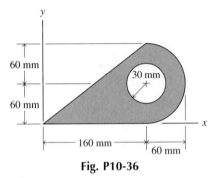

Fig. P10-36

10-37 Determine the second moments of the shaded area shown in Fig. P10-37 with respect to *x*- (horizontal) and *y*- (vertical) axes through the centroid of the area.

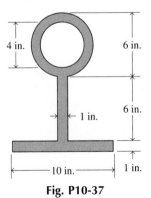

Fig. P10-37

10-38 Four C305 × 45 channels are welded together to form the cross section shown in Fig. P10-38. Determine the second moments of the area with respect to *x*- (horizontal) and *y*- (vertical) axes through the centroid of the area.

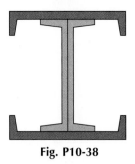

Fig. P10-38

10-39* Two 10×1-in. steel plates are welded to the flanges of an S18 × 70 I-beam as shown in Fig. P10-39. Determine the second moments of the area with respect to x- (horizontal) and y- (vertical) axes through the centroid of the area.

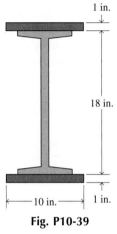

1 in.

18 in.

10 in. ⟶ 1 in.

Fig. P10-39

10-40 Two 250 × 25-mm steel plates and two C254 × 45 channels are welded together to form the cross section shown in Fig. P10-40. Determine the second moments of the area with respect to x- (horizontal) and y- (vertical) axes through the centroid of the area.

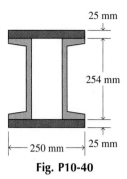

25 mm

254 mm

250 mm ⟶ 25 mm

Fig. P10-40

10-2-5 Mixed Second Moments of Areas

The mixed second moment (commonly called the area product of inertia) dI_{xy} of the element of area dA shown in Fig. 10-13 with respect to the x- and y-axes is defined as the product of the two coordinates of the element multiplied by the area of the element; thus

$$dI_{xy} = xy\,dA$$

The mixed second moment (area product of inertia) of the total area A about the x- and y-axes is the sum of the mixed second moments of the elements of the area; thus,

$$I_{xy} = \int_A xy\,dA \tag{10-12}$$

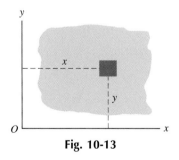

Fig. 10-13

The dimensions of the mixed second moment are the same as for the rectangular or polar second moments, but since the product xy can be either positive or negative, the mixed second moment can be positive, negative, or zero. Recall that rectangular or polar second moments are always positive.

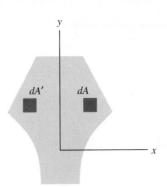

Fig. 10-14

The mixed second moment of an area with respect to any two orthogonal axes is zero when either of the axes is an axis of symmetry. This statement can be demonstrated by means of Fig. 10-14, which is symmetrical with respect to the y-axis. For every element of area dA on one side of the axis of symmetry, there is a corresponding element of area dA' on the opposite side of the axis such that the mixed second moments of dA and dA' will be equal in magnitude but opposite in sign. Thus, they will cancel each other in the summation, and the resulting mixed second moment for the total area will be zero.

The parallel-axis theorem for mixed second moments can be derived from Fig. 10-15 in which the x- and y-axes pass through the centroid C of the area and are parallel to the x'- and y'-axes. The mixed second moment with respect to the x'- and y'-axes is

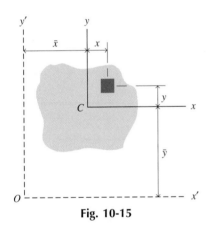

Fig. 10-15

$$
\begin{aligned}
I_{x'y'} &= \int_A x'y' \, dA = \int_A (\bar{x} + x)(\bar{y} + y) \, dA \\
&= \bar{x}\bar{y} \int_A dA + \bar{x} \int_A y \, dA + \bar{y} \int_A x \, dA + \int_A xy \, dA
\end{aligned}
$$

since $\bar{x}$ and $\bar{y}$ are the same for every element of area dA. The second and third integrals in the preceding equation are zero since x and y are centroidal axes. The last integral is the mixed second moment with respect to the centroidal axes. Consequently, the mixed second moment about a pair of axes parallel to a pair of centroidal axes is

$$
I_{x'y'} = I_{xyC} + \bar{x}\bar{y} \, A \tag{10-13}
$$

where the subscript C indicates that the x- and y-axes are centroidal axes. The parallel-axis theorem for mixed second moments (area products of inertia) can be stated as follows: The mixed second moment of an area with respect to any two perpendicular axes x and y in the plane of the area is equal to the mixed second moment of the area with respect to a pair of centroidal axes parallel to the x- and y-axes added to the product of the area and the two centroidal locations from the x- and y-axes. The parallel-axis theorem for mixed second moments is used most frequently in determining mixed second moments for composite areas. Values from Eq. 10-12 for some of the shapes commonly used in these calculations are listed in Table 10-3.

The determination of the mixed second moment (area product of inertia) is illustrated in the next two example problems.

TABLE 10-3   MIXED SECOND MOMENTS OF PLANE AREAS

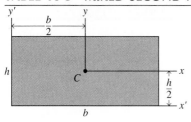

$$I_{xy} = 0$$

$$I_{x'y'} = \frac{b^2 h^2}{4}$$

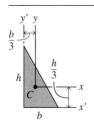

$$I_{xy} = -\frac{b^2 h^2}{72}$$

$$I_{x'y'} = \frac{b^2 h^2}{24}$$

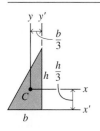

$$I_{xy} = \frac{b^2 h^2}{72}$$

$$I_{x'y'} = -\frac{b^2 h^2}{24}$$

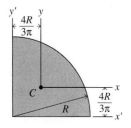

$$I_{xy} = \frac{(9\pi - 32)R^4}{72\pi}$$

$$I_{x'y'} = \frac{R^4}{8}$$

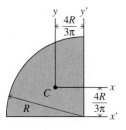

$$I_{xy} = -\frac{(9\pi - 32)R^4}{72\pi}$$

$$I_{x'y'} = -\frac{R^4}{8}$$

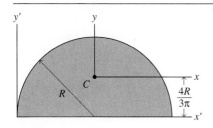

$$I_{xy} = 0$$

$$I_{x'y'} = \frac{2R^4}{3}$$

EXAMPLE PROBLEM 10-8

Determine the mixed second moment (area product of inertia) of the shaded area shown in Fig. 10-16a with respect to the x- and y-axes by using

a. Double integration.
b. The parallel-axis theorem and single integration.

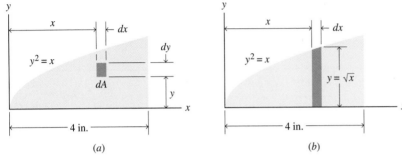

Fig. 10-16

SOLUTION

a. The mixed second moment (area product of inertia) dI_{xy} of the element of area $dA = dy\, dx$ shown in Fig. 10-16a with respect to the x- and y-axes is defined as $dI_{xy} = xy\, dA$. Therefore,

$$I_{xy} = \int_A xy\, dA = \int_0^4 \int_0^{\sqrt{x}} xy\, dy\, dx$$

$$= \int_0^4 \left[\frac{y^2}{2} \right]_0^{\sqrt{x}} x\, dx$$

$$= \int_0^4 \frac{x^2}{2}\, dx = \left[\frac{x^3}{6} \right]_0^4 = 10.67 \text{ in.}^4 \qquad \text{Ans.}$$

b. The mixed second moment (area product of inertia) dI_{xy} of the element of area $dA = y\, dx$ shown in Fig. 10-16b with respect to axes through the centroid of the element parallel to the x- and y-axes is zero. Thus, the mixed second moment of the element with respect to the x- and y-axes is

$$dI_{xy} = dI_{xyC} + \bar{x}\bar{y}\, dA = 0 + x\frac{y}{2}\,(y\, dx) = \frac{x^2}{2}\, dx$$

Therefore,

$$I_{xy} = \int_0^4 dI_{xy} = \int_0^4 \frac{x^2}{2}\, dx = \left[\frac{x^3}{6} \right]_0^4 = 10.67 \text{ in.}^4 \qquad \text{Ans.}$$

Determine the mixed second moment (area product of inertia) of the quadrant of a circle shown in Fig. 10-17 with respect to

a. The x- and y-axes.
b. A pair of axes through the centroid of the area and parallel to the x- and y-axes.

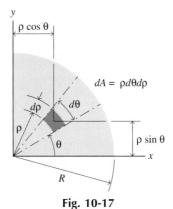

Fig. 10-17

SOLUTION

a. The mixed second moment (area product of inertia) dI_{xy} of the element of area $dA = \rho\, d\theta\, d\rho$ shown in Fig. 10-17 with respect to the x- and y-axes is defined as $dI_{xy} = xy\, dA$. Therefore,

$$dI_{xy} = (\rho \cos \theta)(\rho \sin \theta)(\rho\, d\theta\, d\rho)$$

and

$$I_{xy} = \int_A dI_{xy} = \int_0^R \int_0^{\pi/2} (\rho \cos \theta)(\rho \sin \theta)\, \rho\, d\theta\, d\rho$$

$$= \int_0^R \left[\frac{\sin^2 \theta}{2} \right]_0^{\pi/2} \rho^3\, d\rho$$

$$= \int_0^R \frac{\rho^3}{2}\, d\rho = \left[\frac{\rho^4}{8} \right]_0^R = \frac{R^4}{8} \qquad \text{Ans.}$$

b. Once the mixed second moment is known with respect to a pair of axes, the mixed second moment with respect to a parallel set of axes through the centroid of the area can be found by using the parallel-axis theorem. For the quarter circle, $\bar{x} = \bar{y} = 4R/3\pi$; therefore,

$$I_{xyC} = I_{xy} - \bar{x}\bar{y}\, dA$$

$$= \frac{R^4}{8} - \left(\frac{4R}{3\pi} \right)\left(\frac{4R}{3\pi} \right)\left(\frac{\pi R^2}{4} \right) = \frac{(9\pi - 32)R^4}{72\pi} \qquad \text{Ans.}$$

PROBLEMS

10-41* Determine the mixed second moment of the rectangle shown in Fig. P10-41 with respect to the x- and y-axes.

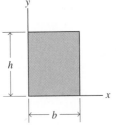

Fig. P10-41

10-42* Determine the mixed second moment of the Z-section shown in Fig. P10-42 with respect to x- and y-axes through the centroid.

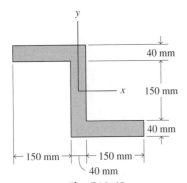

Fig. P10-42

10-43 Determine the mixed second moment of the triangle shown in Fig. P10-43 with respect to

a. The x'- and y'-axes.
b. A pair of xy axes through the centroid of the area and parallel to the x'- and y'-axes.

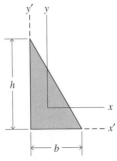

Fig. P10-43

10-44 Determine the mixed second moment of the angle section shown in Fig. P10-44 with respect to

a. The x- and y-axes.
b. A pair of axes through the centroid of the area and parallel to the x- and y-axes.

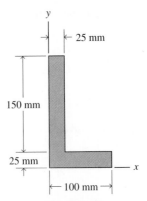

Fig. P10-44

10-45* Determine the mixed second moment of the shaded area shown in Fig. P10-45 with respect to

a. The x- and y-axes.
b. A pair of axes through the centroid of the area and parallel to the x- and y-axes.

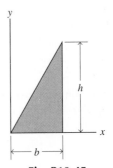

Fig. P10-45

10-46* Determine the mixed second moment of the shaded area shown in Fig. P10-46 with respect to

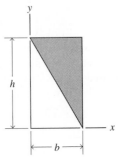

Fig. P10-46

a. The x- and y-axes.
b. A pair of axes through the centroid of the area and parallel to the x- and y-axes.

10-47 Determine the mixed second moment of the shaded area shown in Fig. P10-47 with respect to the x- and y-axes.

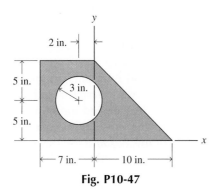

Fig. P10-47

10-48 Determine the mixed second moment of the shaded area shown in Fig. P10-48 with respect to the x- and y-axes.

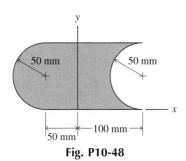

Fig. P10-48

10-49* Determine the mixed second moment of the shaded area shown in Fig. P10-49 with respect to the x- and y-axes.

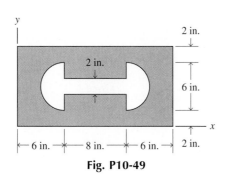

Fig. P10-49

10-50* Determine the mixed second moment of the shaded area shown in Fig. P10-50 with respect to the x- and y-axes.

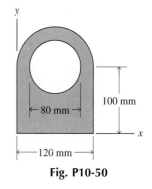

Fig. P10-50

10-51 Determine the mixed second moment of the shaded area shown in Fig. P10-51 with respect to the x- and y-axes.

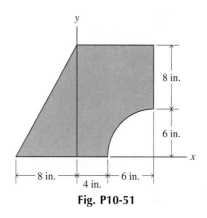

Fig. P10-51

10-52* Determine the mixed second moment of the shaded area shown in Fig. P10-52 with respect to the x- and y-axes.

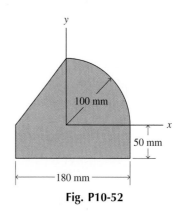

Fig. P10-52

457

10-3   PRINCIPAL SECOND MOMENTS

The second moment of the area A in Fig. 10-18 with respect to the x'-axis through O will, in general, vary with the angle θ. The x- and y-axes used to obtain the polar second moment J_z about a z-axis through O (Eq. 10-2) were any pair of orthogonal axes in the plane of the area passing through O; therefore,

$$J_z = I_x + I_y = I_{x'} + I_{y'}$$

where x' and y' are any pair of orthogonal axes through O. Since the sum of $I_{x'}$ and $I_{y'}$ is a constant, $I_{x'}$ will be maximum and the corresponding $I_{y'}$ will be minimum for one particular value of θ.

The set of axes for which the second moments are maximum and minimum are called the principal axes of the area through point O and are designated as the u- and v-axes. The second moments of the area with respect to these axes are called the principal second moments of the area (principal area moments of inertia) and are designated I_u and I_v. There is only one set of principal axes for any point in an area unless all axes have the same second moment, such as the diameters of a circle. Principal axes are important in mechanics of materials courses, which deal with stresses and deformations in beams and columns.

The principal second moments of an area can be determined by expressing $I_{x'}$ as a function of I_x, I_y, I_{xy}, and θ and setting the derivative of $I_{x'}$ with respect to θ equal to zero to obtain the values of θ that give the maximum and minimum second moments. From Fig. 10-18,

$$dI_{x'} = y'^2\, dA = (y \cos \theta - x \sin \theta)^2 dA$$

Therefore,

$$
\begin{aligned}
I_{x'} &= \int_A dI_{x'} \\
&= \cos^2 \theta \int_A y^2\, dA + \sin^2 \theta \int_A x^2\, dA - 2 \sin \theta \cos \theta \int_A xy\, dA \\
&= I_x \cos^2 \theta + I_y \sin^2 \theta - 2I_{xy} \sin \theta \cos \theta
\end{aligned}
\tag{10-14}
$$

Equation 10-14 can be expressed in terms of the double angle 2θ by using the trigonometric identities

$$\sin 2\theta = 2 \sin \theta \cos \theta$$
$$\cos 2\theta = \cos^2 \theta - \sin^2 \theta$$

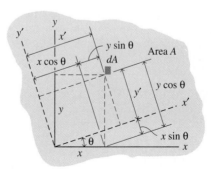

Fig. 10-18

Thus,

$$I_{x'} = \frac{1}{2}(I_x + I_y) + \frac{1}{2}(I_x - I_y)\cos 2\theta - I_{xy}\sin 2\theta \qquad (10\text{-}15)$$

The angle 2θ for which $I_{x'}$ is a maximum (or a minimum) can be obtained by setting the derivative of $I_{x'}$ with respect to θ equal to zero; thus,

$$\frac{dI_{x'}}{d\theta} = -(I_x - I_y)\sin 2\theta - 2I_{xy}\cos 2\theta = 0$$

from which

$$\tan 2\theta_p = -\frac{2I_{xy}}{(I_x - I_y)} \qquad (10\text{-}16)$$

where θ_p represents the two values of θ that locate the principal axes u and v. Equation 10-16 gives two values of $2\theta_p$ that are 180° apart, and thus two values of θ_p that are 90° apart. The principal second moments can be obtained by substituting these values of θ_p in Eq. 10-14. Thus,

$$I_{u,v} = \frac{I_x + I_y}{2} \pm \left[\left(\frac{I_x - I_y}{2}\right)^2 + I_{xy}^2\right]^{1/2} \qquad (10\text{-}17)$$

The mixed second moment (area product of inertia) of the element of area in Fig. 10-18 with respect to the x'- and y'-axes is

$$dI_{x'y'} = x'y'\, dA = (x\cos\theta + y\sin\theta)(y\cos\theta - x\sin\theta)\, dA$$

Therefore,

$$
\begin{aligned}
I_{x'y'} &= \int_A dI_{x'y'} \\
&= \sin\theta\cos\theta \int_A (y^2 - x^2)\, dA + (\cos^2\theta - \sin^2\theta)\int_A xy\, dA \\
&= (I_x - I_y)\sin\theta\cos\theta + I_{xy}(\cos^2\theta - \sin^2\theta) \qquad (10\text{-}18)
\end{aligned}
$$

Or in terms of the double angle 2θ,

$$I_{x'y'} = I_{xy}\cos 2\theta + \frac{1}{2}(I_x - I_y)\sin 2\theta \qquad (10\text{-}19)$$

The mixed second moment $I_{x'y'}$ will be zero when

$$\tan 2\theta = -\frac{2I_{xy}}{(I_x - I_y)} \qquad (a)$$

Equation a is the same as Eq. 10-16. This fact indicates that mixed second moments with respect to principal axes are zero. Since mixed second moments are zero with respect to any axis of symmetry, it follows that any axis of symmetry must be a principal axis for any point on the axis.

The following example problem illustrates the procedure for determining second moments of area with respect to the principal axes.

EXAMPLE PROBLEM 10-10

Determine the maximum and minimum second moments for the triangular area shown in Fig. 10-19a with respect to axes through the centroid of the area.

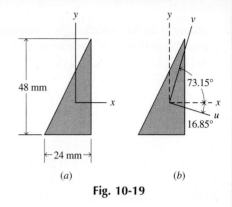

Fig. 10-19

SOLUTION

The second moments I_x, I_y, and I_{xy} can be determined by using the properties listed in Tables 10-1 and 10-4. Thus,

$$I_x = \frac{bh^3}{36} = \frac{24(48)^3}{36} = 73{,}728 \text{ mm}^4$$

$$I_y = \frac{hb^3}{36} = \frac{48(24)^3}{36} = 18{,}432 \text{ mm}^4$$

$$I_{xy} = \frac{b^2h^2}{72} = \frac{(24)^2(48)^2}{72} = 18{,}432 \text{ mm}^4$$

The principal angles θ_p, given by Eq. 10-16, are

$$\theta_p = \frac{1}{2}\tan^{-1}\left[-\frac{2I_{xy}}{I_x - I_y}\right]$$

$$= \frac{1}{2}\tan^{-1}\left[-\frac{2(18432)}{73728 - 18432}\right] = -16.85° \text{ or } 73.15°$$

With $\theta_p = -16.85°$, Eq. 10-14 yields

$$I_{x'} = I_x \cos^2\theta + I_y \sin^2\theta - 2I_{xy}\sin\theta\cos\theta$$
$$= 73{,}728 \cos^2(-16.85°) + 18{,}432 \sin^2(-16.85°)$$
$$- 2(18{,}432)\sin(-16.85°)\cos(-16.85°) = 79{,}309 \text{ mm}^4 = I_{max}$$

With $\theta_p = 73.15°$, Eq. 10-14 yields

$$I_{x'} = I_x \cos^2\theta + I_y \sin^2\theta - 2I_{xy}\sin\theta\cos\theta$$
$$= 73{,}728 \cos^2(73.15°) + 18{,}432 \sin^2(73.15°)$$
$$- 2(18{,}432)\sin(73.15°)\cos(73.15°) = 12{,}851 \text{ mm}^4 = I_{min}$$

Therefore, with respect to the x-axis,

$$I_u = I_{max} = 79{,}300 \text{ mm}^4 \text{ at } \theta_p = -16.85° \qquad \text{Ans.}$$
$$I_v = I_{min} = 12{,}850 \text{ mm}^4 \text{ at } \theta_p = 73.15° \qquad \text{Ans.}$$

The principal second moments can also be determined by using Eq. 10-17:

$$I_{u,v} = \frac{I_x + I_y}{2} \pm \left[\left(\frac{I_x - I_y}{2}\right)^2 + I_{xy}^2\right]^{1/2}$$

$$= \frac{73{,}728 + 18{,}432}{2} \pm \left[\left(\frac{73{,}728 - 18{,}432}{2}\right)^2 + (18{,}432)^2\right]^{1/2}$$

$$= 46{,}080 \pm 33{,}229 = 79{,}309 \text{ mm}^4 \qquad \text{and} \qquad 12{,}851 \text{ mm}^4$$

The orientations of the principal axes are shown in Fig. 10-19b.

PROBLEMS

Problems 10-53–10-56. Determine the maximum and minimum second moments of area with respect to axes through the origin of the *xy*-coordinate system and show the orientations of the principal axes on a sketch.

10-53* For the rectangle shown in Fig. P10-53.

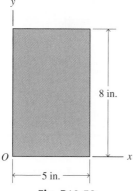

Fig. P10-53

10-54* For the triangle shown in Fig. P10-54.

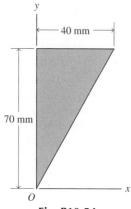

Fig. P10-54

10-55 For the T-section shown in Fig. P10-55.

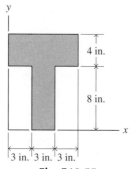

Fig. P10-55

10-56 For the triangle shown in Fig. P10-56.

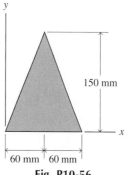

Fig. P10-56

10-57* Determine the maximum and minimum second moments of area for the angle section shown in Fig. P10-57 with respect to axes through the centroid of the area. Show the orientations of the principal axes on a sketch.

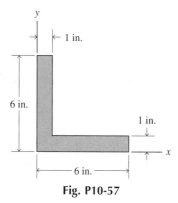

Fig. P10-57

10-58* Determine the maximum and minimum second moments of area for the Z-section shown in Fig. P10-58 with respect to axes through the centroid of the area. Show the orientations of the principal axes on a sketch.

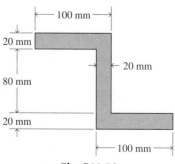

Fig. P10-58

10-59 A composite section consists of a quarter circle and a rectangle with a half circle removed as shown in Fig. P10-59. Determine

a. The second moment $I_{x'}$ for the area.
b. The maximum and minimum second moments for the area with respect to axes through the origin of the xy-coordinate system.

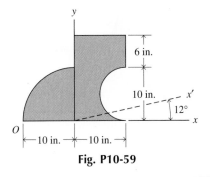

Fig. P10-59

10-60 A composite section consists of a triangle and a rectangle with a circle removed as shown in Fig. P10-60. Determine

a. The second moment $I_{x'}$ for the area.
b. The maximum and minimum second moments for the area with respect to axes through the origin of the xy-coordinate system.

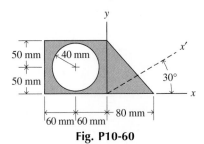

Fig. P10-60

10-3-1   Mohr's Circle for Second Moments of Areas

The equations for locating principal axes and determining maximum and minimum second moments for an area were developed in Section 10-3. The German engineer Otto Mohr (1835–1918) developed a graphic interpretation of these equations that provides a visual picture of the relationships between second moments, mixed moments, and principal moments of an area. This method, commonly called Mohr's circle, involves the construction of a circle in such a manner that the coordinates of each point on the circle represent the rectangular second moment $I_{x'}$ and the mixed second moment $I_{x'y'}$ for one orientation of the x' and y'-axes. The proof that the second moments $I_{x'}$ and $I_{x'y'}$ for a given value of θ can be represented as a point on a circle follows from the general expressions for $I_{x'}$ and $I_{x'y'}$ developed in Section 10-3. Recall Eqs. 10-15 and 10-19 and rewrite in the following form

$$I_{x'} - \frac{I_x + I_y}{2} = \frac{I_x - I_y}{2}\cos 2\theta - I_{xy}\sin 2\theta$$

$$I_{x'y'} = \frac{I_x - I_y}{2}\sin 2\theta + I_{xy}\cos 2\theta$$

Squaring both equations, adding, and simplifying yields

$$\left(I_{x'} - \frac{I_x + I_y}{2}\right)^2 + I_{x'y'}^2 = \left(\frac{I_x - I_y}{2}\right)^2 + I_{xy}^2$$

This is the equation of a circle in terms of the variables $I_{x'}$ and $I_{x'y'}$. The circle is centered on the $I_{x'}$ axis at a distance $(I_x + I_y)/2$ from the $I_{x'y'}$ axis, and the radius of the circle is given by the expression

$$R = \left[\left(\frac{I_x - I_y}{2}\right)^2 + I_{xy}^2\right]^{1/2}$$

Construction of Mohr's circle is initiated by plotting two points A and B as shown in Fig. 10-20a. The coordinates of point A are (I_x, I_{xy}). The coordinates of point B are $(I_y, -I_{xy})$. In this figure, it has been assumed that I_x is greater than I_y and that I_{xy} is positive. Rectangular second moments (area moments of inertia) are plotted along the horizontal axis. Mixed second moments (area products of inertia) are plotted along the vertical axis. Rectangular second moments are always positive; therefore, they are always plotted to the right of the origin. Mixed second moments can be either positive or negative: positive values are plotted above the horizontal axis; negative value are plotted below the horizontal axis. A line connecting A and B is a diameter of Mohr's circle, as shown on Fig. 10-20a. On a physical body, the angle between the x- and y-axes is 90°. On Mohr's circle, points A and B on the circle, which represent quantities associated with the x- and y-axes, are 180° apart. Therefore, all angles on Mohr's circle are double those on the physical body. The diameter intersects the horizontal axis at point C. The distance from the origin O of the coordinate system to the center C of Mohr's circle is

$$OC = \frac{I_x + I_y}{2}$$

The radius R of the circle is

$$R = OA = \left[\left(\frac{I_x - I_y}{2}\right)^2 + I_{xy}^2\right]^{1/2}$$

Maximum and minimum values of $I_{x'}$ occur at points D and E, respectively. At point D:

$$I_{max} = OC + R = \frac{I_x + I_y}{2} + \left[\left(\frac{I_x - I_y}{2}\right)^2 + I_{xy}^2\right]^{1/2} \quad (a)$$

At point E:

$$I_{min} = OC - R = \frac{I_x + I_y}{2} - \left[\left(\frac{I_x - I_y}{2}\right)^2 + I_{xy}^2\right]^{1/2} \quad (b)$$

On Mohr's circle, angle ACD locates I_{max} with respect to I_x. Thus,

$$ACD = \tan^{-1}\frac{I_{xy}}{(I_x - I_y)/2} = \tan^{-1}\frac{2I_{xy}}{(I_x - I_y)} \quad \text{(clockwise)} \quad (c)$$

The rectangular second moment $I_{x'}$ and the mixed second moment $I_{x'y'}$ for an axis passing through a point O in a body and oriented at an angle θ with respect to the x-axis can be obtained by using the Mohr's circle shown in Fig. 10-20b. Thus,

$$\begin{aligned}
I_{x'} &= OC + R\cos(2\theta_p + 2\theta) \\
&= OC + R\cos 2\theta_p \cos 2\theta - R\sin 2\theta_p \sin 2\theta \\
&= \frac{1}{2}(I_x + I_y) + \frac{1}{2}(I_x - I_y)\cos 2\theta - I_{xy}\sin 2\theta \quad (d)
\end{aligned}$$

$$\begin{aligned}
I_{x'y'} &= R\sin(2\theta_p + 2\theta) \\
&= R\cos 2\theta_p \sin 2\theta + R\sin 2\theta_p \cos 2\theta \\
&= \frac{1}{2}(I_x - I_y)\sin 2\theta + I_{xy}\cos 2\theta \quad (e)
\end{aligned}$$

Equations a, b, c, d, and e, obtained using Mohr's circle, are identical to Eqs. 10-15, 10-16, 10-17, and 10-19, obtained by theoretical analysis. Figure 10-21 represents an area and a set of axes for which the data used to construct Fig. 10-20 are valid.

The following example problem illustrates the procedure for locating the principal axes and determining the maximum and minimum second moments of area by using Mohr's circle.

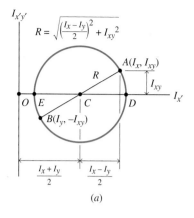

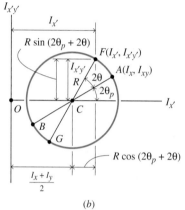

(a)

(b)

Fig. 10-20

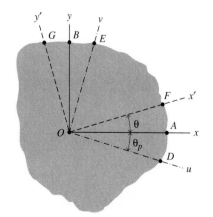

Fig. 10-21

Determine the maximum and minimum second moments for the cross-sectional area of the unequal-leg angle, shown in Fig. 10-22a, with respect to axes through the centroid C of the area.

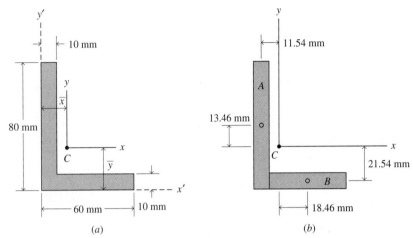

(a)   (b)

Fig. 10-22

SOLUTION

The location of the centroid of the angle with respect to the x'- and y'-axes of Fig. 10-22a are

$$\bar{x} = \frac{60(10)(30) + 70(10)(5)}{60(10) + 70(10)} = 16.54 \text{ mm}$$

$$\bar{y} = \frac{80(10)(40) + 50(10)(5)}{80(10) + 50(10)} = 26.54 \text{ mm}$$

The area is divided into two rectangles A and B. The location of the centroid of each of these areas with respect to the centroid of the composite area is indicated on Fig. 10-22b. The values of I_x, I_y, and I_{xy} for the two areas are obtained by applying the parallel-axis theorem to the expressions for I_{xC}, I_{yC}, and I_{xyC} from Tables 10-1 and 10-3. Thus,

$$I_x = \frac{bh^3}{12} + A(\bar{y})^2$$

$$I_x = I_{Ax} + I_{Bx}$$
$$= \frac{10(80)^3}{12} + 10(80)(13.46)^2 + \frac{50(10)^3}{12} + 50(10)(21.54)^2 = 808(10^3) \text{ mm}^4$$

$$I_y = I_{Ay} + I_{By}$$
$$= \frac{80(10)^3}{12} + 80(10)(11.54)^2 + \frac{10(50)^3}{12} + 10(50)(18.46)^2 = 388(10^3) \text{ mm}^4$$

$$I_{xy} = I_{xyC} + \bar{x}\bar{y} A = 0 + \bar{x}\bar{y} A$$

$$I_{xy} = I_{Axy} + I_{Bxy}$$
$$= -11.54(13.46)(10)(80) + 18.46(-21.54)(50)(10) = -323(10^3) \text{ mm}^4$$

Mohr's circle, shown in Fig. 10-23a, is constructed using these values. Point A (808, −323) is located at I_x and I_{xy} (which is negative) and point B (388, 323) is located at I_y and $-I_{xy}$. A line drawn between points A and B locates point C (the center of the circle). The distance OC is

$$OC = \frac{I_x + I_y}{2} = \frac{(808 + 388)(10^3)}{2} = 598(10^3) \text{ mm}^4$$

The radius of the circle is

$$R = \sqrt{(210)^2 + (323)^2}\,(10^3) = 385(10^3) \text{ mm}^4$$

Thus, as shown in Fig. 10-23b, the principal second moments are

$$I_{max} = OC + R = (598 + 385)(10^3) = 983(10^3) \text{ mm}^4 \qquad \text{Ans.}$$
$$I_{min} = OC - R = (598 - 385)(10^3) = 213(10^3) \text{ mm}^4 \qquad \text{Ans.}$$

Twice the angle from the x-axis to the principal axis with the maximum second moment is shown on Fig. 10-23c as $2\theta_p$. From the information provided on Mohr's circle,

$$\theta_p = \frac{1}{2} \cos^{-1} \frac{210}{385} = 28.5° \quad \text{counterclockwise from the } x\text{-axis}$$

Results of this Mohr circle analysis are shown in Fig. 10-23d. These values were obtained analytically from the geometry of Mohr's circle. They could also have been measured directly from the figure; however, the accuracy of the results obtained by scaling distances from the figure would depend on the scale used and the care employed in constructing the figure.

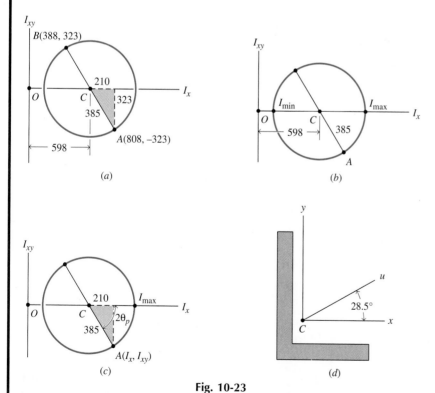

Fig. 10-23

465

PROBLEMS

10-61* Use Mohr's circle to determine the maximum and minimum second moments and the orientations of the principal axes for the composite section shown in Fig. P10-61 with respect to axes through the centroid of the cross section.

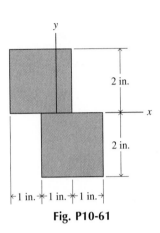

Fig. P10-61

10-62* Use Mohr's circle to determine the maximum and minimum second moments and the orientations of the principal axes for the T-section shown in Fig. P10-62 with respect to axes through the origin of the *xy*-coordinate system.

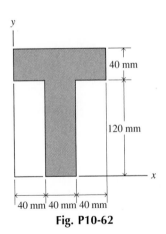

Fig. P10-62

10-63 Use Mohr's circle to determine the maximum and minimum second moments and the orientations of the

principal axes for the triangle shown in Fig. P10-63 with respect to axes through the centroid of the cross section.

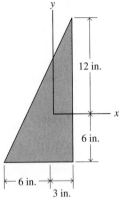

Fig. P10-63

10-64 Use Mohr's circle to determine the maximum and minimum second moments and the orientations of the principal axes for the triangle shown in Fig. P10-64 with respect to axes through the origin of the *xy*-coordinate system.

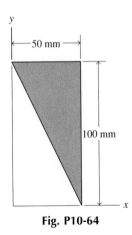

Fig. P10-64

10-65* Use Mohr's circle to determine the maximum and minimum second moments and the orientations of the principal axes for the composite section shown in Fig.

P10-65 with respect to axes through the origin of the xy-coordinate system.

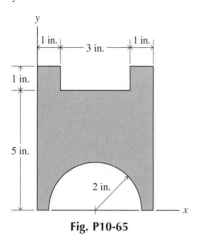

Fig. P10-65

10-66 Use Mohr's circle to determine the maximum and minimum second moments and the orientations of the principal axes for the composite section shown in Fig. P10-66 with respect to axes through the origin of the xy-coordinate system.

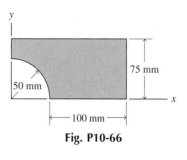

Fig. P10-66

10-4 MOMENTS OF INERTIA

In analyses of the motion of rigid bodies, expressions are often encountered that involve the product of the mass of a small element of the body and the square of its distance from a line of interest. This product is called the second moment of the mass of the element or more frequently the *moment of inertia* of the element. Thus, the moment of inertia dI of an element of mass dm about the axis OO shown in Fig. 10-24 is defined as

$$dI = r^2 \, dm$$

The moment of inertia of the entire body about axis OO is defined as

$$I = \int_m r^2 \, dm \tag{10-20}$$

Fig. 10-24

Since both the mass of the element and the distance squared from the axis to the element are always positive, the moment of inertia of a mass is always a positive quantity.

Moments of inertia have the dimensions of mass multiplied by length squared, ML^2. Common units for the measurement of moment of inertia in the SI system are $\text{kg} \cdot \text{m}^2$. In the U. S. Customary system, force, length, and time are selected as the fundamental quantities and mass has the dimensions FT^2L^{-1}. Therefore, moment of inertia has the units $\text{lb} \cdot \text{s}^2 \cdot \text{ft}$. If the mass of the body W/g is expressed in slugs ($\text{lb} \cdot \text{s}^2/\text{ft}$), the units for measurement of moment of inertia in the U. S. Customary system are $\text{slug} \cdot \text{ft}^2$.

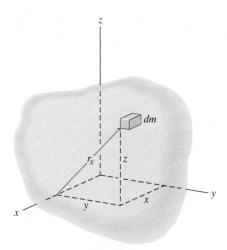

Fig. 10-25

The moments of inertia of a body with respect to an *xyz*-coordinate system can be determined by considering an element of mass as shown in Fig. 10-25. From the definition of moment of inertia,

$$dI_x = r_x^2 \, dm = (y^2 + z^2) \, dm$$

Similar expressions can be written for the *y*- and *z*-axes. Thus,

$$I_x = \int_m r_x^2 \, dm = \int_m (y^2 + z^2) \, dm$$

$$I_y = \int_m r_y^2 \, dm = \int_m (z^2 + x^2) \, dm \qquad (10\text{-}21)$$

$$I_z = \int_m r_z^2 \, dm = \int_m (x^2 + y^2) \, dm$$

10-4-1 Radius of Gyration

The definition of moment of inertia (Eq. 10-20) indicates that the dimensions of moment of inertia are mass multiplied by a length squared. As a result, the moment of inertia of a body can be expressed as the product of the mass *m* of the body and a length *k* squared. This length *k* is defined as the *radius of gyration* of the body. Thus, the moment of inertia *I* of a body with respect to a given line can be expressed as

$$I = mk^2 \qquad \text{or} \qquad k = \sqrt{\frac{I}{m}} \qquad (10\text{-}22)$$

The radius of gyration of the mass of a body with respect to any axis can be viewed as the distance from the axis to the point where the total mass must be concentrated to produce the same moment of inertia with respect to the axis as does the actual (or distributed) mass.

The radius of gyration for masses is very similar to the radius of gyration for areas discussed in Section 10-2-3. The radius of gyration for masses is not the distance from the given axis to any fixed point in the body such as the mass-center. The radius of gyration of the mass of a body with respect to any axis is always greater than the distance from the axis to the mass center of the body. There is no useful physical interpretation for a radius of gyration; it is merely a convenient means of expressing the moment of inertia of the mass of a body in terms of its mass and a length.

10-4-2 Parallel-Axis Theorem for Moments of Inertia

The parallel-axis theorem for moments of inertia is very similar to the parallel-axis theorem for second moments of area discussed in Section 10-2-1. Consider the body shown in Fig. 10-26, which has an xyz-coordinate system with its origin at the mass-center G of the body and a parallel $x'y'z'$-coordinate system with its origin at point O'. Observe in the figure that

$$x' = \bar{x} + x$$
$$y' = \bar{y} + y$$
$$z' = \bar{z} + z$$

The distance d_x between the x'- and x-axes is

$$d_x = \sqrt{\bar{y}^2 + \bar{z}^2}$$

The moment of inertia of the body about an x'-axis that is parallel to the x-axis through the mass-center is by definition

$$I_{x'} = \int_m r_{x'}^2 \, dm = \int_m [(\bar{y} + y)^2 + (\bar{z} + z)^2] \, dm$$

$$= \int_m (y^2 + z^2) \, dm + \bar{y}^2 \int_m dm + 2\bar{y} \int_m y \, dm + \bar{z}^2 \int_m dm + 2\bar{z} \int_m z \, dm$$

However,

$$\int_m (y^2 + z^2) \, dm = I_{xG}$$

and, since the x- and y-axes pass through the mass-center G of the body,

$$\int_m y \, dm = 0 \qquad \int_m z \, dm = 0$$

Therefore,

$$I_{x'} = I_{xG} + (\bar{y}^2 + \bar{z}^2)m = I_{xG} + d_x^2 m$$
$$I_{y'} = I_{yG} + (\bar{z}^2 + \bar{x}^2)m = I_{yG} + d_y^2 m \qquad (10\text{-}23)$$
$$I_{z'} = I_{zG} + (\bar{x}^2 + \bar{y}^2)m = I_{zG} + d_z^2 m$$

Equation 10-23 is the parallel-axis theorem for moments of inertia. The subscript G indicates that the x-axis passes through the mass-center G of the body. Thus, if the moment of inertia of a body with respect to an axis passing through its mass-center is known, the moment of inertia of the body with respect to any parallel axis can be found, without integrating, by use of Eqs. 10-23.

A similar relationship exists between the radii of gyration for the

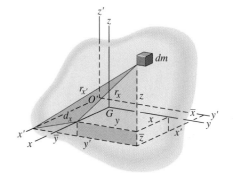

Fig. 10-26

two axes. Thus, if the radii of gyration for the two parallel axes are denoted by k_x and $k_{x'}$, the foregoing equation may be written

$$k_{x'}^2 m = k_{xG}^2 m + d_x^2 m$$

Hence

$$k_{x'}^2 = k_{xG}^2 + d_x^2$$
$$k_{y'}^2 = k_{yG}^2 + d_y^2 \qquad (10\text{-}24)$$
$$k_{z'}^2 = k_{zG}^2 + d_z^2$$

Note: Equations 10-23 and 10-24 are valid only for transfers to or from xyz-axes passing through the mass center of the body. They are not valid for two arbitrary axes.

10-4-3   Moments of Inertia by Integration

When integration methods are used to determine the moment of inertia of a body with respect to an axis, the mass of the body can be divided into elements in various ways. Depending on the way the element is chosen, single, double, or triple integration may be required. The geometry of the body usually determines whether Cartesian or polar coordinates are used. In any case, the elements of mass should always be selected, so that

1. All parts of the element are the same distance from the axis with respect to which the moment of inertia is to be determined.
2. If condition 1 is not satisfied, the element should be selected so that the moment of inertia of the element with respect to the axis about which the moment of inertia of the body is to be found is known. The moment of inertia of the body can then be found by summing the moments of inertia of the elements.
3. If the location of the mass-center of the element is known and the moment of inertia of the element with respect to an axis through its mass-center and parallel to the given axis is known, the moment of inertia of the element can be determined by using the parallel-axis theorem. The moment of inertia of the body can then be found by summing the moments of inertia of the elements.

When triple integration is used, the element always satisfies the first requirement, but this condition is not necessarily satisfied by elements used for single or double integration.

In some instances, a body can be regarded as a system of particles. The moment of inertia of a system of particles with respect to a line of interest is the sum of the moments of inertia of the particles with respect to the given line. Thus, if the masses of the particles of a system are denoted by $m_1, m_2, m_3, \cdots, m_n$, and the distances of the particles from a given line are denoted by $r_1, r_2, r_3, \cdots, r_n$, the moment of inertia of the system can be expressed as

$$I = \sum mr^2 = m_1 r_1^2 + m_2 r_2^2 + m_3 r_3^2 + \cdots + m_n r_n^2$$

Moments of inertia for thin plates are relatively easy to determine. For example, consider the thin plate shown in Fig. 10-27. The plate has a uniform density ρ, a uniform thickness t, and a cross-sectional area A. The moments of inertia about x-, y-, and z-axes are by definition

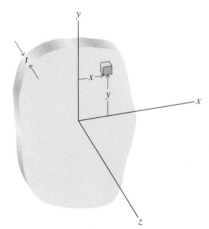

Fig. 10-27

$$I_{xm} = \int_m y^2 \, dm = \int_V y^2 \rho \, dV = \int_A y^2 \rho t \, dA = \rho t \int_A y^2 \, dA = \rho t \, I_{xA}$$

$$I_{ym} = \int_m x^2 \, dm = \int_V x^2 \rho \, dV = \int_A x^2 \rho t \, dA = \rho t \int_A x^2 \, dA = \rho t \, I_{yA} \quad \text{(10-25)}$$

$$I_{zm} = \int_m (x^2 + y^2) \, dm = \rho t \, I_{yA} + \rho t \, I_{xA} = \rho t \, (I_{yA} + I_{xA})$$

where the subscripts m and A denote moments of inertia and second moments of area, respectively. Since the equations for the moments of inertia of thin plates contain the expressions for the second moments of area, the results listed in Table 10-1 for second moments of areas can be used for moments of inertia by simply multiplying the results listed in the table by ρt.

For the general three-dimensional body, moments of inertia with respect to x-, y-, and z-axes are

$$I_x = \int_m r_x^2 \, dm = \int_m (y^2 + z^2) \, dm$$

$$I_y = \int_m r_y^2 \, dm = \int_m (z^2 + x^2) \, dm \quad \text{(10-21)}$$

$$I_z = \int_m r_z^2 \, dm = \int_m (x^2 + y^2) \, dm$$

If the density of the body is uniform, the element of mass dm can be expressed in terms of the element of volume dV of the body as $dm = \rho \, dV$. Equations 10-21 then become

$$I_x = \rho \int_V (y^2 + z^2) \, dV$$

$$I_y = \rho \int_V (z^2 + x^2) \, dV \quad \text{(10-26)}$$

$$I_z = \rho \int_V (x^2 + y^2) \, dV$$

If the density of the body is not uniform, it must be expressed as a function of position and retained within the integral sign.

The specific element of volume to be used depends on the geometry of the body. For the general three-dimensional body, the differential element $dV = dx \, dy \, dz$, which requires a triple integration, is usually used. For bodies of revolution, circular plate elements, which require only a single integration, can be used. For some problems, cylinder elements and polar coordinates are useful. Procedures for determining moments of inertia are illustrated in the following example problems.

Determine the moment of inertia of a homogeneous right circular cylinder with respect to the axis of the cylinder.

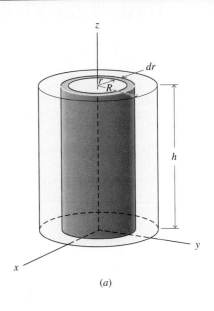

(a)

SOLUTION

The moment of inertia of the cylinder can be determined from the definition of moment of inertia (Eq. 10-20) by selecting a cylindrical tube-type of element as shown in Fig. 10-28a. Thus,

$$dI_{zm} = r^2\,dm = r^2\,(\rho\,dV) = r^2\rho\,(2\pi rh\,dr) = 2\pi\rho hr^3 dr$$

Therefore,

$$I_z = \int_m dI_{zm} = \int_0^R 2\pi\rho hr^3\,dr = \left[\frac{\pi\rho hr^4}{2}\right]_0^R = \frac{1}{2}\pi\rho hR^4$$

Alternatively, a thin circular plate-type of element, such as the one shown in Fig. 10-28b, can be used. The moment of inertia for this type of element is given by Eq. 10-25 as

$$dI_{zm} = \rho t(I_{yA} + I_{xA})$$

Substituting the second moments for a circular area from Table 10-1 yields

$$dI_{zm} = \rho\left(\frac{\pi R^4}{4} + \frac{\pi R^4}{4}\right)dz = \frac{1}{2}\pi\rho R^4\,dz$$

Therefore,

$$I_z = \int_m dI_{zm} = \int_0^h \frac{1}{2}\pi\rho R^4\,dz = \left[\frac{1}{2}\pi\rho R^4\,z\right]_0^h = \frac{1}{2}\pi\rho hR^4$$

The mass of the cylinder is

$$m = \rho V = \rho(\pi R^2 h) = \rho\pi R^2 h$$

Therefore,

$$I_z = \frac{1}{2}(\rho\pi R^2 h)R^2 = \frac{1}{2}m\,R^2 \qquad \text{Ans.}$$

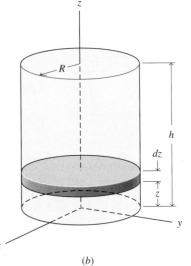

(b)

Fig. 10-28

Determine the moment of inertia for the homogeneous rectangular prism shown in Fig. 10-29a with respect to

a. Axis y through the mass-center of the prism.
b. Axis y' along an edge of the prism.
c. Axis x through the centroid of an end of the prism.

SOLUTION

a. A thin rectangular plate-type of element, such as the one shown in Fig. 10-29b, will be used. The moment of inertia for this type of element is given by Eq. 10-25 as

$$dI_{ym} = \rho t (I_{zA} + I_{xA})$$

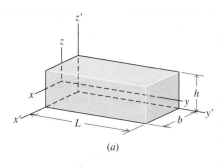

Substituting the second moments for a rectangular area from Table 10-1 yields

$$dI_{ym} = \rho \left(\frac{hb^3}{12} + \frac{bh^3}{12} \right) dy = \rho \frac{bh}{12} (b^2 + h^2) \, dy$$

(a)

Therefore,

$$I_y = \int_m dI_{ym} = \int_0^L \rho \frac{bh}{12} (b^2 + h^2) \, dy$$

$$= \rho \frac{bh}{12} \left[(b^2 + h^2) y \right]_0^L = \frac{\rho bh L}{12} (b^2 + h^2)$$

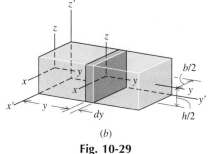

The mass of the prism is

$$m = \rho V = \rho (bhL) = \rho bhL$$

Therefore,

$$I_y = \frac{\rho bhL}{12} (b^2 + h^2) = \frac{1}{12} m (b^2 + h^2) \qquad \text{Ans.}$$

(b)

Fig. 10-29

b. The parallel-axis theorem (Eq. 10-23) can be used to determine the moment of inertia about the y'-axis along an edge of the prism. Thus,

$$I_{y'} = I_{yG} + (\bar{x}^2 + \bar{z}^2) m$$

$$= \frac{1}{12} m (b^2 + h^2) + \left(\frac{b^2}{4} + \frac{h^2}{4} \right) m$$

$$= \frac{1}{3} m (b^2 + h^2) \qquad \text{Ans.}$$

c. The moment of inertia about an x-axis through the mass-center of the thin rectangular plate-type of element shown in Fig. 10-29b is given by Eq. 10-25 as

$$dI_{xm} = \rho t I_{xA}$$

Substituting the second moment for a rectangular area from Table 10-1 yields

$$dI_{xG} = \rho \frac{bh^3}{12} \, dy = \frac{\rho bh^3}{12} \, dy$$

The parallel-axis theorem (Eq. 10-23) with $d_x = y$ then gives the moment of inertia for the thin rectangular plate element about the x-axis shown in Fig. 10-29b as

$$dI_x = dI_{xG} + d_x^2 m = \frac{\rho bh^3}{12} \, dy + y^2 (\rho bh \, dy) = \frac{\rho bh}{12} (h^2 + 12y^2) \, dy$$

$$I_x = \int_m dI_x = \int_0^L \frac{\rho bh}{12} (h^2 + 12y^2) \, dy$$

$$= \frac{\rho bh}{12} \left[h^2 y + 4y^3 \right]_0^L = \frac{\rho bh L}{12} (h^2 + 4L^2)$$

But

$$m = \rho bhL$$

Therefore,

$$I_x = \frac{\rho bh L}{12} (h^2 + 4L^2) = \frac{1}{12} m (h^2 + 4L^2) \qquad \text{Ans.}$$

PROBLEMS

10-67* Determine the moment of inertia of the homogeneous right circular cone, shown in Fig. P10-67, with respect to the axis of the cone.

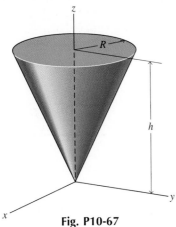

Fig. P10-67

10-68* Determine the moment of inertia of the homogeneous right circular cone, shown in Fig. P10-67, with respect to an axis perpendicular to the axis of the cone at the apex of the cone.

10-69 Determine the moment of inertia of the homogeneous right circular cone, shown in Fig. P10-67, with respect to an axis perpendicular to the axis of the cone at the base of the cone.

10-70 Determine the moment of inertia of a solid homogeneous sphere of radius R with respect to a diameter of the sphere.

10-71* Determine the moment of inertia of a solid homogeneous cylinder of radius R and length L with respect to a diameter in the base of the cylinder.

10-72* Determine the moment of inertia of the solid homogeneous hemisphere shown in Fig. P10-72 with respect to the x-axis shown on the figure.

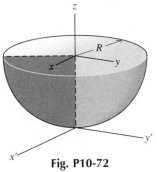

Fig. P10-72

10-73 Determine the moment of inertia of the solid homogeneous hemisphere shown in Fig. P10-72 with respect to the x'-axis shown on the figure.

10-74 Determine the moment of inertia of the solid homogeneous triangular prism shown in Fig. P10-74 with respect to the x-axis shown on the figure.

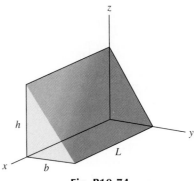

Fig. P10-74

10-75* Determine the moment of inertia of the solid homogeneous triangular prism shown in Fig. P10-74 with respect to a y-axis through the mass center of the prism.

10-76* Determine the moment of inertia of the solid homogeneous tetrahedron shown in Fig. P10-76 with respect to the x-axis shown on the figure.

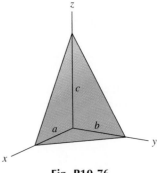

Fig. P10-76

10-77 Determine the moment of inertia of the solid homogeneous tetrahedron shown in Fig. P10-76 with respect to a y-axis through the mass-center of the body.

10-78 A homogeneous solid of revolution is formed by revolving the area shown in Fig. P10-78 around the y-axis. Determine the moment of inertia of the body with respect to the y-axis.

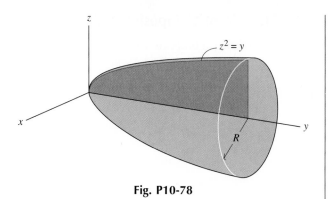

Fig. P10-78

10-79* A homogeneous solid of revolution is formed by revolving the area shown in Fig. P10-78 around the y-axis. Determine the moment of inertia of the body with respect to the x-axis.

10-80* A homogeneous octant of a sphere is formed by rotating the quarter circle shown in Fig. P10-80 for 90° around the z-axis. Determine the moment of inertia of the body with respect to a y-axis through the mass-center of the body.

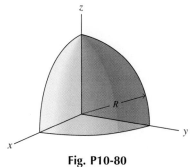

Fig. P10-80

10-81 A homogeneous octant of a cone is formed by rotating the triangle shown in Fig. P10-81 for 90° around the z-axis. Determine the moment of inertia of the body with respect to the y-axis shown on the figure.

10-82 A homogeneous octant of a cone is formed by rotating the triangle shown in Fig. P10-81 for 90° around the z-axis. Determine the moment of inertia of the body with respect to an x-axis through the mass-center of the body.

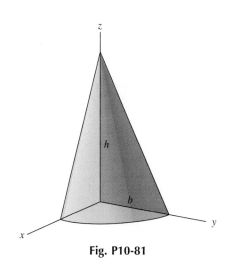

Fig. P10-81

10-4-4 Moment of Inertia of Composite Bodies

Frequently in engineering practice, a body of interest can be broken up into a number of simple shapes, such as cylinders, spheres, plates, and rods, for which the moments of inertia have been evaluated and tabulated. The moment of inertia of the *composite body* with respect to any axis is equal to the sum of the moments of inertia of the separate parts of the body with respect to the specified axis. For example,

TABLE 10-4 MOMENTS OF INERTIA OF COMMON SHAPES

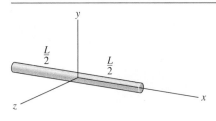

Slender rod

$$I_x = 0$$

$$I_y = I_z = \frac{1}{12}mL^2$$

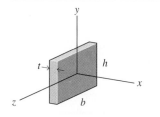

Thin rectangular plate

$$I_x = \frac{1}{12}m(b^2 + h^2)$$

$$I_y = \frac{1}{12}mb^2$$

$$I_z = \frac{1}{12}mh^2$$

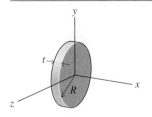

Thin circular plate

$$I_x = \frac{1}{2}mR^2$$

$$I_y = I_z = \frac{1}{4}mR^2$$

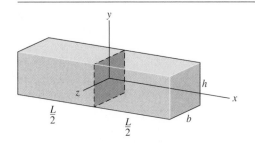

Rectangular prism

$$V = bhL$$

$$I_x = \frac{1}{12}m(b^2 + h^2)$$

$$I_y = \frac{1}{12}m(b^2 + L^2)$$

$$I_z = \frac{1}{12}m(h^2 + L^2)$$

$$I_x = \int_m (y^2 + z^2)\, dm$$

$$= \int_{m_1} (y^2 + z^2)\, dm_1 + \int_{m_2} (y^2 + z^2)\, dm_2 + \cdots + \int_{m_n} (y^2 + z^2)\, dm_n$$

$$= I_{x1} + I_{x2} + I_{x3} + \cdots + I_{xn}$$

When one of the component parts is a hole, its moment of inertia must be subtracted from the moment of inertia of the larger part to obtain the moment of inertia for the composite body. Table 10-4 contains a listing of the moments of inertia for some frequently encountered shapes such as rods, plates, cylinders, spheres, and cones.

TABLE 10-4   (CONTINUED)

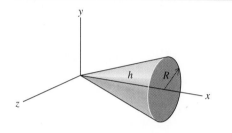

Right circular cone

$$V = \frac{1}{3}\pi R^2 h$$

$$\bar{x} = \frac{3}{4}h$$

$$I_x = \frac{3}{10} mR^2$$

$$I_y = I_z = \frac{3}{20} m(R^2 + 4h^2)$$

$$I_{yG} = I_{zG} = \frac{3}{80} m(4R^2 + h^2)$$

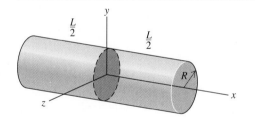

Circular cylinder

$$V = \pi R^2 L$$

$$I_x = \frac{1}{2} mR^2$$

$$I_y = I_z = \frac{1}{12} m(3R^2 + L^2)$$

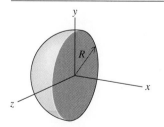

Hemisphere

$$V = \frac{2}{3}\pi R^3$$

$$\bar{x} = \frac{3}{8}R$$

$$I_x = I_y = I_z = \frac{2}{5} mR^2$$

$$I_{yG} = I_{zG} = \frac{83}{320} mR^2$$

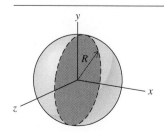

Sphere

$$V = \frac{4}{3}\pi R^3$$

$$I_x = I_y = I_z = \frac{2}{5} mR^2$$

Determine the moment of inertia of the cast-iron flywheel shown in Fig. 10-30 with respect to the axis of rotation of the flywheel. The specific weight of the cast iron is 460 lb/ft^3.

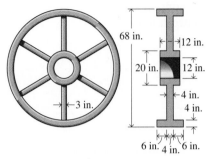

68 in.

20 in.

3 in.

12 in.

12 in.

4 in.

4 in.

6 in. 4 in. 6 in.

Fig. 10-30

SOLUTION

The rim and hub of the flywheel are hollow cylinders and the spokes are rectangular prisms. The density of the cast iron is

$$\rho = \frac{w}{g} = \frac{460}{32.2} = 14.29 \text{ slugs/ft}^3$$

With all dimensions converted to feet, the moment of inertia of the rim is

$$
\begin{aligned}
I_R &= \frac{1}{2} m_o R_o^2 - \frac{1}{2} m_i R_i^2 \\
&= \frac{1}{2} \left[\pi \left(\frac{34}{12} \right)^2 \left(\frac{16}{12} \right) (14.29) \right] \left(\frac{34}{12} \right)^2 - \frac{1}{2} \left[\pi \left(\frac{30}{12} \right)^2 \left(\frac{16}{12} \right) (14.29) \right] \left(\frac{30}{12} \right)^2 \\
&= 1929 - 1169 = 760 \text{ slug} \cdot \text{ft}^2
\end{aligned}
$$

The moment of inertia of the hub is

$$
\begin{aligned}
I_H &= \frac{1}{2} m_o R_o^2 - \frac{1}{2} m_i R_i^2 \\
&= \frac{1}{2} \left[\pi \left(\frac{10}{12} \right)^2 \left(\frac{12}{12} \right) (14.29) \right] \left(\frac{10}{12} \right)^2 - \frac{1}{2} \left[\pi \left(\frac{6}{12} \right)^2 \left(\frac{12}{12} \right) (14.29) \right] \left(\frac{6}{12} \right)^2 \\
&= 10.82 - 1.40 = 9.42 \text{ slug} \cdot \text{ft}^2
\end{aligned}
$$

The moment of inertia of each spoke is

$$
\begin{aligned}
I_S &= I_G + d^2 m \\
&= \frac{1}{12} \left[\frac{3}{12} \left(\frac{4}{12} \right) \left(\frac{20}{12} \right) (14.29) \right] \left[\left(\frac{3}{12} \right)^2 + \left(\frac{20}{12} \right)^2 \right] \\
&\quad + \left(\frac{20}{12} \right)^2 \left[\frac{3}{12} \left(\frac{4}{12} \right) \left(\frac{20}{12} \right) (14.29) \right] \\
&= 0.4698 + 5.5131 = 5.9829 = 5.98 \text{ slug} \cdot \text{ft}^2
\end{aligned}
$$

The total moment of inertia for the flywheel is

$$
\begin{aligned}
I &= I_R + I_H + 6 I_S \\
&= 760 + 9.42 + 6(5.98) = 805 \text{ slug} \cdot \text{ft}^2 \qquad \text{Ans.}
\end{aligned}
$$

PROBLEMS

10-83* A composite body is constructed by attaching a steel ($w = 490$ lb/ft^3) hemisphere to an aluminum ($w = 175$ lb/ft^3) right circular cone as shown in Fig. P10-83. Determine the moment of inertia of the composite body with respect to the y-axis shown on the figure.

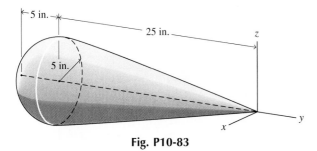

Fig. P10-83

10-84* A composite body consists of a rectangular brass ($\rho = 8.75$ Mg/m^3) block attached to a steel ($\rho = 7.87$ Mg/m^3) cylinder as shown in Fig. P10-84. Determine the moment of inertia of the composite body with respect to the y-axis shown on the figure.

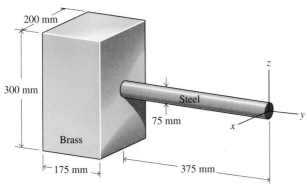

Fig. P10-84

10-85 A composite body is constructed by attaching a steel ($w = 490$ lb/ft^3) hemisphere to an aluminum ($w = 175$ lb/ft^3) right circular cone as shown in Fig. P10-83. Determine the moment of inertia of the composite body with respect to the x-axis shown on the figure.

10-86 A composite body consists of a cylinder attached to a rectangular block as shown in Fig. P10-84. Determine the moment of inertia of the composite body with respect to the x-axis shown on the figure if the body is made of cast iron ($\rho = 7.37$ Mg/m^3).

10-87* Two steel ($w = 490$ lb/ft^3) cylinders and a brass ($w = 546$ lb/ft^3) sphere form the composite body shown in Fig. P10-87. Determine the moment of inertia of the composite body with respect to the x-axis shown on the figure.

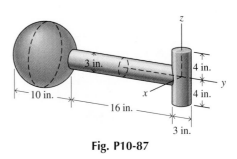

Fig. P10-87

10-88* Determine the moment of inertia of the composite body shown in Fig. P10-88 with respect to the x-axis shown on the figure. The density of the material is 7.87 Mg/m^3.

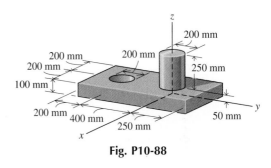

Fig. P10-88

10-89 Two brass ($w = 546$ lb/ft^3) cylinders and a bronze ($w = 553$ lb/ft^3) sphere form the composite body shown in Fig. P10-87. Determine the moment of inertia of the composite body with respect to the y-axis shown on the figure.

10-90 Determine the moment of inertia of the composite body shown in Fig. P10-88 with respect to the y-axis shown on the figure. The density of the material is 2.80 Mg/m^3.

10-91 Two steel ($w = 490$ lb/ft^3) cylinders and an aluminum ($w = 173$ lb/ft^3) sphere form the composite body shown in Fig. P10-87. Determine the moment of inertia of the composite body with respect to the z-axis shown on the figure.

10-92 Determine the moment of inertia of the composite body shown in Fig. P10-88 with respect to the z-axis shown on the figure. The density of the material is 7.87 Mg/m^3.

10-93* Determine the moment of inertia of the composite body shown in Fig. P10-93 with respect to the *y*-axis shown on the figure. The specific weight of the material is 175 lb/ft^3.

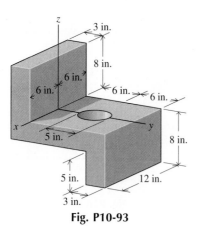

Fig. P10-93

10-94* Determine the moment of inertia of the composite body shown in Fig. P10-94 with respect to the *y*-axis shown on the figure. The density of the material is 7.37 Mg/m^3.

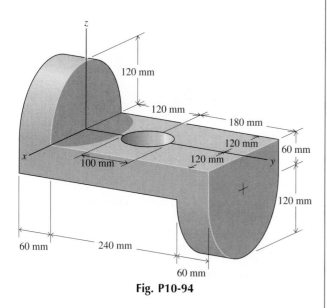

Fig. P10-94

10-95 Determine the moment of inertia of the composite body shown in Fig. P10-93 with respect to the *x*-axis shown on the figure. The specific weight of the material is 546 lb/ft^3.

10-96 Determine the moment of inertia of the composite body shown in Fig. P10-94 with respect to the *x*-axis shown on the figure. The density of the material is 2.77 Mg/m^3.

10-97* Determine the moment of inertia of the composite body shown in Fig. P10-97 with respect to the *y*-axis shown on the figure. The specific weight of the material is 553 lb/ft^3.

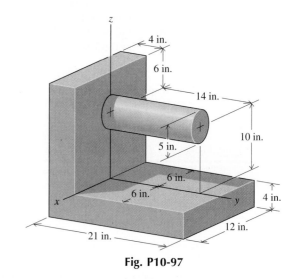

Fig. P10-97

10-98* Determine the moment of inertia of the composite body shown in Fig. P10-98 with respect to the *y*-axis shown on the figure. The density of the material is 7.87 Mg/m^3.

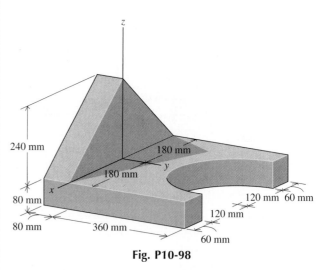

Fig. P10-98

10-99 Determine the moment of inertia of the composite body shown in Fig. P10-97 with respect to the *x*-axis shown on the figure. The specific weight of the material is 490 lb/ft^3.

10-100 Determine the moment of inertia of the composite body shown in Fig. P10-98 with respect to the *x*-axis shown on the figure. The density of the material is 8.75 Mg/m^3.

10-4-5 Product of Inertia

In analyses of the motion of rigid bodies, expressions are sometimes encountered that involve the product of the mass of a small element and the coordinate distances from a pair of orthogonal coordinate planes. This product, which is similar to the mixed second moment of an area, is called the *product of inertia* of the element. For example, the product of inertia of the element shown in Fig. 10-31 with respect to the xz- and yz-planes is by definition

$$dI_{xy} = xy \, dm \qquad (10\text{-}27)$$

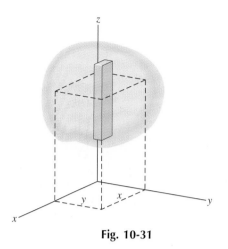

Fig. 10-31

The sum of the products of inertia of all elements of mass of the body with respect to the same orthogonal planes is defined as the product of inertia of the body. The three products of inertia for the body shown in Fig. 10-31 are

$$I_{xy} = \int_m xy \, dm$$

$$I_{yz} = \int_m yz \, dm \qquad (10\text{-}28)$$

$$I_{zx} = \int_m zx \, dm$$

Products of inertia, like moments of inertia, have the dimensions of mass multiplied by a length squared, ML^2. Common units for the measurement of product of inertia in the SI system are $kg \cdot m^2$. In the U. S. Customary system, common units are $slug \cdot ft^2$.

The product of inertia of a body can be positive, negative, or zero, since the two coordinate distances have independent signs. The product of inertia will be positive for coordinates with the same sign and negative for coordinates with opposite signs. The product of inertia will be zero if either of the planes is a plane of symmetry, since pairs of elements on opposite sides of the plane of symmetry will have positive and negative products of inertia that will add to zero in the summation process.

The integration methods used to determine moments of inertia apply equally well to products of inertia. Depending on the way the element is chosen, single, double, or triple integration may be required. Moments of inertia for thin plates were related to second moments of area for the same plate. Likewise, products of inertia can be related to the mixed second moments for the plates. If the plate has a uniform density ρ, a uniform thickness t, and a cross-sectional area A, the products of inertia are by definition

$$I_{xym} = \int_m xy \, dm = \int_V xy \, \rho \, dV = \int_A xy \, \rho t \, dA = \rho t \int_A xy \, dA = \rho t \, I_{xyA}$$

$$I_{yzm} = \int_m yz \, dm = 0 \qquad (10\text{-}29)$$

$$I_{zxm} = \int_m zx \, dm = 0$$

where the subscripts m and A denote products of inertia of mass and mixed second moments of area, respectively. The products of inertia I_{yzm} and I_{zxm} for a thin plate are zero since the x- and y-axes are assumed to lie in the midplane of the plate (plane of symmetry).

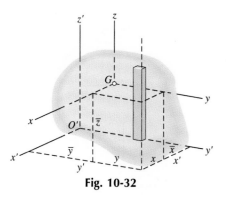

Fig. 10-32

A parallel-axis theorem for products of inertia can be developed that is very similar to the parallel-axis theorem for mixed second moments of area discussed in Section 10-2-5. Consider the body shown in Fig. 10-32, which has an xyz-coordinate system with its origin at the mass-center G of the body and a parallel $x'y'z'$-coordinate system with its origin at point O'. Observe in the figure that

$$x' = \bar{x} + x$$
$$y' = \bar{y} + y$$
$$z' = \bar{z} + z$$

The product of inertia $I_{x'y'}$ of the body with respect to the $x'z'$- and $y'z'$-planes is by definition

$$I_{x'y'} = \int_m x'y' \, dm = \int_m (\bar{x} + x)(\bar{y} + y) \, dm$$
$$= \int_m \bar{x}\bar{y} \, dm + \int_m \bar{x}y \, dm + \int_m \bar{y}x \, dm + \int_m xy \, dm$$

Since $\bar{x}$ and $\bar{y}$ are the same for every element of mass dm,

$$I_{x'y'} = \bar{x}\bar{y}\int_m dm + \bar{x}\int_m y \, dm + \bar{y}\int_m x \, dm + \int_m xy \, dm$$

However,

$$\int_m xy \, dm = I_{xy}$$

and, since the x- and y-axes pass through the mass-center G of the body,

$$\int_m y \, dm = 0 \qquad \int_m z \, dm = 0$$

Therefore,

$$I_{x'y'} = I_{xyG} + \bar{x}\bar{y} \, m$$
$$I_{y'z'} = I_{yzG} + \bar{y}\bar{z} \, m \qquad\qquad (10\text{-}30)$$
$$I_{z'x'} = I_{zxG} + \bar{z}\bar{x} \, m$$

Equations 10-30 are the parallel-axis theorem for products of inertia. The subscript G indicates that the x- and y-axes pass through the mass-center G of the body. Thus, if the product of inertia of a body with respect to a pair of orthogonal planes that pass through its mass-center is known, the product of inertia of the body with respect to any other pair of parallel planes can be found, without integrating, by use of Eqs. 10-30.

Procedures for determining products of inertia are illustrated in the following example problems.

EXAMPLE PROBLEM 10-15

Determine the product of inertia I_{xy} for the homogeneous quarter cylinder shown in Fig. 10-33a.

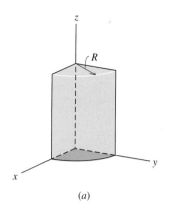

(a)

SOLUTION

All parts of the element of mass dm, shown in Fig. 10-33b, are located at the same distances x and y from the xz- and yz-planes; therefore, the product of inertia dI_{xy} for the element is by definition

$$dI_{xy} = xy \, dm$$

Summing the elements for the entire body yields

$$I_{xy} = \int_m dI_{xy} = \int_m xy \, dm = \int_V xy \, \rho dV$$

$$= \int_0^R \int_0^{\sqrt{R^2-x^2}} \rho xy \, (h \, dy \, dx)$$

$$= \int_0^R \rho hx \left[\frac{y^2}{2}\right]_0^{\sqrt{R^2-x^2}} dx$$

$$= \int_0^R \frac{1}{2} \rho h(R^2 x - x^3) \, dx$$

$$= \frac{1}{2} \rho h \left[\frac{R^2 x^2}{2} - \frac{x^4}{4}\right]_0^R = \frac{1}{2} \rho h \left(\frac{R^4}{4}\right) = \frac{1}{8} \rho R^4 h$$

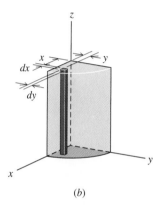

(b)

Alternatively, the thin-plate element shown in Fig. 10-33c could be used to determine I_{xy}. From Eq. 10-29 and the data from Table 10-3,

$$dI_{xym} = \rho t \, dI_{xyA} = \frac{1}{8} \rho R^4 \, dz$$

Therefore,

$$I_{xym} = \rho t \int_A dI_{xyA} = \int_0^h \frac{1}{8} \rho R^4 \, dz = \frac{1}{8} \rho R^4 h$$

Since the mass of the body is

$$m = \rho V = \rho \left(\frac{1}{4} \pi R^2 h\right) = \frac{1}{4} \rho \pi R^2 h$$

the product of inertia I_{xy} can be written as

$$I_{xy} = \frac{1}{2\pi} \left(\frac{1}{4} \rho \pi R^2 h\right) R^2 = \frac{1}{2\pi} mR^2 \qquad \text{Ans.}$$

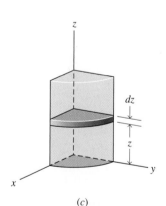

(c)

Fig. 10-33

Determine the products of inertia I_{xy}, I_{yz}, and I_{zx} for the homogeneous flat-plate steel ($\rho = 7870$ kg/m^3) washer shown in Fig. 10-34. The hole is located at the center of the plate.

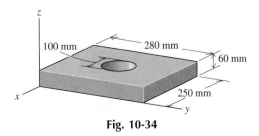

Fig. 10-34

SOLUTION

The products of inertia are zero for the planes of symmetry through the mass-centers of the plate and hole. Since the xy-, yz-, and zx-planes shown in Fig. 10-34 are parallel to these planes of symmetry, the parallel-axis theorem for products of inertia (Eqs. 10-30) can be used to determine the required products of inertia. The masses of the plate, hole, and washer are

$$m_P = \rho V = \rho bht = 7870(0.280)(0.250)(0.060) = 33.05 \text{ kg}$$
$$m_H = \rho V = \rho \pi R^2 t = 7870\pi(0.050)^2(0.060) = 3.71 \text{ kg}$$
$$m_W = m_P - m_H = 33.05 - 3.71 = 29.34 \text{ kg}$$

From Eqs. 10-30,

$$I_{xy} = I_{xyG} + \bar{x}\bar{y}\,m$$
$$= 0 + (-0.125)(0.140)(29.34) = -0.513 \text{ kg} \cdot \text{m}^2 \qquad \text{Ans.}$$

$$I_{yz} = I_{yzG} + \bar{y}\bar{z}\,m$$
$$= 0 + (0.140)(0.030)(29.34) = 0.1232 \text{ kg} \cdot \text{m}^2 \qquad \text{Ans.}$$

$$I_{zx} = I_{zxG} + \bar{z}\bar{x}\,m$$
$$= 0 + (0.030)(-0.125)(29.34) = -0.1100 \text{ kg} \cdot \text{m}^2 \qquad \text{Ans.}$$

PROBLEMS

10-101* Determine the product of inertia I_{xy} for the homogeneous rectangular block shown in Fig. P10-101.

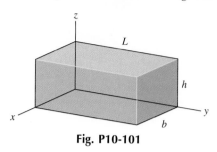

Fig. P10-101

10-102* Determine the product of inertia I_{xy} for the homogeneous octant of a sphere shown in Fig. P10-102.

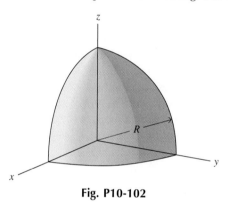

Fig. P10-102

10-103 Determine the products of inertia I_{xy} and I_{yz} for the homogeneous triangular block shown in Fig. P10-103.

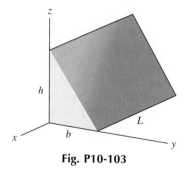

Fig. P10-103

10-104 Determine the products of inertia I_{yz} and I_{zx} for the homogeneous half cylinder shown in Fig. P10-104.

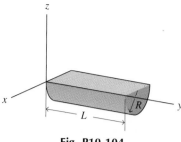

Fig. P10-104

10-105* Determine the products of inertia I_{xy}, I_{yz}, and I_{zx} for the homogeneous steel ($w = 490 \text{ lb/ft}^3$) bracket shown in Fig. P10-105.

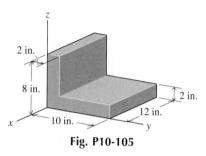

Fig. P10-105

10-106* Determine the products of inertia I_{xy} and I_{zx} for the homogeneous right circular quarter cone shown in Fig. P10-106.

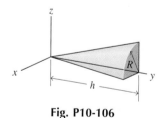

Fig. P10-106

10-107 Determine the products of inertia I_{yz} and I_{zx} for the homogeneous body shown in Fig. P10-107.

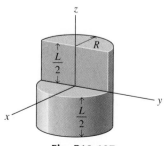

Fig. P10-107

485

10-108 Determine the products of inertia I_{xy}, I_{yz}, and I_{zx} for the homogeneous steel ($\rho = 7870$ kg/m³) block shown in Fig. 10-108.

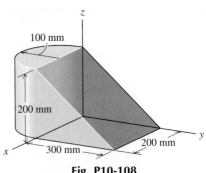

Fig. P10-108

10-5 PRINCIPAL MOMENTS OF INERTIA

In some instances, in the dynamic analysis of bodies, principal axes, and maximum and minimum moments of inertia, which are similar to maximum and minimum second moments of an area, must be determined. Again, the problem is one of transforming known or easily calculated moments and products of inertia with respect to one coordinate system (such as an xyz-coordinate system along the edges of a rectangular prism) to a second $x'y'z'$-coordinate system through the same origin O but inclined with respect to the xyz-system.

For example, consider the body shown in Fig. 10-35, where the x'-axis is oriented at angles $\theta_{x'x}$, $\theta_{x'y}$, and $\theta_{x'z}$ with respect to the x-, y-, and z-axes, respectively. The moment of inertia $I_{x'}$ is by definition

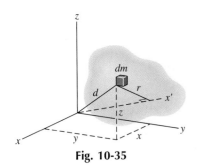

Fig. 10-35

$$I_{x'} = \int_m r^2 \, dm$$

The distance d from the origin of coordinates to the element dm is given by the expression

$$d^2 = x^2 + y^2 + z^2 = x'^2 + y'^2 + z'^2 = x'^2 + r^2$$

Therefore,

$$r^2 = x^2 + y^2 + z^2 - x'^2$$

and since

$$x' = x \cos \theta_{x'x} + y \cos \theta_{x'y} + z \cos \theta_{x'z}$$
$$r^2 = x^2 + y^2 + z^2 - (x \cos \theta_{x'x} + y \cos \theta_{x'y} + z \cos \theta_{x'z})^2$$

Recall that

$$\cos^2 \theta_{x'x} + \cos^2 \theta_{x'y} + \cos^2 \theta_{x'z} = 1$$

Therefore,

$$r^2 = (x^2 + y^2 + z^2)(\cos^2 \theta_{x'x} + \cos^2 \theta_{x'y} + \cos^2 \theta_{x'z})$$
$$- (x \cos \theta_{x'x} + y \cos \theta_{x'y} + z \cos \theta_{x'z})^2$$

which reduces to

$$r^2 = (y^2 + z^2) \cos^2 \theta_{x'x} + (z^2 + x^2) \cos^2 \theta_{x'y} + (x^2 + y^2) \cos^2 \theta_{x'z}$$
$$- 2xy \cos \theta_{x'x} \cos \theta_{x'y} - 2yz \cos \theta_{x'y} \cos \theta_{x'z} - 2zx \cos \theta_{x'z} \cos \theta_{x'x}$$

Therefore,

$$I_{x'} = \int_m r^2 \, dm = \cos^2 \theta_{x'x} \int_m (y^2 + z^2) \, dm + \cos^2 \theta_{x'y} \int_m (z^2 + x^2) \, dm$$

$$+ \cos^2 \theta_{x'z} \int_m (x^2 + y^2) \, dm - \cos \theta_{x'x} \cos \theta_{x'y} \int_m 2xy \, dm$$

$$- \cos \theta_{x'y} \cos \theta_{x'z} \int_m 2yz \, dm - \cos \theta_{x'z} \cos \theta_{x'x} \int_m 2zx \, dm$$

From Eqs. 10-21 and 10-28,

$$I_x = \int_m (y^2 + z^2) \, dm \qquad I_{xy} = \int_m xy \, dm$$

$$I_y = \int_m (z^2 + x^2) \, dm \qquad I_{yz} = \int_m yz \, dm$$

$$I_z = \int_m (x^2 + y^2) \, dm \qquad I_{zx} = \int_m zx \, dm$$

Therefore,

$$I_{x'} = I_x \cos^2 \theta_{x'x} + I_y \cos^2 \theta_{x'y} + I_z \cos^2 \theta_{x'z} - 2I_{xy} \cos \theta_{x'x} \cos \theta_{x'y}$$
$$- 2I_{yz} \cos \theta_{x'y} \cos \theta_{x'z} - 2I_{zx} \cos \theta_{x'z} \cos \theta_{x'x} \qquad (10\text{-}31a)$$

In a similar fashion the product of inertia

$$I_{x'y'} = \int_m x'y' \, dm$$

can be expressed in terms of I_x, I_y, I_z, I_{xy}, I_{yz}, and I_{zx} as

$$I_{x'y'} = -I_x \cos \theta_{x'x} \cos \theta_{y'x} - I_y \cos \theta_{x'y} \cos \theta_{y'y} - I_z \cos \theta_{x'z} \cos \theta_{y'z}$$
$$+ I_{xy}(\cos \theta_{x'x} \cos \theta_{y'y} + \cos \theta_{x'y} \cos \theta_{y'x})$$
$$+ I_{yz}(\cos \theta_{x'y} \cos \theta_{y'z} + \cos \theta_{x'z} \cos \theta_{y'y})$$
$$+ I_{zx}(\cos \theta_{x'z} \cos \theta_{y'x} + \cos \theta_{x'x} \cos \theta_{y'z}) \qquad (10\text{-}31b)$$

If the original xyz-axes are principal axes (such as those shown for the figures in Table 10-4),

$$I_{xy} = I_{yz} = I_{zx} = 0$$

and Eqs. 10-31 reduce to

$$I_{x'} = I_x \cos^2 \theta_{x'x} + I_y \cos^2 \theta_{x'y} + I_z \cos^2 \theta_{x'z} \qquad (10\text{-}32a)$$

and

$$I_{x'y'} = -I_x \cos \theta_{x'x} \cos \theta_{y'x} - I_y \cos \theta_{x'y} \cos \theta_{y'y}$$
$$- I_z \cos \theta_{x'z} \cos \theta_{y'z} \qquad (10\text{-}32b)$$

Equation 10-31a for moments of inertia is the three-dimensional equivalent of Eq. 10-14 for second moments of area. By using a similar but much more complicated procedure than the one used with Eq. 10-14 to locate principal axes and determine maximum and minimum second moments of area, principal axes can be located and maximum and minimum moments of inertia can be determined. The procedure yields the following equations:

$$(I_x - I_P) \cos \theta_{Px} - I_{xy} \cos \theta_{Py} - I_{zx} \cos \theta_{Pz} = 0$$
$$(I_y - I_P) \cos \theta_{Py} - I_{yz} \cos \theta_{Pz} - I_{xy} \cos \theta_{Px} = 0 \qquad (10\text{-}33)$$
$$(I_z - I_P) \cos \theta_{Pz} - I_{zx} \cos \theta_{Px} - I_{yz} \cos \theta_{Py} = 0$$

This set of equations has a nontrivial solution only if the determinant of the coefficients of the direction cosines is equal to zero. Expansion of the determinant yields the following cubic equation for determining the principal moments of inertia of the body for the particular origin of coordinates being used:

$$I_P^3 - (I_x + I_y + I_z)I_P^2 + (I_x I_y + I_y I_z + I_z I_x - I_{xy}^2 - I_{yz}^2 - I_{zx}^2)I_P$$
$$- (I_x I_y I_z - I_x I_{yz}^2 - I_y I_{zx}^2 - I_z I_{xy}^2 - 2I_{xy}I_{yz}I_{zx}) = 0 \qquad \text{(10-34)}$$

Equation 10-34 yields three values I_1, I_2, and I_3 for the principal moments of inertia. One value is the maximum moment of inertia of the body for the origin of coordinates being used, a second value is the minimum moment of inertia of the body for the origin of coordinates being used, and the third value is an intermediate value of the moment of inertia of the body that has no particular significance.

The direction cosines for the principal inertia axes can be obtained by substituting the three values I_1, I_2, and I_3 obtained from Eq. 10-34, in turn, into Eqs. 10-33 and using the additional relation

$$\cos^2 \theta_{Px} + \cos^2 \theta_{Py} + \cos^2 \theta_{Pz} = 1$$

Equations 10-33 and 10-34 are valid for bodies of any shape. The procedure for locating principal axes and determining maximum and minimum moments of inertia is illustrated in the following example problem.

Locate the principal axes and determine the maximum and minimum moments of inertia for the rectangular steel ($w = 490$ lb/ft^3) block shown in Fig. 10-36.

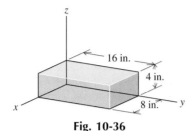

Fig. 10-36

SOLUTION

The moments and products of inertia for the block are given by Eqs. 10-23 and 10-30 as

$$I_x = I_{xG} + (\overline{y}^2 + \overline{z}^2)m \qquad I_{xy} = I_{xyG} + \overline{x}\,\overline{y}\, m$$
$$I_y = I_{yG} + (\overline{z}^2 + \overline{x}^2)m \qquad I_{yz} = I_{yzG} + \overline{y}\,\overline{z}\, m$$
$$I_z = I_{zG} + (\overline{x}^2 + \overline{y}^2)m \qquad I_{zx} = I_{zxG} + \overline{z}\,\overline{x}\, m$$

The mass of the block is

$$m = \rho V = \frac{w}{g}bhL = \frac{490}{32.2}\left(\frac{8}{12}\right)\left(\frac{4}{12}\right)\left(\frac{16}{12}\right) = 4.509 \text{ slugs}$$

Thus, from the results listed in Table 10-4,

$$I_x = \frac{1}{12}m(b^2 + h^2) + \left[\left(\frac{h}{2}\right)^2 + \left(\frac{b}{2}\right)^2\right]m$$
$$= \frac{1}{3}m(b^2 + h^2) = \frac{1}{3}(4.509)\left[\left(\frac{8}{12}\right)^2 + \left(\frac{4}{12}\right)^2\right] = 0.835 \text{ slug} \cdot \text{ft}^2$$

$$I_y = \frac{1}{12}m(b^2 + L^2) + \left[\left(\frac{b}{2}\right)^2 + \left(\frac{L}{2}\right)^2\right]m$$
$$= \frac{1}{3}m(b^2 + L^2) = \frac{1}{3}(4.509)\left[\left(\frac{8}{12}\right)^2 + \left(\frac{16}{12}\right)^2\right] = 3.340 \text{ slug} \cdot \text{ft}^2$$

$$I_z = \frac{1}{12}m(h^2 + L^2) + \left[\left(\frac{h}{2}\right)^2 + \left(\frac{L}{2}\right)^2\right]m$$
$$= \frac{1}{3}m(h^2 + L^2) = \frac{1}{3}(4.509)\left[\left(\frac{4}{12}\right)^2 + \left(\frac{16}{12}\right)^2\right] = 2.839 \text{ slug} \cdot \text{ft}^2$$

$$I_{xy} = 0 + \left(\frac{L}{2}\right)\left(\frac{h}{2}\right)m = \frac{1}{4}mLh = \frac{1}{4}(4.509)\left(\frac{16}{12}\right)\left(\frac{4}{12}\right) = 0.501 \text{ slug} \cdot \text{ft}^2$$

$$I_{yz} = 0 + \left(\frac{h}{2}\right)\left(\frac{b}{2}\right)m = \frac{1}{4}mhb = \frac{1}{4}(4.509)\left(\frac{4}{12}\right)\left(\frac{8}{12}\right) = 0.251 \text{ slug} \cdot \text{ft}^2$$

$$I_{zx} = 0 + \left(\frac{b}{2}\right)\left(\frac{L}{2}\right)m = \frac{1}{4}mbL = \frac{1}{4}(4.509)\left(\frac{8}{12}\right)\left(\frac{16}{12}\right) = 1.002 \text{ slug} \cdot \text{ft}^2$$

Once the moments and products of inertia have been determined, the principal moments of inertia can be determined by using Eq. 10-34. Thus,

$$I_P^3 - (I_x + I_y + I_z)I_P^2 + (I_xI_y + I_yI_z + I_zI_x - I_{xy}^2 - I_{yz}^2 - I_{zx}^2)I_P$$
$$- (I_xI_yI_z - I_xI_{yz}^2 - I_yI_{zx}^2 - I_zI_{xy}^2 - 2I_{xy}I_{yz}I_{zx}) = 0$$

Substituting values for the moments and products of inertia gives

$$I_P^3 - 7.014I_P^2 + 13.324I_P - 3.548 = 0$$

which has the solution

$$I_1 = I_{max} = 3.451 \text{ slug} \cdot \text{ft}^2 \qquad \text{Ans.}$$
$$I_2 = I_{int} = 3.246 \text{ slug} \cdot \text{ft}^2$$
$$I_3 = I_{min} = 0.317 \text{ slug} \cdot \text{ft}^2 \qquad \text{Ans.}$$

The principal directions are obtained by substituting the principal moments of inertia, in turn, into Eqs. 10-33. With $I_P = I_1 = I_{max} = 3.451 \text{ slug} \cdot \text{ft}^2$:

$$(I_x - I_P)\cos\theta_{Px} - I_{xy}\cos\theta_{Py} - I_{zx}\cos\theta_{Pz} = 0$$
$$(I_y - I_P)\cos\theta_{Py} - I_{yz}\cos\theta_{Pz} - I_{xy}\cos\theta_{Px} = 0$$
$$(I_z - I_P)\cos\theta_{Pz} - I_{zx}\cos\theta_{Px} - I_{yz}\cos\theta_{Py} = 0$$

$$-2.616\cos\theta_{1x} - 0.501\cos\theta_{1y} - 1.002\cos\theta_{1z} = 0$$
$$-0.501\cos\theta_{1x} - 0.111\cos\theta_{1y} - 0.251\cos\theta_{1z} = 0 \qquad (a)$$
$$-1.002\cos\theta_{1x} - 0.251\cos\theta_{1y} - 0.612\cos\theta_{1z} = 0$$

Equations a together with the required relationship for direction cosines

$$\cos^2\theta_{1x} + \cos^2\theta_{1y} + \cos^2\theta_{1z} = 1$$

have the solution

$$\cos\theta_{1x} = 0.0891 \quad \text{or} \quad \theta_{1x} = 84.9°$$
$$\cos\theta_{1y} = -0.9643 \qquad\qquad \theta_{1y} = 164.6° \qquad \text{Ans.}$$
$$\cos\theta_{1z} = 0.2495 \qquad\qquad \theta_{1z} = 75.6°$$

With $I_P = I_2 = I_{int} = 3.246 \text{ slug} \cdot \text{ft}^2$:

$$-2.411\cos\theta_{2x} - 0.501\cos\theta_{2y} - 1.002\cos\theta_{2z} = 0$$
$$-0.501\cos\theta_{2x} + 0.094\cos\theta_{2y} - 0.251\cos\theta_{2z} = 0$$
$$-1.002\cos\theta_{2x} - 0.251\cos\theta_{2y} - 0.407\cos\theta_{2z} = 0$$
$$\cos^2\theta_{2x} + \cos^2\theta_{2y} + \cos^2\theta_{2z} = 1$$

the solution becomes

$$\cos\theta_{2x} = -0.4105 \quad \text{or} \quad \theta_{2x} = 114.2°$$
$$\cos\theta_{2y} = 0.1925 \qquad\qquad \theta_{2y} = 78.9° \qquad \text{Ans.}$$
$$\cos\theta_{2z} = 0.8913 \qquad\qquad \theta_{2z} = 27.0°$$

With $I_P = I_3 = I_{min} = 0.317 \text{ slug} \cdot \text{ft}^2$:

$$0.518\cos\theta_{3x} - 0.501\cos\theta_{3y} - 1.002\cos\theta_{3z} = 0$$
$$-0.501\cos\theta_{3x} + 3.023\cos\theta_{3y} - 0.251\cos\theta_{3z} = 0$$
$$-1.002\cos\theta_{3x} - 0.251\cos\theta_{3y} + 2.522\cos\theta_{3z} = 0$$
$$\cos^2\theta_{3x} + \cos^2\theta_{3y} + \cos^2\theta_{3z} = 1$$

the solution becomes

$$\cos\theta_{3x} = 0.9075 \quad \text{or} \quad \theta_{3x} = 24.8°$$
$$\cos\theta_{3y} = 0.1818 \qquad\qquad \theta_{3y} = 79.5° \qquad \text{Ans.}$$
$$\cos\theta_{3z} = 0.3786 \qquad\qquad \theta_{3z} = 67.8°$$

Thus, the unit vectors associated with the three principal directions are

$$\mathbf{n}_1 = 0.0891\mathbf{i} - 0.9643\mathbf{j} + 0.2495\mathbf{k} \qquad \mathbf{n}_1 \cdot \mathbf{n}_2 = 0$$
$$\mathbf{n}_2 = -0.4105\mathbf{i} + 0.1925\mathbf{j} + 0.8913\mathbf{k} \qquad \mathbf{n}_2 \cdot \mathbf{n}_3 = 0$$
$$\mathbf{n}_3 = 0.9075\mathbf{i} + 0.1818\mathbf{j} + 0.3786\mathbf{k} \qquad \mathbf{n}_3 \cdot \mathbf{n}_1 = 0$$

which verifies that the three principal axes are orthogonal.

PROBLEMS

10-109* Locate the principal axes and determine the maximum and minimum moments of inertia for the triangular steel (w = 490 lb/ft^3) block shown in Fig. P10-103 if b = 8 in., h = 6 in., and L = 10 in.

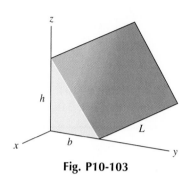

Fig. P10-103

10-110* Locate the principal axes and determine the maximum and minimum moments of inertia for the steel (ρ = 7870 kg/m^3) half cylinder shown in Fig. P10-104 if R = 100 mm and L = 150 mm.

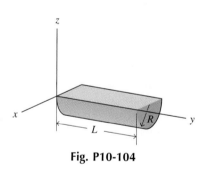

Fig. P10-104

10-111 Locate the principal axes and determine the maximum and minimum moments of inertia for the steel (w = 490 lb/ft^3) angle bracket shown in Fig. P10-105.

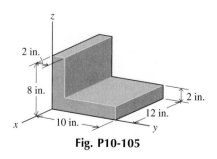

Fig. P10-105

10-112 Locate the principal axes and determine the maximum and minimum moments of inertia for the steel (ρ = 7870 kg/m^3) washer shown in Fig. 10-34.

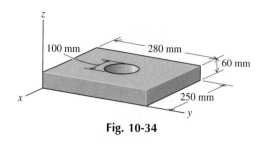

Fig. 10-34

10-113* Locate the principal axes and determine the maximum and minimum moments of inertia for the aluminum (w = 175 lb/ft^3) cylinder shown in Fig. P10-107 if R = 4 in. and L = 6 in.

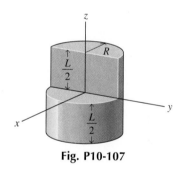

Fig. P10-107

10-114* Locate the principal axes and determine the maximum and minimum moments of inertia for the brass (ρ = 8750 kg/m^3) block shown in Fig. P10-108.

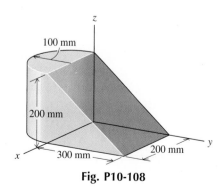

Fig. P10-108

491

10-115 Locate the principal axes and determine the maximum and minimum moments of inertia for the cast iron ($w = 460$ lb/ft^3) tetrahedron shown in Fig. P10-76 if $a = 6$ in., $b = 8$ in., and $c = 10$ in.

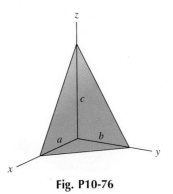

Fig. P10-76

10-116 Locate the principal axes and determine the maximum and minimum moments of inertia for the brass ($\rho = 8750$ kg/m^3) sphere shown in Fig. P10-116. The radius of the sphere is 200 mm. The surface of the sphere is tangent to the three coordinate planes.

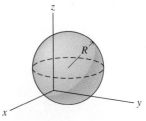

Fig. P10-116

SUMMARY

Previously, the centroid of an area A was located by using the integrals $\int_A y\, dA$ and $\int_A x\, dA$, which are the first moments of the area with respect to the x- and y-axes, respectively. Integrals of the form $\int_A x^2\, dA$ and $\int_A xy\, dA$ yield quantities known as second moments of the area and mixed second moments of the area, respectively.

When the second moment of an area has been determined with respect to a given axis, the second moment with respect to a parallel axis can be obtained by means of the parallel-axis theorem. If one of the axes (say the x-axis) passes through the centroid of the area, the second moment of the area about a parallel x'-axis is given by Eq. 10-4 as

$$I_{x'} = I_{xC} + \bar{y}^2 A$$

Similarly, from Eqs. 10-4, 10-5, and 10-13,

$$I_{y'} = I_{yC} + \bar{x}^2 A \qquad J_{z'} = J_{zC} + (\bar{x}^2 + \bar{y}^2)A \qquad I_{x'y'} = I_{xyC} + \bar{x}\bar{y}\, A$$

The parallel-axis theorem is valid only for transfers to or from centroidal axes.

The second moment of an area A with respect to an x'-axis through the origin O of an xy-coordinate system will, in general, vary with the angle θ between the x- and x'-axes. The set of x'- and y'-axes for which the second moments are maximum and minimum are called principal axes. Second moments of the area with respect to these axes, called the principal second moments, are designated I_u and I_v. In terms of the second moments I_x, I_y, and the mixed second moment I_{xy}, the principal second moments I_u and I_v are

$$I_{u,v} = \frac{I_x + I_y}{2} \pm \left[\left(\frac{I_x - I_y}{2} \right)^2 + I_{xy}^2 \right]^{1/2} \qquad (10\text{-}17)$$

The angle θ_p for which second moments are maximum or minimum are determined from the expression

$$\tan 2\theta_p = -\frac{2I_{xy}}{(I_x - I_y)} \qquad (10\text{-}16)$$

In analyses of the motion of rigid bodies, second moments of the mass of an element are often encountered. These second moments are commonly known as moments of inertia and products of inertia. Moments of inertia and products of inertia of a body with respect to an xyz-coordinate system are given by Eqs. 10-21 and 10-28 as

$$I_x = \int_m (y^2 + z^2)\, dm \qquad I_{xy} = \int_m xy\, dm$$

$$I_y = \int_m (z^2 + x^2)\, dm \qquad I_{yz} = \int_m yz\, dm$$

$$I_z = \int_m (x^2 + y^2)\, dm \qquad I_{zx} = \int_m zx\, dm$$

For a body which has an xyz-coordinate system with its origin at the mass-center G of the body and a parallel $x'y'z'$-coordinate system with its origin at point O', the parallel-axis theorem for moments and products of inertia, Eqs. 10-23 and 10-30, yield

$$I_{x'} = I_{xG} + (\bar{y}^2 + \bar{z}^2)\, m \qquad I_{x'y'} = I_{xyG} + \bar{x}\bar{y}\, m$$
$$I_{y'} = I_{yG} + (\bar{z}^2 + \bar{x}^2)\, m \qquad I_{y'z'} = I_{yzG} + \bar{y}\bar{z}\, m$$
$$I_{z'} = I_{zG} + (\bar{x}^2 + \bar{y}^2)\, m \qquad I_{z'x'} = I_{zxG} + \bar{z}\bar{x}\, m$$

Equations 10-23 and 10-30 are valid only for transfers to or from axes passing through the mass-center G of the body.

For some dynamic analyses of bodies, maximum and minimum moments of inertia must be determined. Maximum and minimum moments of inertia can be determined and principal axes can be located by using the following equations:

$$(I_x - I_P) \cos \theta_{Px} - I_{xy} \cos \theta_{Py} - I_{zx} \cos \theta_{Pz} = 0$$
$$(I_y - I_P) \cos \theta_{Py} - I_{yz} \cos \theta_{Pz} - I_{xy} \cos \theta_{Px} = 0 \qquad (10\text{-}33)$$
$$(I_z - I_P) \cos \theta_{Pz} - I_{zx} \cos \theta_{Px} - I_{yz} \cos \theta_{Py} = 0$$

Equations 10-33 have a solution if the determinant of the coefficients of the cosine terms is equal to zero. Expansion of the determinant yields the following cubic equation for determining the principal moments of inertia of the body for the particular origin of coordinates being used:

$$I_P^3 - (I_x + I_y + I_z)I_P^2 + (I_x I_y + I_y I_z + I_z I_x - I_{xy}^2 - I_{yz}^2 - I_{zx}^2)I_P$$
$$- (I_x I_y I_z - I_x I_{yz}^2 - I_y I_{zx}^2 - I_z I_{xy}^2 - 2I_{xy} I_{yz} I_{zx}) = 0 \qquad (10\text{-}34)$$

Equation 10-34 yields three values I_1, I_2, and I_3 for the principal moments of inertia. One value is the maximum moment of inertia of the body for the origin of coordinates being used, the second value is the minimum moment of inertia, and the third value is an intermediate value of the moment of inertia that has no particular significance. The direction cosines for the principal axes are obtained by substituting the three values I_1, I_2, and I_3 obtained from Eq. 10-34, in turn, into Eqs. 10-33 and using the additional relation

$$\cos^2 \theta_{Px} + \cos^2 \theta_{Py} + \cos^2 \theta_{Pz} = 1$$

REVIEW PROBLEMS

10-117* Determine the second moment of area for the shaded region shown in Fig. P10-117

a. With respect to the x-axis.
b. With respect to the y-axis.

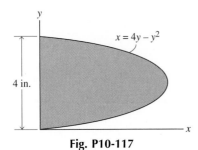

Fig. P10-117

10-118* Determine the radii of gyration of the shaded area shown in Fig. P10-118 with respect to

a. The x- and y-axes shown on the figure.
b. Horizontal and vertical centroidal axes.

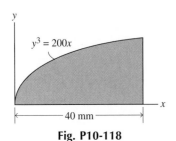

Fig. P10-118

10-119 For the shaded composite area shown in Fig. P10-119, determine

a. The second moment of the area with respect to the x-axis.
b. The second moment of the area with respect to a centroidal x-axis.

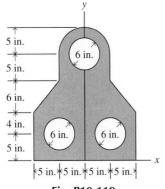

Fig. P10-119

10-120 For the shaded composite area shown in Fig. P10-120, determine

a. The mixed second moment of the area with respect to the x- and y-axes.
b. The maximum and minimum second moments of the area with respect to axes through the origin of the xy-coordinate system and their orientations with respect to the x-axis.

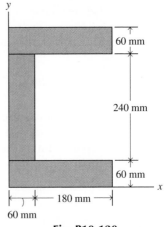

Fig. P10-120

10-121* Determine the moment of inertia about the y-axis for the frustrum of a homogeneous cone shown in Fig. P10-121.

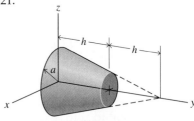

Fig. P10-121

10-122* Determine the product of inertia I_{xy} for the homogeneous tetrahedron shown in Fig. P10-122.

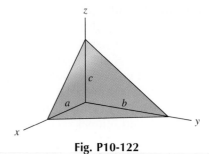

Fig. P10-122

494

10-123 Determine the moment of inertia about an axis that passes through points A and B of the homogeneous circular cylinder shown in Fig. P10-123. Line AB lies in the xy-plane and passes through the mass-center G of the cylinder.

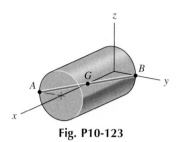

Fig. P10-123

10-124 A composite body consists of a homogeneous sphere and a homogeneous bent bar as shown in Fig. P10-124. The mass of the sphere is 35 kg. The masses of sections OA and AB of the bar are 16 kg and 8 kg, respectively. Determine

a. The moment of inertia I_z of the composite body.
b. The product of inertia I_{yz} of the composite body.

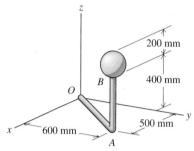

Fig. P10-124

Computer Problems

C10-125 For the T-section shown in Fig. P10-125, the second moments of area are $I_x = 4160$ in.^{4}, $I_y = 1476$ in.^{4}, and $I_{xy} = 2052$ in.^{4}. Compute the second moments of area relative to the $x'y'$-axes for various angles θ ($-90° \leq \theta \leq 90°$) and create a graph using the second moments $I_{x'}$ and $I_{y'}$ for the horizontal coordinates and the mixed second moments $I_{x'y'}$ for the vertical coordinates. Then, for each angle θ, plot the points $[(I_{x'}), (I_{x'y'})]$ and $[(I_{y'}), (-I_{x'y'})]$. On the graph, clearly label the points corresponding to $\theta = 0°$, 30°, 60°, $-28.41°$, and 61.6°. (Note that the last two angles correspond to the principal axes for the area.)

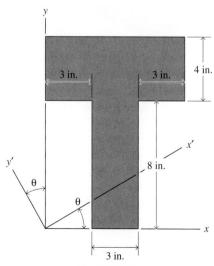

Fig. P10-125

C10-126 In Section 10-5 it was shown that the principal moments of inertia are the roots of the cubic equation

$$I_p^3 - aI_p^2 + bI_p - c = 0 \qquad (10\text{-}34)$$

in which

$$a = I_x + I_y + I_z$$
$$b = I_x I_y + I_y I_z + I_z I_x - I_{xy}^2 - I_{yz}^2 - I_{zx}^2$$
$$c = I_x I_y I_z - I_x I_{yz}^2 - I_y I_{zx}^2 - I_z I_{xy}^2 - 2I_{xy}I_{yz}I_{zx}$$

Write a program to compute and print the three principal moments of inertia ($I_1 \geq I_2 \geq I_3$) and the maximum products of inertia (I_{12}, I_{23}, and I_{31}) by the following method:

a. First solve Eq. 10-34 for one of the principal moments, say I_1, by using the Newton–Raphson method (see Appendix D).
b. Next, since I_1 is a solution of Eq. 10-34, the equation can be written

$$(I_p - I_1)(I_p^2 + AI_p - B) = 0 \qquad (a)$$

Multiplying out the terms in Eq. a and comparing them with the terms of Eq. 10-34 shows that

$$A = I_1 - a \qquad B = b + I_1(I_1 - a)$$

c. Next solve the quadratic equation part of Eq. a to get

$$I_2, I_3 = \frac{-A \pm \sqrt{A^2 - 4B}}{2}$$

d. Finally, renumber the principal moments (if necessary) so that $I_1 \geq I_2 \geq I_3$ and compute the maximum products of inertia I_{12}, I_{23}, and I_{31}.

Use the program to compute the principal moments of inertia and maximum products of inertia for the following sets of data:

i. $I_x = .0835 \text{ slug} \cdot \text{ft}^2$ $I_{xy} = .0501 \text{ slug} \cdot \text{ft}^2$
$I_y = .3340 \text{ slug} \cdot \text{ft}^2$ $I_{yz} = .1002 \text{ slug} \cdot \text{ft}^2$
$I_z = .2839 \text{ slug} \cdot \text{ft}^2$ $I_{zx} = .0251 \text{ slug} \cdot \text{ft}^2$

ii. $I_x = .2781 \text{ kg} \cdot \text{m}^2$ $I_{xy} = -.0590 \text{ kg} \cdot \text{m}^2$
$I_y = .3709 \text{ kg} \cdot \text{m}^2$ $I_{yz} = .0787 \text{ kg} \cdot \text{m}^2$
$I_z = .1854 \text{ kg} \cdot \text{m}^2$ $I_{zx} = -.1391 \text{ kg} \cdot \text{m}^2$

iii. $I_x = .2446 \text{ slug} \cdot \text{ft}^2$ $I_{xy} = -.1957 \text{ slug} \cdot \text{ft}^2$
$I_y = .5773 \text{ slug} \cdot \text{ft}^2$ $I_{yz} = .0587 \text{ slug} \cdot \text{ft}^2$
$I_z = .6458 \text{ slug} \cdot \text{ft}^2$ $I_{zx} = -.1468 \text{ slug} \cdot \text{ft}^2$

iv. $I_x = .8239 \text{ kg} \cdot \text{m}^2$ $I_{xy} = -.5130 \text{ kg} \cdot \text{m}^2$
$I_y = .6635 \text{ kg} \cdot \text{m}^2$ $I_{yz} = .1232 \text{ kg} \cdot \text{m}^2$
$I_z = 1.4169 \text{ kg} \cdot \text{m}^2$ $I_{zx} = -.1100 \text{ kg} \cdot \text{m}^2$

v. $I_x = 1.534 \text{ slug} \cdot \text{ft}^2$ $I_{xy} = -.4940 \text{ slug} \cdot \text{ft}^2$
$I_y = 2.787 \text{ slug} \cdot \text{ft}^2$ $I_{yz} = .1174 \text{ slug} \cdot \text{ft}^2$
$I_z = 3.257 \text{ slug} \cdot \text{ft}^2$ $I_{zx} = -.3520 \text{ slug} \cdot \text{ft}^2$

vi. $I_x = 1.573 \text{ kg} \cdot \text{m}^2$ $I_{xy} = .408 \text{ kg} \cdot \text{m}^2$
$I_y = 1.760 \text{ kg} \cdot \text{m}^2$ $I_{yz} = .867 \text{ kg} \cdot \text{m}^2$
$I_z = 1.900 \text{ kg} \cdot \text{m}^2$ $I_{zx} = .625 \text{ kg} \cdot \text{m}^2$

C10-127 The direction cosines for the principal inertia axes are obtained by solving the system of equations (Eqs. 10-33)

$$(I_x - I_p) l_p \quad - I_{xy} m_p \quad - I_{zx} n_p = 0 \quad (a)$$
$$- I_{xy} l_p + (I_y - I_p) m_p \quad - I_{yz} n_p = 0 \quad (b)$$
$$- I_{zx} l_p \quad - I_{yz} m_p + (I_z - I_p) n_p = 0 \quad (c)$$

and using the additional relation

$$l_p^2 + m_p^2 + n_p^2 = 1 \quad (d)$$

where I_p ($p = 1, 2, 3$) is one of the principal moments obtained from Eq. 10-34 and $l_p = \cos \theta_{px}$, $m_p = \cos \theta_{py}$, and $n_p = \cos \theta_{pz}$. Write a program to compute and print the three direction cosines l_p, m_p, and n_p for each of the princi-

pal moments I_p ($p = 1, 2, 3$). In writing the program, you should note the following:

a. When the principal moment I_p used in Eqs. a, b, and c is different from the other two principal moments (a single root of Eq. 10-34), then Eqs. a, b, and c are not independent: one of the equations is a combination of the other two. Equation d must be used along with the two independent equations to get a solution for the direction cosines.

b. When the principal moment I_p used in Eqs. a, b, and c is the same as another value of principal moment (a double root of Eq. 10-34), then only one of the three equations is unique. In this case, only Eqs. a and d can be used to solve for the direction cosines. The problem statement is completed by requiring that the unit vectors defined by the three sets of direction cosines must be perpendicular.

c. Although not all three of the direction cosines can be zero for a particular principal moment I_p, it is quite possible that one or two of the direction cosines will be zero.

Use the program to compute the direction cosines for the principal inertia axes for the same sets of data as in Problem C10-126.

C10-128 A common approximation for the moment of inertia of a cylinder about an axis parallel to the axis of the cylinder is

$$I \cong ma^2$$

where a is the distance between the two axes (Fig. P10-128). Plot the percent error in using this approximation as a function of a/r ($1 \leq a/r \leq 6$).

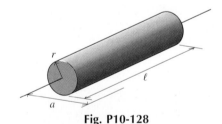

Fig. P10-128

11

METHOD OF
VIRTUAL WORK

Virtual work principles can be used to determine joint forces and to
investigate the stability of equilibrium for various positions of complicated
mechanisms.

11-1   INTRODUCTION

Static equilibrium problems in Chapter 6 (Equilibrium of Rigid Bodies), Chapter 7 (Trusses, Frames, and Machines), and Chapter 8 (Internal Forces in Structural Members) were solved by using the problem-solving method outlined in Chapter 1. Once the problem was clearly identified, free-body diagrams (one or more as required for the specific problem) were drawn to identify all forces (applied forces and support reactions) involved in the problem. The free-body diagram was then used with the appropriate equations of equilibrium to solve for the required unknowns. All of these types of problems can also be solved by using a method based on work and potential energy concepts. The method provides a convenient means for determining an unknown force whenever displacements of applied forces can be determined in terms of the displacement of an unknown force. Since most equilibrium problems involve bodies at rest, the displacements involved in the process are virtual[1] (imaginary or fictitious) displacements. Thus, the method has become known as the method of virtual displacements or the method of virtual work. The method is often useful when systems of connected bodies are studied. Frequently, when the equations of equilibrium are used to solve such problems, the system must be dismembered in order to determine a particular reaction component. It is not necessary to investigate the internal forces of a system when the method of virtual work is used.

The method of virtual work, as applied to the solution of equilibrium problems, is developed in the following sections.

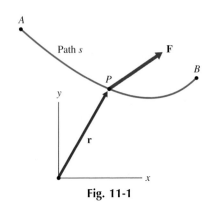

Fig. 11-1

11-2   DEFINITION OF WORK AND VIRTUAL WORK

In everyday life, the word *work* is applied to any form of activity that requires the exertion of muscular or mental effort. In mechanics, the term is used in a much more restricted sense.

11-2-1   Work of a Force

A particle P moves under the action of a force $\mathbf{F}$ from point A to point B along a curved path s, as shown in Fig. 11-1. The position of the particle along the path is specified by using the position vector $\mathbf{r}$. As the particle moves an infinitesimal distance from point 1 to point 2, the position vector changes from $\mathbf{r}_1$ to $\mathbf{r}_2$ as shown in Fig. 11-2. The motion of the particle, as it moves from point 1 to point 2, is described by the vector $d\mathbf{s} = \mathbf{r}_2 - \mathbf{r}_1$. The vector $d\mathbf{s}$ is known as a linear displacement. Since the arc length ds along the curve is infinitesimal, the direction of the vector $d\mathbf{s}$ is tangent to the path s and the magnitude of the vector $d\mathbf{s}$ is the length ds.

By definition, the work dU done by force $\mathbf{F}$ on the particle as it travels from point 1 to point 2 is the dot product of the two vectors $\mathbf{F}$ and $d\mathbf{s}$. Thus,

$$dU = \mathbf{F} \cdot d\mathbf{s} = |\mathbf{F}||d\mathbf{s}| \cos \alpha = F \, ds \cos \alpha \qquad (11\text{-}1)$$

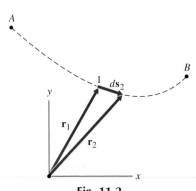

Fig. 11-2

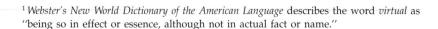

[1] *Webster's New World Dictionary of the American Language* describes the word *virtual* as "being so in effect or essence, although not in actual fact or name."

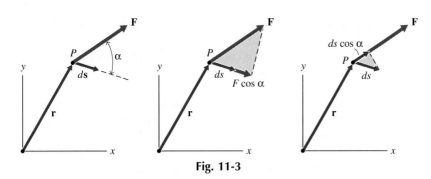

Fig. 11-3

As shown in Fig. 11-3, the increment of work dU can be viewed either as the component of the force in the direction of the displacement multiplied by the magnitude of the displacement [$(F \cos \alpha) \, ds$], or alternatively, as the magnitude of the force multiplied by the component of the displacement in the direction of the force [$F(ds \cos \alpha)$]. Since work is defined as the scalar product of two vectors, work is a scalar quantity with only magnitude and algebraic sign. When the sense of the displacement and the sense of the force component in the direction of the displacement are the same ($0 \leq \alpha < 90°$), the work done by the force is positive. When the sense of the displacement and the sense of the force component in the direction of the displacement are opposite ($90° < \alpha \leq 180°$), the work done by the force is negative. When the direction of the force is perpendicular to the direction of the displacement ($\alpha = 90°$), the component of the force in the direction of the displacement is zero; therefore, the work done by the force is zero.

Since work is defined as the product of a force and a displacement, the units of work are obtained by multiplying a unit of force by a unit of length. In the U. S. Customary system, the primary unit for work is the ft·lb. In the SI system, the unit of work is called a joule (1 J = 1 N·m).

When the force **F** acting on the particle shown in Fig. 11-1 varies in either magnitude or direction, Eq. 11-1 is valid only for an infinitesimal change in position. The total work done by the force as the particle moves from point A to point B is obtained by integrating Eq. 11-1 along the path of the particle. Thus,

$$U = \int_A^B dU = \int_A^B \mathbf{F} \cdot d\mathbf{s} = \int_A^B F \cos \alpha \, ds \qquad (11\text{-}2)$$

In Eq. 11-2, ds is the magnitude of the infinitesimal displacement $d\mathbf{s}$ and α is the angle between **F** and $d\mathbf{s}$. Equation 11-2 can be used to compute the work done by any force **F** if relationships can be established between F, α, and s. Since the dot product of the vector quantities **F** and $d\mathbf{s}$ is a scalar quantity, ordinary scalar integration can be used to determine the work done. Also, if more than one force acts on the particle or body, the total work done is the sum of the work done by the individual forces.

For the special case of a force **F** (see Fig. 11-4a) that is constant in both magnitude and direction as the particle moves along path s from point A to point B, Eq. 11-2 indicates that

$$U = \int_A^B dU = \int_A^B \mathbf{F} \cdot d\mathbf{s} = F\int_A^B \cos \alpha \, ds = F\int_A^B dn = Fd \qquad (11\text{-}3)$$

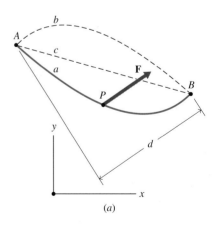

(a)

(b)

Fig. 11-4

where dn is the component of ds in the direction of $\mathbf{F}$ (Fig. 11-4b) and d is the distance between A and B in the $\mathbf{n}$-direction. Equation 11-3 indicates that the work done by a constant force does not depend on the path traveled by the particle. The work done by the constant force $\mathbf{F}$ is the same if the particle moves from point A to point B along the curved path a shown in Fig. 11-4a or along the curved path b or the straight line path c. The fact that the work done by a constant force is independent of the path is useful when computing the work done by the weight of a body. The work is simply the product of the weight W of the body and the vertical displacement h of the center of gravity. If the displacement is toward the center of the Earth (in the same direction as the gravity force), the work is positive; if it is away from the center of the Earth (opposite the direction of the gravity force), the work is negative.

Examples of forces that do work when a body moves from one position to another include the weight of the body, friction between the body and other surfaces, and externally applied loads. Examples of forces that do no work include forces at fixed points ($ds = 0$) or forces acting in a direction perpendicular to the displacement ($\cos \alpha = 0$).

11-2-2   Work of a Couple

The work done by a force $\mathbf{F}$ during an infinitesimal linear displacement $d\mathbf{s}$ of a particle was previously defined as

$$dU = \mathbf{F} \cdot d\mathbf{s} = |\mathbf{F}||d\mathbf{s}| \cos \alpha = F \, ds \cos \alpha \tag{11-1}$$

The work done by a couple $\mathbf{C}$, as it rotates through an infinitesimal angle $d\boldsymbol{\theta}$, can be obtained by applying Eq. 11-1 to the two forces that compose the couple. Infinitesimal angular displacements are vector quantities, but finite angular displacements are not. Therefore, infinitesimal angular displacements must be used for all work determinations involving angular movements. Thus,

$$dU = \mathbf{F}_1 \cdot d\mathbf{s}_1 + \mathbf{F}_2 \cdot d\mathbf{s}_2 \tag{11-4}$$

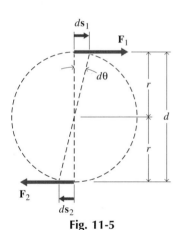

Fig. 11-5

As the couple rotates through an infinitesimal angle $d\boldsymbol{\theta}$, the two forces $\mathbf{F}_1$ and $\mathbf{F}_2$ experience infinitesimal linear displacements $d\mathbf{s}_1$ and $d\mathbf{s}_2$ in the direction of the forces, as shown in Fig. 11-5. The magnitudes of the displacements are

$$ds_1 = ds_2 = r \, d\theta$$

Since the two forces of the couple have the same magnitude F, the work done by the forces is

$$\begin{aligned} dU &= F_1 \, ds_1 + F_2 \, ds_2 \\ &= F \, (r \, d\theta) + F \, (r \, d\theta) \\ &= F \, (2r) \, d\theta = Fd \, d\theta = M \, d\theta \end{aligned} \tag{11-5}$$

where $M = Fd$ is the moment of the couple.

The work done during a finite angular displacement θ can be obtained by integrating Eq. 11-5. Thus,

$$U = \int dU = \int_0^\theta M \, d\theta \tag{11-6}$$

If the moment M of the couple is constant,

$$U = \int_0^\theta M \, d\theta = M \int_0^\theta d\theta = M\theta \qquad (11\text{-}7)$$

When a body is subjected to a linear displacement while being acted on by a couple, the work done by the couple is

$$dU = \mathbf{F}_1 \cdot d\mathbf{s}_1 + \mathbf{F}_2 \cdot d\mathbf{s}_2 = \mathbf{F}_1 \cdot d\mathbf{s}_1 + (-\mathbf{F}_1) \cdot d\mathbf{s}_1 = 0$$

since $\mathbf{F}_2 = -\mathbf{F}_1$ and $d\mathbf{s}_2 = d\mathbf{s}_1$. Thus, a couple does no work during translation of a body on which it is acting. If a body is simultaneously translated and rotated, the couple does work only as a result of the rotation. A couple does work when it turns through an angle in the plane of the couple (about the axis of the couple). The work is positive if the angular displacement is in the same direction as the sense of rotation of the couple and negative if the displacement is in the opposite direction. No work is done if the couple is translated or rotated about an axis parallel to the plane of the couple.

If the body rotates in space, the component of the infinitesimal angular displacement $d\theta$ in the direction of the couple $\mathbf{C}$ is required. For this case, the work done is determined by using the dot product relationship,

$$dU = \mathbf{C} \cdot d\boldsymbol{\theta} \qquad \text{or} \qquad dU = M \, d\theta \cos\alpha \qquad (11\text{-}8)$$

where M is the magnitude of the moment of the couple, $d\theta$ is the magnitude of the infinitesimal angular displacement, and α is the angle between $\mathbf{C}$ and $d\boldsymbol{\theta}$.

Since work is a scalar quantity, the work done on a rigid body by a system of external forces and couples is the algebraic sum of work done by the individual forces and couples.

11-2-3 Virtual Work

When a body being acted on by a force $\mathbf{F}$ moves through an infinitesimal linear displacement $d\mathbf{r}$ as described in Section 11-2-1, the body is not in equilibrium. In studying the equilibrium of bodies by the method of virtual work, it is necessary to introduce fictitious displacements that are called virtual displacements. An infinitesimal virtual linear displacement will be represented by the first-order differential $\delta\mathbf{s}$ rather than $d\mathbf{s}$. The work done by a force $\mathbf{F}$ acting on a body during a virtual displacement $\delta\mathbf{s}$ is called virtual work δU and is represented mathematically as

$$\delta U = \mathbf{F} \cdot \delta\mathbf{s} \qquad \text{or} \qquad \delta U = F \, \delta s \cos\alpha \qquad (11\text{-}9)$$

where F and δs are the magnitudes of the force $\mathbf{F}$ and virtual displacement $\delta\mathbf{s}$, respectively, and α is the angle between $\mathbf{F}$ and $\delta\mathbf{s}$.

A virtual displacement may also be a rotation of the body. The virtual work done by a couple $\mathbf{C}$ during an infinitesimal virtual angular displacement $\delta\boldsymbol{\theta}$ of the body is

$$\delta U = \mathbf{C} \cdot \delta\boldsymbol{\theta} \qquad \text{or} \qquad \delta U = M \, \delta\theta \cos\alpha \qquad (11\text{-}10)$$

where M and $\delta\theta$ are the magnitudes of the couple $\mathbf{C}$ and virtual displacement $\delta\boldsymbol{\theta}$, respectively, and α is the angle between $\mathbf{C}$ and $\delta\boldsymbol{\theta}$. Since the infinitesimal virtual displacements δs and $\delta\theta$ in Eqs. 11-9 and 11-10 refer to fictitious movements, the equations cannot be integrated.

A 500-lb block A is held in equilibrium on an inclined surface with a cable and weight system and a force **P** as shown in Fig. 11-6a. When the force **P** is removed, block A slides down the incline at a constant velocity for a distance of 10 ft. The coefficient of friction between block A and the inclined surface is 0.20. Determine

a. The work done by the cable on block A.
b. The work done by gravity on block A.
c. The work done by the surface of the incline on block A.
d. The total work done by all forces on block A.
e. The work done by the cable on block B.
f. The work done by gravity on block B.
g. The total work done by all forces on block B.

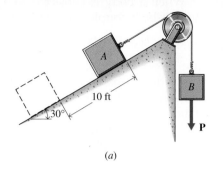

(a)

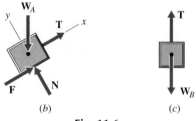

(b)   (c)

Fig. 11-6

SOLUTION

Free-body diagrams for blocks A and B are shown in Fig. 11-6b. Four forces act on block A: the cable tension T, the weight W_A, and the normal and frictional forces N and F at the surface of contact between the block and the inclined surface. Two forces act on block B: the cable tension T and the weight W_B. Since the blocks move at a constant velocity, they are in equilibrium and the equilibrium equations applied to block A yield

$$N = W_A \cos 30° = 500 \cos 30° = 433 \text{ lb}$$
$$F = \mu N = 0.20(433) = 86.6 \text{ lb}$$
$$T = W_A \sin 30° - F = 500 \sin 30° - 86.6 = 163.4 \text{ lb}$$

The equilibrium equations applied to block B yield

$$W_B = T = 163.4 \text{ lb}$$

Since all the forces are constant in both magnitude and direction during the movements of the blocks, the work can be computed by using Eq. 11-3 in the form

$$U = \int_A^B dU = \int_A^B \mathbf{F} \cdot d\mathbf{s} = F \int_A^B \cos \alpha \, ds = F s \cos \alpha$$

Therefore, for block A

a. $U_T = T s \cos \alpha = 163.4(10) \cos 180° = -1634 \text{ ft} \cdot \text{lb}$ Ans.
b. $U_W = W_A s \cos \alpha = 500(10) \cos 60° = 2500 \text{ ft} \cdot \text{lb}$ Ans.

Alternatively,

   $U_W = W_A h = 500(10 \sin 30°) = 2500 \text{ ft} \cdot \text{lb}$
c. $U_F = F s \cos \alpha = 86.6(10) \cos 180° = -866 \text{ ft} \cdot \text{lb}$ Ans.
   $U_N = N s \cos \alpha = 433(10) \cos 90° = 0$
d. $U_{\text{total}} = U_T + U_W + U_F + U_N = -1634 + 2500 - 866 + 0 = 0$ Ans.

and for block B

e. $U_T = T s \cos \alpha = 163.4(10) \cos 0° = 1634 \text{ ft} \cdot \text{lb}$ Ans.
f. $U_W = W_B s \cos \alpha = 163.4(10) \cos 180° = -1634 \text{ ft} \cdot \text{lb}$ Ans.
g. $U_{\text{total}} = U_T + U_W = -1634 + 1634 = 0$ Ans.

A constant couple $\mathbf{C} = 25\mathbf{i} + 35\mathbf{j} - 50\mathbf{k}$ N·m acts on a rigid body. The unit vector associated with the fixed axis of rotation of the body for an infinitesimal angular displacement $d\boldsymbol{\theta}$ is $\mathbf{e}_\theta = 0.667\mathbf{i} + 0.333\mathbf{j} + 0.667\mathbf{k}$. Determine the work done on the body by the couple during an angular displacement of 2.5 rad.

SOLUTION

The magnitude (moment M) of the couple is

$$M = \sqrt{(25)^2 + (35)^2 + (-50)^2} = 65.95 \text{ N·m}$$

The unit vector associated with the couple is

$$\mathbf{e}_C = \frac{+25}{65.95}\mathbf{i} + \frac{+35}{65.95}\mathbf{j} + \frac{-50}{65.95}\mathbf{k} = 0.379\mathbf{i} + 0.531\mathbf{j} - 0.758\mathbf{k}$$

The cosine of the angle between the axis of the couple and the axis of rotation of the body is

$$\cos \alpha = \mathbf{e}_C \cdot \mathbf{e}_\theta$$
$$= (0.379\mathbf{i} + 0.531\mathbf{j} - 0.758\mathbf{k}) \cdot (0.667\mathbf{i} + 0.333\mathbf{j} + 0.667\mathbf{k})$$
$$= -0.0760$$

Therefore,

$$\alpha = 94.36°$$

The work done by the couple during the finite rotation of 2.5 rad can be obtained by using Eq. 11-8. Thus,

$$dU = M \cos \alpha \, d\theta = 65.95(-0.0760) \, d\theta = -5.0122 \, d\theta$$
$$U = \int_0^{2.5} dU = \int_0^{2.5} (M \cos \alpha) \, d\theta$$
$$= -5.0122 \int_0^{2.5} d\theta = -12.53 \text{ N·m} \qquad \text{Ans.}$$

Alternatively,

$$U = \int_0^{2.5} \mathbf{C} \cdot d\boldsymbol{\theta}$$
$$= \int_0^{2.5} (25\mathbf{i} + 35\mathbf{j} - 50\mathbf{k}) \cdot (0.667\mathbf{i} + 0.333\mathbf{j} + 0.667\mathbf{k}) \, d\theta$$
$$= \int_0^{2.5} -5.02 \, d\theta = -12.55 \text{ N·m} \qquad \text{Ans.}$$

The slight difference between this answer and the previous answer is due to the way the numbers were rounded off before being multiplied together.

PROBLEMS

11-1* Determine the work done by a locomotive in drawing a freight train for one mile at constant speed on a level track if the locomotive exerts a constant force of 6000 lb on the train.

11-2* A horse tows a canal boat with a rope that makes an angle of 10° with the towpath. If the tension in the rope is 800 N, how much work does the horse do while pulling the boat for 300 m along the canal?

11-3* A box is dragged across a floor by using a rope that makes an angle of 30° with the horizontal. Determine the work done if the tension in the rope is 50 lb and the horizontal distance moved is 12 ft.

11-4* A child on a sled is pulled by using a rope that makes an angle of 40° with the horizontal. Determine the work done if the tension in the rope is 75 N and the horizontal distance moved is 30 m.

11-5 A 175-lb man climbs a flight of stairs 12 ft high. Determine

a. The work done by the man.
b. The work done on the man by gravity.

11-6 A box with a mass of 600 kg is dragged up an incline 12 m long and 4 m high by using a cable that is parallel to the incline. The force in the cable is 2500 N. Determine

a. The work done on the box by the cable.
b. The work done on the box by gravity.

11-7* A 100-lb block is pushed at constant speed for a distance of 20 ft along a level floor by a force that makes an angle of 35° with the horizontal, as shown in Fig. P11-7. The coefficient of friction between the block and the floor is 0.35. Determine

a. The work done on the block by the force.
b. The work done on the block by gravity.
c. The work done on the block by the floor.

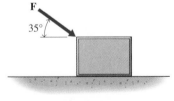

Fig. P11-7

11-8* A block with a mass of 100 kg slides down an inclined surface that is 5 m long and makes an angle of 25° with the horizontal. A man pushes horizontally on the block so that it slides down the incline at a constant speed. The coefficient of friction between the block and the inclined surface is 0.15. Determine

a. The work done on the block by the man.
b. The work done on the block by gravity.
c. The work done on the block by the inclined surface.

11-9 A crate weighing 300 lb is supported by a rope that is 40 ft long. A man pushes the crate 8 ft horizontally and holds it there. Determine

a. The work done on the crate by the man.
b. The work done on the crate by the rope.
c. The work done on the crate by gravity.

11-10 A steel bar of uniform cross section is 2 m long and has a mass of 25 kg. The bar is supported in a vertical position by a horizontal pin at the top end of the bar. Determine the work done on the bar by gravity as the bar rotates 60° about the pin in a vertical plane.

11-11* A constant force acting on a particle can be expressed in Cartesian vector form as $\mathbf{F} = 8\mathbf{i} - 6\mathbf{j} + 2\mathbf{k}$ lb. Determine the work done by the force on the particle if the displacement of the particle can be expressed in Cartesian vector form as $\mathbf{s} = 5\mathbf{i} + 4\mathbf{j} + 6\mathbf{k}$ ft.

11-12* A constant force acting on a particle can be expressed in Cartesian vector form as $\mathbf{F} = 3\mathbf{i} + 5\mathbf{j} - 4\mathbf{k}$ N. Determine the work done by the force on the particle if the displacement of the particle can be expressed in Cartesian vector form as $\mathbf{s} = 4\mathbf{i} - 2\mathbf{j} + 3\mathbf{k}$ m.

11-13 A constant couple $\mathbf{C} = 120\mathbf{i} + 75\mathbf{j} - 150\mathbf{k}$ ft·lb acts on a rigid body. The unit vector associated with the fixed axis of rotation of the body for an infinitesimal angular displacement $d\boldsymbol{\theta}$ is $\mathbf{e}_\theta = 0.600\mathbf{i} + 0.300\mathbf{j} - 0.742\mathbf{k}$. Determine the work done on the body by the couple during an angular displacement of 0.75 rad.

11-14 A constant couple $\mathbf{C} = 200\mathbf{i} + 300\mathbf{j} + 350\mathbf{k}$ N·m acts on a rigid body. The unit vector associated with the fixed axis of rotation of the body for an infinitesimal angular displacement $d\boldsymbol{\theta}$ is $\mathbf{e}_\theta = 0.250\mathbf{i} + 0.350\mathbf{j} - 0.903\mathbf{k}$. Determine the work done on the body by the couple during an angular displacement of 1.5 rad.

11-3 PRINCIPLE OF VIRTUAL WORK AND EQUILIBRIUM

The principle of virtual work can be stated as follows:

> If the virtual work done by all external forces (or couples) acting on a particle, a rigid body, or a system of connected rigid bodies with ideal (frictionless) connections and supports is zero for all virtual displacements of the system, the system is in equilibrium.

The principle of virtual work can be expressed mathematically as

$$\delta U = \sum_{i=1}^{m} \mathbf{F}_i \cdot \delta\mathbf{s}_i + \sum_{j=1}^{n} \mathbf{C}_j \cdot \delta\boldsymbol{\theta}_j = 0 \qquad (11\text{-}11)$$

11-3-1 Equilibrium of a Particle

Consider the particle shown in Fig. 11-7, which is acted on by several forces $\mathbf{F}_1, \mathbf{F}_2, \cdots, \mathbf{F}_n$. The work done on the particle by these forces during an arbitrary virtual displacement $\delta\mathbf{s}$ is

$$\begin{aligned}\delta U &= \mathbf{F}_1 \cdot \delta\mathbf{s} + \mathbf{F}_2 \cdot \delta\mathbf{s} + \cdots + \mathbf{F}_n \cdot \delta\mathbf{s} \\ &= (\mathbf{F}_1 + \mathbf{F}_2 + \cdots + \mathbf{F}_n) \cdot \delta\mathbf{s} = \Sigma\mathbf{F} \cdot \delta\mathbf{s} = \mathbf{R} \cdot \delta\mathbf{s} \qquad (11\text{-}12)\end{aligned}$$

where $\mathbf{R}$ is the resultant of the forces acting on the particle.

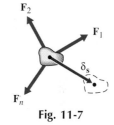

Fig. 11-7

Expressing the resultant $\mathbf{R}$ and the virtual displacement $\delta\mathbf{s}$ in Cartesian vector form and computing the vector scalar product yields

$$\begin{aligned}\delta U = \mathbf{R} \cdot \delta\mathbf{s} &= (\Sigma F_x\mathbf{i} + \Sigma F_y\mathbf{j} + \Sigma F_z\mathbf{k}) \cdot (\delta x\mathbf{i} + \delta y\mathbf{j} + \delta z\mathbf{k}) \\ &= \Sigma F_x\,\delta x + \Sigma F_y\,\delta y + \Sigma F_z\,\delta z \qquad (11\text{-}13)\end{aligned}$$

Applying the principle of virtual work by combining Eqs. 11-11 and 11-13 yields

$$\delta U = \mathbf{R} \cdot \delta\mathbf{s} = \Sigma F_x\,\delta x + \Sigma F_y\,\delta y + \Sigma F_z\,\delta z = 0 \qquad (11\text{-}14)$$

By considering virtual displacements (δx, δy, and δz) taken one at a time in each of the three mutually perpendicular coordinate directions,

$$\Sigma F_x = 0 \qquad \Sigma F_y = 0 \qquad \Sigma F_z = 0$$

Thus, the virtual work equation $\delta U = 0$ is simply an alternative statement of the equilibrium equations for a particle. The principle of virtual work does not simplify the solution of problems involving equilibrium of a particle since the equations $\delta U = 0$ and $\Sigma F = 0$ are equivalent.

11-3-2 Equilibrium of a Rigid Body

If a rigid body is in equilibrium, all particles forming the body must be in equilibrium. Therefore, according to the principle of virtual work, the total virtual work of all forces, both internal and external, acting on all of the particles must be zero. Since the internal forces between particles occur as equal, opposite, collinear pairs, the work done by each pair of forces sum to zero during any virtual displacement of the rigid body. Thus, only the external forces do work during any virtual displacement of the body. Any system of forces acting on a rigid body can be replaced by a resultant force $\mathbf{R}$ and a resultant couple $\mathbf{C}$. Therefore, the work done on a rigid body by the external forces during an arbitrary linear virtual displacement δs and an arbitrary angular virtual displacement $\delta\theta$ is

$$\delta U = \mathbf{R} \cdot \delta\mathbf{s} + \mathbf{C} \cdot \delta\boldsymbol{\theta} \qquad (11\text{-}15)$$

Expressing the resultant force $\mathbf{R}$, the resultant couple $\mathbf{C}$, and virtual displacements $\delta\mathbf{r}$ and $\delta\boldsymbol{\theta}$ in Cartesian vector form and computing the vector scalar products yields

$$
\begin{aligned}
\delta U &= \mathbf{R} \cdot \delta\mathbf{s} + \mathbf{C} \cdot \delta\boldsymbol{\theta} \\
&= (\Sigma F_x \mathbf{i} + \Sigma F_y \mathbf{j} + \Sigma F_z \mathbf{k}) \cdot (\delta x \mathbf{i} + \delta y \mathbf{j} + \delta z \mathbf{k}) \\
&\quad + (\Sigma M_x \mathbf{i} + \Sigma M_y \mathbf{j} + \Sigma M_z \mathbf{k}) \cdot (\delta\theta_x \mathbf{i} + \delta\theta_y \mathbf{j} + \delta\theta_z \mathbf{k}) \\
&= \Sigma F_x\, \delta x + \Sigma F_y\, \delta y + \Sigma F_z\, \delta z + \Sigma M_x\, \delta\theta_x + \Sigma M_y\, \delta\theta_y + \Sigma M_z\, \delta\theta_z
\end{aligned}
$$
$$(11\text{-}16)$$

Applying the principle of virtual work by combining Eqs. 11-11 and 11-16 yields

$$
\begin{aligned}
\delta U &= \Sigma F_x\, \delta x + \Sigma F_y\, \delta y + \Sigma F_z\, \delta z \\
&\quad + \Sigma M_x\, \delta\theta_x + \Sigma M_y\, \delta\theta_y + \Sigma M_z\, \delta\theta_z = 0 \quad (11\text{-}17)
\end{aligned}
$$

By considering virtual linear displacements (δx, δy, and δz) and virtual angular displacements ($\delta\theta_x$, $\delta\theta_y$, and $\delta\theta_z$) taken one at a time,

$$
\begin{array}{ccc}
\Sigma F_x = 0 & \Sigma F_y = 0 & \Sigma F_z = 0 \\
\Sigma M_x = 0 & \Sigma M_y = 0 & \Sigma M_z = 0
\end{array}
$$

Therefore, the virtual work equation $\delta U = 0$ is simply an alternative statement of the equilibrium equations for a rigid body. The principle of virtual work does not simplify the solution of problems involving equilibrium of a single rigid body since the equation $\delta U = 0$ is equivalent to the equilibrium equations $\Sigma F = 0$ and $\Sigma M = 0$.

11-3-3   Equilibrium of an Ideal System of Connected Rigid Bodies

The principle of virtual work can also be used to study systems of connected rigid bodies. Frequently it is possible to solve such problems by using the complete system rather than individual free-body diagrams of each member of the system.

When the system remains connected during the virtual displacement, only the work of the forces external to the system need be considered, since the net work done by the internal forces at connections between members during any virtual displacement is zero because the forces exist as equal, opposite, collinear pairs. Such a condition exists when the connection is a smooth pin, a smooth roller, or an inextensible link or cable. When the reaction exerted by a support is to be determined, the restraint is replaced by a force and the body is given a virtual displacement with a component in the direction of the force. The virtual work done by the reaction and all other forces acting on the body is computed. If several forces are to be determined, the system of bodies can be given a series of separate virtual displacements in which only one of the unknown forces does virtual work during each displacement.

The problems in this chapter will be limited to a single degree of freedom, systems for which the virtual displacements of all points can be expressed in terms of a single variable (displacement).

A beam is loaded and supported as shown in Fig. 11-8a. Use the method of virtual work to determine the reaction at support B. Neglect the weight of the beam.

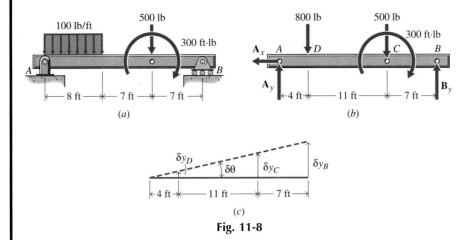

(a)   (b)

(c)

Fig. 11-8

SOLUTION

A free-body diagram for the beam is shown in Fig. 11-8b. The 100-lb/ft distributed load has been replaced by its resultant $R = w\ell = 100(8) = 800$ lb. The beam is given a counterclockwise virtual angular displacement $\delta\theta$ about support A, which produces the virtual linear displacements $\delta y_B = 22\, \delta\theta$, $\delta y_C = 15\, \delta\theta$, and $\delta y_D = 4\, \delta\theta$ at support B and load points C and D, respectively, as shown in the displacement diagram (Fig. 11-8c). The virtual work done as a result of these linear and angular virtual displacements is given by Eqs. 11-9 and 11-10 as

$$\delta U = F\, \delta s \cos\alpha \quad\text{and}\quad \delta U = M\, \delta\theta \cos\alpha$$

Thus,

$$\delta U_B = B_y\, \delta y_B \cos 0° = B_y\, (22\, \delta\theta) \cos 0° = 22 B_y\, \delta\theta$$
$$\delta U_C = F_C\, \delta y_C \cos 180° = 500\, (15\, \delta\theta) \cos 180° = -7500\, \delta\theta$$
$$\delta U_D = F_D\, \delta y_D \cos 180° = 800\, (4\, \delta\theta) \cos 180° = -3200\, \delta\theta$$
$$\delta U_M = M\, \delta\theta \cos 180° = 300\, \delta\theta \cos 180° = -300\, \delta\theta$$

The total work on the beam is zero when the beam is in equilibrium. Thus,

$$\delta U_{total} = \delta U_B + \delta U_C + \delta U_D + \delta U_M = (22 B_y - 7500 - 3200 - 300)\, \delta\theta = 0$$

Since $\delta\theta \neq 0$

$$(22 B_y - 7500 - 3200 - 300) = 0 \qquad \mathbf{B}_y = 500\ \text{lb}\uparrow \qquad\qquad \text{Ans.}$$

Solutions to problems of this type are much simpler if the equations of equilibrium are used. As an example, for this problem,

$$+\!\!\curvearrowleft\Sigma M_A = 0 \qquad B_y(22) - 500(15) - 300 - 800(4) = 0$$
$$\mathbf{B}_y = 500\ \text{lb}\uparrow \qquad\qquad\qquad\qquad \text{Ans.}$$

The slender bar shown in Fig. 11-9a is 7.2 m long and has a mass of 100 kg. The bar rests against smooth surfaces at supports *A* and *B*. Use the method of virtual work to determine the magnitude of the force **F** required to maintain the bar in the equilibrium position shown in the figure.

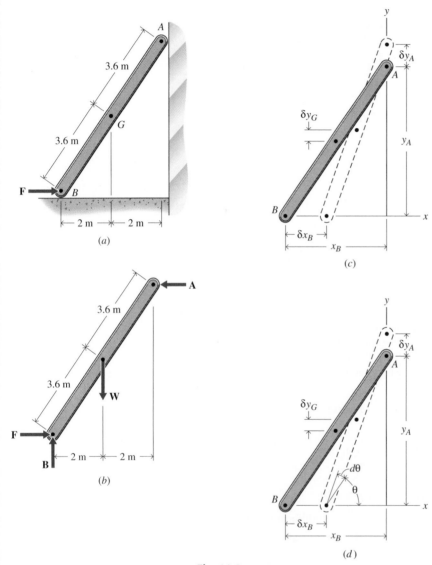

Fig. 11-9

SOLUTION

A free-body diagram of the bar is shown in Fig. 11-9b and a virtual displacement diagram is shown in Fig. 11-9c. The virtual displacements of interest can be expressed in terms of the displacement δx_B in the direction of the force **F**. Thus,

$$(x_B - \delta x_B)^2 + (y_A + \delta y_A)^2 = x_B^2 + y_A^2 = L^2 \qquad (a)$$

from which

$$x_B^2 - 2x_B\delta x_B + \delta x_B^2 + y_A^2 + 2y_A\delta y_A + \delta y_A^2 = x_B^2 + y_A^2 \qquad (b)$$

Since the virtual displacements may be considered very small, the terms δx_B^2 and δy_A^2 can be neglected and Eq. b simplifies to

$$\delta y_A = \frac{x_B}{y_A} \, \delta x_B$$

The y-component of the virtual displacement of the center of gravity is

$$\delta y_G = \frac{1}{2} \, \delta y_A = \frac{1}{2} \frac{x_B}{y_A} \, \delta x_B$$

For the given geometry

$$y_A = \sqrt{L^2 - x_B^2} = \sqrt{(7.2)^2 - (4)^2} = 5.987 \text{ m}$$

Therefore,

$$\delta y_G = \frac{1}{2} \frac{x_B}{y_A} \, \delta x_B = \frac{1}{2} \frac{4}{5.987} \, \delta x_B = 0.3341 \, \delta x_B$$

Forces **A** and **B** undergo no displacements in the directions of the forces as a result of displacement δx_B; therefore, they do no work. The work of the force **F** and the work of the weight **W** on the bar as a result of displacement δx_B are as follows:

Since F and δx_B are in the same direction,

$$\delta U_F = F \, \delta x_B$$

Since W and δy are in opposite directions,

$$\delta U_W = W \, (-\delta y_G) = -0.3341 \, W \, \delta x_B$$

For the bar to be in equilibrium, the total work on the bar must be zero. Thus,

$$\delta U_{\text{total}} = \delta U_F + \delta U_W = F \, \delta x_B - 0.3341 \, W \, \delta x_B = (F - 0.3341W) \, \delta x_B = 0$$

Since $\delta x_B \neq 0$,

$$F - 0.3341W = 0$$
$$F = 0.3341W = 0.3341(mg) = 0.3341(100)(9.81) = 328 \text{ N} \qquad \text{Ans.}$$

The relationships between the virtual displacements can also be obtained by using an angular displacement $\delta\theta$ as shown in Fig. 11-9d. Thus,

$$x_B = -L \cos \theta \quad \text{and} \quad \delta x_B = L \sin \theta \, \delta\theta = 0.8315L \, \delta\theta$$
$$y_A = L \sin \theta \quad \text{and} \quad \delta y_A = L \cos \theta \, \delta\theta = 0.5556L \, \delta\theta$$
$$\delta y_G = \frac{1}{2} \, \delta y_A = 0.2778L \, \delta\theta$$

$$\delta U_F = F \, \delta x_B = F(0.8315L \, \delta\theta) = 0.8315FL \, \delta\theta$$
$$\delta U_W = W(-\delta y_G) = W(-0.2778L \, \delta\theta) = -0.2778WL \, \delta\theta$$
$$\delta U_{\text{total}} = \delta U_F + \delta U_W = (0.8315F - 0.2778W) \, L \, \delta\theta = 0$$

Since L and $\delta\theta \neq 0$

$$0.8315F - 0.2778W = 0$$
$$F = 0.3341W = 0.3341(mg) = 0.3341(100)(9.81) = 328 \text{ N} \qquad \text{Ans.}$$

A system of pin-connected bars supports a system of loads as shown in Fig. 11-10a. Use the method of virtual work to determine the horizontal component of the reaction at support G. Neglect the weights of the bars.

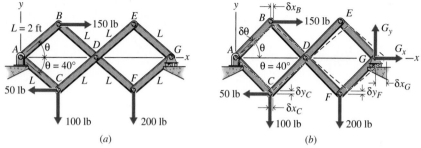

(a)   (b)

Fig. 11-10

SOLUTION

A virtual displacement diagram of the system with the loads superimposed is shown in Fig. 11-10b. The virtual displacements δx and δy at each of the load points and at the support G can be expressed in terms of a virtual angular displacement $\delta \theta$ about support A. Thus,

$$x_B = L \cos \theta \qquad \delta x_B = -L \sin \theta \, \delta \theta = -2(\sin 40°) \, \delta \theta = -1.2856 \, \delta \theta$$
$$x_C = L \cos \theta \qquad \delta x_C = -L \sin \theta \, \delta \theta = -2(\sin 40°) \, \delta \theta = -1.2856 \, \delta \theta$$
$$x_G = 4L \cos \theta \qquad \delta x_G = -4L \sin \theta \, \delta \theta = -4(2)(\sin 40°) \, \delta \theta = -5.1423 \, \delta \theta$$

$$y_C = -L \sin \theta \qquad \delta y_C = -L \cos \theta \, \delta \theta = -2(\cos 40°) \, \delta \theta = -1.5321 \, \delta \theta$$
$$y_F = -L \sin \theta \qquad \delta y_F = -L \cos \theta \, \delta \theta = -2(\cos 40°) \, \delta \theta = -1.5321 \, \delta \theta$$

The virtual work done by each of the forces can then be computed by carefully noting the directions of the forces and the directions of the associated virtual displacements. Thus,

$$\delta U_{Bx} = B_x(\delta x_B) = 150(-1.2856 \, \delta \theta) = -192.84 \, \delta \theta$$
$$\delta U_{Cx} = C_x(\delta x_C) = 50(+1.2856 \, \delta \theta) = +64.28 \, \delta \theta$$
$$\delta U_{Cy} = C_y(\delta y_C) = 100(+1.5321 \, \delta \theta) = +153.21 \, \delta \theta$$
$$\delta U_{Fy} = F_y(\delta y_F) = 200(+1.5321 \, \delta \theta) = +306.42 \, \delta \theta$$
$$\delta U_{Gx} = G_x(\delta x_G) = G_x(-5.1423 \, \delta \theta) = -5.1423 G_x \, \delta \theta$$

Force G_y does not undergo a virtual displacement in the direction of the force as a result of the virtual angular displacement $\delta \theta$; therefore, it does no work. For the system to be in equilibrium, the total work on the system must be zero. Thus

$$\delta U_{\text{total}} = \delta U_{Bx} + \delta U_{Cx} + \delta U_{Cy} + \delta U_{Fy} + \delta U_{Gx}$$
$$= (-192.84 + 64.28 + 153.21 + 306.42 - 5.1423 G_x) \, \delta \theta$$
$$= (331.07 - 5.1423 G_x) \, \delta \theta = 0$$

Since $\delta \theta \neq 0$,

$$331.07 - 5.1423 G_x = 0$$
$$G_x = \frac{331.07}{5.1423} = 64.4 \text{ lb} \qquad \qquad \text{Ans.}$$

The two-bar mechanism shown in Fig. 11-11a is used to support a body W, which has a mass of 50 kg. A couple $\mathbf{C} = 1500$ N·m is also applied to member CDE as shown in the figure. Use the method of virtual work to determine the magnitude of the force exerted by the link at support A to maintain the mechanism in equilibrium in the position shown in the figure. Neglect the weights of the bars.

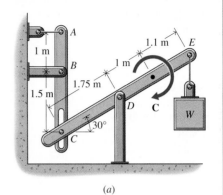

(a)

SOLUTION

A free-body diagram of the mechanism is shown in Fig. 11-11b and a virtual displacement diagram of the mechanism is shown in Fig. 11-11c. The magnitudes of the virtual displacements δx_A and δy_E can be expressed in terms of a virtual angular displacement $\delta\theta$ about support B. Thus,

$$\delta x_A = 1\ \delta\theta = 1.000\ \delta\theta$$

Note from the small diagram shown in Fig. 11-11d that

$$1.5\ \delta\theta = 1.75\ \delta\phi \sin 30°$$

therefore,

$$\delta\phi = \frac{1.5\ \delta\theta}{1.75(0.5000)} = 1.7143\ \delta\theta$$

From the small diagram shown in Fig. 11-11e,

$$\delta y_E = 2.1\ \delta\phi \cos 30° = 2.1(1.7143\ \delta\theta) \cos 30° = 3.118\ \delta\theta$$

Forces $\mathbf{B}_x$, $\mathbf{B}_y$, $\mathbf{D}_x$, and $\mathbf{D}_y$ undergo no virtual displacements in the directions of the forces as a result of the virtual angular displacement $\delta\theta$; therefore, they do no work. The work of the force $\mathbf{A}$, the couple $\mathbf{C}$, and the weight $\mathbf{W}$ on the mechanism, as a result of the virtual angular displacement $\delta\theta$, are

$$\delta U_A = A\ \delta x_A \cos 180° = A(1.000\ \delta\theta)(-1) = -1.000A\ \delta\theta$$
$$\delta U_C = C\ \delta\phi \cos 0° = 1500(1.7143\ \delta\theta)(1) = 2571\ \delta\theta$$
$$\delta U_W = W\ \delta y_E \cos 0° = mg\ \delta y_E \cos 0°$$
$$= 50(9.81)(3.118\ \delta\theta)(1) = 1529\ \delta\theta$$

The total work must be zero for the mechanism to be in equilibrium. Thus,

$$\delta U_{\text{total}} = \delta U_A + \delta U_C + \delta U_W = -1.000A\ \delta\theta + 2571\ \delta\theta + 1529\ \delta\theta$$
$$= (-1.000A + 2571 + 1529)\ \delta\theta = 0$$

Since $\delta\theta \neq 0$,

$$-1.000A + 2571 + 1529 = 0$$

Therefore,

$$A_x = 4100\ \text{N} = 4.10\ \text{kN} \qquad\qquad \text{Ans.}$$

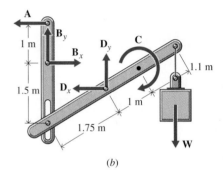

(b)

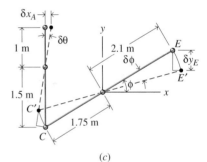

(c)

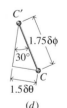

(d)

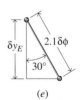

(e)

Fig. 11-11

PROBLEMS

Use the method of virtual work to solve the following problems.

11-15* Two carts carry boxes A and B as shown in Fig. P11-15. The two carts are connected by an inextensible cable, which passes over an ideal pulley. Determine the angle θ for equilibrium if the weights of boxes A and B are 500 and 1000 lb, respectively.

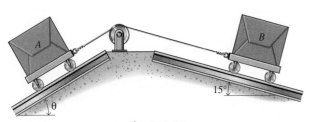

Fig. P11-15

11-16* The cable and pulley system shown in Fig. P11-16 is in equilibrium. If the mass of block B is 200 kg, determine the mass of block A.

Fig. P11-16

11-17* Two spheres are supported with slender rods as shown in Fig. P11-17. Determine the angle θ for equilibrium if the weights of spheres A and B are 25 and 10 lb, respectively.

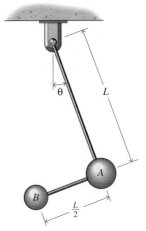

Fig. P11-17

11-18* A three-bar frame is loaded and supported as shown in Fig. P11-18. Determine the angle θ for equilibrium if $\mathbf{P} = 500$ N, $\mathbf{M} = 350$ N·m, $L = 1$ m, and the masses of blocks A and B are 50 and 75 kg, respectively. The weights of the bars are negligible.

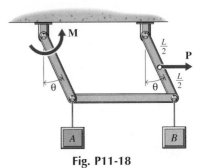

Fig. P11-18

11-19 The two-bar linkage shown in Fig. P11-19 is in equilibrium. The collars at supports A and C slide on

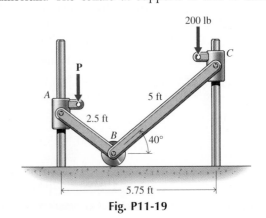

Fig. P11-19

512

smooth rods. Bars *AB* and *BC* have negligible weight. Determine the magnitude of force **P**.

11-20 Two blocks are suspended on a continuous inextensible cord as shown in Fig. P11-20. Determine the angle θ for equilibrium if the masses of blocks *A* and *B* are 50 and 40 kg, respectively.

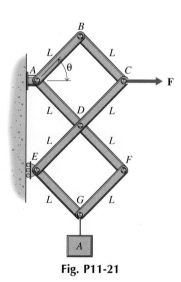

Fig. P11-20

11-21* A frame is loaded and supported as shown in Fig. P11-21. If the weight of block *A* is 500 lb and angle θ equals 45°, determine the magnitude of force **F** required to maintain equilibrium.

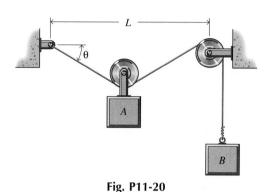

Fig. P11-21

11-22* A frame is loaded and supported as shown in Fig. P11-22. Determine the magnitude of the force **F** required to maintain equilibrium if *P* = 250 N and θ = 45°.

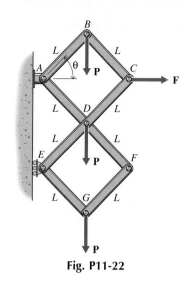

Fig. P11-22

11-23 A frame is loaded and supported as shown in Fig. P11-22. Determine the angle θ for equilibrium if *F* = 750 lb and *P* = 500 lb.

11-24 A frame is loaded and supported as shown in Fig. P11-21. Determine the angle θ for equilibrium if *F* = 800 N and the mass of block *A* is 100 kg.

11-25* A beam is loaded and supported as shown in Fig. P11-25. Determine the reaction at support *B*. Neglect the weight of the beam.

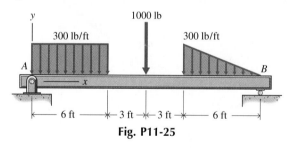

Fig. P11-25

11-26* A beam is loaded and supported as shown in Fig. P11-26. Determine the reaction at support *A*. Neglect the weight of the beam.

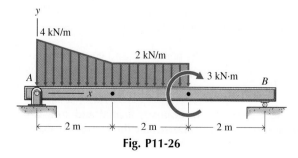

Fig. P11-26

513

11-27* The slender bar shown in Fig. P11-27 weighs 100 lb. If $P = 50$ lb and $M = 0$, determine the magnitude of the force **F** required to maintain the bar in the equilibrium position shown in the figure.

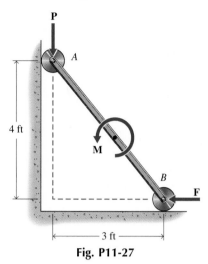

Fig. P11-27

11-28* Determine the horizontal and vertical components of the reaction at support C of the two-bar frame shown in Fig. P11-28.

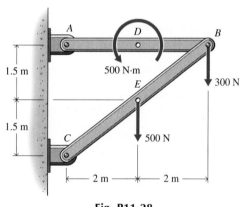

Fig. P11-28

11-29 The slender bar shown in Fig. P11-27 weighs 150 lb. If $P = 75$ lb and $M = 50$ ft · lb, determine the magnitude of the force **F** required to maintain the bar in the equilibrium position shown in the figure.

11-30 Determine the horizontal and vertical components of the reaction at support A of the two-bar frame shown in Fig. P11-28.

11-31* A system of beams is loaded and supported as shown in Fig. P11-31. Determine the reaction at support C. Neglect the weights of the beams.

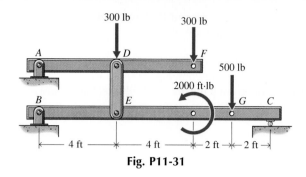

Fig. P11-31

11-32* A system of beams is loaded and supported as shown in Fig. P11-32. Determine the reaction at support A. Neglect the weights of the beams.

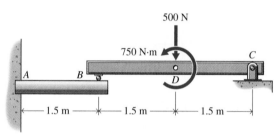

Fig. P11-32

11-33 Determine the force transmitted by member CD of the truss shown in Fig. P11-33.

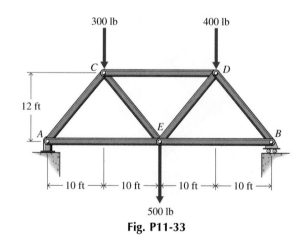

Fig. P11-33

11-34 Determine the force transmitted by member BC of the truss shown in Fig. P11-34.

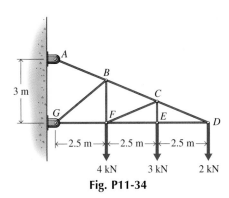

Fig. P11-34

11-35* Determine the horizontal and vertical components of the reaction at support B of the two-bar mechanism shown in Fig. P11-35.

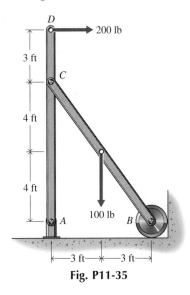

Fig. P11-35

11-36* Determine the force transmitted by member CD of the A frame shown in Fig. P11-36.

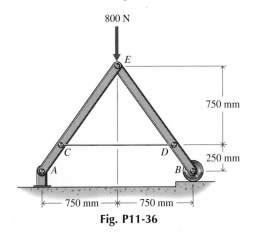

Fig. P11-36

11-37 Determine the horizontal and vertical components of the reaction at support B of the three-bar mechanism shown in Fig. P11-37.

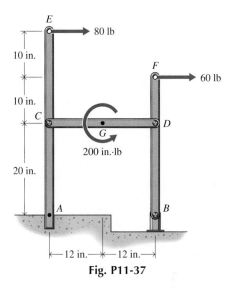

Fig. P11-37

11-38 Determine the horizontal and vertical components of the reaction at support B of the frame shown in Fig. P11-38.

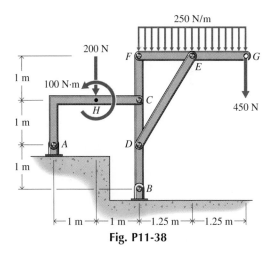

Fig. P11-38

11-39* Determine the force and moment components of the reaction at support A of the three-bar mechanism shown in Fig. P11-37.

11-40* Determine the horizontal and vertical components of the reaction at support A of the frame shown in Fig. P11-38.

11-41 Determine the force transmitted by cable CD of the two-bar frame shown in Fig. P11-41.

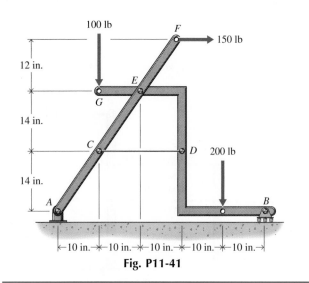

100 lb

F

150 lb

12 in.

E

G

14 in.

C

D 200 lb

14 in.

A

B

←10 in.→*←10 in.→*←10 in.→*←10 in.→*←10 in.→

Fig. P11-41

11-42 The mass of block A shown in Fig. P11-42 is 250 kg. Determine the magnitude of the force F required to support the block in the equilibrium position shown in the figure.

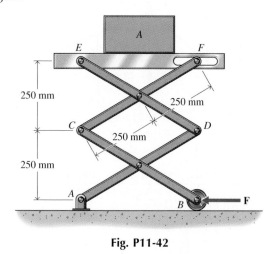

E A F

250 mm

250 mm

C D

250 mm

250 mm

A

B F

Fig. P11-42

11-4   POTENTIAL ENERGY AND EQUILIBRIUM

The potential energy of a body is a measure of its capacity for doing work. The potential energy can be defined quantitatively as the amount of work that the body is capable of doing against outside forces as it passes from a given position or configuration to some standard or reference position or configuration. A compressed spring is capable of doing work by virtue of the relative positions (configuration) of its particles. Potential energy is expressed in the same units as work: in joules (J) if SI units are used and in ft·lb or in.·lb if U.S. Customary units are used.

Conservative Force Systems   When the work done by a body depends only on the initial and final configurations of the system and not at all on the paths taken by the parts of the system while moving from the initial state to the final state, the system is a conservative force system. Conservative force systems occur frequently in engineering. A common example is that of a system consisting of the Earth and an elevated body (rigid or deformable). The work done on the body during any displacement is equal to the Earth pull (weight) of the body times the vertical displacement of the center of gravity of the body. Any intermediate positions occupied by the body as it moves from its initial position to its final position have no effect on net work done on the body.

Another example is that of an elastic body such as a spring. If the body is elastic, the energy possessed by the body in a given strained condition (that is, for a given configuration of its particles) is independent of the movements of the particles that occurred while the body was being put in the given strained condition. The standard configuration may be arbitrarily chosen, but, for convenience, it is usually chosen so that the potential energy of the body is initially positive or zero. Thus, in the case of the Earth and an elevated body, the Earth is considered fixed and the standard configuration occurs when the body is in contact with the Earth.

Nonconservative Force Systems   The most common case of a non-conservative system is one in which the system does work against frictional forces. Any rigid body under the action of a force system in which friction does not occur (or may be considered negligible) is a conservative system provided no change in state or condition of the body except that of position or configuration takes place.

11-4-1   Elastic Potential Energy

A deformable body that changes shape under load but resumes its original shape when the loads are removed is known as an elastic body. The spring shown in Fig. 11-12a is an example of an elastic body that is widely used in engineering applications. When a tensile force is applied to the spring, the length of the spring increases. Similarly, when a compressive force is applied to the spring, the length decreases. The magnitude F of the force that must be applied to the spring to stretch it by an amount s is given by the expression

$$F = ks \tag{11-18}$$

where s is the deformation of the spring from its unloaded position and k is a constant known as the modulus of the spring. Any spring whose behavior is governed by Eq. 11-18 is an ideal linear elastic spring.

The work done in stretching an ideal spring from an initial unstretched position to a stretched position s can be determined from Eq. 11-1. Since the force F and displacement s are in the same direction, the angle α is zero and Eq. 11-1 becomes

$$U = \int_0^s ks \, ds = \frac{1}{2} ks^2 \tag{11-19}$$

For this case, the work done by the force F as it stretches the spring can be represented as the shaded triangular area under the load versus deformation curve shown in Fig. 11-12b. This area also represents the elastic potential energy V_e stored in the spring as a result of the change in shape of the spring.

In a similar manner, the work done in stretching an ideal spring from an initial position s_1 to a further extended position s_2 can be determined from Eq. 11-1. Thus,

$$U = V_e = \int_{s_1}^{s_2} ks \, ds = \frac{1}{2} k(s_2^2 - s_1^2) \tag{11-20}$$

The work done by the force F, in this case, as it stretches the spring can be represented as the shaded trapezoidal area under the load versus deformation curve shown in Fig. 11-12c. It is important to note here that Eq. 11-19 is valid only if the deflection of the spring is measured from its undeformed position.

As a spring is being deformed (either stretched or compressed), the force on the spring and the displacement are in the same direction;

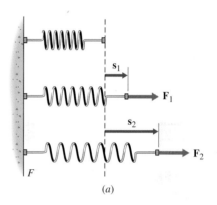

(a)

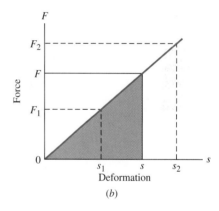

(b)

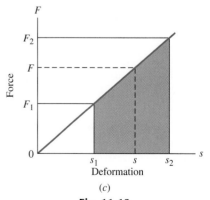

(c)

Fig. 11-12

therefore, the work done on the spring is positive, which increases its potential energy. If a spring is initially deformed and gradually released, the force and displacement are in opposite directions, the work is negative, and the potential energy decreases. The force of a spring on a body is opposite to the force of the body on the spring. Since both the body and the end of the spring have the same displacement, the work done by the spring on the body results in an equal decrease in the potential energy of the spring.

A torsional spring, which is used to resist rotation rather than linear displacement, can also store and release potential energy. The magnitude of the torque (twisting moment) T that must be applied to a torsional spring to produce a rotation θ is given by the expression

$$T = k\theta \tag{11-21}$$

where T is the torque, θ is the angular deformation of the spring from its unloaded position, and k is a constant known as the modulus of the spring. The work U done by the torque T, which is equal to the potential energy V_e stored in the spring, is given by the expression

$$U = V_e = \int_0^\theta T\, d\theta = \int_0^\theta k\theta\, d\theta = \frac{1}{2}k\theta^2 \tag{11-22}$$

which is analogous to the expression for the linear spring.

During a virtual displacement δs of a spring, the virtual work δU done on the spring and the change in virtual elastic potential energy δV_e of the spring are given by the expression

$$\delta U = \delta V_e = F\, \delta s = ks\, \delta s \tag{11-23}$$

11-4-2 Gravitational Potential Energy

The work done by the weight of a body is obtained by multiplying the the weight W of the body by the vertical displacement dh of the center of gravity of the body. Thus,

$$U = W\, dh \tag{11-24}$$

If the displacement is toward the center of the Earth, the work is positive; if it is away from the center of the Earth, the work is negative. The work is independent of the path followed by the body in getting from its initial position to its final position; it depends only on the vertical displacement h of the center of gravity of the body. Thus,

$$U = V_g = \int_0^h W\, dh = Wh = mgh \tag{11-25}$$

The gravitational potential energy possessed by the body is available to do work as the body returns to a lower position. The work done by the weight W as it displaces downward will be positive. This downward movement will produce a decrease in the gravitational potential energy of the body equal to the work done. Thus,

$$\Delta U = W\, \Delta h = -\Delta V_g \tag{11-26}$$

During a virtual displacement δh of the body, the virtual work δU done by the weight of the body, and the change in virtual gravitational potential energy δV_g of the body are given by the expression

$$\delta U = -\delta V_g = W\, \delta h \tag{11-27}$$

11-4-3   The Principle of Potential Energy

Since the work done by a linear spring on a body equals the negative of the change in elastic potential energy of the spring and the work done by the weight W of a body equals the negative of the change in gravitational potential energy of the body, the virtual-work equation can be written as

$$\delta U + \delta V_e + \delta V_g = \delta U + \delta V = 0 \qquad \text{or} \qquad \delta U = -\delta V \quad (11\text{-}28)$$

where $V = V_e + V_g$ stands for the total potential energy of the system and δU represents the virtual work done on the system during a virtual displacement by all external forces other than spring forces and gravitational forces. Thus, the principle of virtual work can be restated as follows:

> A system is in equilibrium if the virtual work done on the system by the external forces equals the change in elastic and gravitational potential energy of the system for all possible virtual displacements.

System Having One Degree of Freedom   For a system whose position can be defined in terms of a single variable (say s), we can write

$$V = V(s) \qquad \delta V = \frac{dV}{ds} \delta s$$

Since $\delta s \neq 0$, the principle of virtual work for equilibrium of the system, as represented by Eq. 11-28, yields

$$\frac{dV}{ds} = 0 \qquad\qquad (11\text{-}29)$$

Thus, the derivative of the potential energy of a conservative system in equilibrium with respect to its position variable is zero. This property is referred to as the principle of potential energy. A system whose position can be described by a single variable is known as a single-degree-of-freedom system.

System Having *n* Degrees of Freedom   If the position of a system depends on several independent variables, the total potential energy stored in the system will be a function of the n independent variables. Thus

$$V = V(s_1, s_2, \cdots, s_n)$$

The change in potential energy δV of such a system is obtained by using the "chain rule" of differential calculus. Thus

$$\delta V = \frac{\partial V}{\partial s_1} \delta s_1 + \frac{\partial V}{\partial s_2} \delta s_2 + \cdots + \frac{\partial V}{\partial s_n} \delta s_n = 0$$

Since $\delta s_1, \delta s_2, \cdots, \delta s_n$ are independent and not equal to zero,

$$\frac{\partial V}{\partial s_1} = 0 \qquad \frac{\partial V}{\partial s_2} = 0 \qquad \cdots \qquad \frac{\partial V}{\partial s_n} = 0 \qquad (11\text{-}30)$$

Equations 11-30 are mathematical expressions of the principle of potential energy for a conservative system with n degrees of freedom.

11-5   STABILITY OF EQUILIBRIUM

The principle of potential energy used in Section 11-4 to determine the equilibrium position of a body or a system of bodies can also be used to determine the state of equilibrium of the body or the system of bodies. The three states of equilibrium (stable, neutral, and unstable) are illustrated in Fig. 11-13.

11-5-1   Stable Equilibrium

The ball in Fig. 11-13a is in a stable equilibrium position because gravity will cause the ball to return to the bottom of the pit once any perturbing force is removed. In this state of stable equilibrium, the work done on the ball during any small displacement is negative since the weight of the ball acts opposite to a component of the displacement; therefore, the potential energy of the ball increases. When the disturbing force is removed, the weight of the ball does positive work as the ball returns to its position of stable equilibrium at the bottom of the pit. Thus, it is evident that a configuration of stable equilibrium occurs at a point of minimum potential energy. Mathematically, the potential energy is minimum for a single-degree-of-freedom system when the second derivative of the potential energy with respect to the displacement variable is positive. Thus, for stable equilibrium

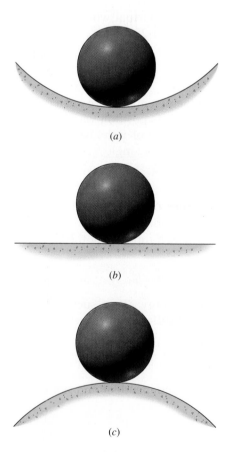

(a)

(b)

(c)

Fig. 11-13

$$\frac{dV}{ds} = 0 \quad \text{and} \quad \frac{d^2V}{ds^2} > 0 \tag{11-31}$$

11-5-2   Neutral Equilibrium

The ball shown in Fig. 11-13b is in a neutral equilibrium position on the horizontal plane because it will remain at any new position to which it is displaced once the disturbing force is removed. In this state of neutral equilibrium, no work is done on the ball during the displacement since the weight of the ball acts perpendicular to the direction of the displacement; therefore, the potential energy of the ball remains constant. Thus, the conditions for neutral equilibrium are

$$\frac{dV}{ds} = 0 \quad \text{and} \quad \frac{d^2V}{ds^2} = \frac{d^3V}{ds^3} = \cdots = 0 \tag{11-32}$$

11-5-3   Unstable Equilibrium

The ball shown in Fig. 11-13c is in an unstable equilibrium position at the top of a hill because, if perturbed, gravity will cause it to move away from its original equilibrium position. In this state of unstable equilibrium, the work done on the ball during any small displacement is positive since the weight of the ball acts in the same direction as a component of the displacement; therefore, the potential energy of the ball decreases. When the disturbing force is removed, the potential energy continues to decrease as the weight of the body causes the body to continue to move away from the position of unstable equilibrium. Thus, a configuration of unstable equilibrium occurs at a point of maximum potential energy. Mathematically, the potential energy is maximum when the second derivative of the potential energy with respect to the displacement variable is negative. Thus, for unstable equilibrium

$$\frac{dV}{ds} = 0 \quad \text{and} \quad \frac{d^2V}{ds^2} < 0 \tag{11-33}$$

If both the first and the second derivatives of V are zero, the sign of a higher order derivative must be used to determine the type of equilibrium. The equilibrium will be stable if the order of the first nonzero derivative is even and the value of the derivative is positive. The equilibrium will be neutral if all derivatives are zero. In all other cases the equilibrium will be unstable.

If the system possesses several degrees of freedom, the potential energy V depends on several variables, and a more advanced treatment involving the theory of functions of several variables is required to determine whether V is minimum. Therefore, stability criteria for multiple-degree-of-freedom systems is beyond the scope of this text.

The use of the potential energy of a system for the determination of static equilibrium and stability is illustrated in the following example problems.

Two bars (L = 2 ft) are used to support a weight (W = 100 lb) as shown in Fig. 11-14a. Determine the value of k for equilibrium at θ = 30°. Assume that the weights of the bars are negligible and that the spring is unstretched when θ = 0°. Show that the equilibrium is stable.

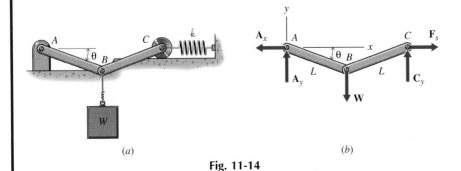

(a) (b)

Fig. 11-14

SOLUTION

A free-body diagram for the system is shown in Fig. 11-14b. From the diagram it is evident that

$$y_B = -L \sin \theta \quad \text{and} \quad x_C = 2L \cos \theta$$

Therefore, the potential energies of the weight, spring, and system with respect to a reference position at θ = 0° are

$$V_W = Wy_B = W(-L \sin \theta) = -WL \sin \theta$$
$$V_S = \frac{1}{2}ks^2 = \frac{1}{2}k(2L - x_C)^2 = \frac{1}{2}k(2L - 2L \cos \theta)^2 = 2kL^2(1 - \cos \theta)^2$$
$$V = V_W + V_S = -WL \sin \theta + 2kL^2(1 - \cos \theta)^2$$

For equilibrium (Eq. 11-29)

$$\frac{dV}{d\theta} = -WL \cos \theta + 4kL^2(1 - \cos \theta) \sin \theta = 0$$

Solving for k yields,

$$k = \frac{W}{4L\,(1 - \cos \theta)\tan \theta} = \frac{100}{4(2)(1 - \cos 30°)\tan 30°} = 161.6 \text{ lb/ft} \quad \text{Ans.}$$

From Eq. 11-31

$$\frac{d^2V}{d\theta^2} = WL \sin \theta + 4kL^2 \cos \theta + 4kL^2 \sin^2 \theta - 4kL^2 \cos^2 \theta$$

With W = 100 lb, L = 2 ft, k = 161.6 lb/ft, and θ = 30°:

$$\frac{d^2V}{d\theta^2} = +1046 > 0$$

Therefore, the equilibrium is stable.

EXAMPLE PROBLEM 11-8

A mass ($m = 100$ kg) and a spring ($k = 50$ kN/m) are attached to a crank, which rotates about a frictionless pin at support O as shown in Fig. 11-15a. The spring is unstretched when $\theta = 0°$. Determine the angle θ for equilibrium and show that the equilibrium is stable.

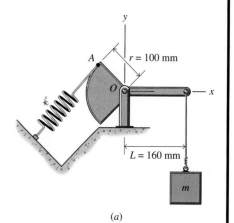

(a)

SOLUTION

The potential energies of the mass, spring, and system with respect to a reference position at $\theta = 0°$ (see Fig. 11-15b) are

$$V_m = mg(-h) = mg(-L \sin \theta) = -mgL \sin \theta$$

$$V_S = \frac{1}{2}ks^2 = \frac{1}{2}k(r\theta)^2 = \frac{1}{2}kr^2\theta^2$$

$$V = V_m + V_S = -mgL \sin \theta + \frac{1}{2}kr^2\theta^2$$

For equilibrium (Eq. 11-29)

$$\frac{dV}{d\theta} = -mgL \cos \theta + kr^2\theta = 0$$

Therefore,

$$\frac{\cos \theta}{\theta} = \frac{kr^2}{mgL} = \frac{50(10^3)(0.100)^2}{100(9.81)(0.160)} = 3.1855 \qquad (a)$$

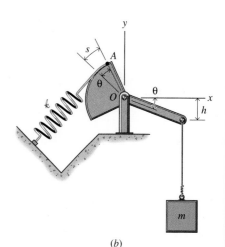

(b)

Fig. 11-15

Equation a can be solved by using trial and error or by using a numerical method. Solving by using the Newton–Raphson method (see Appendix D) yields

$$f(\theta) = \cos \theta - 3.1855 \, \theta = 0$$
$$f'(\theta) = -\sin \theta - 3.1855 = 0$$
$$\delta = -\frac{f(\theta)}{f'(\theta)} = \frac{\cos \theta - 3.1855 \, \theta}{\sin \theta + 3.1855}$$
$$\theta_{n+1} = \theta_n + \delta(\theta_n) \qquad \theta \text{ in radians}$$

n	θ_n	$\delta(\theta_n)$
0	0	0.3139
1	0.3139	−0.0140
2	0.2999	$9.94(10^{-6})$

Thus,

$$\theta = 0.2999 \text{ rad} = 17.18° \qquad \text{Ans.}$$

From Eq. 11.31

$$\frac{d^2V}{d\theta^2} = mgL \sin \theta + kr^2$$

$$= 100(9.81)(0.160)(\sin 17.18°) + 50(10^3)(0.100)^2$$
$$= +546 > 0$$

Therefore, the equilibrium is stable. $\qquad$ Ans.

A system of bars and springs is used to support a weight W as shown in Fig. 11-16a. Determine the value of W for equilibrium and investigate the stability of equilibrium. The weights of the bars are negligible and the springs are unstretched when $\theta = 0°$. Assume that the spring forces remain horizontal.

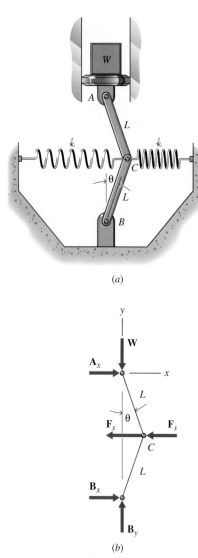

(a)

(b)

Fig. 11-16

SOLUTION

A free-body diagram for the system is shown in Fig. 11-16b. From the diagram it is evident that

$$x_C = L \sin \theta \quad \text{and} \quad y_A = 2L - 2L \cos \theta$$

Therefore, the potential energies of the weight, springs, and system with respect to a reference position at $\theta = 0°$ are

$$V_W = W y_A = W(2L - 2L \cos \theta) = -2WL(1 - \cos \theta)$$

$$V_S = \frac{1}{2} k s^2 = \frac{1}{2} k(x_C)^2 = \frac{1}{2} k(L \sin \theta)^2 = \frac{1}{2} k L^2 \sin^2 \theta$$

$$V = V_W + 2V_S = -2WL(1 - \cos \theta) + k L^2 \sin^2 \theta$$

For equilibrium (Eq. 11-29)

$$\frac{dV}{d\theta} = -2WL \sin \theta + 2k L^2 \sin \theta \cos \theta = 0$$

or

$$\sin \theta \, (-W + k L \cos \theta) = 0$$

Thus, the system is in equilibrium when

$$\sin \theta = 0 \qquad \text{or} \qquad \theta = 0$$
$$-W + k L \cos \theta = 0 \qquad \text{or} \qquad W = k L \cos \theta$$

From Eq. 11-31

$$\frac{d^2V}{d\theta^2} = -2WL \cos \theta - 2k L^2 \sin^2 \theta + 2k L^2 \cos^2 \theta$$

For $\theta = 0°$

$$\frac{d^2V}{d\theta^2} = -2WL + 2k L^2 = 2L(-W + k L)$$

With $W < k L$

$$\frac{d^2V}{d\theta^2} > 0 \qquad \text{the equilibrium is stable}$$

With $W = k L$

$$\frac{d^2V}{d\theta^2} = 0 \qquad \text{the equilibrium is neutral}$$

With $W > k L$

$$\frac{d^2V}{d\theta^2} < 0 \qquad \text{the equilibrium is unstable}$$

For $W = k L \cos \theta$

$$\frac{d^2V}{d\theta^2} = -2WL \cos \theta - 2k L^2 \sin^2 \theta + 2k L^2 \cos^2 \theta$$

$$= -2WL\left(\frac{W}{k L}\right) - 2k L^2\left[1 - \left(\frac{W}{k L}\right)^2\right] + 2k L^2\left(\frac{W}{k L}\right)^2 = \frac{2}{k}(W^2 - k^2 L^2)$$

With $W > k L$

$$\frac{d^2V}{d\theta^2} > 0 \qquad \text{the equilibrium is stable}$$

With $W = k L$

$$\frac{d^2V}{d\theta^2} = 0 \qquad \text{the equilibrium is neutral}$$

With $W < k L$

$$\frac{d^2V}{d\theta^2} < 0 \qquad \text{the equilibrium is unstable}$$

PROBLEMS

11-43* A body ($W = 120$ lb) is supported by a flexible cable and a spring ($k = 80$ lb/in.) as shown in Fig. P11-43. The pulley over which the cable passes is weightless and frictionless. If a force **P** of 100 lb is slowly applied to the body, what is the change in potential energy of

a. The spring?
b. The system consisting of the body, the cable, and the spring?

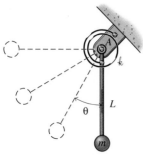

Fig. P11-43

11-44* A sphere ($m = 10$ kg) is supported at the end of a slender rod ($L = 300$ mm) as shown in Fig. P11-44. The torsional spring ($k = 300$ N·m/rad at support A is neutral when $\theta = 0°$. Determine the potential energy of the system at angles of 30°, 60°, and 90°.

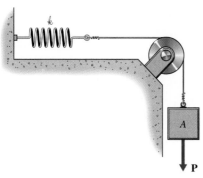

Fig. P11-44

11-45 A moment **M** is slowly increased as it is applied to the disk shown in Fig. P11-45. Determine the change in potential energy of the system with respect to a reference at $\theta = 0°$ as the disk is rotated to the position $\theta = 45°$. The spring ($k = 10$ lb/in.) is unstretched at $\theta = 0°$.

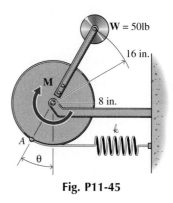

Fig. P11-45

11-46 A force **F** is slowly increased as it is applied to bar AB of Fig. P11-46. Determine the change in potential energy of the system with respect to a reference at $\theta = 0°$ as the bar is rotated to the position $\theta = 30°$. The bar is in equilibrium at $\theta = 0°$ with $F = 0$. The spring constant k equals 500 N/m.

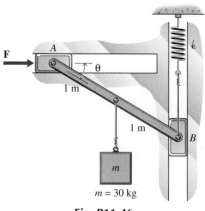

Fig. P11-46

11-47* The uniform bar AB shown in Fig. P11-47 weigh 30 lb. If the spring is unstretched when $\theta = 0°$, determine the angle θ for equilibrium if $k = 15$ lb/ft and $L = 3$ ft.

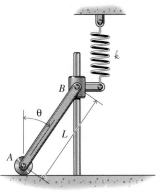

Fig. P11-47

11-48* The uniform bar AB shown in Fig. P11-48 has a mass of 10 kg. The spring is unstretched when $\theta = 0°$. If $k = 50$ N/m and $L = 2$ m, determine the angle θ for equilibrium.

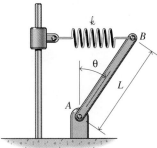

Fig. P11-48

11-49 The uniform bar AB shown in Fig. P11-49 weighs 100 lb. If the spring has an unstretched length of 5 ft, determine the spring constant k required for equilibrium at the position indicated in the figure.

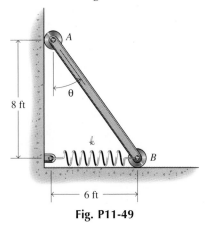

Fig. P11-49

11-50 The uniform bar AB shown in Fig. P11-50 has a mass of 20 kg and a length of 2 m. If the spring is un-

stretched at an angle $\theta = 30°$, determine the spring constant k for equilibrium at an angle $\theta = 60°$.

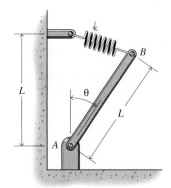

Fig. P11-50

11-51* A two-bar mechanism supports a weight ($W = 100$ lb) as shown in Fig. P11-51. The length of bar AB is 3 ft, the unstretched length of the spring is 1.5 ft, and the spring stiffness $k = 30$ lb/ft. Determine the angle θ for equilibrium of the mechanism. Assume that the weights of the bars are negligible.

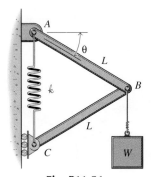

Fig. P11-51

11-52* A two-bar mechanism supports a body ($m = 75$ kg) as shown in Fig. P11-52. The unstretched length of the spring is 600 mm. Determine the spring stiffness k required to limit the angle θ for equilibrium to 35°. Assume that the masses of the bars are negligible.

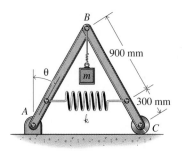

Fig. P11-52

11-53* The spring in Fig. P11-53 has an unstretched length of 2 ft. Determine the angle θ for equilibrium when $W_D = W_E = W_F = 80$ lb, $L = 2$ ft, and $k = 350$ lb/ft. Assume that the weights of the bars are negligible.

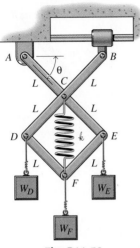

Fig. P11-53

11-54* The two springs in Fig. P11-54 are identical and have unstretched lengths of 0.5 m. Determine the spring constant k required for equilibrium at an angle $\theta = 30°$ if $m_A = m_B = m_C = 20$ kg and $L = 1.5$ m. Assume that the masses of the bars are negligible.

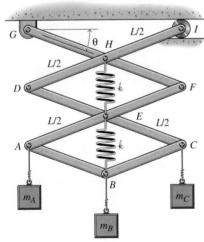

Fig. P11-54

11-55 Solve Problem 11-53 if $W_D = 75$ lb, $W_E = 100$ lb, and $W_F = 90$ lb.

11-56 Solve Problem 11-54 if $m_A = 10$ kg, $m_B = 25$ kg, and $m_C = 15$ kg.

11-57* The spring in Fig. P11-57 is unstretched when $\theta = \theta_0 = 45°$. Find the angle θ for equilibrium if $W = 20$ lb, $L =$

18 in., $a = 8$ in., and $k = 12$ lb/in. Assume that the weights of the bars are negligible.

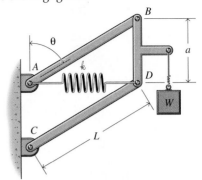

Fig. P11-57

11-58* The spring in Fig. P11-58 is unstretched when $\theta = \theta_0 = 15°$. Find the angle θ for equilibrium if $m_B = 60$ kg, $m_D = 25$ kg, $L = 1$ m, and $k = 750$ N/m. Assume that the masses of the bars are negligible.

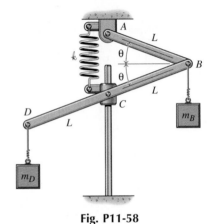

Fig. P11-58

11-59 The spring in Fig. P11-59 has an unstretched length of 2 ft. Find the spring constant k required for equilibrium at an angle $\theta = 30°$ if $W_B = 300$ lb, $W_D = 500$ lb, and $L = 4$ ft. Assume that the weights of the bars are negligible.

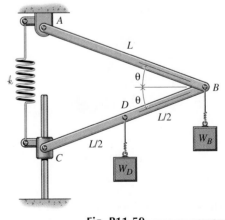

Fig. P11-59

11-60 The spring in Fig. P11-60 is unstretched when $\theta = 0°$. Find the spring constant k required for equilibrium at an angle $\theta = 25°$ if $m_B = 50$ kg, $m_D = 25$ kg, $L = 1.5$ m. Assume that the masses of the bars are negligible.

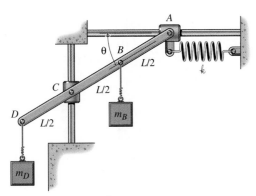

Fig. P11-60

11-61* A system of bars and a spring are used to support a weight W as shown in Fig. P11-61. The weights of the bars are negligible and the spring is unstretched when $\theta = 25°$. Determine the angle θ for equilibrium if $W = 50$ lb, $L = 12$ in., and $k = 12$ lb/in.

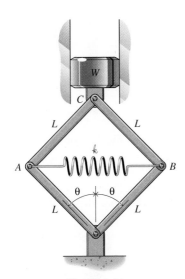

Fig. P11-61

11-62* Two bars and a torsional spring are used to support a mass m as shown in Fig. P11-62. The masses of the bars are negligible and the spring is neutral when $\theta = 0°$. Determine the angle θ for equilibrium if $m = 25$ kg, $L = 1.25$ m, and $k = 150$ N·m/rad.

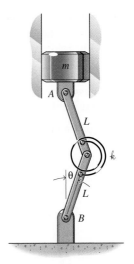

Fig. P11-62

11-63 A block is supported by the three-bar frame shown in Fig. P11-63. The weights of the bars are negligible. The spring can act in either tension or compression and is unstretched when $\theta = 0°$. Determine the maximum weight W for the block, in terms of k and L, if the system is to be in stable equilibrium in the position shown in the figure.

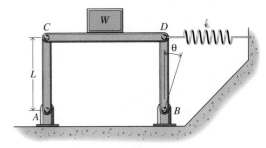

Fig. P11-63

11-64 A block of mass m is supported by the three-bar frame shown in Fig. P11-64. The masses of the bars are negligible and the torsional springs are neutral when $\theta = 0°$. Determine the minimum stiffness k for the springs, in terms of mass m and length L, if the system is to be in stable equilibrium in the position shown in the figure.

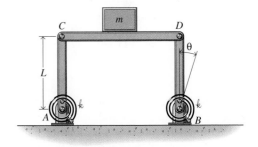

Fig. P11-64

11-65* Bar *AB* of Fig. P11-65 with weight *W* and length *L* has a uniform cross section. Rotation of the bar is resisted by a torsional spring with spring constant k. The spring is neutral when $\theta = 0°$. Determine

a. The minimum stiffness k for the spring, in terms of *W* and *L*, if the bar is to be in stable equilibrium when $\theta = 0°$.

b. The equilibrium angle θ and the stability of the equilibrium if *W* = 25 lb, *L* = 3 ft, and k = 35 lb/ft.

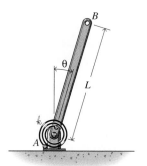

Fig. P11-65

11-66* The two springs shown in Fig. P11-66 can act in either tension or compression and are unstretched when $\theta = 0°$. Determine the possible positions of equilibrium and the stability of equilibrium in each of these positions if *m* = 30 kg, *L* = 1 m, and k = 150 N/m.

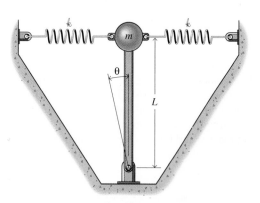

Fig. P11-66

11-67 The two springs shown in Fig. P11-67 can act in either tension or compression and are unstretched when $\theta = 0°$. Determine the possible positions of equilibrium and the stability of equilibrium in each of these positions if *W* = 50 lb, *L* = 4 ft, and k = 28 lb/ft.

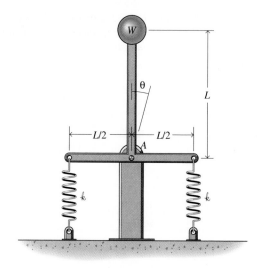

Fig. P11-67

11-68 The two springs shown in Fig. P11-68 can act in either tension or compression and are unstretched when $\theta = 0°$. Determine the possible positions of equilibrium and the stability of equilibrium in each of these positions if *m* = 10 kg, *L* = 2 m, and k = 68 N/m.

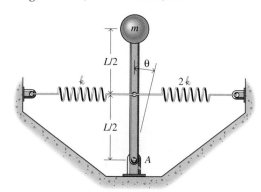

Fig. P11-68

11-69* Bar *AB* of Fig. P11-69 with weight *W* and length *L* has a uniform cross section. The spring has an unstretched

Fig. P11-69

530

length L. Determine the possible positions of equilibrium and the stability of equilibrium in each of these positions if $W = 20$ lb, $L = 2$ ft, and $k = 28$ lb/ft.

11-70   The weight of bar BC of Fig. P11-70 is negligible. The spring has an unstretched length L. Determine the possible positions of equilibrium and the stability of equilibrium in each of these positions if $m = 50$ kg, $L = 1.5$ m, and $k = 600$ N/m.

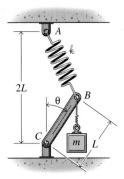

Fig. P11-70

SUMMARY

Static equilibrium problems are normally solved by isolating the body of interest, drawing a free-body diagram, and solving the equations $\mathbf{R} = \mathbf{C} = \mathbf{0}$ for the desired unknowns. A second method for solving static equilibrium problems makes use of the principle of virtual work. The work dU done by a force $\mathbf{F}$ is defined as the magnitude of the force multiplied by the component of an infinitesimal linear displacement in the direction of the force. Thus,

$$dU = \mathbf{F} \cdot d\mathbf{s} = |\mathbf{F}||d\mathbf{s}| \cos \alpha = F \, ds \cos \alpha \qquad (11\text{-}1)$$

The work done by a couple $\mathbf{C}$ is determined from the work done by the two equal forces that compose the couple during an infinitesimal angular displacement. The two forces experience infinitesimal linear displacements $ds_1 = ds_2 = r \, d\theta$; therefore,

$$dU = 2F(r \, d\theta) = M \, d\theta \qquad (11\text{-}5)$$

When a body being acted on by a force $\mathbf{F}$ moves through an infinitesimal linear displacement $d\mathbf{s}$, the body is not in equilibrium. In studying the equilibrium of bodies by the method of virtual work, fictitious displacements called virtual displacements ($\delta\mathbf{s}$ rather than $d\mathbf{s}$) are introduced. The work done by a force $\mathbf{F}$ acting on a body during a virtual displacement $\delta\mathbf{s}$ is called virtual work δU and is represented mathematically as

$$\delta U = \mathbf{F} \cdot \delta\mathbf{s} \qquad \text{or} \qquad \delta U = F \, \delta s \cos \alpha \qquad (11\text{-}9)$$
$$\delta U = \mathbf{C} \cdot \delta\boldsymbol{\theta} \qquad \text{or} \qquad \delta U = M \, \delta\theta \cos \alpha \qquad (11\text{-}10)$$

The principle of virtual work states that a particle, a rigid body, or a system of connected rigid bodies is in equilibrium if the virtual work done by all external forces and couples acting on the system is zero for all possible virtual displacements of the system.

The principle of virtual work can be expressed mathematically as:

$$\delta U = \sum_{i=1}^{m} \mathbf{F}_i \cdot \delta\mathbf{s}_i + \sum_{j=1}^{n} \mathbf{C}_j \cdot \delta\boldsymbol{\theta}_j = 0 \qquad (11\text{-}11)$$

531

If the virtual work done by all external forces acting on the body for all possible virtual displacements of the body is zero, then

$$\delta U = \Sigma F_x\, \delta_x + \Sigma F_y\, \delta y + \Sigma F_z\, \delta z + \Sigma M_x\, \delta\theta_x + \Sigma M_y\, \delta\theta_y + \Sigma M_z\, \delta\theta_z = 0$$

$$(11\text{-}17)$$

Applying the virtual work principle by taking virtual linear and angular displacements one at a time yields expressions that are identical to the six scalar equations of equilibrium; namely,

$$\Sigma F_x = 0 \qquad \Sigma F_y = 0 \qquad \Sigma F_z = 0$$
$$\Sigma M_x = 0 \qquad \Sigma M_y = 0 \qquad \Sigma M_z = 0$$

Therefore, the virtual work equation $\delta U = 0$ is simply an alternative statement of the equilibrium equations for a rigid body.

The potential energy of a body is a measure of its capacity for doing work. A deformed elastic member (such as a spring) stores energy V_e, which is potentially available to do work on some other body. Similarly, an elevated body possesses gravitational potential energy V_g, which is available to do work on some other body. In terms of potential energy, the virtual-work equation can be written as

$$\delta U + \delta V_e + \delta V_g = 0 \qquad \text{or} \qquad \delta U = -(\delta V_e + \delta V_g) = -\delta V \quad (11\text{-}28)$$

Equation 11-28 indicates that the virtual work done by the external forces on a system in equilibrium is equal to the change in elastic and gravitational potential energy of the system for all possible virtual displacements consistent with the constraints.

For a system whose position can be defined in terms of a single variable (single-degree-of-freedom system),

$$V = V(s) \qquad \delta V = \frac{dV}{ds}\, \delta s$$

Since $\delta s \neq 0$, Eq. 11-28 indicates that the system is in equilibrium when

$$\frac{dV}{ds} = 0 \qquad\qquad (11\text{-}29)$$

The potential energy method can be extended to determine whether the equilibrium is stable or unstable. Mathematically, the potential energy is minimum for a single-degree-of-freedom system when the second derivative of the potential energy with respect to the displacement variable is positive. For stable equilibrium (at a position of minimum potential energy)

$$\frac{dV}{ds} = 0 \qquad \text{and} \qquad \frac{d^2V}{ds^2} > 0 \qquad\qquad (11\text{-}31)$$

For unstable equilibrium (at a position of maximum potential energy)

$$\frac{dV}{ds} = 0 \qquad \text{and} \qquad \frac{d^2V}{ds^2} < 0 \qquad\qquad (11\text{-}32)$$

If both the first and the second derivatives of the potential energy V are zero, the sign of a higher order derivative must be used to determine the type of equilibrium. The equilibrium will be stable if the first nonzero derivative is even and positive. The equilibrium will be neutral if all derivatives are zero. In all other cases the equilibrium will be unstable. If the system possesses several degrees of freedom, the potential energy V depends on several variables, and a more advanced treatment involving the theory of functions of several variables is required to determine when the potential energy V is minimum. Stability criteria for such systems is beyond the scope of this text.

REVIEW PROBLEMS

Use the method of virtual work to solve the following problems.

11-71* A system of beams supports three concentrated loads as shown in Fig. P11-71. Determine the reaction at support B.

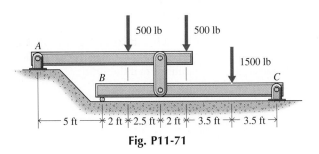

Fig. P11-71

11-72* A pin-connected truss is loaded and supported as shown in Fig. P11-72. Determine

a. The reaction at support E.
b. The force in member CD.

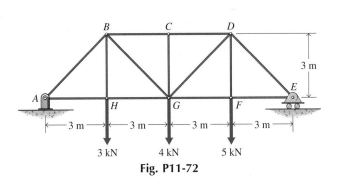

Fig. P11-72

11-73 Determine the equilibrium position (angles θ_1 and θ_2) for the system of pin-connected bars shown in Fig. P11-73. Bars AB and BC weigh 50 and 30 lb, respectively, and $P = 20$ lb.

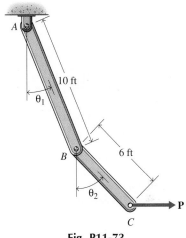

Fig. P11-73

11-74 The masses of bar AB, bar BC, and body W of Fig. P11-74 are 20 kg, 20 kg, and 10 kg, respectively. Determine the magnitude of force P if the system is in equilibrium in the position shown.

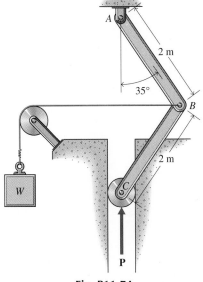

Fig. P11-74

11-75* A slider crank mechanism with a spring attached to the slider is shown in Fig. P11-75. Bars AB and BC weigh 10 and 20 lb, respectively. The spring has a modulus k of 50 lb/in. and is unstretched when $\theta = 0°$. If the system is in equilibrium when $\theta = 30°$, determine the magnitude of the moment **M** that must be applied to crank AB. Friction between the slider and the horizontal surfaces is negligible.

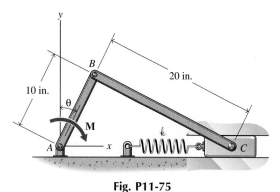

10 in.

20 in.

Fig. P11-75

11-77 With force **P** applied, the bars of Fig. P11-77 are in equilibrium and the spring ($k = 25$ lb/in.) is unstretched. The weights of bar AB, bar BC, and block W are 20 lb, 25 lb, and 50 lb, respectively. Determine the new equilibrium position of the system if force **P** is removed.

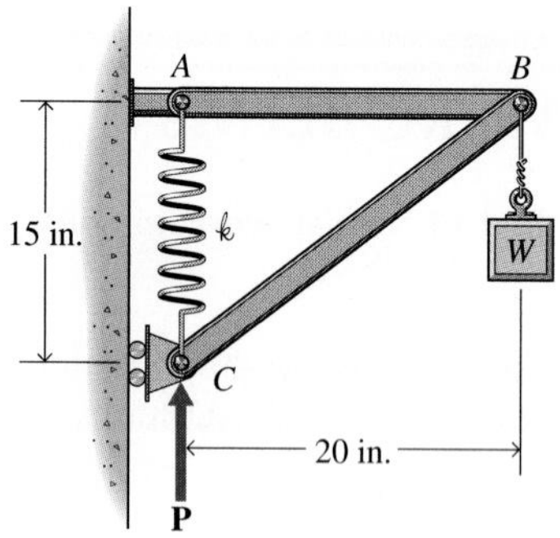

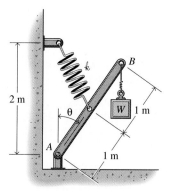

15 in.

20 in.

Fig. P11-77

11-76* The three-bar frame shown in Fig. P11-76 supports a block W, which has a mass of 200 kg. The spring AC is unstretched when $\theta = 0°$. The three bars have negligible masses. If the system is in equilibrium when $\theta = 30°$, determine

a. The modulus k of the spring.
b. The stability of this equilibrium position.

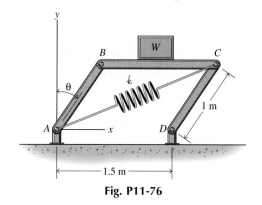

1 m

1.5 m

Fig. P11-76

11-78 The uniform bar AB shown in Fig. P11-78 has a mass of 20 kg and the block W has a mass of 50 kg. Determine the angle θ for equilibrium if the spring ($k = 5$ kN/m) is unstretched when $\theta = 30°$.

2 m

1 m

1 m

Fig. P11-78

VECTOR OPERATIONS

Physical quantities of interest in mechanics such as mass, force, time, and distance, can be represented by two kinds of quantities: scalars and vectors. The mathematical operations required for their use in statics and dynamics problems will be presented in this appendix.

A-1-1 Scalar Quantities

Scalar quantities can be completely described with a magnitude. Examples of scalar quantities in mechanics are mass, density, length, area, volume, speed, energy, time, and temperature. In this book and in the companion text on dynamics, symbols representing scalar quantities will be printed in lightface italic type A. In mathematical operations, scalars follow the rules of elementary algebra.

A-1-2 Vector Quantities

A vector quantity has both a magnitude and a direction (line of action and sense) and obeys the parallelogram law of addition, which will be described in Section A-2 of this Appendix. Examples of vector quantities in mechanics are force, displacement, velocity, and acceleration. A vector quantity can be represented graphically by using a directed line segment (arrow), as shown in Fig. A-1. The magnitude of the quantity is represented by the length of the arrow. The direction is specified by using the angle θ between the arrow and some known reference direction. The sense is indicated by the arrowhead at the tip of the line segment. Symbols representing vector quantities will be printed in boldface type $\mathbf{A}$ to distinguish them from scalar quantities. Symbols representing magnitudes $|\mathbf{A}|$ of vector quantities will be printed in lightface italic type A. In all handwritten work, it is important to distinguish between scalar and vector quantities since, in mathematical operations vector quantities do not follow the rules of elementary alge-

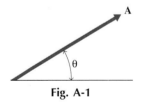

Fig. A-1

bra. A small arrow over the symbol for a vector quantity $\vec{A}$ is often used in handwritten work to take the place of boldface type in printed material.

Vectors can be classified into three types: free, sliding, or fixed.

1. A free vector has a specific magnitude, slope, and sense, but its line of action does not pass through a unique point in space.
2. A sliding vector has a specific magnitude, slope, and sense, and its line of action passes through a unique point in space. The point of application of a sliding vector can be anywhere along its line of action.
3. A fixed vector has a specific magnitude, slope, and sense, and its line of action passes through a unique point in space. The point of application of a fixed vector is confined to a fixed point on its line of action.

A-2   ELEMENTARY VECTOR OPERATIONS

A-2-1   Addition of Vectors

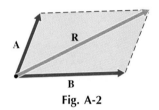

Fig. A-2

Two free vectors **A** and **B** can be translated so that their tails meet at a common point, as shown in Fig. A-2. The parallelogram law of vector addition states that the two vectors **A** and **B** are equivalent to a vector **R**, which is the diagonal of a parallelogram constructed by using vectors **A** and **B** as the two adjacent sides (see Fig. A-2). The vector **R** is the diagonal that passes through the point of intersection of vectors **A** and **B**. Vector **R** is known as the resultant of the two vectors **A** and **B**. Vector quantities must obey this parallelogram law of addition. The resultant **R** can be represented mathematically by the vector equation

$$\mathbf{R} = \mathbf{A} + \mathbf{B} \tag{A-1}$$

The plus sign used in conjunction with vector quantities **A** and **B** (boldface type) indicates vector (parallelogram law) addition not scalar (algebraic) addition.

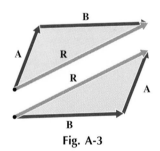

Fig. A-3

It is obvious from Fig. A-2 that the two vectors **A** and **B** can be added in a head-to-tail fashion, as shown in the top part of Fig. A-3 to obtain the vector sum **R**. Similarly, the two vectors can be added in the manner shown in the bottom part of Fig. A-3 to obtain **R**, since the two triangles shown in Fig. A-3 are the two different halves of the parallelogram shown in Fig. A-2. The procedure of adding vectors in a head-to-tail fashion is known as the triangle law. Note from Fig. A-3 that the order of addition of the vectors does not affect their sum; therefore,

$$\mathbf{A} + \mathbf{B} = \mathbf{B} + \mathbf{A} \tag{A-2}$$

Equation A-2 establishes the fact that vector addition is commutative.

The sum of three or more vectors $\mathbf{A} + \mathbf{B} + \mathbf{C} + \cdots$ can be obtained by first adding the vectors **A** and **B**, and then adding the vector **C** to the vector sum $(\mathbf{A} + \mathbf{B})$. The procedure, which is illustrated in Fig. A-4, can be expressed mathematically as

$$\mathbf{A} + \mathbf{B} + \mathbf{C} = (\mathbf{A} + \mathbf{B}) + \mathbf{C} \tag{A-3}$$

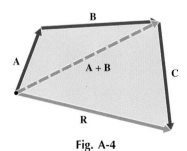

Fig. A-4

Similarly, the sum of four vectors can be obtained by adding the fourth vector to the sum of the first three. Thus, the sum of any number of

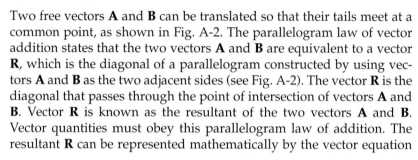

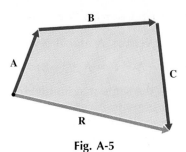

Fig. A-5

vectors can be obtained by applying the parallelogram law, as expressed mathematically by Eq. A-3, to successive pairs of vectors until all the given vectors are replaced by a single resultant vector **R**.

The sum of three or more vectors **A** + **B** + **C** + ··· can also be obtained by repeated application of the triangle law. In Fig. A-5 the sum of three vectors is obtained directly by connecting the three vectors, in sequence, in a tip-to-tail fashion. The resultant **R** is the vector that connects the tail of the first vector with the tip of the last one. The procedure, illustrated in Fig. A-5, can be applied to any number of vectors and is known as the polygon rule for vector addition. The sketch of the vectors connected in a tip-to-tail fashion is known as a vector polygon. As illustrated in Fig. A-6, the resultant **R** is independent of the order in which the vectors **A**, **B**, and **C** are introduced into the vector polygon. This fact, which can be expressed mathematically as

$$\mathbf{A} + \mathbf{B} + \mathbf{C} = (\mathbf{A} + \mathbf{B}) + \mathbf{C} = \mathbf{A} + (\mathbf{B} + \mathbf{C}) \qquad \text{(A-4)}$$

shows that vector addition is associative.

The resultant **R** of two or more vectors can be determined graphically by drawing either the parallelogram or the vector polygon to scale. In practice, the magnitude R of the resultant is determined algebraically by applying the cosine law for a general triangle, as illustrated in Fig. A-7. Thus,

$$R = \sqrt{A^2 + B^2 - 2AB \cos \theta} \qquad \text{(A-5)}$$

The angles between resultant **R** and vectors **A** and **B** can be determined by using the law of sines for a general triangle. Thus, for the triangle shown in Fig. A-7,

$$\frac{\sin \alpha}{A} = \frac{\sin \beta}{B} = \frac{\sin \theta}{R} \qquad \text{(A-6)}$$

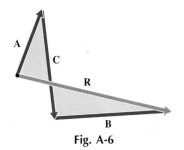

Fig. A-6

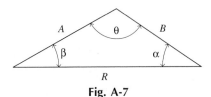

Fig. A-7

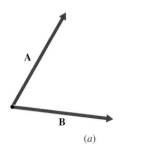

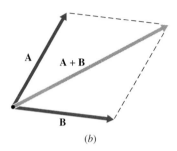

(a)

(b)

A + B

A + (−B)

−B

(c)

Fig. A-8

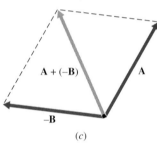

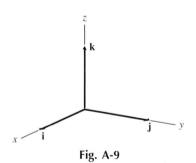

A-2-2 Subtraction of Vectors

The difference between vectors **A** and **B** is defined by the relation

$$\mathbf{A} - \mathbf{B} = \mathbf{A} + (-\mathbf{B}) \tag{A-7}$$

The vector $-\mathbf{B}$ has the same magnitude and direction angle θ as vector **B** but is of opposite sense. Two vectors **A** and **B** are shown in Fig. A-8a. Their sum and difference are illustrated in Figs. A-8b and A-8c, respectively. A zero or null vector is obtained when a vector is subtracted from itself; that is,

$$\mathbf{A} - \mathbf{A} = 0 \tag{A-8}$$

It is important to realize that the zero that occurs on the right-hand side of this vector equation must be a vector.

The minus sign used in conjunction with vectors **A** and **B** (boldface type) indicates vector (parallelogram law) subtraction and not scalar subtraction. Since vector subtraction is defined as a special case of vector addition, all of the rules for vector addition discussed in Section A-2-1 apply to vector subtraction.

A-2-3 Multiplication of Vectors by Scalars

The product of a vector **A** and a scalar m is a vector $m\mathbf{A}$, whose magnitude is mA (a positive number) and whose direction has the same sense as **A** if m is positive and a sense opposite to **A** if m is negative. Operations involving the products of scalars m and n and vectors **A** and **B** include the following:

$$\begin{aligned}
(m + n)\mathbf{A} &= m\mathbf{A} + n\mathbf{A} \\
m(\mathbf{A} + \mathbf{B}) &= m\mathbf{A} + m\mathbf{B} \\
m(n\mathbf{A}) &= (mn)\mathbf{A} = n(m\mathbf{A})
\end{aligned} \tag{A-9}$$

A-3 VECTORS REFERRED TO RECTANGULAR CARTESIAN COORDINATES

A-3-1 Unit Vectors

A vector of unit magnitude is called a unit vector. Unit vectors directed along the x-, y-, and z-axes of a Cartesian coordinate system are normally given the symbols **i**, **j**, and **k**, respectively. The sense of these unit vectors is indicated analytically by using a plus sign if the unit vector points in a positive x-, y-, or z-direction and a minus sign if the unit vector points in a negative x-, y-, or z-direction. The unit vectors shown in Fig. A-9 are positive. In this text, the symbol **e** with an appropriate subscript is used to denote a unit vector in a direction other than a coordinate direction.

Any vector quantity can be written as the product of its magnitude (a positive number) and a unit vector in the direction of the given vector. Thus, for a vector **A** in the positive n direction

$$\mathbf{A} = |\mathbf{A}|\mathbf{e}_n = A\mathbf{e}_n \tag{A-10}$$

For a vector **B** in the negative n direction

$$\mathbf{B} = |\mathbf{B}|(-\mathbf{e}_n) = -B\mathbf{e}_n \tag{A-11}$$

Fig. A-9

It is obvious from Eq. A-10 that the unit vector $\mathbf{e}_n$ can be expressed as

$$\mathbf{e}_n = \frac{\mathbf{A}}{|\mathbf{A}|} = \frac{\mathbf{A}}{A} \qquad \text{(A-12)}$$

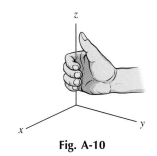

Fig. A-10

The Cartesian coordinate axes shown in Fig. A-9 are arranged as a right-hand system. In a right-hand system, if the fingers of the right hand are curled about the z-axis in a direction from the positive x-axis toward the positive y-axis, then the thumb points in the positive z-direction, as shown in Fig. A-10. All the vector relationships developed in the remainder of this appendix will utilize a right-hand coordinate system.

A-3-2 Cartesian Components of a Vector

The process of adding two or more vectors to obtain a resultant vector **R** was discussed in Section A-2-1. The reverse process of resolving a given vector **A** into two or more components will now be discussed. The vector **A** can be resolved into any number of components provided the components sum to the given vector **A** by the parallelogram law. Mutually perpendicular components, called rectangular components, along the x-, y-, and z-coordinate axes are most useful. Any vector **A** can be resolved into rectangular components $\mathbf{A}_x$, $\mathbf{A}_y$, and $\mathbf{A}_z$ by constructing a rectangular parallelepiped (see Fig. A-11) with vector **A** as the diagonal and rectangular components $\mathbf{A}_x$, $\mathbf{A}_y$, and $\mathbf{A}_z$ along the x-, y-, and z-axes, respectively, as the edges. The three components $\mathbf{A}_x$, $\mathbf{A}_y$, and $\mathbf{A}_z$ can be written in Cartesian vector form, as illustrated in Fig. A-12, by using Eq. A-10. Thus

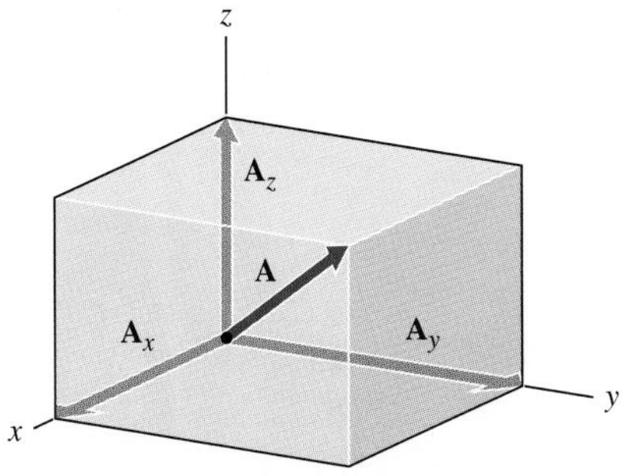

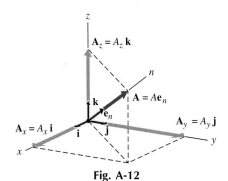

Fig. A-11

$$\mathbf{A}_x = A_x \mathbf{i} \qquad \mathbf{A}_y = A_y \mathbf{j} \qquad \mathbf{A}_z = A_z \mathbf{k} \qquad \text{(A-13)}$$

Furthermore, from Eqs. A-1, A-10, and A-13,

$$\mathbf{A} = \mathbf{A}_x + \mathbf{A}_y + \mathbf{A}_z = A_x \mathbf{i} + A_y \mathbf{j} + A_z \mathbf{k} = A\mathbf{e}_n \qquad \text{(A-14)}$$

The magnitudes of the rectangular components are related to the magnitude of vector **A** by the expressions

$$A_x = A \cos \theta_x \qquad A_y = A \cos \theta_y \qquad A_z = A \cos \theta_z \qquad \text{(A-15)}$$

The terms $\cos \theta_x$, $\cos \theta_y$, and $\cos \theta_z$ are called the direction cosines of the vector **A**. These cosines are also the direction cosines of the unit vector $\mathbf{e}_n$ since its line of action, as illustrated in Fig. A-13, coincides with the line of action of vector **A**.

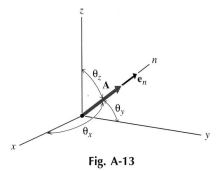

Fig. A-12

The magnitude of vector **A** can be expressed in terms of the magnitudes of its rectangular components by the Pythagorean theorem as

$$A = \sqrt{A_x^2 + A_y^2 + A_z^2} \qquad \text{(A-16)}$$

Also, from Eq. A-15, the direction cosines of a vector **A** can be expressed in terms of the magnitude of vector **A** and the magnitudes of its rectangular components as

$$\cos \theta_x = \frac{A_x}{A} \qquad \cos \theta_y = \frac{A_y}{A} \qquad \cos \theta_z = \frac{A_z}{A} \qquad \text{(A-17)}$$

Fig. A-13

Finally, the unit vector $\mathbf{e}_n$ can be written in terms of vector $\mathbf{A}$ and its rectangular components by using Eqs. A-14 and A-15. Thus,

$$\mathbf{e}_n = \frac{\mathbf{A}}{A} = \frac{A_x}{A}\mathbf{i} + \frac{A_y}{A}\mathbf{j} + \frac{A_z}{A}\mathbf{k}$$
$$= \cos\theta_x\mathbf{i} + \cos\theta_y\mathbf{j} + \cos\theta_z\mathbf{k} \qquad \text{(A-18)}$$

Since the magnitude of $\mathbf{e}_n = 1$

$$\cos^2\theta_x + \cos^2\theta_y + \cos^2\theta_z = 1 \qquad \text{(A-19)}$$

A-3-3   Position Vectors

A directed line segment $\mathbf{r}$, known as a position vector, can be used to locate a point in space relative to the origin of a coordinate system or relative to another point in space. For example, consider the two points A and B shown in Fig. A-14. The location of point A can be specified by using the position vector $\mathbf{r}_A$ drawn from the origin of coordinates to point A. Position vector $\mathbf{r}_A$ can be written in Cartesian vector form as

$$\mathbf{r}_A = x_A\mathbf{i} + y_A\mathbf{j} + z_A\mathbf{k}$$

Similarly, the location of point B with respect to the origin of coordinates can be specified by using the position vector $\mathbf{r}_B$, which can be written in Cartesian vector form as

$$\mathbf{r}_B = x_B\mathbf{i} + y_B\mathbf{j} + z_B\mathbf{k}$$

Finally, the position of point B with respect to point A can be specified by using the position vector $\mathbf{r}_{B/A}$ where the subscript B/A indicates B with respect to A. Observe in Fig. A-14 that

$$\mathbf{r}_B = \mathbf{r}_A + \mathbf{r}_{B/A}$$

Therefore,

$$\begin{aligned}
\mathbf{r}_{B/A} &= \mathbf{r}_B - \mathbf{r}_A \\
&= (x_B\mathbf{i} + y_B\mathbf{j} + z_B\mathbf{k}) - (x_A\mathbf{i} + y_A\mathbf{j} + z_A\mathbf{k}) \\
&= (x_B - x_A)\mathbf{i} + (y_B - y_A)\mathbf{j} + (z_B - z_A)\mathbf{k} \qquad \text{(A-20)}
\end{aligned}$$

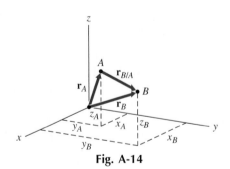

Fig. A-14

Determine the magnitude and direction of the position vector from point A to point B if the coordinates of points A and B are $(3, 4, 5)$ and $(6, -3, -2)$, respectively.

SOLUTION

The position vector $\mathbf{r}_{B/A}$ is given by Eq. A-20 as

$$\mathbf{r}_{B/A} = (x_B - x_A)\mathbf{i} + (y_B - y_A)\mathbf{j} + (z_B - z_A)\mathbf{k}$$
$$= (6 - 3)\mathbf{i} + (-3 - 4)\mathbf{j} + (-2 - 5)\mathbf{k}$$
$$= 3\mathbf{i} - 7\mathbf{j} - 7\mathbf{k}$$

The magnitude of $\mathbf{r}_{B/A}$ is determined by using Eq. A-22. Thus,

$$r_{B/A} = \sqrt{(x_B - x_A)^2 + (y_B - y_A)^2 + (z_B - z_A)^2}$$
$$= \sqrt{(3)^2 + (-7)^2 + (-7)^2} = 10.34 \qquad \text{Ans.}$$

The direction is determined by using Eqs. A-23. Thus,

$$\theta_x = \cos^{-1} \frac{x_B - x_A}{r_{B/A}} = \cos^{-1} \frac{3}{10.34} = 73.1° \qquad \text{Ans.}$$

$$\theta_y = \cos^{-1} \frac{y_B - y_A}{r_{B/A}} = \cos^{-1} \frac{-7}{10.34} = 132.6° \qquad \text{Ans.}$$

$$\theta_z = \cos^{-1} \frac{z_B - z_A}{r_{B/A}} = \cos^{-1} \frac{-7}{10.34} = 132.6° \qquad \text{Ans.}$$

A-4   ADDITION OF CARTESIAN VECTORS

The vector operations of addition and subtraction are greatly simplified when more than two vectors are involved if the vectors are expressed in Cartesian vector form. For example, consider the two vectors **A** and **B**, which can be written as

$$\mathbf{A} = A_x\mathbf{i} + A_y\mathbf{j} + A_z\mathbf{k}$$
$$\mathbf{B} = B_x\mathbf{i} + B_y\mathbf{j} + B_z\mathbf{k}$$

The sum of the two vectors is

$$\mathbf{R} = \mathbf{A} + \mathbf{B} = (A_x\mathbf{i} + A_y\mathbf{j} + A_z\mathbf{k}) + (B_x\mathbf{i} + B_y\mathbf{j} + B_z\mathbf{k})$$
$$= (A_x + B_x)\mathbf{i} + (A_y + B_y)\mathbf{j} + (A_z + B_z)\mathbf{k} \qquad \text{(A-21)}$$

Thus, the resultant vector **R** has components that represent the scalar sums of the rectangular components of **A** and **B**. The process represented by Eq. A-21 can be extended to any number of vectors. Thus,

$$\mathbf{R}_x = (A_x + B_x + C_x + \cdots)\mathbf{i} = R_x\mathbf{i}$$
$$\mathbf{R}_y = (A_y + B_y + C_y + \cdots)\mathbf{j} = R_y\mathbf{j}$$
$$\mathbf{R}_z = (A_z + B_z + C_z + \cdots)\mathbf{k} = R_z\mathbf{k}$$
$$\mathbf{R} = \mathbf{R}_x + \mathbf{R}_y + \mathbf{R}_z = R_x\mathbf{i} + R_y\mathbf{j} + R_z\mathbf{k}$$

The magnitude and direction of the resultant **R** are then obtained by using Eqs. A-16 and A-17. Thus,

$$R = \sqrt{(R_x)^2 + (R_y)^2 + (R_z)^2} \qquad \text{(A-22)}$$

$$\theta_x = \cos^{-1}\frac{R_x}{R} \qquad \theta_y = \cos^{-1}\frac{R_y}{R} \qquad \theta_z = \cos^{-1}\frac{R_z}{R} \qquad \text{(A-23)}$$

EXAMPLE PROBLEM A-2

Determine the magnitude and direction of the resultant of the following three vectors $\mathbf{A} = 3\mathbf{i} + 7\mathbf{j} + 8\mathbf{k}$, $\mathbf{B} = 4\mathbf{i} - 5\mathbf{j} + 3\mathbf{k}$, and $\mathbf{C} = 2\mathbf{i} + 3\mathbf{j} - 4\mathbf{k}$. Express the resultant $\mathbf{R}$ and the unit vector $\mathbf{e}_R$ associated with the resultant in Cartesian vector form.

SOLUTION

The x-, y-, and z-components of the resultant $\mathbf{R}$ are

$$R_x = A_x + B_x + C_x = 3 + 4 + 2 = 9$$
$$R_y = A_y + B_y + C_y = 7 - 5 + 3 = 5$$
$$R_z = A_z + B_z + C_z = 8 + 3 - 4 = 7$$

The magnitude of the resultant is determined by using Eq. A-22. Thus,

$$R = \sqrt{(R_x)^2 + (R_y)^2 + (R_z)^2} = \sqrt{(9)^2 + (5)^2 + (7)^2} = 12.45 \qquad \text{Ans.}$$

The direction is determined by using Eqs. A-23. Thus,

$$\theta_x = \cos^{-1}\frac{R_x}{R} = \cos^{-1}\frac{9}{12.45} = 43.7° \qquad \text{Ans.}$$

$$\theta_y = \cos^{-1}\frac{R_y}{R} = \cos^{-1}\frac{5}{12.45} = 66.3° \qquad \text{Ans.}$$

$$\theta_z = \cos^{-1}\frac{R_z}{R} = \cos^{-1}\frac{7}{12.45} = 55.8° \qquad \text{Ans.}$$

The resultant can be expressed in Cartesian vector form as

$$\mathbf{R} = R_x\mathbf{i} + R_y\mathbf{j} + R_z\mathbf{k} = 9\mathbf{i} + 5\mathbf{j} + 7\mathbf{k} \qquad \text{Ans.}$$

The unit vector $\mathbf{e}_R$ associated with the resultant $\mathbf{R}$ is given by Eq. A-18 as

$$\mathbf{e}_R = \frac{\mathbf{R}}{R} = \frac{9\mathbf{i} + 5\mathbf{j} + 7\mathbf{k}}{12.45} = 0.723\mathbf{i} + 0.402\mathbf{j} + 0.562\mathbf{k} \qquad \text{Ans.}$$

A-5   MULTIPLICATION OF CARTESIAN VECTORS

Two types of multiplication involving vector quantities are performed; namely, the scalar or dot product (written as $\mathbf{A} \cdot \mathbf{B}$) and the vector or cross product (written as $\mathbf{A} \times \mathbf{B}$). These two types of multiplication have entirely different properties and are used for different purposes.

A-5-1   Dot or Scalar Product

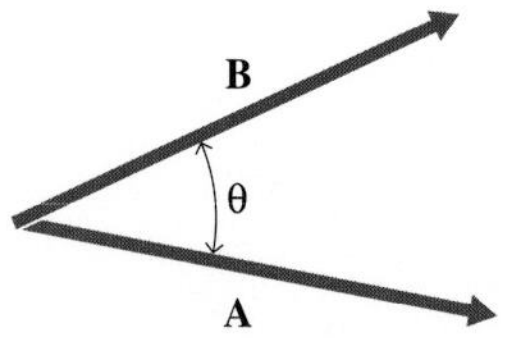

The dot or scalar product of two intersecting vectors is defined as the product of the magnitudes of the vectors and the cosine of the angle between them. Thus, by definition, for the vectors $\mathbf{A}$ and $\mathbf{B}$ shown in Fig. A-15,

$$\mathbf{A} \cdot \mathbf{B} = \mathbf{B} \cdot \mathbf{A} = AB \cos \theta \tag{A-24}$$

where $0° \leq \theta \leq 180°$. This type of vector multiplication yields a scalar, not a vector. For $0° \leq \theta < 90°$, the scalar is positive. For $90° < \theta \leq 180°$, the scalar is negative. When $\theta = 90°$, the two vectors are perpendicular and the scalar is zero.

The dot product can be used to obtain the rectangular scalar component of a vector in a specified direction. For example, the rectangular scalar component of vector $\mathbf{A}$ along the x-axis is

$$A_x = \mathbf{A} \cdot \mathbf{i} = A(1) \cos \theta_x = A \cos \theta_x$$

Similarly, the rectangular scalar component of vector $\mathbf{A}$ in a direction n (see Fig. A-16a) is

$$A_n = \mathbf{A} \cdot \mathbf{e}_n = A \cos \theta_n \tag{A-25}$$

where $\mathbf{e}_n$ is the unit vector associated with the direction n. The vector

Fig. A-15

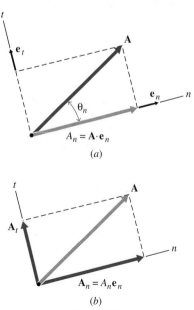

$$A_n = \mathbf{A} \cdot \mathbf{e}_n$$

(a)

$$\mathbf{A}_n = A_n \mathbf{e}_n$$

(b)

Fig. A-16

component of **A** in the direction n (see Fig. A-16b) is then given by Eq. A-10 as

$$\mathbf{A}_n = (\mathbf{A} \cdot \mathbf{e}_n)\mathbf{e}_n \tag{A-26}$$

The component of vector **A** perpendicular to the direction n lies in the plane containing **A** and n, as shown in Fig. A-16b, and can be obtained from the expression

$$\mathbf{A}_t = \mathbf{A} - \mathbf{A}_n \tag{a}$$

Once an expression for $\mathbf{A}_t$ has been obtained by using Eq. a, the magnitude and direction of $\mathbf{A}_t$ can be determined by using Eqs. A-22 and A-18.

If two vectors **A** and **B** are written in Cartesian vector form, the scalar product becomes

$$\begin{aligned}
\mathbf{A} \cdot \mathbf{B} &= (A_x\mathbf{i} + A_y\mathbf{j} + A_z\mathbf{k}) \cdot (B_x\mathbf{i} + B_y\mathbf{j} + B_z\mathbf{k}) \\
&= A_xB_x(\mathbf{i} \cdot \mathbf{i}) + A_xB_y(\mathbf{i} \cdot \mathbf{j}) + A_xB_z(\mathbf{i} \cdot \mathbf{k}) \\
&\quad + A_yB_x(\mathbf{j} \cdot \mathbf{i}) + A_yB_y(\mathbf{j} \cdot \mathbf{j}) + A_yB_z(\mathbf{j} \cdot \mathbf{k}) \\
&\quad + A_zB_x(\mathbf{k} \cdot \mathbf{i}) + A_zB_y(\mathbf{k} \cdot \mathbf{j}) + A_zB_z(\mathbf{k} \cdot \mathbf{k})
\end{aligned}$$

Since **i**, **j**, and **k** are orthogonal,

$$\mathbf{i} \cdot \mathbf{j} = \mathbf{j} \cdot \mathbf{k} = \mathbf{k} \cdot \mathbf{i} = (1)(1)\cos 90° = 0$$
$$\mathbf{i} \cdot \mathbf{i} = \mathbf{j} \cdot \mathbf{j} = \mathbf{k} \cdot \mathbf{k} = (1)(1)\cos 0° = 1$$

Therefore,

$$\mathbf{A} \cdot \mathbf{B} = A_xB_x + A_yB_y + A_zB_z \tag{A-27}$$

Note also the special case

$$\mathbf{A} \cdot \mathbf{A} = A^2 \cos 0° = A^2 = A_x^2 + A_y^2 + A_z^2$$

which verifies Eq. A-16 obtained by using the Pythagorean theorem.

By combining Eqs. A-24 and A-27, an expression for the angle between two vectors is obtained. Thus,

$$\mathbf{A} \cdot \mathbf{B} = A_xB_x + A_yB_y + A_zB_z = AB \cos \theta$$

or

$$\cos \theta = \frac{\mathbf{A} \cdot \mathbf{B}}{AB} = \frac{A_xB_x + A_yB_y + A_zB_z}{AB} \tag{A-28}$$

In mechanics, the scalar product is used to determine the rectangular component of a vector (force, moment, velocity, acceleration, etc.) along a line and to find the angle between two vectors (two forces, a force and a line, a force and an acceleration, etc.).

EXAMPLE PROBLEM A-3

Determine the rectangular components of vector $\mathbf{A} = -312\mathbf{i} + 72\mathbf{j} - 228\mathbf{k}$ parallel and perpendicular to a line whose direction is given by the unit vector $\mathbf{e}_n = -0.60\mathbf{i} + 0.80\mathbf{j}$. Express the unit vector $\mathbf{e}_t$ associated with the perpendicular component of vector $\mathbf{A}$ in Cartesian vector form.

SOLUTION

The magnitude of the component of vector $\mathbf{A}$ parallel to the line is given by Eq. A-27 as

$$A_n = \mathbf{A} \cdot \mathbf{e}_n = (-312\mathbf{i} + 72\mathbf{j} - 228\mathbf{k}) \cdot (-0.60\mathbf{i} + 0.80\mathbf{j})$$
$$= -312(-0.60) + 72(0.80) - 228(0) = 244.8$$

Thus, from Eq. A-26

$$\mathbf{A}_n = (\mathbf{A} \cdot \mathbf{e}_n)\mathbf{e}_n = 244.8(-0.60\mathbf{i} + 0.80\mathbf{j})$$
$$= -146.9\mathbf{i} + 195.8\mathbf{j} \qquad \text{Ans.}$$

From Eq. A-21

$$\mathbf{A} = \mathbf{A}_n + \mathbf{A}_t$$

Therefore,

$$\mathbf{A}_t = \mathbf{A} - \mathbf{A}_n = (-312\mathbf{i} + 72\mathbf{j} - 228\mathbf{k}) - (-146.9\mathbf{i} + 195.8\mathbf{j})$$
$$= -165.1\mathbf{i} - 123.8\mathbf{j} - 228\mathbf{k} \qquad \text{Ans.}$$

The magnitude of vector $\mathbf{A}_t$ is determined by using Eq. A-22. Thus,

$$A_t = |\mathbf{A}_t| = \sqrt{(-165.1)^2 + (-123.8)^2 + (-228)^2} = 307.5$$

The unit vector $\mathbf{e}_t$ is given by Eq. A-18 as

$$\mathbf{e}_t = \frac{\mathbf{A}_t}{A_t} = \frac{-165.1\mathbf{i} - 123.8\mathbf{j} - 228\mathbf{k}}{307.5}$$
$$= -0.537\mathbf{i} - 0.403\mathbf{j} - 0.741\mathbf{k} \qquad \text{Ans.}$$

EXAMPLE PROBLEM A-4

Determine the angle θ between the vectors

$$\mathbf{A} = 8\mathbf{i} + 9\mathbf{j} + 7\mathbf{k} \qquad \text{and} \qquad \mathbf{B} = 6\mathbf{i} - 5\mathbf{j} + 3\mathbf{k}$$

SOLUTION

The magnitudes of vectors $\mathbf{A}$ and $\mathbf{B}$ are determined by using Eq. A-22. Thus,

$$A = \sqrt{(A_x)^2 + (A_y)^2 + (A_z)^2} = \sqrt{(8)^2 + (9)^2 + (7)^2} = 13.93$$
$$B = \sqrt{(B_x)^2 + (B_y)^2 + (B_z)^2} = \sqrt{(6)^2 + (-5)^2 + (3)^2} = 8.37$$

The dot product of vectors $\mathbf{A}$ and $\mathbf{B}$ is given by Eq. A-27 as

$$\mathbf{A} \cdot \mathbf{B} = A_x B_x + A_y B_y + A_z B_z = 8(6) + 9(-5) + 7(3) = 24$$

The angle θ between the two vectors is given by Eq. A-28 as

$$\theta = \cos^{-1}\frac{\mathbf{A} \cdot \mathbf{B}}{AB} = \cos^{-1}\frac{24}{13.93(8.37)} = 78.1° \qquad \text{Ans.}$$

A-5-2   Cross or Vector Product

The cross or vector product of two intersecting vectors **A** and **B**, by definition, yields a vector **C** that has a magnitude that is the product of the magnitudes of vectors **A** and **B** and the sine of the angle θ between them and a direction that is perpendicular to the plane containing the vectors **A** and **B**. Thus, by definition, for the vectors **A** and **B** shown in Fig. A-17,

$$\mathbf{C} = \mathbf{A} \times \mathbf{B} = (AB \sin \theta) \, \mathbf{e}_C \qquad \text{(A-29)}$$

where $0 \le \theta \le 180°$ and $\mathbf{e}_C$ is a unit vector in a direction perpendicular to the plane containing the vectors **A** and **B**. The sense of $\mathbf{e}_C$ is obtained by using the right-hand rule, that is, if the fingers of the right hand are curled from **A** toward **B** about an axis perpendicular to the plane containing vectors **A** and **B** then the thumb points in the direction of vector **C**. Because of this definition of the unit vector $\mathbf{e}_C$, the cross-product is not commutative. In fact,

$$\mathbf{A} \times \mathbf{B} = -\mathbf{B} \times \mathbf{A}$$

It is apparent from Eq. A-29, however, that

$$\mathbf{A} \times s\mathbf{B} = s(\mathbf{A} \times \mathbf{B})$$

and by combining Eqs. A-29 and the law of addition that

$$\mathbf{A} \times (\mathbf{B} + \mathbf{C}) = (\mathbf{A} \times \mathbf{B}) + (\mathbf{A} \times \mathbf{C})$$

If two vectors **A** and **B** are written in Cartesian vector form, the cross-product becomes,

$$
\begin{aligned}
\mathbf{A} \times \mathbf{B} &= (A_x \mathbf{i} + A_y \mathbf{j} + A_z \mathbf{k}) \times (B_x \mathbf{i} + B_y \mathbf{j} + B_z \mathbf{k}) \\
&= A_x B_x (\mathbf{i} \times \mathbf{i}) + A_x B_y (\mathbf{i} \times \mathbf{j}) + A_x B_z (\mathbf{i} \times \mathbf{k}) \\
&\quad + A_y B_x (\mathbf{j} \times \mathbf{i}) + A_y B_y (\mathbf{j} \times \mathbf{j}) + A_y B_z (\mathbf{j} \times \mathbf{k}) \\
&\quad + A_z B_x (\mathbf{k} \times \mathbf{i}) + A_z B_y (\mathbf{k} \times \mathbf{j}) + A_z B_z (\mathbf{k} \times \mathbf{k})
\end{aligned} \qquad (a)
$$

Since **i**, **j**, and **k** are orthogonal,

$$
\begin{aligned}
\mathbf{i} \times \mathbf{i} &= [(1)(1) \sin 0°] \mathbf{k} = \mathbf{0} \\
\mathbf{i} \times \mathbf{j} &= [(1)(1) \sin 90°] \mathbf{k} = \mathbf{k}
\end{aligned}
$$

Similarly,

$$
\begin{array}{lll}
\mathbf{i} \times \mathbf{i} = \mathbf{0} & \mathbf{j} \times \mathbf{i} = -\mathbf{k} & \mathbf{k} \times \mathbf{i} = \mathbf{j} \\
\mathbf{i} \times \mathbf{j} = \mathbf{k} & \mathbf{j} \times \mathbf{j} = \mathbf{0} & \mathbf{k} \times \mathbf{j} = -\mathbf{i} \\
\mathbf{i} \times \mathbf{k} = -\mathbf{j} & \mathbf{j} \times \mathbf{k} = \mathbf{i} & \mathbf{k} \times \mathbf{k} = \mathbf{0}
\end{array} \qquad \text{(A-30)}
$$

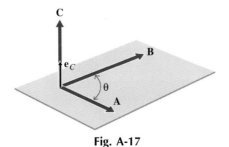

Fig. A-17

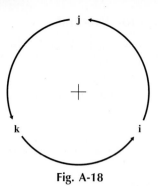

Fig. A-18

Equations A-30 can be represented graphically by arranging the unit vectors **i**, **j**, and **k** in a circle in a counterclockwise order, as shown in Fig. A-18. The product of two unit vectors will be positive if they follow each other in counterclockwise order, and negative if they follow each other in clockwise order. Substituting Eqs. A-30 into Eq. *a* yields

$$
\begin{aligned}
\mathbf{A}\times\mathbf{B} &= A_x B_y \mathbf{k} + A_x B_z(-\mathbf{j}) + A_y B_x(-\mathbf{k}) \\
&\quad + A_y B_z \mathbf{i} + A_z B_x \mathbf{j} + A_z B_y(-\mathbf{i}) \\
&= (A_y B_z - A_z B_y)\mathbf{i} + (A_z B_x - A_x B_z)\mathbf{j} \qquad \text{(A-31)} \\
&\quad + (A_x B_y - A_y B_x)\mathbf{k}
\end{aligned}
$$

which is the expanded form of the determinant

$$
\mathbf{A}\times\mathbf{B} = \begin{vmatrix} \mathbf{i} & \mathbf{j} & \mathbf{k} \\ A_x & A_y & A_z \\ B_x & B_y & B_z \end{vmatrix} = \mathbf{C} \qquad \text{(A-32)}
$$

Note carefully the arrangement of terms in the determinant that places the unit vectors **i**, **j**, **k** in the first row, the components A_x, A_y, A_z of **A** in the second row, and the components B_x, B_y, B_z of **B** in the third row. If the two bottom rows of the determinant are interchanged, the sign of the determinant will change. Thus,

$$
\mathbf{A}\times\mathbf{B} = -\mathbf{B}\times\mathbf{A}
$$

In mechanics, the cross-product is used to find the moment of a force about a point. The direction of the moment vector is the axis about which the force tends to rotate a body and the magnitude of the moment is the strength of the tendency to rotate the body.

EXAMPLE PROBLEM A-5

If $\mathbf{A} = -3.75\mathbf{i} - 2.50\mathbf{j} + 1.50\mathbf{k}$ and $\mathbf{B} = 32\mathbf{i} + 44\mathbf{j} + 64\mathbf{k}$, determine the magnitude and direction of the vector $\mathbf{C}$ if $\mathbf{C} = \mathbf{A} \times \mathbf{B}$. Express the unit vector $\mathbf{e}_C$ associated with the vector $\mathbf{C}$ in Cartesian vector form.

SOLUTION

The vector cross-product $\mathbf{A} \times \mathbf{B}$ is obtained by using Eq. A-32. Thus,

$$\mathbf{C} = \mathbf{A} \times \mathbf{B} = \begin{vmatrix} \mathbf{i} & \mathbf{j} & \mathbf{k} \\ A_x & A_y & A_z \\ B_x & B_y & B_z \end{vmatrix} = \begin{vmatrix} \mathbf{i} & \mathbf{j} & \mathbf{k} \\ -3.75 & -2.50 & 1.50 \\ 32 & 44 & 64 \end{vmatrix}$$

$$= -226\mathbf{i} + 288\mathbf{j} - 85\mathbf{k} \qquad \text{Ans.}$$

The magnitude of vector $\mathbf{C}$ is determined by using Eq. A-22. Thus,

$$C = |\mathbf{C}| = \sqrt{C_x^2 + C_y^2 + C_z^2} = \sqrt{(-226)^2 + (288)^2 + (-85)^2} = 376 \quad \text{Ans.}$$

The direction is determined by using Eqs. A-23. Thus,

$$\theta_x = \cos^{-1}\frac{C_x}{|\mathbf{C}|} = \cos^{-1}\frac{-226}{376} = 126.9° \qquad \text{Ans.}$$

$$\theta_y = \cos^{-1}\frac{C_y}{|\mathbf{C}|} = \cos^{-1}\frac{288}{376} = 40.0° \qquad \text{Ans.}$$

$$\theta_z = \cos^{-1}\frac{C_z}{|\mathbf{C}|} = \cos^{-1}\frac{-85}{376} = 103.1° \qquad \text{Ans.}$$

The unit vector $\mathbf{e}_C$ is given by Eq. A-18 as

$$\mathbf{e}_C = \frac{\mathbf{C}}{|\mathbf{C}|} = \frac{-226\mathbf{i} + 288\mathbf{j} - 85\mathbf{k}}{376}$$

$$= -0.601\mathbf{i} + 0.766\mathbf{j} - 0.226\mathbf{k} \qquad \text{Ans.}$$

A-5-3   Triple Scalar Product

The triple scalar product involves the dot product of vector **A** and the cross-product of vectors **B** and **C**. The triple scalar product is written as

$$\mathbf{A} \cdot (\mathbf{B} \times \mathbf{C}) \qquad \text{or} \qquad (\mathbf{B} \times \mathbf{C}) \cdot \mathbf{A} \qquad \text{(A-33)}$$

Expressing vectors **A**, **B**, and **C** in Cartesian vector form and expanding yields

$$
\begin{aligned}
\mathbf{A} \cdot (\mathbf{B} \times \mathbf{C}) &= (A_x \mathbf{i} + A_y \mathbf{j} + A_z \mathbf{k}) \cdot \\
&\quad [B_x \mathbf{i} + B_y \mathbf{j} + B_z \mathbf{k}) \times (C_x \mathbf{i} + C_y \mathbf{j} + C_z \mathbf{k})] \\
&= A_x(B_y C_z - B_z C_y) + A_y(B_z C_x - B_x C_z) \\
&\quad + A_z(B_x C_y - B_y C_x)
\end{aligned}
\qquad \text{(A-34)}
$$

which is the expanded form of the determinant

$$\mathbf{A} \cdot (\mathbf{B} \times \mathbf{C}) = \begin{vmatrix} A_x & A_y & A_z \\ B_x & B_y & B_z \\ C_x & C_y & C_z \end{vmatrix} \qquad \text{(A-35)}$$

This product is a scalar; hence the name triple scalar product. However, the vectors cannot be indiscriminately interchanged since vector cross-products are not commutative. It is true, however, that

$$\mathbf{A} \cdot (\mathbf{B} \times \mathbf{C}) = (\mathbf{A} \times \mathbf{B}) \cdot \mathbf{C} = (\mathbf{B} \times \mathbf{C}) \cdot \mathbf{A} = \mathbf{B} \cdot (\mathbf{C} \times \mathbf{A})$$

Since the cross-product is used to find the moment of a force about a point and the scalar product is used to find the component of a vector along a line, the triple scalar product represents the component of the moment of a force along a line. That is, the triple scalar product represents the tendency of a force to rotate a body about a line.

A-5-4   Triple Vector Product

The triple vector product involves the cross product of a vector **A** with the result of a cross-product of vectors **B** and **C**. Thus, the triple vector product is written as

$$\mathbf{A} \times (\mathbf{B} \times \mathbf{C})$$

If the vectors are written in Cartesian vector form, Eq. A-31 gives the cross product (**B** × **C**) as

$$
\begin{aligned}
\mathbf{B} \times \mathbf{C} &= (B_x \mathbf{i} + B_y \mathbf{j} + B_z \mathbf{k}) \times (C_x \mathbf{i} + C_y \mathbf{j} + C_z \mathbf{k}) \\
&= (B_y C_z - B_z C_y)\mathbf{i} + (B_z C_x - B_x C_z)\mathbf{j} + (B_x C_y - B_y C_x)\mathbf{k}
\end{aligned}
$$

Similarly, the Eq. A-31 gives the cross product **A** × (**B** × **C**) as

$$
\begin{aligned}
\mathbf{A} \times (\mathbf{B} \times \mathbf{C}) &= (A_x \mathbf{i} + A_y \mathbf{j} + A_z \mathbf{k}) \times [(B_y C_z - B_z C_y)\mathbf{i} \\
&\quad + (B_z C_x - B_x C_z)\mathbf{j} + (B_x C_y - B_y C_x)\mathbf{k}] \\
&= [A_y(B_x C_y - B_y C_x) - A_z(B_z C_x - B_x C_z)]\mathbf{i} \\
&\quad + [A_z(B_y C_z - B_z C_y) - A_x(B_x C_y - B_y C_x)]\mathbf{j} \\
&\quad + [A_x(B_z C_x - B_x C_z) - A_y(B_y C_z - B_z C_y)]\mathbf{k} \\
&= (A_y C_y + A_z C_z)B_x \mathbf{i} + (A_z C_z + A_x C_x)B_y \mathbf{j} \\
&\quad + (A_x C_x + A_y C_y)B_z \mathbf{k} - (A_y B_y + A_z B_z)C_x \mathbf{i} \\
&\quad - (A_z B_z + A_x B_x)C_y \mathbf{j} - (A_x B_x + A_y B_y)C_z \mathbf{k}
\end{aligned}
$$

Adding and subtracting the vector

$$A_x B_x C_x \mathbf{i} + A_y B_y C_y \mathbf{j} + A_z B_z C_z \mathbf{k}$$

and regrouping yields

$$\mathbf{A} \times (\mathbf{B} \times \mathbf{C}) = (A_x C_x + A_y C_y + A_z C_z)(B_x \mathbf{i} + B_y \mathbf{j} + B_z \mathbf{k}) \\ - (A_x B_x + A_y B_y + A_z B_z)(C_x \mathbf{i} + C_y \mathbf{j} + C_z \mathbf{k})$$

However, from Eq. A-27,

$$A_x C_x + A_y C_y + A_z C_z = \mathbf{A} \cdot \mathbf{C} \\ A_x B_x + A_y B_y + A_z B_z = \mathbf{A} \cdot \mathbf{B}$$

Therefore

$$\mathbf{A} \times (\mathbf{B} \times \mathbf{C}) = (\mathbf{A} \cdot \mathbf{C})\mathbf{B} - (\mathbf{A} \cdot \mathbf{B})\mathbf{C} \qquad \text{(A-36)}$$

In a similar fashion it can be established that

$$(\mathbf{A} \times \mathbf{B}) \times \mathbf{C} = (\mathbf{A} \cdot \mathbf{C})\mathbf{B} - (\mathbf{B} \cdot \mathbf{C})\mathbf{A} \qquad \text{(A-37)}$$

The triple vector product is used in dynamics to determine the acceleration of points in a rotating body.

Determine the scalar triple product $\mathbf{A} \cdot (\mathbf{B} \times \mathbf{C})$ and the vector triple product $\mathbf{A} \times (\mathbf{B} \times \mathbf{C})$ for the vectors

$$\mathbf{A} = 3\mathbf{i} + 5\mathbf{j} + 8\mathbf{k}$$
$$\mathbf{B} = 4\mathbf{i} - 5\mathbf{j} + 3\mathbf{k}$$
$$\mathbf{C} = 2\mathbf{i} + 3\mathbf{j} - 4\mathbf{k}$$

SOLUTION

The scalar triple product $\mathbf{A} \cdot (\mathbf{B} \times \mathbf{C})$ can be expressed in determinate form as given by Eq. A-35. Thus,

$$\mathbf{A} \cdot (\mathbf{B} \times \mathbf{C}) = \begin{vmatrix} A_x & A_y & A_z \\ B_x & B_y & B_z \\ C_x & C_y & C_z \end{vmatrix} = \begin{vmatrix} 3 & 5 & 8 \\ 4 & -5 & 3 \\ 2 & 3 & -4 \end{vmatrix}$$

$$= 3(20 - 9) - 5(-16 - 6) + 8(12 + 10)$$
$$= 319 \qquad \qquad \text{Ans.}$$

The triple vector product $\mathbf{A} \times (\mathbf{B} \times \mathbf{C})$ is given by Eq. A-36 as

$$\mathbf{A} \times (\mathbf{B} \times \mathbf{C}) = (\mathbf{A} \cdot \mathbf{C})\mathbf{B} - (\mathbf{A} \cdot \mathbf{B})\mathbf{C}$$
$$= [(3\mathbf{i} + 5\mathbf{j} + 8\mathbf{k}) \cdot (2\mathbf{i} + 3\mathbf{j} - 4\mathbf{k})](4\mathbf{i} - 5\mathbf{j} + 3\mathbf{k})$$
$$\quad - [(3\mathbf{i} + 5\mathbf{j} + 8\mathbf{k}) \cdot (4\mathbf{i} - 5\mathbf{j} + 3\mathbf{k})](2\mathbf{i} + 3\mathbf{j} - 4\mathbf{k})$$
$$= [3(2) + 5(3) + 8(-4)](4\mathbf{i} - 5\mathbf{j} + 3\mathbf{k})$$
$$\quad - [3(4) + 5(-5) + 8(3)](2\mathbf{i} + 3\mathbf{j} - 4\mathbf{k})$$
$$= -11(4\mathbf{i} - 5\mathbf{j} + 3\mathbf{k}) - 11(2\mathbf{i} + 3\mathbf{j} - 4\mathbf{k})$$
$$= -66\mathbf{i} + 22\mathbf{j} + 11\mathbf{k} \qquad \qquad \text{Ans.}$$

CENTROIDS OF VOLUMES, AREAS, AND LINES

TABLE B-1 CENTROID LOCATIONS FOR A FEW COMMON LINE SEGMENTS AND AREAS

Circular arc

$L = 2r\alpha$

$\bar{x} = \dfrac{r \sin \alpha}{\alpha}$

$\bar{y} = 0$

Circular sector

$A = r^2\alpha$

$\bar{x} = \dfrac{2r \sin \alpha}{3\alpha}$

$\bar{y} = 0$

Quarter circular arc

$L = \dfrac{\pi r}{2}$

$\bar{x} = \dfrac{2r}{\pi}$

$\bar{y} = \dfrac{2r}{\pi}$

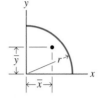

Quadrant of a circle

$A = \dfrac{\pi r^2}{4}$

$\bar{x} = \dfrac{4r}{3\pi}$

$\bar{y} = \dfrac{4r}{3\pi}$

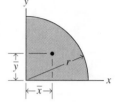

Semicircular arc

$L = \pi r$

$\bar{x} = r$

$\bar{y} = \dfrac{2r}{\pi}$

Semicircular area

$A = \dfrac{\pi r^2}{2}$

$\bar{x} = r$

$\bar{y} = \dfrac{4r}{3\pi}$

Rectangular area

$A = bh$

$\bar{x} = \dfrac{b}{2}$

$\bar{y} = \dfrac{h}{2}$

Quadrant of an ellipse

$A = \dfrac{\pi ab}{4}$

$\bar{x} = \dfrac{4a}{3\pi}$

$\bar{y} = \dfrac{4b}{3\pi}$

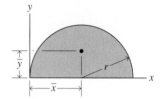

Triangular area

$A = \dfrac{bh}{2}$

$\bar{x} = \dfrac{2b}{3}$

$\bar{y} = \dfrac{h}{3}$

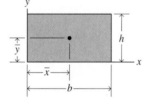

Parabolic spandrel

$A = \dfrac{bh}{3}$

$\bar{x} = \dfrac{3b}{4}$

$\bar{y} = \dfrac{3h}{10}$

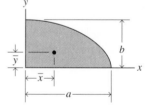

Triangular area

$A = \dfrac{bh}{2}$

$\bar{x} = \dfrac{a + b}{3}$

$\bar{y} = \dfrac{h}{3}$

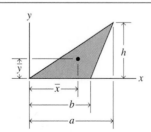

Quadrant of a parabola

$A = \dfrac{2bh}{3}$

$\bar{x} = \dfrac{5b}{8}$

$\bar{y} = \dfrac{2h}{5}$

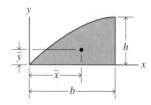

TABLE B-2 CENTROID LOCATIONS FOR A FEW COMMON VOLUMES

Rectangular parallelepiped

$V = abc$

$\bar{x} = \dfrac{a}{2}$

$\bar{y} = \dfrac{b}{2}$

$\bar{z} = \dfrac{c}{2}$

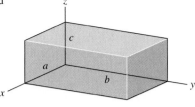

Rectangular tetrahedron

$V = \dfrac{abc}{6}$

$\bar{x} = \dfrac{a}{4}$

$\bar{y} = \dfrac{b}{4}$

$\bar{z} = \dfrac{c}{4}$

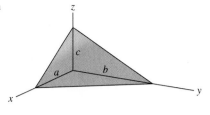

Circular cylinder

$V = \pi r^2 L$

$\bar{x} = 0$

$\bar{y} = \dfrac{L}{2}$

$\bar{z} = 0$

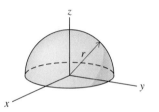

Semicylinder

$V = \dfrac{\pi r^2 L}{2}$

$\bar{x} = 0$

$\bar{y} = \dfrac{L}{2}$

$\bar{z} = \dfrac{4r}{3\pi}$

Hemisphere

$V = \dfrac{2\pi r^3}{3}$

$\bar{x} = 0$

$\bar{y} = 0$

$\bar{z} = \dfrac{3r}{8}$

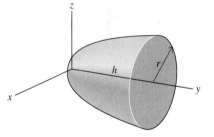

Paraboloid

$V = \dfrac{\pi r^2 h}{2}$

$\bar{x} = 0$

$\bar{y} = \dfrac{2h}{3}$

$\bar{z} = 0$

Right circular cone

$V = \dfrac{\pi r^2 h}{3}$

$\bar{x} = 0$

$\bar{y} = \dfrac{3h}{4}$

$\bar{z} = 0$

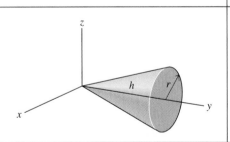

Half cone

$V = \dfrac{\pi r^2 h}{6}$

$\bar{x} = 0$

$\bar{y} = \dfrac{3h}{4}$

$\bar{z} = \dfrac{r}{\pi}$

SECOND MOMENTS AND MOMENTS OF INERTIA

TABLE C-1 SECOND MOMENTS OF PLANE AREAS

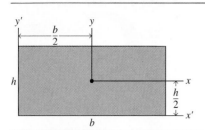

$$I_x = \frac{bh^3}{12}$$

$$I_{x'} = \frac{bh^3}{3}$$

$$A = bh$$

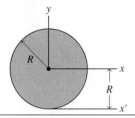

$$I_x = \frac{bh^3}{36}$$

$$I_{x'} = \frac{bh^3}{12}$$

$$A = \frac{1}{2}bh$$

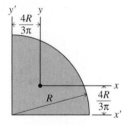

$$I_x = \frac{\pi R^4}{4}$$

$$I_{x'} = \frac{5\pi R^4}{4}$$

$$A = \pi R^2$$

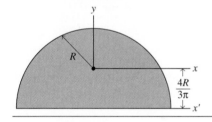

$$I_x = \frac{\pi R^4}{8} - \frac{8R^4}{9\pi}$$

$$I_y = \frac{\pi R^4}{8}$$

$$I_{x'} = \frac{\pi R^4}{8}$$

$$A = \frac{1}{2}\pi R^2$$

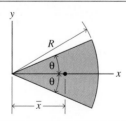

$$I_x = \frac{\pi R^4}{16} - \frac{4R^4}{9\pi}$$

$$I_{x'} = \frac{\pi R^4}{16}$$

$$A = \frac{1}{4}\pi R^2$$

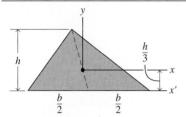

$$I_x = \frac{R^4}{4}\left(\theta - \frac{1}{2}\sin 2\theta\right)$$

$$I_y = \frac{R^4}{4}\left(\theta + \frac{1}{2}\sin 2\theta\right)$$

$$\bar{x} = \frac{2}{3}\frac{R\sin\theta}{\theta}$$

$$A = \theta R^2$$

TABLE C-2   MIXED SECOND MOMENTS OF PLANE AREAS

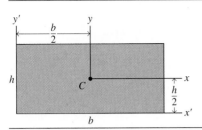

$$I_{xy} = 0$$

$$I_{x'y'} = \frac{b^2h^2}{4}$$

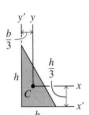

$$I_{xy} = -\frac{b^2h^2}{72}$$

$$I_{x'y'} = \frac{b^2h^2}{24}$$

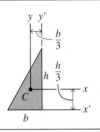

$$I_{xy} = \frac{b^2h^2}{72}$$

$$I_{x'y'} = -\frac{b^2h^2}{24}$$

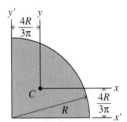

$$I_{xy} = \frac{(9\pi - 32)R^4}{72\pi}$$

$$I_{x'y'} = \frac{R^4}{8}$$

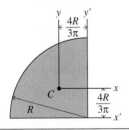

$$I_{xy} = -\frac{(9\pi - 32)R^4}{72\pi}$$

$$I_{x'y'} = -\frac{R^4}{8}$$

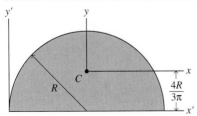

$$I_{xy} = 0$$

$$I_{x'y'} = \frac{2R^4}{3}$$

TABLE C-3   MOMENTS OF INERTIA OF COMMON SHAPES

Slender rod

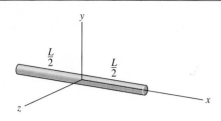

$$I_x = 0$$

$$I_y = I_z = \frac{1}{12}mL^2$$

Thin rectangular plate

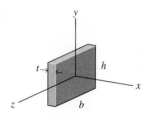

$$I_x = \frac{1}{12}m(b^2 + h^2)$$

$$I_y = \frac{1}{12}mb^2$$

$$I_z = \frac{1}{12}mh^2$$

Thin circular plate

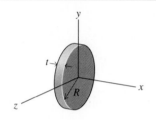

$$I_x = \frac{1}{2}mR^2$$

$$I_y = I_z = \frac{1}{4}mR^2$$

Rectangular prism

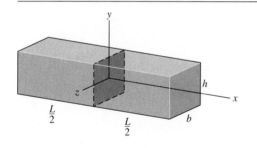

$$V = bhL$$

$$I_x = \frac{1}{12}m(b^2 + h^2)$$

$$I_y = \frac{1}{12}m(b^2 + L^2)$$

$$I_z = \frac{1}{12}m(h^2 + L^2)$$

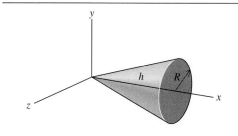

Right circular cone

$$V = \frac{1}{3}\pi R^2 h$$

$$\bar{x} = \frac{3}{4}h$$

$$I_x = \frac{3}{10}mR^2$$

$$I_y = I_z = \frac{3}{20}m(R^2 + 4h^2)$$

$$I_{yG} = I_{zG} = \frac{3}{80}m(4R^2 + h^2)$$

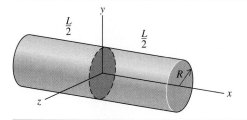

Circular cylinder

$$V = \pi R^2 L$$

$$I_x = \frac{1}{2}mR^2$$

$$I_y = I_z = \frac{1}{12}m(3R^2 + L^2)$$

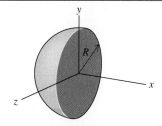

Hemisphere

$$V = \frac{2}{3}\pi R^3$$

$$\bar{x} = \frac{3}{8}R$$

$$I_x = I_y = I_z = \frac{2}{5}mR^2$$

$$I_{yG} = I_{zG} = \frac{83}{320}mR^2$$

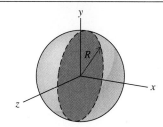

Sphere

$$V = \frac{4}{3}\pi R^3$$

$$I_x = I_y = I_z = \frac{2}{5}mR^2$$

COMPUTATIONAL METHODS

D-1 INTRODUCTION

The purpose of this appendix is to provide a few simple numerical methods that can be used to help solve (to take some of the drudgery out of the solution of) mechanics problems. No attempt has been made to provide a collection of computer programs to be used for the solution of the various types of problems encountered in a course in Statics. The problems designated as computer problems are basically simple applications of the elementary principles. They simply require solving the same problem over and over again as some parameter in the problem varies. Although they could be solved by hand or with a programmable calculator, these problems are most conveniently solved using a computer—either a microcomputer programmed in BASIC or a mainframe computer programmed in FORTRAN. In any event, the parametric study is intended to display characteristics of the problems that cannot be seen by solving for just one particular value. For example, Problem C3-35 examines the relationship between the tension in a cable and the amount of sag of the cable.

This appendix is not designed to teach the student all there is to know about numerical methods. The numerical methods presented here are purposely kept simple: simple for understanding and simple for use. Much more sophisticated methods exist. Students interested in these more sophisticated methods should take a course in numerical methods and/or see some of the references listed at the end of this appendix.

This appendix addresses three types of problems encountered in various mechanics problems in general and Statics problems in particular:

1. Nonlinear equations (root solving)
2. Systems of linear equations
3. Numerical integration

One or two simple methods are presented for the solution of each of these types of problems. The methods presented are purposely kept simple so that they can be used in hand calculation with a calculator or with a programmable calculator as well as with a computer.

In each case, a simple program is included to demonstrate the use of the numerical method. Versions are supplied in both BASIC and FORTRAN. A companion disk is not included, as the programs are short enough to type in without much effort. The programs are not elegant; like the numerical methods they illustrate, they have been kept as simple as possible so that they will be easy to understand, easy to modify, and easy to customize, and also so that they will run on the widest possible selection of computers. Students will probably want to modify and enhance these programs to improve the transfer of data into the programs and to improve the format of the results output by the programs. Students may also want to modify the programs to take advantage of special features of their individual computers, such as graphical output.

D-2   NONLINEAR EQUATIONS

Problems in mechanics (Statics) often require the solution of nonlinear equations, such as

$$x^3 - 7.014x^2 + 13.324x - 3.548 = 0 \tag{D-1a}$$

These problems are sometimes stated in the form: Find the zeros or roots of the function

$$f(x) = x^3 - 7.014x^2 + 13.324x - 3.548 \tag{D-1b}$$

(that is, find the values of x that make $f(x) = 0$). Therefore, they are sometimes called *root solving* problems. Equation D-1 is a typical equation encountered in the problem of finding the maximum moment of inertia of a body (Eq. D-1a is taken directly from Example Problem 10-17).

While such equations can be solved by trial and error (simply guessing values until the left-hand side of the equation is nearly zero), there exist simple, systematic ways to solve such problems. Two such methods—the *Newton–Raphson* method and the *Method of False Position*—will be discussed here.

D-2-1   Newton–Raphson Method

The Newton–Raphson method of root solving is an iterative method that for most functions converges very quickly. In fact, if the initial guess is reasonably close to the correct value, the Newton–Raphson method often converges within three or four iterations. Because of its quick convergence and ease of use, the Newton–Raphson method is the method of choice in most cases.

At each step n in the iteration process, the Newton–Raphson method uses the tangent line to the curve at the point x_n (Fig. D-1) to estimate the location of the root. The slope of the tangent line at x_n is just the derivative of the function evaluated at x_n

$$\text{slope} = f'(x_n)$$

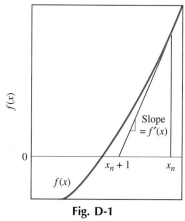

Fig. D-1

But from the geometry of Fig. D-1, the slope is also given by

$$\text{slope} = \frac{f(x_n) - 0}{x_n - x_{n+1}}$$

Setting these two expressions equal and solving for x_{n+1} gives

$$x_{n+1} = x_n - \frac{f(x_n)}{f'(x_n)} \tag{D-2}$$

Equation D-2 is used iteratively to get improved estimates of the location of the root. A rough graph of the function $f(x)$ should be used to get a first estimate of the root location, x_0.

Solve Eq. D-1

$$x^3 - 7.014x^2 + 13.324x - 3.548 = 0$$

using the Newton–Raphson Method to a relative accuracy of 0.001 percent.

SOLUTION

The function

$$f(x) = x^3 - 7.014x^2 + 13.324x - 3.548$$

is a cubic polynomial so the equation $f(x) = 0$ should have three roots. A rough sketch of the function (Fig. D-2) indicates that the three roots are near the points $x = 0$, $x = 3$, and $x = 3.5$. Starting from the initial point $x_0 = 3.5$, Eq. D-3 is then used to generate the next several points

$$x_1 = x_0 - \frac{x_0^3 - 7.014x_0^2 + 13.324x_0 - 3.548}{3x_0^2 - 14.028x_0 + 13.324}$$

$$= 3.5 - \frac{0.0395}{0.9760} = 3.4595$$

$$x_2 = 3.4595 - \frac{0.0056}{0.6986} = 3.4514$$

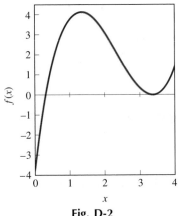

Fig. D-2

and so on. The process is repeated until the relative error

$$\left| \frac{x_{n+1} - x_n}{x_{n+1}} \right|$$

(the difference between successive approximations and the current approximation) is less than or equal to 0.00001 (0.001 percent). The result of this process is shown in Fig. D-3a. The column labeled DL is just the difference between successive iterations (the second term in the iteration formula)

$$DL = -\frac{f(x_n)}{f'(x_n)} = x_{n+1} - x_n$$

After just three iterations, the root is located as $x = 3.4511$. Similar iterations for starting values of 0 and 3.0 give the other two roots $x = 0.31670$ (Fig. D-3b) and $x = 3.24619$ (Fig. D-3c), respectively.

Xo = 3.50000 Error = 0.00001			Xo = 0.00000 Error = 0.00001			Xo = 3.00000 Error = 0.00001		
Xn	DL	ABS(DL/X1)	Xn	DL	ABS(DL/X1)	Xn	DL	ABS(DL/X1)
3.50000	−0.04047	0.01170	0.00000	0.26629	1.00000	3.00000	0.16932	0.05342
3.45953	−0.00808	0.00234	0.26629	0.04882	0.15492	3.16932	0.06170	0.01910
3.45145	−0.00034	0.00010	0.31510	0.00160	0.00504	3.23102	0.01426	0.00439
3.45111	−0.00000	0.00000	0.31670	0.00000	0.00001	3.24528	0.00090	0.00028
						3.24619	0.00000	0.00000

The root is 3.45111   The root is 0.31670

The root is 3.24619

(a)   (b)   (c)

Fig. D-3

Simple programs in BASIC and FORTRAN to solve nonlinear equations using the Newton–Raphson method are given in Programs D-1a and D-1b, respectively. The function statements defining the function and its derivative in lines 100 and 110 must be changed for the particular problem being solved.

D-2-2 Method of False Position

Although it is not as sophisticated as the Newton–Raphson method, the Method of False Position is a good method to use on functions for which the Newton–Raphson method has difficulty. The Newton–Raphson method has trouble when the derivative $f'(x)$ goes to zero at or near the root being located. This commonly happens when a function has two roots at the same value of x as, for example, $f(x) = (x + 1)(x - 1)^2$. Also, the Newton–Raphson method can be difficult to use for complex functions whose derivative $f'(x)$ may not be easily calculated. Therefore, an alternate method of solution is desirable and the Method of False Position is a good one.

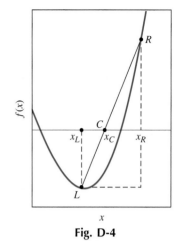

Fig. D-4

The Method of False Position is a systematic method of narrowing down the region in which a root exists. For example, Fig. D-4 shows a graph of a typical function $f(x)$ as a function of x. Point L lies on the left side of the point where $f(x) = 0$ and point R lies on the right side. Note that for two such points L and R, which bracket a simple root, $f(x_L)f(x_R) < 0$, always. A rough graph of the function $f(x)$ may be needed to find appropriate initial points L and R.

Construct point C—the point where the straight line joining point

```
100 DEF FNY (X) = X*X*X − 7.014*X*X + 13.324*X − 3.548
110 DEF FNYP (X) = 3*X*X − 14.028*X + 13.324
120 PRINT "Enter Xo = ";
130 INPUT X0
140 PRINT "Enter Error = ";
150 INPUT ER
160 CLS
170 PRINT USING "    Xo = ###,###.#####"; X0
180 PRINT USING " Error =        ##.#####"; ER
190 PRINT "————————————————————————————"
200 PRINT "    Xn        DL      ABS(DL/X1) "
210 PRINT "————————————————————————————"
220    DL = −FNY(X0) / FNYP(X0)
230    X1 = X0 + DL
240    PE = ABS(DL / X1)
250    PRINT USING "###,###.#####   ###,###.#####   ###.#####"; X0; DL; PE
260    IF ABS(DL / X1) < ER THEN 290
270      X0 = X1
280      GOTO 220
290 PRINT "————————————————————————————"
300 PRINT USING "The root is ###,###.#####"; X1
310 END
```

Program D-1a. BASIC program listing for performing the Newton–Raphson method.

```
100 Y (X) = X*X*X − 7.014*X*X + 13.324*X − 3.548
110 YP (X) = 3*X*X − 14.028*X + 13.324
    PRINT *, ' Enter Xo = '
    READ *, X0
    PRINT *, ' Enter Error = '
    READ *, ER
    PRINT
    PRINT 1, X0
  1 FORMAT ('     Xo = ', F13.5)
    PRINT 2, ER
  2 FORMAT ('   Error =      ',F8.5)
    PRINT *, '——————————————————————————————.'
    PRINT *, '     Xn          DL      ABS(DL/X1) '
    PRINT *, '——————————————————————————————'
220    DL = −Y(X0) / YP(X0)
    X1 = X0 + DL
    PE = ABS(DL / X1)
    PRINT 3, X0, DL, PE
  3 FORMAT (1X,F13.5,4X,F13.5,4X,F9.5)
    IF (ABS(DL/X1) .LT. ER) THEN
        GOTO 290
      ELSE
        X0 = X1
        GOTO 220
      END IF
290 PRINT *, ' ——————————————————————————————.'
    PRINT 4, X1
  4 FORMAT (' The root is ',F13.5)
    CALL EXIT
    END
```

Program D-1b. FORTRAN program listing for performing the Newton–
Raphson method.

L and point R goes through zero—and use it as an estimate of the root.
Point C can be found by similar triangles

$$\frac{f(x_R)}{x_R - x_C} = \frac{f(x_R) - f(x_L)}{x_R - x_L} \tag{D-3}$$

Rearranging and solving for x_C gives

$$x_C = x_R - \frac{f(x_R)(x_R - x_L)}{f(x_R) - f(x_L)} \tag{D-4}$$

However, since $f(x)$ is not a straight line, $f(x_C)$ will not be zero. If
$f(x_C)f(x_R) < 0$, then x_C is on the left side of the root. In this case, point L
is moved to point C and the process is repeated. If $f(x_C)f(x_L) < 0$, then
x_C is on the right side of the root. In this case, point R is moved to point
C and the process is repeated. The process ends when point C becomes
the same as one of the end points (within the limits of numerical
roundoff error). Then point C is the desired root.

EXAMPLE PROBLEM D-2

Solve

$$200 = \frac{T_0}{5}\left[\cosh\frac{5(800)}{2T_0} - 1\right]$$

(from Example Problem 8-10) using the Method of False Position to a relative accuracy of 0.001 percent.

SOLUTION

First, rewrite the equation in the form of Eq. D-1

$$f(x) = x\left[\cosh\frac{2000}{x} - 1\right] - 1000 = 0$$

Based on a rough sketch of $f(x)$ (Fig. D-5), the initial points $x_L = 2000$ and $x_R = 2500$ are chosen. Note that $f(x_L) = 86.1614$, $f(x_R) = -156.4125$, and $f(x_L)f(x_R) < 0$. Point C is constructed using Eq. D-4, giving

$$x_C = 2500 - \frac{(-156.4125)(2500 - 2000)}{(-156.4125) - (86.1614)} = 2177.5982$$

Since $f(x_C) = -15.1522$, the product $f(x_C)f(x_R) > 0$ and the product $f(x_C)f(x_L) < 0$. Therefore, point R is moved to point C and the process is repeated. The process ends when the relative error, either

$$\frac{x_R - x_C}{x_C} \quad \text{or} \quad \frac{x_C - x_L}{x_C}$$

is less than or equal to 0.00001. The result of this process is shown in Fig. D-6. After just five iterations, the root is located as $x = 2148.642$.

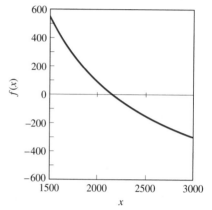

Fig. D-5

Error = 0.00001

XL = 2,000.00000	F(XL) = 86.16138
XR = 2,500.00000	F(XR) = −156.41248
XC = 2,177.59814	F(XC) = −15.15216
XL = 2,000.00000	F(XL) = 86.16138
XR = 2,177.59814	F(XR) = −15.15216
XC = 2,151.03711	F(XC) = −1.27262
XL = 2,000.00000	F(XL) = 86.16138
XR = 2,151.03711	F(XR) = −1.27262
XC = 2,148.83862	F(XC) = −0.10550
XL = 2,000.00000	F(XL) = 86.16138
XR = 2,148.83862	F(XR) = −0.10550
XC = 2,148.65649	F(XC) = −0.00865
XL = 2,000.00000	F(XL) = 86.16138
XR = 2,148.65649	F(XR) = −0.00865
XC = 2,148.64160	F(XC) = −0.00072

The root is 2,148.64200

Fig. D-6

```
 90 DEF FNCOSH(X) = (EXP(X) + EXP(−X))/2
100 DEF FNY(X) = X*(FNCOSH(2000/X) − 1) − 1000
110 PRINT "Enter XL = ";
120 INPUT XL
130 PRINT "ENTER XR = ";
140 INPUT XR
150 PRINT "Enter Error = ";
160 INPUT ER
170 CLS
180 PRINT USING " Error = ###,###.#####"; ER
190 PRINT "—————————————————————————"
200 XC = XR − FNY(XR)*(XR − XL)/(FNY(XR) − FNY(XL))
210 PRINT USING " XL = ###,###.#####    F(XL) = ###,###.#####"; XL; FNY(XL)
220 PRINT USING " XR = ###,###.#####    F(XR) = ###,###.#####"; XR; FNY(XR)
230 PRINT USING " XC = ###,###.#####    F(XC) = ###,###.#####"; XC; FNY(XC)
240    IF FNY(XC)*FNY(XL) < = 0 THEN 320
250    IF FNY(XC)*FNY(XR) < = 0 THEN 290
260 PRINT
270 PRINT "*** ERROR ***"
280 END
290 IF ABS((XC − XL)/XC) < ER THEN 350
300    XL = XC
310    GOTO 190
320 IF ABS((XC − XR)/XC) < ER THEN 350
330    XR = XC
340    GOTO 190
350 PRINT "—————————————————————————"
360 PRINT USING "     The root is ###,###.#####"; XC
370 END
```

Program D-2a. BASIC program listing for performing the Method of False Position.

Simple programs in BASIC and FORTRAN to solve nonlinear equations are given in Programs D-2a and D-2b, respectively. The function statement in line 100 must be changed for the particular problem being solved.

D-3 SYSTEMS OF LINEAR EQUATIONS

Many problems in mechanics require the solution of a system of linear equations, such as

$$5x + 3y + 4z = 23 \qquad \text{(D-5a)}$$
$$2x + 1y + 1z = \ 7 \qquad \text{(D-5b)}$$
$$1x + 3y + 5z = 22 \qquad \text{(D-5c)}$$

One possible scheme for solving such equations would be to:

First, use Eq. D-5a to eliminate x from Eqs. D-5b and D-5c. For example, subtract 2/5 times Eq. D-5a from Eq. D-5b and subtract 1/5 times Eq. D-5a from Eq. D-5c. (If the coefficient of x in Eq. D-5a were zero, the equations must first be reordered so that the coefficient is not zero.)

Next, use the resulting Eq. D-5b to eliminate y from Eqs. D-5a and D-5c.

Finally, use Eq. D-5c to eliminate z from Eqs. D-5a and D-5b.

```
100 Y(X) = X*(COSH(2000/X) − 1) − 1000
    PRINT *, ' Enter XL = '
    READ *, XL
    PRINT *, ' Enter XR = '
    READ *, XR
    PRINT *, ' Enter Error = '
    READ *, ER
    PRINT 1, ER
  1 FORMAT ('  Error = ',F13.5)
190 PRINT *, ' ——————————————————————————————.'
    XC = XR − Y(XR)*(XR − XL)/(Y(XR) − Y(XL))
    PRINT 2, XL, Y(XL)
  2 FORMAT ('  XL = ',F13.5,'  F(XL) = ',F13.5)
    PRINT 3, XR, Y(XR)
  3 FORMAT ('  XR = ',F13.5,'  F(XR) = ',F13.5)
    PRINT 4, XC, Y(XC)
  4 FORMAT ('  XC = ',F13.5,'  F(XC) = ',F13.5)
    IF (Y(XC)*Y(XL) .LE. 0) GOTO 320
    IF (Y(XC)*Y(XR) .LE. 0) GOTO 290
    PRINT
    PRINT *, ' *** ERROR ***'
    CALL EXIT
290 IF (ABS((XC − XL)/XC) .LT. ER) GOTO 350
    XL = XC
    GOTO 190
320 IF (ABS((XC − XR)/XC) .LT. ER) GOTO 350
    XR = XC
    GOTO 190
350 PRINT *, ' ——————————————————————————————.'
    PRINT 5, XC
  5 FORMAT ('     The root is ',F13.5)
    CALL EXIT
    END
```

Program D-2*b*. FORTRAN program listing for performing the Method of False Position.

At this point, Eq. D-5*a* will give the value of x; Eq. D-5*b*, the value of y; and Eq. D-5*c*, the value of z.

The procedure described above is called the Gauss–Jordan method for the solution of systems of linear equations. Since the letters representing the unknowns (x, y, and z) serve no function in the solution scheme except to hold the coefficients in their proper places, the procedure is conveniently carried out on a matrix of the coefficients

$$\begin{bmatrix} 5 & 3 & 4 & 23 \\ 0 & 1 & 1 & 7 \\ 1 & 3 & 5 & 22 \end{bmatrix} \tag{D-6}$$

where the rows of the matrix represent the equations being solved. The first column of each row contains the coefficients of x, the second column contains the coefficients of y, the third column contains the coefficients of z, and the last column contains the right-hand side of the equations. The rows of the matrix are to be multiplied by constants and added to other rows in the same manner that the equations were in the procedure described above.

Solve the system of linear equations (Eqs. D-5a, D-5b, and D-5c) using the Gauss–Jordan method.

SOLUTION

The equations are first written in matrix form as in Eq. D-6. Then 2/5 times the first row (equation) is subtracted from the second row (equation), and 1/5 times the first row (equation) is subtracted from the third row (equation) to give

$$\begin{bmatrix} 5 & 3 & 4 & 23 \\ 0 & -0.2 & -0.6 & -2.2 \\ 0 & 2.4 & 4.2 & 17.4 \end{bmatrix}$$

Then, 15 times the second row is added to the first row and 12 times the second row is added to the third row to give

$$\begin{bmatrix} 5 & 0 & -5.0 & -10.0 \\ 0 & -0.2 & -0.6 & -2.2 \\ 0 & 0 & -3.0 & -9.0 \end{bmatrix}$$

Finally, 5/3 times the third row is subtracted from the first row and 1/5 times the third row is subtracted from the second row to give

$$\begin{bmatrix} 5 & 0 & 0 & 5 \\ 0 & -0.2 & 0 & -0.4 \\ 0 & 0 & -3.0 & -9.0 \end{bmatrix}$$

The answer is more easily interpreted if the first row is divided by 5; the second row by -0.2, and the third row by -3, which gives the matrix

$$\begin{bmatrix} 1 & 0 & 0 & 1 \\ 0 & 1 & 0 & 2 \\ 0 & 0 & 1 & 3 \end{bmatrix}$$

The rows of this matrix represent the three equations

$$x = 1 \qquad y = 2 \qquad z = 3$$

which is the solution to the original system of equations.

```
100 DATA 3
110 DATA 5, 3, 4, 23
120 DATA 2, 1, 1,  7
130 DATA 1, 3, 5, 22
140 DIM A(20,21)
150 GOSUB 310
160 FOR I = 1 TO N
170    GOSUB 500
180    GOSUB 720
190    NEXT I
200 PRINT
210 PRINT
220 PRINT "The solution is:"
230 PRINT
240 FOR I = 1 TO N
250    PRINT "x(";I;") = "; A(I,N+1)
260    NEXT I
270 END
280 REM
290 REM   Read the input matrix from DATA statements
300 REM
310 READ N
320 FOR I = 1 TO N
330    FOR J = 1 TO N+1
340       READ A(I,J)
350       NEXT J
360    NEXT I
370 CLS
380 PRINT "The input matrix is:"
390 PRINT
400 FOR I = 1 TO N
410    FOR J = 1 TO N+1
420       PRINT A(I,J),
430       NEXT J
440    PRINT
450    NEXT I
460 RETURN
470 REM
480 REM   Search for row with largest element in column I
490 REM
500 TEMP = ABS(A(I,I))
510 KT = I
520 FOR K = I TO N
530    TT = ABS(A(K,I))
540    IF TT < = TEMP THEN 570
550       KT = K
560       TEMP = TT
570    NEXT K
580 IF KT = I THEN 680
590 REM
600 REM   Interchange rows if necessary to make the
610 REM   'pivot' element as large as possible
620 REM
630 FOR K = 1 TO N+1
640    TEMP = A(I,K)
650    A(I,K) = A(KT,K)
660    A(KT,K) = TEMP
670    NEXT K
680 RETURN
690 REM
700 REM   'Normalize' the pivot row
710 REM
720 PV = A(I,I)
730 FOR K = I TO N+1
740    A(I,K) = A(I,K)/PV
750    NEXT K
760 REM
770 REM   Eliminate all entries from the Ith column
780 REM   except the pivot element which has been
790 REM   normalized to 1
800 REM
810 FOR K = 1 TO N
820    IF K = I THEN 870
830    PV = A(K,I)
840    FOR KK = I TO N+1
850       A(K,KK) = A(K,KK) − PV*A(I,KK)
860       NEXT KK
870    NEXT K
880 RETURN
```

Program D-3a. BASIC program listing for solving a system of linear equations using the Gauss–Jordan elimination method.

Simple programs in BASIC and FORTRAN to solve systems of linear equations using the Gauss–Jordan elimination procedure are given in Programs D-3a and D-3b, respectively. The number of equations being solved and the coefficients of the matrix are included in data statements (lines 100–130) and must be changed for the specific problem being solved. A desirable modification of these programs would be to have these values entered directly from the keyboard for small numbers of equations or read from a file for large numbers of equations.

```
      REAL A(20,21)                                        END IF
100 DATA N/3/                                       570    CONTINUE
110 DATA A(1,1),A(1,2),A(1,3),A(1,4)/5, 3, 4, 23/          IF (KT .EQ. I) GOTO 680
120 DATA A(2,1),A(2,2),A(2,3),A(2,4)/2, 1, 1,  7/   C        Interchange rows if necessary to make the 'pivot'
130 DATA A(3,1),A(3,2),A(3,3),A(3,4)/1, 3, 5, 22/   C        element as large as possible
      PRINT *, ' The input matrix is:'                     DO 670 K = 1, N+1
      PRINT *, '  '                                          TEMP = A(I,K)
      DO 450 I = 1, N                                        A(I,K) = A(KT,K)
        PRINT *, (A(I,J), J = 1, N+1)                        A(KT,K) = TEMP
450   CONTINUE                                       670    CONTINUE
      DO 190 I = 1, N                                680 RETURN
        CALL PIVOT(N,A,I)                                  END
        CALL ELIM(N,A,I)                                   SUBROUTINE ELIM(N,A,I)
190   CONTINUE                                             REAL A(20,21)
      PRINT *, '  '                                 C        'Normalize' the pivot row
      PRINT *, '  '                                          PV = A(I,I)
      PRINT *, ' The solution is:'                          DO 750 K = I, N+1
      PRINT *, '  '                                           A(I,K) = A(I,K)/PV
      DO 260 I = 1, N                                750    CONTINUE
        PRINT *, 'X(', I, ') = ', A(I,N+1)           C        Eliminate all entries from the Ith column
260   CONTINUE                                       C        except the pivot element which has been
      CALL EXIT                                      C        normalized to 1
      END                                                   DO 870 K = 1, N
      SUBROUTINE PIVOT(N,A,I)                                 IF (K .NE. I) THEN
      REAL A(20,21)                                             PV = A(K,I)
C        Search for row with largest element in column I        DO 860 KK = I, N+1
      TEMP = ABS(A(I,I))                                          A(K,KK) = A(K,KK) - PV*A(I,KK)
      KT = I                                          860        CONTINUE
      DO 570 K = I, N                                          END IF
        TT = ABS(A(K,I))                             870    CONTINUE
        IF (TT .GT. TEMP) THEN                       880 RETURN
          KT = K                                         END
          TEMP = TT
```

Program D-3b.   FORTRAN program listing for solving a system of linear equations using the Gauss–Jordan elimination method.

D-4   NUMERICAL INTEGRATION

Most of the functions to be integrated in Statics problems are polynomials or other simple functions and are easily evaluated analytically. Sometimes, however, the function to be integrated will be complex enough to require advanced integration techniques. At other times, the function to be integrated will not be given explicitly. Instead, the function may be given by experimentally determined values at a few points. In these last two cases, numerical methods may be useful in evaluating the integrals.

In the simplest form, numerical integration derives from the physical interpretation of an integral as the area under a curve. The area may be approximated using several rectangles, trapezoids, or other simple shapes whose area is easily determined. The approximate value of the integral is obtained by adding together the areas of the several pieces. The method to be described here uses trapezoids to approximate the area under the curve; hence the name: Trapezoidal Rule.

For example, the value of the integral

$$I = \int_a^b f(x)\,dx \qquad \text{(D-7)}$$

is represented by the shaded area in Fig. D-7a. Approximating this area by one large trapezoid of width $h = b - a$, as in Fig. D-7b, gives

$$I \cong T_1 = \frac{h}{2}[f_a + f_b] \qquad \text{(D-8)}$$

where $f_a = f(a)$ and $f_b = f(b)$. This approximation is obviously in error by the amount of the shaded area in Fig. D-7b between the top of the trapezoid and the curve. The amount of error can be reduced, however, by using two trapezoids of width $h = (b - a)/2$, as in Fig. D-7c, and adding their areas together

$$I \cong T_2 = \frac{h}{2}[f_a + f_1] + \frac{h}{2}[f_1 + f_b] = \frac{h}{2}[f_a + f_b + 2f_1] \qquad \text{(D-9)}$$

where $f_1 = f(x_1)$ and $x_1 = a + h$. The tops of the two trapezoids follow the curve more closely than did the single trapezoid. The error in the approximation T_2 (represented by the shaded area in Fig. D-7c) is less than the error in the approximation T_1.

Continuing with this logic and dividing the interval into N (trapezoids) of equal width h gives the Trapezoidal Rule approximation

$$I \cong T_n = \frac{h}{2}\left[f_0 + f_n + 2\sum f_i\right] \qquad \text{(D-10)}$$

where

$$f_0 = f(a) \qquad f_n = f(b) \qquad h = \frac{b - a}{n}$$
$$f_i = f(x_i) = f(a + ih) \qquad i = 1, 2, \cdots, n - 1$$

As the number of panels used is increased, the width of the panels will decrease, the tops of the trapezoids will more closely fit the function being integrated, and the total error of the approximation will be reduced. It can be shown that the error in using the Trapezoidal Rule is approximately proportional to h^2. Therefore, reducing h by a factor of 2 should reduce the error by a factor of 4. Typically, the integral is evaluated several times using smaller and smaller panel widths until the value for two different panel widths is nearly the same. This value then is taken to be the value of the integral.

For experimental data given only at discrete points (possibly not equal-spaced), the Trapezoidal Rule is applied to each pair of points and the values are added together. In this case, however, it is usually not possible to vary the number of panels and get an estimate of the error. Instead, a sketch of the function may be used to get a rough estimate of how well the Trapezoidal Rule approximates the function (see Example Problem D-5). If it appears that the Trapezoidal Rule does not accurately represent the function, then either an interpolation procedure could be used to generate additional points at a large number of equal-spaced values as required for the Trapezoidal Rule or a more accurate integration procedure must be used.

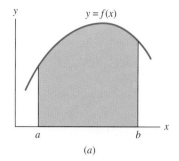

(a)

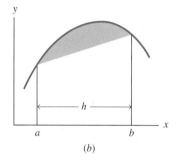

(b)

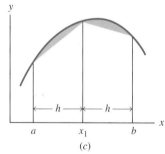

(c)

Fig. D-7

Evaluate the integral

$$\int_0^2 xe^{-x^2} dx$$

using the Trapezoidal Rule to an estimated relative error of 0.1 percent.

SOLUTION

Evaluating the integral using Eq. D-10 and a single panel gives

$$T_1 = \frac{2}{2}[0 + 2e^{-4}] = 0.0366$$

whereas using two panels gives

$$T_2 = \frac{1}{2}[0 + 2 e^{-4} + 2(e^{-1})] = 0.3862$$

The absolute error (just the difference between the current estimate and the correct value) is estimated to be

$$E_{abs} = |T_2 - T_1| = 0.3496$$

and the relative error (the ratio of the absolute error and the correct value) is estimated to be

$$E_{rel} = \left| \frac{T_2 - T_1}{T_2} \right| \times 100 = 90.5\%$$

Using four panels gives

$$T_4 = \frac{0.5}{2}[0 + 2 e^{-4} + 2(0.5e^{-0.25} + e^{-1} + 1.5e^{-2.25})] = 0.4668$$

and the relative error is estimated to be

$$E_{rel} = \left| \frac{0.4668 - 0.3862}{0.4668} \right| \times 100 = 17.28\%$$

Continuing using more and more panels until the desired accuracy has been reached gives the results in Fig. D-8. Using 64 panels, the integral is evaluated as 0.4908 with an estimated error of 0.06 percent. (Notice in Fig. D-8 that reducing the width of the panels by a factor of 2 reduced the error by approximately a factor of 4 in accordance with the error estimate mentioned earlier).

No of Panels	Value of Integral	% rel Error
1	0.036631	
2	0.386195	90.514820
4	0.466847	17.275900
8	0.484937	3.730329
16	0.489371	0.906150
32	0.490475	0.224978
64	0.490750	0.056155
128	0.490819	0.014032

Note: The result is 0.4908 with an estimated error of 0.1 percent.

Fig. D-8 Evaluation of the integral $\int_0^2 xe^{-x^2} dx$, using the Trapezoidal Rule (Eq. D-10) and varying numbers of panels.

```
100 DEF FNY(X) = X*EXP(−X*X)
110 A = 0
120 B = 2
130 CLS
140 N = 1
150 H = B − A
160 T = H*(FNY(A) + FNY(B))/2
170 PRINT "————————————————-"
180 PRINT "No of    Value of      % rel"
190 PRINT "Panels   Integral     Error"
200 PRINT "————————————————-"
210 PRINT USING " #### ###.######"; N, T
220 FOR K = 1 TO 7
230    TOLD = T
240    H = H/2
250    T = FNY(A) + FNY(B)
260    N = (B − A)/H
270    FOR I = 1 TO N − 1
280      X1 = A + I*H
290      T = T + 2*FNY(X1)
300    NEXT I
310    T = T*H/2
320    ER = 100*ABS((T − TOLD)/T)
330    PRINT USING " ####   ###.######   ###.######"; N, T, ER
340    NEXT K
350 PRINT "————————————————-"
```

Program D-4a. BASIC program listing for evaluating integrals using the Trapezoidal Rule.

```
100 Y(X) = X*EXP(−X*X)                      H = H/2
110 A = 0                                   T = Y(A) + Y(B)
120 B = 2                                   N = (B − A)/H
    N = 1                                   DO 300 I = 1, N − 1
    H = B − A                                 X1 = A + I*H
    T = H*(Y(A) + Y(B))/2                      T = T + 2*Y(X1)
    PRINT *, '————————————————-'   300   CONTINUE
    PRINT *, 'No of    Value of      % rel'    T = T*H/2
    PRINT *, 'Panels   Integral     Error'     ER = 100*ABS((T − TOLD)/T)
    PRINT *, '————————————————-'         PRINT 210, N, T, ER
    PRINT 210, N, T                     340   CONTINUE
210 FORMAT (3X,I4,2(3X,F10.6))               PRINT *, ' ————————————————-'
    DO 340 K = 1, 7                          CALL EXIT
      TOLD = T                               END
```

Program D-4b. FORTRAN program listing for evaluating integrals using the Trapezoidal Rule.

A listing of the BASIC and FORTRAN programs used to generate the output of Fig. D-8 are given in Programs D-4a and D-4b, respectively. The function statement and integration limits in lines 100–120 need to be changed for the function being integrated.

A circular plate is used to distribute the weight carried by a column (Fig. D-9). Because the plate is not rigid, the pressure between the plate and the floor is not constant. If pressure sensors on the bottom of the plate indicate the following radial pressure distribution,

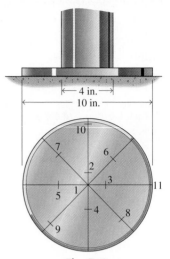

Fig. D-9

Radial Position (in.)	Pressure (psi)	Radial Position (in.)	Pressure (psi)
0.0	21.6	3.0	2.4
0.5	21.8	3.5	1.8
1.0	21.7	4.0	1.4
1.5	21.8	4.5	1.0
2.0	13.6	5.0	0.9
2.5	3.5		

determine the total weight carried by the plate.

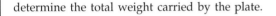

SOLUTION

The total weight W carried by the plate is

$$W = \int_0^5 p(2\pi r \, dr) = 2\pi \int_0^5 (pr) \, dr$$

Since the data are given at equal-spaced values of r, the Trapezoidal Rule (Eq. D-10) will be used directly with $N = 10$ and $h = \dfrac{(5-0)}{10} = 0.5$ to get

$$
\begin{aligned}
W = 2\pi\frac{0.5}{2}\{ & (0)(21.6) + (5.0)(0.9) \\
& + 2[(0.5)(21.8) + (1.0)(21.7) + (1.5)(21.8) \\
& + (2.0)(13.6) + (2.5)(3.5) + (3.0)(2.4) \\
& + (3.5)(1.8) + (4.0)(1.4) + (4.5)(1.0)]\} \\
= & \ 399.3 \text{ lb}
\end{aligned}
$$

The function being integrated, pr, is plotted in Fig. D-10 as a function of r. The solid line indicates the area used by the Trapezoidal Rule, whereas the dotted line indicates the expected true function shape. The expected error is the shaded region between the two curves. It appears that the Trapezoidal Rule has underestimated the integral by a small amount (perhaps 1 or 2 percent). If a more accurate value of the integral were needed, a more accurate integration method would be needed.

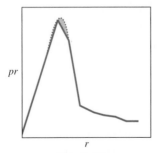

Fig. D-10

D-5   FURTHER READING

1. Chapra, S. C., and R. P. Canale (1985). *Numerical Methods for Engineers with Personal Computer Applications*, McGraw-Hill, New York.

2. Cheney, W., and D. Kincaid (1980). *Numerical Mathematics and Computing*, Brooks/Cole, Monterey, Calif.

3. James, M. L., G. M. Smith, and J. C. Wolford (1985). *Applied Numerical Methods for Digital Computation*, 3rd ed., Harper & Row, New York.

4. Johnston, R. L. (1982). *Numerical Methods: A Software Approach*, Wiley, New York.

5. Mathews, J. H. (1987). *Numerical Methods for Computer Science, Engineering and Mathematics*, Prentice Hall, Englewood Cliffs, N.J.

6. Shoup, T. E. (1983). *Numerical Methods for the Personal Computer*, Prentice Hall, Englewood Cliffs, N.J.

ANSWERS TO SELECTED PROBLEMS

Chapter 1

1-1 $m = 15.54$ slug

1-2 $W = 5.64$ kN

1-3 **a.** $W = 179.7$ lb
   b. $W = 179.5$ lb

1-4 $W = 561$ N

1-9 $W = 20.7$ lb

1-10 $r = 3.38(10^6)$ m

1-11 $g = 27.8$ ft/s^2

1-12 $F = 1.984(10^{20})$ N

1-15 $W = 165.4$ lb

1-16 $m = 227$ kg

1-19 $\rho = 692$ kg/m^3

1-20 $\gamma = 0.320$ lb/in.^{3}

1-23 $V = 7.21$ L

1-24 $FC = 23.5$ mi/gal

1-27 **a.** $d = 128.7$ km
   b. $d = 37.0$ km
   c. $d = 73.2$ m

1-28 $c_p = 6000$ ft $\cdot$ lb/slug $\cdot$ °R

1-29 $J = L^4$

1-30 $E = M/LT^2$

1-33 $a = L; b = L^2; w = L^3$

1-34 $A = L; B = L^2; C = L^3; D = L^7$

1-35 **a.** 0.015 (%$D = -2.36$%)
   b. 0.035 (%$D = +0.751$%)
   c. 0.057 (%$D = +0.666$%)

1-36 **a.** 0.84 (%$D = +0.301$%)
   b. 0.47 (%$D = -0.617$%)
   c. 0.66 (%$D = -0.673$%)

1-38 **a.** 26.4 (%$D = +0.01997$%)
   b. 74.8 (%$D = -0.0390$%)
   c. 55.3 (%$D = -0.0665$%)

1-41 **a.** 63,750 (%$D = +0.00585$%)
   b. 27,380 (%$D = -0.01037$%)
   c. 55,130 (%$D = +0.0001451$%)

1-44 **a.** $F = 1.984(10^{20})$ N
   b. $F = 3.55(10^{22})$ N

1-45 **a.** $r = 2.56(10^7)$ ft
   b. $r = 3.62(10^7)$ ft

1-48 $\mu 2.51(10^{-5})$ lb $\cdot$ s/ft^2

1-49 $A = 259$ hectare

Chapter 2

2-1 $\mathbf{R} = 150$ lb ↗ @ 36.9°

2-2 $\mathbf{R} = 98.8$ N ↗ @ 28.3°

2-5 $\mathbf{R} = 100.1$ lb ↗ @ 50.5°

2-6 $\mathbf{R} = 151.7$ N ↗ @ 30.5°

2-9 $\mathbf{R} = 1361$ lb ↗ @ 43.7°

2-10 $\mathbf{R} = 115.3$ kN ↗ @ 67.2°

2-13 $\mathbf{R} = 23.2$ kip ↗ @ 71.9°

2-14 $\mathbf{R} = 99.0$ kN ↖ @ 93.4°

2-17 $F_u = 582$ lb
   $F_v = 718$ lb

2-18 $F_u = 964$ N
   $F_v = 750$ N

2-21 $F_u = 181.8$ lb
   $F_v = 214$ lb

2-22 $F_u = 1026$ N
   $F_v = 798$ N

2-25 $\mathbf{F}_u = 53.0$ kip ↘ @ 45.0°
   $\mathbf{F}_v = 39.5$ kip ↖ @ 161.6°

2-26 $\mathbf{F}_u = 92.9$ kN ↙ @ 157.4°
   $\mathbf{F}_v = 107.1$ kN ↘ @ 36.9°

2-29 $F_x = 866$ lb
   $F_v = 500$ lb

2-30 $F_x = 338$ N
   $F_y = 725$ N

2-33 **a.** $F_{1x} = 171.0$ lb
   $F_{1y} = 470$ lb
   $F_{2x} = 650$ lb
   $F_{2y} = -375$ lb
   b. $F_{1x'} = -211$ lb
   $F_{1y'} = 453$ lb
   $F_{2x'} = 724$ lb
   $F_{2y'} = 194.1$ lb

2-34 **a.** $F_{1x} = -514$ N
   $F_{1y} = 613$ N
   $F_{2x} = 940$ N
   $F_{2y} = 342$ N
   b. $F_{1x'} = -752$ N
   $F_{1y'} = 274$ N
   $F_{2x'} = 643$ N
   $F_{2y'} = 766$ N

2-37 **a.** $F_x = 5.00$ kip
   $F_y = 3.42$ kip
   $F_z = 7.95$ kip
   b. $\mathbf{F} = 5.00$ **i** $+ 3.42$ **j** $+ 7.95$ **k** kip

2-38 **a.** $F_x = 3.88$ kN
   $F_y = -9.64$ kN
   $F_z = 10.81$ kN
   b. $\mathbf{F} = 3.88$ **i** $- 9.64$ **j** $+ 10.81$ **k** kN

2-41 **a.** $\theta_x = 64.9$°
   $\theta_y = 45.0$°
   $\theta_z = 55.6$°
   b. $F_x = 339$ lb
   $F_y = 566$ lb
   $F_z = 453$ lb
   c. $\mathbf{F} = 339$ **i** $+ 566$ **j** $+ 453$ **k** lb

2-42 **a.** $\theta_x = 136.7°$
$\theta_y = 119.0°$
$\theta_z = 61.0°$
b. $F_x = -36.4$ kN
$F_y = -24.3$ kN
$F_z = 24.3$ kN
c. $\mathbf{F} = -36.4\ \mathbf{i} - 24.3\ \mathbf{j} + 24.3\ \mathbf{k}$ kN

2-43 **a.** $F_{1x} = -557$ lb
$F_{1y} = 278$ lb
$F_{1z} = 650$ lb
b. $\mathbf{F}_1 = -557\ \mathbf{i} + 278\ \mathbf{j} + 650\ \mathbf{k}$ lb
c. $F_n = 773$ lb
d. $\alpha = 30.7°$

2-44 **a.** $F_{1x} = 25.4$ kN
$F_{1y} = -15.21$ kN
$F_{1z} = 5.07$ kN
b. $\mathbf{F}_1 = 25.4\ \mathbf{i} - 15.21\ \mathbf{j} + 5.07\ \mathbf{k}$ kN
c. $F_n = 25.8$ kN
d. $\alpha = 30.6°$

2-47 $\mathbf{R} = 639$ lb ↗ @ $10.02°$

2-48 $\mathbf{R} = 7.23$ kN ↗ @ $26.1°$

2-51 $\mathbf{R} = 5.22$ kip ↖ @ $121.0°$

2-52 $\mathbf{R} = 15.77$ kN ↗ @ $78.8°$

2-55 $R = 52.9$ kip
$\theta_x = 66.7°$
$\theta_y = 62.1°$
$\theta_z = 37.7°$

2-57 $R = 1640$ lb
$\theta_x = 49.7°$
$\theta_y = 55.9°$
$\theta_z = 58.8°$

2-58 $R = 32.6$ kN
$\theta_x = 72.0°$
$\theta_y = 42.8°$
$\theta_z = 52.7°$

2-61 $F_3 = 81.2$ lb
$R = 125.9$ lb

2-62 $R = 2.71$ kN
$\theta_{x'} = 182.1°$

2-65 $F_x = 394$ lb
$F_y = 308$ lb
$F_u = 367$ lb
$F_y = 205$ lb

2-66 **a.** $R = 52.5$ N
$\theta_x = 63.6°$
b. $F_u = 41.9$ N
$F_v = 18.45$ N

Chapter 3

3-1 $F_2 = 220$ lb
$F_3 = 269$ lb

3-2 $F_3 = 4.58$ kN
$F_4 = 9.52$ kN

3-5 $F_4 = 677$ lb @ $\theta = 69.1°$

3-6 $F_4 = 4.78$ kN @ $\theta = 32.2°$

3-9 $F_A = 25.9$ lb
$F_B = 36.6$ lb

3-10 $F_N = 90.5$ N
$\theta = 53.1°$

3-13 $\theta = 10.89°$

3-14 $F_A = 4.16$ kN
$F_B = 3.92$ kN
$F_C = 1.387$ kN

3-17 $F_1 = 160.0$ lb
$F_2 = 90.0$ lb
$T = 120.0$ lb
$\theta = 36.9°$

3-19 $F_4 = 149.2$ lb
$\theta_x = 115.4°$
$\theta_y = 147.0°$
$\theta_z = 70.4°$

3-20 $F_4 = 8.66$ kN
$\theta_x = 144.6°$
$\theta_y = 109.8°$
$\theta_z = 118.0°$

3-23 $T_A = 2.34$ kip
$T_B = 2.47$ kip
$T_C = 2.70$ kip

3-24 $T_A = 452$ N
$T_B = 329$ N
$T_C = 694$ N

3-27 $T_A = 1200$ lb
$T_B = 1170$ lb
$T_C = 685$ lb

3-28 $T_A = 1790$ N
$T_B = 964$ N
$T_C = 2620$ N

3-29 $A_H = 283$ lb
$A_V = 603$ lb
$B_H = 283$ lb
$B_V = 547$ lb

3-30 $F_A = 467$ N
$F_B = 902$ N
$\mathbf{F}_C = 234$ N ↗ @ $2.90°$

3-33 $T_A = 348$ lb
$T_B = 171.0$ lb
$T_C = 317$ lb

Chapter 4

4-1 **a.** $\mathbf{M}_O = 15.00$ in. · kip ↻
$\mathbf{M}_A = 15.00$ in. · kip ↻
$\mathbf{M}_B = 10.00$ in. · kip ↺
$\mathbf{M}_C = 10.00$ in. · kip ↺
b. $\mathbf{M}_O = 7.50$ in. · kip ↺
$\mathbf{M}_A = 4.50$ in. · kip ↻
$\mathbf{M}_B = 9.00$ in. · kip ↻
$\mathbf{M}_C = 7.50$ in. · kip ↺

4-2   **a.** $\mathbf{M}_O = 22.5\ \text{N} \cdot \text{m} \downarrow$
              $\mathbf{M}_A = 5.63\ \text{N} \cdot \text{m} \downarrow$
              $\mathbf{M}_B = 13.13\ \text{N} \cdot \text{m} \downarrow$
              $\mathbf{M}_C = 22.5\ \text{N} \cdot \text{m} \downarrow$
   b. $\mathbf{M}_O = 21.9\ \text{N} \cdot \text{m} \downarrow$
              $\mathbf{M}_A = 21.9\ \text{N} \cdot \text{m} \downarrow$
              $\mathbf{M}_B = 31.3\ \text{N} \cdot \text{m} \downarrow$
              $\mathbf{M}_C = 31.3\ \text{N} \cdot \text{m} \downarrow$

4-3   **a.** $\mathbf{M}_O = 100.0\ \text{in.} \cdot \text{lb} \downarrow$
   b. $\mathbf{M}_O = 525\ \text{in.} \cdot \text{lb} \downarrow$

4-4   **a.** $\mathbf{M}_B = 75.0\ \text{N} \cdot \text{m} \downarrow$
   b. $\mathbf{M}_A = 40.0\ \text{N} \cdot \text{m} \downarrow$
   c. $\mathbf{M}_B = 14.14\ \text{N} \cdot \text{m} \downarrow$

4-7   **a.** $\mathbf{M}_C = 6000\ \text{in.} \cdot \text{lb} \downarrow$
   b. $\mathbf{M}_B = 3600\ \text{in.} \cdot \text{lb} \downarrow$
   c. $\mathbf{M}_B = 750\ \text{in.} \cdot \text{lb} \downarrow$
   d. $\mathbf{M}_E = 400\ \text{in.} \cdot \text{lb} \downarrow$

4-8   **a.** $\mathbf{M}_B = 25.2\ \text{N} \cdot \text{m} \downarrow$
   b. $\mathbf{M}_A = 66.0\ \text{N} \cdot \text{m} \downarrow$
   c. $\mathbf{M}_C = 36.4\ \text{N} \cdot \text{m} \downarrow$
   d. $\mathbf{M}_E = 41.1\ \text{N} \cdot \text{m} \downarrow$

4-11 $\mathbf{M}_A = 3.40\ \text{in.} \cdot \text{kip} \downarrow$

4-12 $\mathbf{M}_A = 80.8\ \text{N} \cdot \text{m} \downarrow$

4-15   **a.** $\mathbf{M}_A = 6.00\ \text{in.} \cdot \text{kip} \downarrow$
   b. $\mathbf{M}_B = 12.60\ \text{in.} \cdot \text{kip} \downarrow$

4-16   **a.** $\mathbf{M}_O = 0.707\ \text{kN} \cdot \text{m} \downarrow$
   b. $\mathbf{M}_A = 1.358\ \text{kN} \cdot \text{m} \downarrow$

4-19   **a.** $\mathbf{M}_A = 5.06\ \text{in.} \cdot \text{kip} \downarrow$
   b. $\mathbf{M}_B = 5.22\ \text{in.} \cdot \text{kip} \downarrow$

4-20   **a.** $\mathbf{M}_A = 1.129\ \text{kN} \cdot \text{m} \downarrow$
   b. $\mathbf{M}_B = 1.344\ \text{kN} \cdot \text{m} \downarrow$

4-23 $\mathbf{M}_O = 2.24\ \text{in.} \cdot \text{kip} \downarrow$

4-24 $\mathbf{M}_O = 1.587\ \text{kN} \cdot \text{m} \downarrow$

4-27   **a.** $\mathbf{M}_A = 6.00\ \text{in.} \cdot \text{kip} \downarrow$
   b. $\mathbf{M}_B = 12.60\ \text{in.} \cdot \text{kip} \downarrow$

4-28   **a.** $\mathbf{M}_O = 0.707\ \text{kN} \cdot \text{m} \downarrow$
   b. $\mathbf{M}_A = 1.358\ \text{kN} \cdot \text{m} \downarrow$

4-31   **a.** $\mathbf{M}_A = 5.06\ \text{in.} \cdot \text{kip} \downarrow$
   b. $\mathbf{M}_B = 5.22\ \text{in.} \cdot \text{kip} \downarrow$

4-32   **a.** $\mathbf{M}_A = 1.129\ \text{kN} \cdot \text{m} \downarrow$
   b. $\mathbf{M}_B = 1.344\ \text{kN} \cdot \text{m} \downarrow$

4-35   **a.** $\mathbf{M}_O = 4.30\ \text{in.} \cdot \text{kip} \downarrow$
   b. $\mathbf{M}_O = 5.55\ \text{in.} \cdot \text{kip} \downarrow$
   c. $\mathbf{M}_D = 9.97\ \text{in.} \cdot \text{kip} \downarrow$

4-36   **a.** $\mathbf{M}_O = 31.4\ \text{N} \cdot \text{m} \downarrow$
   b. $\mathbf{M}_O = 15.00\ \text{N} \cdot \text{m} \downarrow$
   c. $\mathbf{M}_D = 2.83\ \text{N} \cdot \text{m} \downarrow$

4-39 $\mathbf{M}_B = -3.36\ \mathbf{i} - 0.450\ \mathbf{j} - 2.88\ \mathbf{k}\ \text{in.} \cdot \text{kip}$

4-40 $\mathbf{M}_B = 0\ \mathbf{i} - 140\ \mathbf{j} - 210\ \mathbf{k}\ \text{N} \cdot \text{m}$

4-43   **a.** $\mathbf{M}_B = 2.40\ \mathbf{i} + 4.20\ \mathbf{j} + 7.98\ \mathbf{k}\ \text{in.} \cdot \text{kip}$
            $M_B = 9.33\ \text{in.} \cdot \text{kip}$
   b. $\theta_x = 75.1°$
            $\theta_y = 63.2°$
            $\theta_z = 31.2°$

4-44   **a.** $\mathbf{M}_B = -89.9\ \mathbf{i} + 112.3\ \mathbf{j} - 179.7\ \mathbf{k}\ \text{N} \cdot \text{m}$
            $M_B = 230\ \text{N} \cdot \text{m}$
   b. $\theta_x = 113.0°$
            $\theta_y = 60.8°$
            $\theta_z = 141.3°$

4-47 $\mathbf{M}_B = -4.35\ \mathbf{i} + 2.37\ \mathbf{j} - 0.510\ \mathbf{k}\ \text{in.} \cdot \text{kip}$

4-48 $\mathbf{M}_B = -190.4\ \mathbf{i} + 220\ \mathbf{j} + 214\ \mathbf{k}\ \text{N} \cdot \text{m}$

4-51 $M_{BC} = 1216\ \text{in.} \cdot \text{lb}$

4-52 $M_{BC} = -41.2\ \text{N} \cdot \text{m}$

4-55   **a.** $M_{OC} = 10.35\ \text{in.} \cdot \text{kip}$
   b. $M_{DE} = -3.63\ \text{in.} \cdot \text{kip}$

4-56   **a.** $M_{OC} = 96.1\ \text{N} \cdot \text{m}$
   b. $M_{DE} = 67.1\ \text{N} \cdot \text{m}$

4-59 $M_{OC} = -2.19\ \text{in.} \cdot \text{kip}$

4-60 $M_{OB} = -68.4\ \text{N} \cdot \text{m}$

4-65 $\mathbf{M} = 1299\ \mathbf{k}\ \text{in.} \cdot \text{lb}$
       $d = 8.66\ \text{in.}$

4-66 $\mathbf{M} = 55.3\ \mathbf{k}\ \text{N} \cdot \text{m}$
       $d = 221\ \text{mm}$

4-67 $\mathbf{M} = 380\ \mathbf{i} - 190.0\ \mathbf{j}\ \text{ft} \cdot \text{lb}$
       $d = 5.08\ \text{ft}$

4-68 $\mathbf{M} = 33.0\ \mathbf{i} + 3.75\ \mathbf{j} + 14.25\ \mathbf{k}\ \text{N} \cdot \text{m}$
       $d = 153.2\ \text{mm}$

4-71 $\mathbf{C} = 2.50\ \mathbf{k}\ \text{in.} \cdot \text{kip}$

4-72 $\mathbf{C} = 33.0\ \mathbf{k}\ \text{N} \cdot \text{m}$

4-75 $C = 887\ \text{in.} \cdot \text{lb}$
       $\theta_x = 105.7°$
       $\theta_y = 109.8°$
       $\theta_z = 154.3°$

4-77 $\mathbf{F} = 125.0\ \mathbf{i} + 217\ \mathbf{j}\ \text{lb}$
       $\mathbf{C} = 650\ \mathbf{k}\ \text{ft} \cdot \text{lb}$

4-78 $\mathbf{F} = 410\ \mathbf{i} + 287\ \mathbf{j}\ \text{N}$
       $\mathbf{C} = -124.9\ \mathbf{k}\ \text{N} \cdot \text{m}$

4-81   **a.** $\mathbf{F} = -499\ \mathbf{i} + 599\ \mathbf{j}\ \text{lb}$
            $\mathbf{C} = -11.98\ \mathbf{i} - 9.99\ \mathbf{j} + 14.98\ \mathbf{k}\ \text{in.} \cdot \text{kip}$
   b. $\theta_x = 123.7°$
            $\theta_y = 117.5°$
            $\theta_z = 46.2°$

4-82   **a.** $\mathbf{F} = -440\ \mathbf{i} + 400\ \mathbf{k}\ \text{N}$
            $\mathbf{C} = 96.0\ \mathbf{i} - 88.0\ \mathbf{j} + 105.7\ \mathbf{k}\ \text{N} \cdot \text{m}$
   b. $\theta_x = 55.1°$
            $\theta_y = 121.7°$
            $\theta_z = 51.0°$

4-87 $\mathbf{R} = 95.0\ \mathbf{j}\ \text{lb}$
       $\mathbf{C} = -380\ \mathbf{k}\ \text{in.} \cdot \text{lb}$

4-88 $\mathbf{R} = -11.44\ \mathbf{i} + 91.7\ \mathbf{j}\ \text{N}$
       $\mathbf{C} = -12.06\ \mathbf{k}\ \text{N} \cdot \text{m}$

4-91 $\mathbf{R} = 796\ \text{lb} \nearrow\ @\ 51.1°$
       $x = 4.86\ \text{in.}$

4-92 $\mathbf{R} = 585\ \text{N} \nearrow\ @\ 70.0°$
       $d = 170.9\ \text{mm}$

4-95 $\mathbf{R} = 127.5\ \text{lb} \nearrow\ @\ 8.53°$
       $\mathbf{C} = 4.50\ \text{in.} \cdot \text{kip} \downarrow$

4-96	$\mathbf{R} = 39.1 \text{ N} \nearrow$ @ $39.8°$		
	$\mathbf{C} = 65.0 \text{ N} \cdot \text{m} \downarrow$		

4-96 $\mathbf{R} = 39.1 \text{ N} \nearrow$ @ $39.8°$
$\mathbf{C} = 65.0 \text{ N} \cdot \text{m} \downarrow$

4-99 a. $\mathbf{R} = 2.13 \text{ kip} \swarrow$ @ $116.7°$
b. $d_R = 6.28 \text{ ft}$

4-100 a. $\mathbf{R} = 12.37 \text{ kN} \searrow$ @ $76.0°$
b. $d_R = 3.23 \text{ m}$

4-101 $\mathbf{R} = 150.0 \text{ k lb}$
$x_R = 5.87 \text{ ft}$
$y_R = 6.93 \text{ ft}$

4-102 $\mathbf{R} = 75.0 \text{ j N}$
$x_R = -2.60 \text{ m}$
$y_R = -4.89 \text{ m}$

4-105 $F_1 = 100.0 \text{ lb}$
$F_2 = 100.0 \text{ lb}$
$F_3 = 60.0 \text{ lb}$

4-106 $F_1 = 50.0 \text{ N}$
$F_2 = 160.0 \text{ N}$
$F_3 = 90.0 \text{ N}$

4-109 $\mathbf{R} = 238 \mathbf{i} - 47.0 \mathbf{j} + 238 \mathbf{k} \text{ lb}$
$\mathbf{C} = -953 \mathbf{j} - 1071 \mathbf{k} \text{ ft} \cdot \text{lb}$

4-110 $\mathbf{R} = 465 \mathbf{i} + 61.6 \mathbf{j} + 474 \mathbf{k} \text{ N}$
$\mathbf{C} = 600 \mathbf{i} - 41.1 \mathbf{j} - 123.2 \mathbf{k} \text{ N} \cdot \text{m}$

4-113 $\mathbf{R} = -800 \mathbf{i} + 750 \mathbf{j} + 600 \mathbf{k} \text{ lb}$
$\mathbf{C} = 1500 \mathbf{i} - 1000 \mathbf{j} + 2700 \mathbf{k} \text{ ft} \cdot \text{lb}$

4-114 $\mathbf{R} = -16.77 \mathbf{i} + 66.3 \mathbf{j} + 187.6 \mathbf{k} \text{ N}$
$\mathbf{C} = -125.1 \mathbf{i} - 62.6 \mathbf{j} + 66.3 \mathbf{k} \text{ N} \cdot \text{m}$

4-117 $\mathbf{R} = 75.0 \mathbf{i} + 40.0 \mathbf{j} + 50.0 \mathbf{k} \text{ lb}$
$\mathbf{C} = 900 \mathbf{i} + 720 \mathbf{j} - 1350 \mathbf{k} \text{ ft} \cdot \text{lb}$

4-118 $\mathbf{R} = -20.6 \mathbf{i} + 195.7 \mathbf{j} - 241 \mathbf{k} \text{ N}$
$\mathbf{C} = -168.8 \mathbf{i} + 80.0 \mathbf{j} + 33.7 \mathbf{k} \text{ N} \cdot \text{m}$

4-121 $\mathbf{R} = 238 \mathbf{i} - 47.0 \mathbf{j} + 238 \mathbf{k} \text{ lb}$
$\mathbf{C}_{\parallel} = -433 \mathbf{i} + 85.4 \mathbf{j} - 432 \mathbf{k} \text{ ft} \cdot \text{lb}$
$x_R = 0.359 \text{ ft}$
$y_R = 2.61 \text{ ft}$

4-122 $\mathbf{R} = 465 \mathbf{i} + 61.6 \mathbf{j} - 474 \mathbf{k} \text{ N}$
$\mathbf{C}_{\parallel} = 350 \mathbf{i} + 46.4 \mathbf{j} - 357 \mathbf{k} \text{ N} \cdot \text{m}$
$x_R = 1.815 \text{ m}$
$y_R = 2.47 \text{ m}$

4-125 a. $\mathbf{M} = 11.47 \mathbf{k} \text{ ft} \cdot \text{kip}$
b. $\mathbf{M} = -86.8 \mathbf{k} \text{ ft} \cdot \text{kip}$

4-126 a. $\mathbf{M}_O = -177.4 \mathbf{i} + 35.5 \mathbf{j} - 71.0 \mathbf{k} \text{ N} \cdot \text{m}$
b. $M_{OD} = 169.3 \text{ N} \cdot \text{m}$

4-129 $\mathbf{R} = -130.0 \text{ k lb}$
$x_R = 4.15 \text{ ft}$
$y_R = 29.2 \text{ ft}$

4-130 $\mathbf{C} = 750 \mathbf{i} + 306 \mathbf{j} + 231 \mathbf{k} \text{ N} \cdot \text{m}$

4-133 a. $\mathbf{R} = 309 \text{ lb} \swarrow$ @ $119.1°$
b. $d = 1.619 \text{ ft}$

4-134 a. $\mathbf{R} = 450 \text{ k N}$
$\mathbf{C} = 67.5 \mathbf{i} + 202 \mathbf{j} \text{ N} \cdot \text{m}$
b. $M_{OA} = 202 \text{ N} \cdot \text{m}$
$M_{BA} = 47.7 \text{ N} \cdot \text{m}$
$M_{BC} = 67.5 \text{ N} \cdot \text{m}$

Chapter 5

5-1 $\bar{x} = 2.09 \text{ in.}$
$\bar{y} = 7.83 \text{ in.}$
$\bar{z} = 1.739 \text{ in.}$

5-2 $\bar{x} = 53.3 \text{ mm}$
$\bar{y} = 77.8 \text{ mm}$
$\bar{z} = 160.0 \text{ mm}$

5-3 $\bar{x} = 7.39 \text{ in.}$
$\bar{y} = 9.02 \text{ in.}$
$\bar{z} = 4.07 \text{ in.}$

5-4 $\bar{x} = 225 \text{ mm}$
$\bar{y} = 263 \text{ mm}$
$\bar{z} = 173.6 \text{ mm}$

5-7 $\bar{x} = -0.1667 \text{ ft}$
$\bar{y} = 1.167 \text{ ft}$
$\bar{z} = 5.33 \text{ ft}$

5-8 $\bar{x} = 1.250 \text{ m}$
$\bar{y} = 1.438 \text{ m}$
$\bar{z} = -0.813 \text{ m}$

5-9 $\bar{x} = 4.00 \text{ in.}$
$\bar{y} = 2.67 \text{ in.}$

5-10 $\bar{x} = 133.3 \text{ mm}$
$\bar{y} = 100.0 \text{ mm}$

5-13 $\bar{x} = 4a/3\pi$
$\bar{y} = 4b/3\pi$

5-14 $\bar{x} = 3b/4$
$\bar{y} = 3b/20$

5-17 $\bar{x} = (\pi - 2)L/\pi$
$\bar{y} = \pi a/8$

5-18 $\bar{x} = 25.0 \text{ mm}$
$\bar{y} = 10.00 \text{ mm}$

5-21 $\bar{x} = 2a/5$
$\bar{y} = a/2$

5-22 $\bar{x} = 24.3 \text{ mm}$
$\bar{y} = 60.7 \text{ mm}$

5-25 $\bar{x} = 10.94 \text{ in.}$
$\bar{y} = 7.43 \text{ in.}$

5-26 $\bar{x} = 91.2 \text{ mm}$
$\bar{y} = 62.0 \text{ mm}$

5-29 $\bar{x} = a/4$
$\bar{y} = b/4$
$\bar{z} = c/4$

5-30 $\bar{x} = r/\pi$
$\bar{y} = r/\pi$
$\bar{z} = h/4$

5-33 $\bar{x} = 1.250 \text{ in.}$
$\bar{y} = \bar{z} = 0$

5-34 $\bar{x} = 5b/8$
$\bar{y} = \bar{z} = 0$

5-37 $\bar{x} = 2L/3$
$\bar{y} = \bar{z} = 0$

5-38 $\bar{x} = \bar{y} = 0$
$\bar{z} = 4h/5$

5-41	$\bar{x} = 3.18$ in.
	$\bar{y} = 6.68$ in.
5-42	$\bar{x} = 65.5$ mm
	$\bar{y} = 190.2$ mm
5-45	$\bar{x} = 4.44$ in.
	$\bar{y} = 8.64$ in.
	$\bar{z} = 2.67$ in.
5-47	$\bar{x} = 1.500$ in.
	$\bar{y} = 2.00$ in.
5-48	$\bar{x} = 22.0$ mm
	$\bar{y} = 60.0$ mm
5-51	$\bar{x} = 0$
	$\bar{y} = 4.00$ in.
5-52	$\bar{x} = 75.5$ mm
	$\bar{y} = 87.2$ mm
5-55	$\bar{x} = 3.75$ in.
	$\bar{y} = 7.73$ in.
5-56	$\bar{x} = 38.0$ mm
	$\bar{y} = 156.2$ mm
5-59	$\bar{x} = 6.44$ in.
	$\bar{y} = 4.03$ in.
5-60	$\bar{x} = 61.4$ mm
	$\bar{y} = 35.7$ mm
5-63	$\bar{x} = 0.430$ in.
	$\bar{y} = 0.430$ in.
5-64	$\bar{y} = -64.3$ mm
5-65	**a.** $\bar{x}_C = 2.40$ in.
	$\bar{y}_C = 3.55$ in.
	$\bar{z}_C = 1.494$ in.
	b. $\bar{x}_G = 3.22$ in.
	$\bar{y}_G = 3.76$ in.
	$\bar{z}_G = 0.794$ in.
5-66	**a.** $\bar{x}_C = 0$
	$\bar{y}_C = 0$
	$\bar{z}_C = 229$ mm
	b. $\bar{x}_G = 0$
	$\bar{y}_G = 0$
	$\bar{z}_G = 197.0$ mm
5-69	$\bar{x}_G = 10.86$ in.
	$\bar{y}_G = 9.95$ in.
	$\bar{z}_G = 3.50$ in.
5-70	$\bar{x}_G = 0$
	$\bar{y}_G = 165.0$ mm
	$\bar{z}_G = 0$
5-73	$A = 1276$ in.^{2}
	$V = 1679$ in.^{3}
5-74	$A = 1.274(10^6)$ mm^2
	$V = 56.2(10^6)$ mm^3
5-77	$V = 846$ in.^{3}
5-78	$V = 23.8(10^6)$ mm^3
5-81	$V = 4.87$ in.^{3}
5-82	$V = 393(10^3)$ mm^3

5-85	$R = 1350$ lb
	$d = 3.83$ ft
5-86	$R = 4.35$ kN
	$d = 4.41$ m
5-89	$R = 2375$ lb
	$d = 3.24$ ft
5-90	$R = 2.25$ kN
	$d = 2.67$ m
5-93	$R = 2500$ lb
	$d = 10.50$ ft
5-94	$R = 955$ N
	$d = 1.090$ m
5-97	$R = 2.50(10^6)$ lb
5-98	$R = 63.7$ MN
5-101	$R = 390$ lb
	$d_P = 1.667$ ft
5-102	$R = 3.92$ MN
	$d_P = 4.58$ m
5-105	$R = 12.62$ kip
	$d_P = 3.27$ ft
5-106	$R = 79.8$ kN
	$d_P = 0.496$ m
5-109	$\bar{x} = 0.900$ in.
	$\bar{y} = 0.900$ in.
5-110	$\bar{x} = 30.0$ mm
	$\bar{y} = 16.00$ mm
5-111	$\bar{x} = 0$
	$\bar{y} = 11.62$ in.
5-114	$A = 2.53(10^6)$ mm^2
	$V = 142.9(10^6)$ mm^3
5-115	$V = 196.9$ in.^{3}
5.118	$\mathbf{R} = 3.90$ kN$\downarrow$
	$d = 3.28$ m

Chapter 6

6-27	$\mathbf{A} = 300$ lb$\uparrow$
	$\mathbf{M}_A = 805$ ft $\cdot$ lb $\downarrow$
6-28	$\mathbf{A} = 366$ N $\nearrow$ @ 14.82°
	$\mathbf{B} = 500$ N $\nwarrow$ @ 135.0°
6-31	$\mathbf{A} = 1240$ lb$\uparrow$
	$\mathbf{B} = 1160$ lb$\uparrow$
6-32	$\mathbf{A} = 2.00$ kN$\uparrow$
	$\mathbf{M}_A = 11.00$ kN $\cdot$ m $\downarrow$
6-35	$\mathbf{A} = 2750$ lb$\uparrow$
	$\mathbf{B} = 2740$ lb$\uparrow$
6-36	$\mathbf{A} = 8.26$ kN$\uparrow$
	$\mathbf{M}_A = 37.4$ kN $\cdot$ m $\downarrow$
6-39	$\mathbf{A} = 61.8$ lb $\nearrow$ @ 76.0°
	$\mathbf{B} = 15.00$ lb$\leftarrow$
6-40	$\mathbf{A} = 621$ N$\uparrow$
	$\mathbf{B} = 300$ N $\nwarrow$ @ 143.1°

6-43 $\mathbf{A} = 80.9$ lb $\nearrow$ @ 44.5°
       $\mathbf{B} = 63.3$ lb↑

6-44   a. $\mathbf{A} = 1152$ N $\nearrow$ @ 61.9°
          $\mathbf{C} = 709$ N $\nwarrow$ @ 140.0°
      b. $\mathbf{D} = 1355$ N $\nwarrow$ @ 113.7°
          $\mathbf{M}_D = 815$ N·m $\downarrow$

6-47 $T = 100$ lb

6-48 $T = 429$ N

6-51 $\mathbf{A} = 150.3$ lb $\nwarrow$ @ 146.2°
       $\mathbf{B} = 250$ lb $\nearrow$ @ 60.0°

6-52 $\mathbf{A} = 803$ N $\nwarrow$ @ 118.4°
       $\mathbf{B} = 855$ N $\nearrow$ @ 63.4°

6-55 $\mathbf{R}_A = 50$ **k** lb
       $\mathbf{C}_A = 1150$ **i** + 350 **j** in.·lb

6-56 $\mathbf{R}_A = 2890$ **i** − 705 **j** + 44.7 **k** N
       $T_B = 1754$ N
       $T_C = 2040$ N

6-59 $\mathbf{R}_A = 1036$ **j** + 364 **k** lb
       $T_B = 643$ lb
       $T_C = 573$ lb

6-60 $\mathbf{R}_A = -0.536$ **i** + 3.00 **j** + 1.403 **k** kN
       $\mathbf{B} = 2.00$ **i** kN
       $T_C = 3.51$ kN

6-63 $\mathbf{P} = -150.0$ **k** lb
       $\mathbf{R}_A = 928$ **i** − 116.7 **j** lb
       $\mathbf{R}_B = -278$ **i** − 200 **j** + 267 **k** lb

6-64 $\mathbf{P} = -750$ **k** N
       $\mathbf{R}_A = 250$ **i** + 500 **k** N
       $\mathbf{R}_B = 500$ **i** + 250 **k** N

6-67 $\mathbf{R}_A = 64.6$ **i** + 129.2 **j** + 92.3 **k** lb
       $\mathbf{C}_A = -1185$ **i** + 2649 **k** in.·lb
       $T_C = 180.2$ lb

6-68 $\mathbf{R}_A = 613$ **i** − 1073 **j** + 920 **k** N
       $\mathbf{R}_B = -460$ **j** N
       $T_C = 1761$ N

6-72 $\mathbf{A} = 5.48$ kN $\nwarrow$ @ 132.7°
       $\mathbf{B} = 5.26$ kN↑

6-73 $\mathbf{P} = 168.7$ lb $\nearrow$ @ 30.0°

6-76 $\mathbf{A} = 442$ N↑
       $\mathbf{B} = 308$ N↑

6-77 $\mathbf{T}_A = 195.3$ **k** lb
       $\mathbf{T}_B = 78.1$ **k** lb
       $T_C = 227$ **k** lb

6-80 $\mathbf{R}_C = 6.67$ **j** + 2.50 **k** kN
       $T = 3.66$ kN

Chapter 7

7-1 $T_{AB} = 1000$ lb T
      $T_{AC} = 1500$ lb T
      $T_{BC} = 1732$ lb C

7-2 $T_{AB} = 4.50$ kN T
      $T_{AC} = 6.75$ kN T
      $T_{BC} = 7.80$ kN C

7-5 $T_{AB} = 1128$ lb C
      $T_{AC} = 205$ lb T
      $T_{BC} = 564$ lb T

7-8 $T_{AB} = 2.50$ kN C
      $T_{AD} = 2.17$ kN T
      $T_{BD} = 5.00$ kN T
      $T_{BC} = 4.33$ kN C
      $T_{CD} = 2.17$ kN T

7-10 $T_{AB} = 1.400$ kN C
       $T_{AD} = 3.84$ kN T
       $T_{BD} = 4.00$ kN T
       $T_{BC} = 4.80$ kN C
       $T_{CD} = 3.84$ kN T

7-11 $T_{AB} = 681$ lb C
       $T_{AD} = 161.9$ lb T
       $T_{BD} = 924$ lb C
       $T_{BC} = 1143$ lb C
       $T_{CD} = 1320$ lb T

7-14 $T_{AB} = 3.70$ kN C (9% larger)
       $T_{AD} = 1.394$ kN T (72% larger)
       $T_{BD} = 5.03$ kN C (9% larger)
       $T_{BC} = 6.21$ kN C (9% larger)
       $T_{CD} = 7.17$ kN T (9% larger)

7-15 $T_{AB} = T_{FG} = 2290$ lb C
       $T_{AC} = T_{EG} = 2320$ lb C
       $T_{BC} = T_{EF} = 0$
       $T_{BD} = T_{DF} = 907$ lb C
       $T_{CD} = T_{DE} = 2320$ lb C

7-17 $T_{AB} = 5000$ lb T
       $T_{AE} = 1000$ lb T
       $T_{BC} = 3330$ lb T
       $T_{BD} = 2000$ lb T
       $T_{BE} = 1667$ lb C
       $T_{CD} = 2670$ lb C
       $T_{DE} = 2670$ lb C

7-19 $T_{CG} = 0$
       $T_{FG} = 750$ lb T

7-20 $T_{CG} = 2.08$ kN C
       $T_{FG} = 11.25$ kN T

7-22 $T_{CG} = 2.49$ kN C
       $T_{FG} = 13.30$ kN T

7-23 $T_{BC} = 400$ lb C
       $T_{BG} = 200$ lb C
       $T_{CG} = 178.9$ lb T

7-26 $T_{BC} = 10.61$ kN C
       $T_{CD} = 10.61$ kN C
       $T_{CK} = 15.00$ kN T

7-29 $T_{DE} = 1732$ lb C
       $T_{DF} = 1400$ lb T
       $T_{EF} = 1000$ lb T

7-30 $T_{AB} = 2.36$ kN C
       $T_{AC} = 5.89$ kN C
       $T_{AH} = 3.53$ kN T
       $T_{BC} = 0$
       $T_{CD} = 3.53$ kN C

$T_{CG} = T_{CH} = 0$
$T_{DE} = 0$
$T_{DF} = 5.89$ kN C
$T_{DG} = 0$
$T_{EF} = 2.36$ kN C
$T_{FG} = 3.53$ kN T
$T_{GH} = 3.53$ kN T

7-44 $T_{EJ} = 3.89$ kN C
$T_{HJ} = 9.20$ kN T

7-45 $T_{CD} = 2730$ lb C
$T_{EF} = 2660$ lb T

7-48 $T_{BC} = 7.81$ kN T
$T_{BG} = 16.88$ kN C
$T_{GH} = 12.19$ kN T

7-51 $T_{CD} = 1617$ lb C
$T_{CE} = 400$ lb C
$T_{EF} = 924$ lb T

7-54 $T_{BC} = 25.4$ kN C
$T_{BF} = 5.18$ kN C

7-55 $T_{CD} = 10{,}000$ lb C
$T_{CE} = 0$

7-57 $T_{BC} = 556$ lb C
$T_{EF} = 445$ lb T

7-60 $T_{CD} = 30.0$ kN C
$T_{CE} = 2.89$ kN T
$T_{FG} = 24.5$ kN T

7-62 $T_{CD} = 34.1$ kN C
$T_{DG} = 8.00$ kN C
$T_{EG} = 1.463$ kN T

7-63 $T_{EG} = 1000$ lb T
$T_{FG} = 3610$ lb T
$T_{FH} = 4000$ lb T

7-67 $T_{CD} = 129.0$ lb T
$T_{DG} = 215$ lb T
$T_{EG} = 215$ lb C

7-70 $T_{CH} = 2.07$ kN T
$T_{DF} = 2.21$ kN C
$T_{EF} = 5.65$ kN C

7-71 $P_{\max} = 1000$ lb

7-74 $P_{\max} = 50.0$ kN

7-76 $\mathbf{A} = -750\,\mathbf{i} + 469\,\mathbf{j} + 375\,\mathbf{k}$ N
$\mathbf{B} = -938\,\mathbf{j}$ N
$\mathbf{C} = 750\,\mathbf{i} + 469\,\mathbf{j} + 375\,\mathbf{k}$ N
$T_{AB} = T_{BC} = 530$ N C
$T_{AD} = T_{CD} = 600$ N C
$T_{BD} = 1200$ N T

7-77 $\mathbf{A} = 50\,\mathbf{i} + 191.7\,\mathbf{j} + 93.8\,\mathbf{k}$ lb
$\mathbf{B} = -100.0\,\mathbf{i}$ lb
$\mathbf{C} = 50\,\mathbf{i} - 141.7\,\mathbf{j} + 56.3\,\mathbf{k}$ lb
$T_{AB} = 114.6$ lb C
$T_{AD} = 114.6$ lb C
$T_{BC} = 52.1$ lb C
$T_{BD} = 141.4$ lb T
$T_{CD} = 114.6$ lb C

7-80 $\mathbf{A} = -75.0\,\mathbf{j}$ N
$\mathbf{B} = 200\,\mathbf{i} + 656\,\mathbf{j} + 650\,\mathbf{k}$ N
$\mathbf{C} = 150.0\,\mathbf{j}$ N
$\mathbf{F} = -731\,\mathbf{j}$ N
$T_{AB} = 100.0$ N C
$T_{AD} = 125.0$ N T
$T_{AF} = 0$
$T_{BC} = 100.0$ N T
$T_{BD} = 656$ N C
$T_{BF} = 650$ N C
$T_{CE} = 180.3$ N C
$T_{CF} = 0$
$T_{DE} = 338$ N C
$T_{DF} = 406$ N T
$T_{EF} = 586$ N T

7-83 $\mathbf{A} = 75.0\,\mathbf{k}$ lb
$\mathbf{B} = 112.5\,\mathbf{i} + 390\,\mathbf{j} + 725\,\mathbf{k}$ lb
$\mathbf{C} = -112.5\,\mathbf{i} - 575\,\mathbf{k}$ lb
$T_{BD} = 0$
$T_{EG} = 767$ lb C
$T_{FG} = 608$ lb T

7-84 $\mathbf{B} = 474$ N $\nearrow$ @ $71.6°$
$\mathbf{C} = 636$ N $\swarrow$ @ $135.0°$

7-85 $\mathbf{B} = 48.2$ lb $\nearrow$ @ $57.4°$
$\mathbf{C} = 85.7$ lb $\swarrow$ @ $135.0°$

7-88 $\mathbf{A} = 200$ N $\nwarrow$ @ $179.1°$
$\mathbf{B} = 292$ N $\nearrow$ @ $36.9°$
$\mathbf{C} = 294$ N $\swarrow$ @ $142.6°$

7-91 $\mathbf{A} = 114.6$ lb $\nwarrow$ @ $139.1°$
$\mathbf{D} = 70.3$ lb $\nwarrow$ @ $145.3°$
$\mathbf{E} = 144.0$ lb $\searrow$ @ $53.0°$
$\mathbf{G} = 57.8$ lb $\rightarrow$

7-93 $\mathbf{A} = 166.7$ lb $\searrow$ @ $63.4°$
$\mathbf{B} = 424$ lb $\nwarrow$ @ $135.0°$
$\mathbf{C} = 335$ lb $\searrow$ @ $26.6°$

7-94 $\mathbf{A} = 35.1$ N $\rightarrow$
$\mathbf{B} = 53.5$ N $\nwarrow$ @ $130.9°$
$\mathbf{C} = 40.5$ N $\downarrow$
$\mathbf{E} = 35.1$ N $\rightarrow$

7-96 $w = 327$ N/m

7-99 $\mathbf{A} = 70.5$ lb $\searrow$ @ $7.13°$
$\mathbf{B} = 61.3$ lb $\downarrow$
$\mathbf{C} = 99.0$ lb $\nwarrow$ @ $135.0°$
$\mathbf{M}_C = 245$ ft·lb $\downarrow$

7-101 $\mathbf{A} = 0$
$\mathbf{B} = 145.8$ lb $\uparrow$
$\mathbf{C} = 500$ lb $\downarrow$
$\mathbf{D} = 250$ lb $\uparrow$
$\mathbf{E} = 104.2$ lb $\uparrow$

7-104 $F_1/F_2 = 3.00$
$\mathbf{A} = 0$
$\mathbf{B} = 800$ N $\uparrow$
$\mathbf{C} = 400$ N $\uparrow$
$\mathbf{F}_1 = 1200$ N $\downarrow$

7-107 $\mathbf{B} = 14.14$ lb $\nearrow$ @ 45.0°
      $\mathbf{C} = 32.5$ lb $\leftarrow$
      $\mathbf{D} = 24.6$ lb $\searrow$ @ 24.0°

7-108 $\mathbf{B} = 13.75$ N$\uparrow$
      $\mathbf{D} = 8.75$ N$\downarrow$

7-111 $P = 138.6$ lb
      $\mathbf{A} = 72.6$ lb $\searrow$ @ 77.3°
      $\mathbf{B} = 138.6$ lb $\nwarrow$ @ 120.0°
      $\mathbf{C} = 72.6$ lb $\searrow$ @ 42.7°

7-113 $T_{AB} = 2510$ lb T
      $T_{AG} = 112.2$ lb T
      $T_{BC} = 1740$ lb T
      $T_{BF} = 375$ lb T
      $T_{BG} = 1230$ lb C
      $T_{CD} = 1250$ lb T
      $T_{CE} = 0$
      $T_{CF} = 772$ lb C
      $T_{DE} = 1030$ lb C
      $T_{EF} = 1030$ lb C
      $T_{FG} = 1800$ lb C

7-114 $T_{AB} = 12.02$ kN C
      $T_{AJ} = 7.12$ kN T
      $T_{BC} = 12.02$ kN C
      $T_{BJ} = 0$
      $T_{CD} = 16.00$ kN C
      $T_{CI} = 11.67$ kN T
      $T_{CJ} = 0$
      $T_{DE} = 16.00$ kN C
      $T_{DI} = 8.00$ kN C
      $T_{EF} = 13.62$ kN C
      $T_{EH} = 0$
      $T_{EI} = 10.56$ kN T
      $T_{FG} = 13.62$ kN C
      $T_{FH} = 0$
      $T_{GH} = 8.07$ kN T
      $T_{HI} = 8.07$ kN T
      $T_{IJ} = 7.12$ kN T

7-117 $T_{AB} = 133.4$ lb T
      $T_{AC} = 800$ lb T
      $T_{AD} = 224$ lb C
      $T_{AE} = 211$ lb T
      $T_{BD} = 166.7$ lb T
      $T_{BE} = 250$ lb C
      $T_{BF} = 316$ lb C
      $T_{CD} = 400$ lb T
      $T_{CE} = 300$ lb T
      $T_{DE} = 361$ lb C
      $T_{DF} = 300$ lb T
      $T_{EF} = 0$

7-118   **a.** $\mathbf{D} = 6.91$ kN $\searrow$ @ 49.4°
        b. $\mathbf{A} = 2.65$ kN $\searrow$ @ 8.13°
             $\mathbf{E} = 4.28$ kN $\nwarrow$ @ 127.9°

7-121   **a.** $\mathbf{T} = 43.1$ lb $\nearrow$ @ 64.2°
        b. $\mathbf{A} = 130.8$ lb $\nearrow$ @ 82.8°
        c. $\mathbf{C} = 35.9$ lb $\nwarrow$ @ 169.6°

7-122   **a.** $P = 214$ N
        b. $P = 462$ N
        c. $P = 800$ N

Chapter 8

8-1   **a.** $F_{AB} = 35.0$ kip T
            $F_{BC} = 2.00$ kip C
            $F_{CD} = 18.00$ kip T

8-2   **a.** $F_{AB} = 25.0$ kN C
            $F_{BC} = 50.0$ kN T
            $F_{CD} = 50.0$ kN C

8-5   **a.** $T_{AB} = 80.0$ ft $\cdot$ kip $\lcurvearrowdown$
            $T_{BC} = 20.0$ ft $\cdot$ kip $\rcurvearrowdown$
            $T_{CD} = 20.0$ ft $\cdot$ kip $\lcurvearrowdown$
            $T_{DE} = 45.0$ ft $\cdot$ kip $\lcurvearrowdown$

8-6   **a.** $T_{BC} = 500$ N $\cdot$ m $\lcurvearrowdown$
            $T_{CD} = 400$ N $\cdot$ m $\lcurvearrowdown$
            $T_{DE} = 250$ N $\cdot$ m $\lcurvearrowdown$

8-10   **a.** $T_{\max} = 14.00$ kN $\cdot$ m

8-11 $P = 85.1$ lb
      $V = 206$ lb
      $M = 789$ ft $\cdot$ lb

8-12 $P = 1.543$ kN
      $V = 2.57$ kN
      $M = 0.300$ kN $\cdot$ m

8-15   **a.** $P = 877$ lb
            $V = 280$ lb
            $M = 700$ ft $\cdot$ lb
        b. $P = 750$ lb
            $V = 13.36$ lb
            $M = 1000$ ft $\cdot$ lb

8-16   **a.** $P = 8.66$ kN
            $V = 5.00$ kN
            $M = 0.866$ kN $\cdot$ m
        b. $P = 10.00$ kN
            $V = 0$
            $M = 1.000$ kN $\cdot$ m
        c. $P = 7.07$ kN
            $V = 7.07$ kN
            $M = 0.707$ kN $\cdot$ m

8-19   **a.** $P = 2080$ lb
            $V = 1562$ lb
            $M = 5000$ ft $\cdot$ lb
        b. $P = 520$ lb
            $V = 80$ lb
            $M = 2080$ ft $\cdot$ lb

8-20   **a.** $P = 1248$ N
            $V = 69.3$ N
            $M = 25.0$ N $\cdot$ m
        b. $P = 750$ N
            $V = 500$ N
            $M = 100.0$ N $\cdot$ m

8-23 $V = -300x + 1800$ lb
      $M = -150x^2 + 1800x - 4500$ ft $\cdot$ lb

8-24 $V = -2x + 9$ kN
$M = -x^2 + 9x - 21$ kN $\cdot$ m

8-27 $V = -250x + 2800$ lb
$M = -125x^2 + 2800x - 6000$ ft $\cdot$ lb

8-28 $V = -18x + 76.5$ kN
$M = -9x^2 + 76.5x - 45$ kN $\cdot$ m

8-31 **a.** $V = -1000x + 5500$ lb
$M = -500x^2 + 5500x$ ft $\cdot$ lb
b. $V = -6500$ lb
$M = -6500x - 72{,}000$ ft $\cdot$ lb
c. $V_{max} = -6500$ lb; $12 \le x \le 16$ ft
$M_{max} = -32{,}000$ ft $\cdot$ lb at $x = 16$ ft

8-32 **a.** $V = 67.5$ kN
$M = 67.5x$ kN $\cdot$ m
b. $V = 37.5$ kN
$M = 37.5x + 60$ kN $\cdot$ m
c. $V = -45x + 217.5$ kN
$M = -22.5x^2 + 217.5x - 300$ kN $\cdot$ m
d. $V_{max} = -142.5$ kN at $x = 8$ m
$M_{max} = 226$ kN $\cdot$ m at $x = 4.83$ m

8-35 $V_{4ft} = 6185$ lb and 1185 lb
$V_{9ft} = 1185$ lb and -6815 lb
$M_{4ft} = 24{,}740$ ft $\cdot$ lb
$M_{9ft} = 30{,}665$ ft $\cdot$ lb

8-36 $V_{4m} = 9.09$ kN and -0.91 kN
$V_{8m} = -0.91$ kN and -30.9 kN
$V_{11m} = -30.9$ kN and 20.0 kN
$M_{4m} = 36.4$ kN $\cdot$ m
$M_{8m} = 32.7$ kN $\cdot$ m
$M_{11m} = -60.0$ kN $\cdot$ m

8-39 $V_{4ft} = 2000$ lb
$M_{4ft} = 2000$ ft $\cdot$ lb and -4000 ft $\cdot$ lb

8-40 $V_{2m} = 4$ kN
$M_{2m} = 2$ kN $\cdot$ m and -4 kN $\cdot$ m

8-43 $V_0 = -1000$ lb and 1600 lb
$V_{10ft} = -400$ lb and -1000 lb
$M_0 = -4000$ ft $\cdot$ lb
$M_{10ft} = 2000$ ft $\cdot$ lb and 5000 ft $\cdot$ lb

8-44 $V_0 = 0$ and 21 kN
$V_{6m} = -15$ kN and 3 kN
$M_0 = -24$ kN $\cdot$ m
$M_{6m} = -6$ kN $\cdot$ m

8-47 $V_0 = -1000$ lb and 1750 lb
$V_{6ft} = -250$ lb and -2250 lb
$V_{8ft} = -2250$ lb and 1000 lb
$M_0 = 0$
$M_{6ft} = 6500$ ft $\cdot$ lb and 2500 ft $\cdot$ lb
$M_{8ft} = -2000$ ft $\cdot$ lb

8-48 $V_0 = -6$ kN and 9.33 kN
$V_{2m} = 9.33$ kN
$V_{4m} = -0.67$ kN and -3.67 kN
$M_0 = -4$ kN $\cdot$ m
$M_{2m} = 14.67$ kN $\cdot$ m and 8.67 kN $\cdot$ m
$M_{4m} = 17.33$ kN $\cdot$ m

8-57 **a.** $\mathbf{A}_x = 1810$ lb$\leftarrow$
$\mathbf{A}_y = 350$ lb$\uparrow$
$\mathbf{D}_x = 1810$ lb$\rightarrow$
$\mathbf{D}_y = 850$ lb$\uparrow$
b. $y_B = 1.933$ ft
$y_C = 2.35$ ft
c. $L = 30.7$ ft

8-58 **a.** $\mathbf{A}_x = 22.8$ kN$\leftarrow$
$\mathbf{A}_y = 7.60$ kN$\uparrow$
$\mathbf{D}_x = 22.8$ kN$\rightarrow$
$\mathbf{D}_y = 8.40$ kN$\uparrow$
b. $T_{AB} = 24.0$ kN
$T_{BC} = 22.9$ kN
$T_{CD} = 24.3$ kN
c. $L = 15.62$ m

8-61 **a.** $\mathbf{A}_x = 1236$ lb$\leftarrow$
$\mathbf{A}_y = 350$ lb$\uparrow$
$\mathbf{D}_x = 1236 \rightarrow$
$\mathbf{D}_y = 850$ lb$\uparrow$
b. $y_B = 2.83$ ft
$y_C = 3.44$ ft
c. $L = 31.5$ ft

8-62 **a.** $\mathbf{A}_x = 29.6$ kN$\leftarrow$
$\mathbf{A}_y = 18.60$ kN$\uparrow$
$\mathbf{E}_x = 29.6$ kN$\rightarrow$
$\mathbf{E}_y = 13.40$ kN$\uparrow$
b. $y_B = 1.882$ m
$y_C = 2.77$ m
$y_D = 1.356$ m
c. $L = 16.13$ m

8-65 **a.** $\mathbf{A}_x = 4460$ lb$\leftarrow$
$\mathbf{A}_y = 1070$ lb$\uparrow$
$\mathbf{D}_x = 3860$ lb$\rightarrow$
$\mathbf{D}_y = 1930$ lb$\uparrow$
b. $T_{AB} = 4590$ lb
$T_{BC} = 4560$ lb
$T_{CD} = 4320$ lb
c. $y_B = 6.92$ ft
d. $L = 31.5$ ft

8-66 **a.** $\mathbf{A}_x = 24.2$ kN$\leftarrow$
$\mathbf{A}_y = 22.0$ kN$\uparrow$
$\mathbf{D}_x = 39.2$ kN$\rightarrow$
$\mathbf{D}_y = 7.98$ kN$\uparrow$
b. $y_B = 6.37$ m
$y_C = 7.04$ m
c. $L = 32.7$ m

8-69 **a.** $T_{max} = 606$ kip
b. $L = 616$ ft

8-70 **a.** $a = 12.58$ m
b. $L = 15.42$ m

8-73 **a.** $a = 229$ ft
b. $L = 247$ ft

8-74 **a.** $\theta_x = 13.88°$
b. $h = 24.7$ m
c. $L = 404$ m

9-57 a. $P = 1726$ lb
  b. System is in equilibrium
  c. $P = 469$ lb
  d. $\phi_{max} = 22.6°$

9-58 a. $P = 1360$ N
  b. System is in equilibrium
  c. $P = 420$ N
  d. $\phi_{max} = 38.6°$

9-60 a. $P = 3230$ N
  b. System is in equilibrium
  c. $P = 572$ N
  d. $\phi_{max} = 33.4°$

9-63 a. $P = 1077$ lb
  b. $P = 30.1$ lb
  c. $\phi_{max} = 11.31°$

9-64 a. $P = 673$ N
  b. $P = 0$
  c. $\phi_{max} = 19.29°$

9-66 a. $P = 1789$ N
  b. $P = 138.4$ N
  c. $\phi_{max} = 16.70°$

9-69 a. $P = 1341$ lb
  b. $P = 30.1$ lb
  c. $\phi_{max} = 11.31°$

9-71 a. $P = 2420$ lb
  b. $P = 667$ lb
  c. $\phi_{max} = 8.53°$

9-72 a. $P = 2050$ N
  b. $P = 138.4$ N
  c. $\phi_{max} = 16.70°$

9-74 a. $P = 2080$ lb
  b. $P = 0$
  c. $\phi_{max} = 19.75°$

9-75 a. $P = 1387$ lb
  b. $P = 0$
  c. $\phi_{max} = 17.02°$

9-77 a. $P = 2520$ lb
  b. $P = 263$ lb
  c. $\phi_{max} = 14.24°$

9-80 a. $P = 208$ N
  b. $\phi_{max} = 25.3°$
  c. $\phi_{min} = 70.3°$

9-83 a. $P = 252$ lb
  b. $\phi_{max} = 30.8°$
  c. $\phi_{min} = 75.8°$

9-84 a. $P = 219$ N
  b. $\phi_{max} = 22.6°$
  c. $\phi_{min} = 67.6°$

9-87 a. $P = 134.1$ lb
  b. $\phi_{max} = 30.7°$
  c. $\phi_{min} = 78.7°$

9-88 a. $P = 80.1$ N
  b. $\phi_{max} = 22.7°$
  c. $\phi_{min} = 70.7°$

9-89 a. $P = 242$ lb
  b. $\phi_{max} = 33.5°$
  c. $\phi_{min} = 81.5°$

9-92 a. $P = 353$ N
  b. $\phi_{max} = 11.43°$
  c. $\phi_{min} = 70.3°$

9-93 a. $P = 225$ lb
  b. $\phi_{max} = 14.15°$
  c. $\phi_{min} = 73.0°$

9-96 a. $P = 387$ N
  b. $\phi_{max} = 8.76°$
  c. $\phi_{min} = 67.6°$

9-99 $W_{min} = 43.9$ lb

9-103 $M = 9.52$ in. · lb
Clamp will stay in place

9-104 $M = 8.59$ N · m
No moment needed to release

9-107 a. $L_{max} = 2.20$ in.
  b. $L = 1.368$ in.

9-110 $M_{min} = 0.890$ N · m (tighten)
$M_{min} = 0.095$ N · m (loosen)

9-111 $M = 0.0750$ in. · lb

9-112 $M = 1.516$ N · m

9-113 a. $P = 54.7$ lb
  b. $P = 53.3$ lb
  c. $P = 45.7$ lb
  d. $P = 46.9$ lb

9-118 $77.6 \leq W \leq 82.4$ N

9-120 a. $M = 3.00$ M · m
  b. $M = 2.25$ M · m

9-122 $M_{min} = 2.99$ N · m (raise)
$M_{min} = 2.72$ N · m (lower)

9-123 $P = 2.67$ lb

9-125 $P = 2.59$ lb

9-128 a. $P_{min} = 1347$ N
  b. $P_{min} = 3460$ N

9-129 $P_{min} = 58.6$ lb

9-131 $T_{min} = 10.49$ lb

9-134 $T_{max} = 190.3$ N · m

9-135 a. $\mu = 0.350$
  b. Brake is not self locking

9-138 $P_{min} = 159.0$ N

9-141 $d = 10.95$ in.

9-142 $P_{max} = 392$ N

9-144 $m_{max} = 38.0$ kg

9-145 a. System is in equilibrium
  b. $P_{max} = 1988$ lb

9-148 $m_{min} = 21.5$ kg

9-149 a. $T_{max} = 8340$ in. · lb
  b. $T_{max} = 632$ in. · lb

9-152 $T_{max} = 10.93$ kN · m

9-153 $F = 5.43$ lb

9-154 $a = 4.80$ mm

9-156	$F = 198.8$ N
9-159	$a = 0.25$ in.
9-162	$F = 722$ N
	Ratio = 0.380
9-163	$F = 81.8$ lb
	Ratio = 1.422
9-165	$\mu_{\min} = 1/3$
9-166	**a.** $M = 94.7$ N·m
	b. $M = 123.8$ N·m
9-169	$P = 211$ lb
9-170	$T = 21.1$ kN
	$C_A = 1.460$ kN·m
	$C_B = 0.450$ kN·m

Chapter 10

10-1	**a.** $I_x = bh^3/12$
	b. $I_{xC} = bh^3/36$
10-2	$I_{yC} = hb^3/48$
10-5	**a.** $I_x = 2/9$
	b. $I_y = \pi^2/4 - 2$
10-6	**a.** $I_x = \pi R^4/8$
	b. $I_{xC} = \dfrac{\pi R^4}{8} - \dfrac{8R^4}{9\pi}$
10-9	**a.** $I_x = 5\pi R^4/4$
	b. $J_z = 3\pi R^4/2$
10-10	**a.** $I_x = 8\pi$
	b. $I_y = 32\pi$
10-13	$I_x = 2bh^3/7$
10-14	$I_y = 4a^4/33$
10-17	**a.** $k_x = 3.46$ in.
	$k_y = 1.732$ in.
	b. $k_{xC} = 1.732$ in.
	$k_{yC} = 0.866$ in.
10-18	**a.** $k_x = 51.0$ mm
	$k_y = 81.6$ mm
	b. $k_{xC} = 29.4$ mm
	$k_{yC} = 47.1$ mm
10-21	$k_z = 6.80$ in.
10-22	$k_{zC} = 11.25$ mm
10-25	$k_z = 7.35$ in.
10-27	$I_{xC} = 56.0$ in.^{4}
	$I_{yC} = 11.00$ in.^{4}
10-28	$I_{xC} = 116.5(10^6)$ mm^4
	$I_{yC} = 318(10^6)$ mm^4
10-31	$I_x = 4810$ in.^{4}
	$I_y = 20,200$ in.^{4}
10-32	$I_x = 127.4(10^6)$ mm^4
	$I_y = 343(10^6)$ mm^4
10-35	**a.** $I_x = 5510$ in.^{4}
	$I_y = 20,900$ in.^{4}
	b. $I_{xC} = 1572$ in.^{4}
	$I_{yC} = 4810$ in.^{4}

10-36	**a.** $I_x = 37.7(10^6)$ mm^4
	$I_y = 246(10^6)$ mm^4
	b. $I_{xC} = 13.02(10^6)$ mm^4
	$I_{yC} = 34.5(10^6)$ mm^4
10-39	$I_{xC} = 2730$ in.^{4}
	$I_{yC} = 190.8$ in.^{4}
10-41	$I_{xy} = b^2h^2/4$
10-42	$I_{xyC} = -108.3(10^6)$ mm^4
10-45	**a.** $I_{xy} = b^2h^2/8$
	b. $I_{xyC} = b^2h^2/72$
10-46	**a.** $I_{xy} = 5b^2h^2/24$
	b. $I_{xyC} = -b^2h^2/72$
10-49	$I_{xy} = 7790$ in.^{4}
10-50	$I_{xy} = 48.4(10^6)$ mm^4
10-52	$I_{xy} = 7.58(10^6)$ mm^4
10-53	$I_{\max} = 1070$ in.^{4} @ $-28.5°$
	$I_{\min} = 116.2$ in.^{4} @ $61.5°$
10-54	$I_{\max} = 3.72(10^6)$ mm^4 @ $-16.32°$
	$I_{\min} = 0.0861(10^6)$ mm^4 @ $73.7°$
10-57	$I_{\max} = 55.9$ in.^{4} @ $45.0°$
	$I_{\min} = 15.00$ in.^{4} @ $-45.0°$
10-58	$I_{\max} = 18.41(10^6)$ mm^4 @ $42.9°$
	$I_{\min} = 2.37(10^6)$ mm^4 @ $-47.1°$
10-61	$I_{\max} = 12.67$ in.^{4} @ $26.6°$
	$I_{\min} = 2.67$ in.^{4} @ $-63.4°$
10-62	$I_{\max} = 148.6(10^6)$ mm^4 @ $-28.2°$
	$I_{\min} = 10.13(10^6)$ mm^4 @ $61.8°$
10-65	$I_{\max} = 398$ in.^{4} @ $-38.5°$
	$I_{\min} = 48.1$ in.^{4} @ $51.5°$
10-67	$I_z = 3mR^2/10$
10-68	$I_y = 3m(R^2 + 4h^2)/20$
10-71	$I_y = m(3R^2 + 4L^2)/12$
10-72	$I_x = 2mR^2/5$
10-75	$I_{yG} = m(2h^2 + 3L^2)/36$
10-76	$I_x = m(c^2 + b^2)/10$
10-79	$I_x = mR^2(1 + 3R^2)/6$
10-80	$I_{yG} = 19mR^2/160$
10-83	$I_y = 0.267$ slug·ft^2
10-84	$I_y = 1.004$ kg·m^2
10-87	$I_x = 19.22$ slug·ft^2
10-88	$I_x = 22.9$ kg·m^2
10-93	$I_y = 0.555$ slug·ft^2
10-94	$I_y = 0.464$ kg·m^2
10-97	$I_y = 8.32$ slug·ft^2
10-98	$I_y = 1.616$ kg·m^2
10-101	$I_{xy} = mbL/4$
10-102	$I_{xy} = 2mR^2/5\pi$
10-105	$I_{xy} = -0.493$ slug·ft^2
	$I_{yz} = 0.1174$ slug·ft^2
	$I_{zx} = -0.352$ slug·ft^2

10-106 $I_{xy} = 4mRh/5\pi$
$I_{zx} = 3mR^2/10\pi$

10-109 $I_{max} = 0.695 \text{ slug} \cdot \text{ft}^2$
$\theta_x = 68.6°$
$\theta_y = 81.5°$
$\theta_z = 23.1°$
$I_{int} = 0.665 \text{ slug} \cdot \text{ft}^2$
$\theta_x = 71.0°$
$\theta_y = 25.2°$
$\theta_z = 105.9°$
$I_{min} = 0.1078 \text{ slug} \cdot \text{ft}^2$
$\theta_x = 29.2°$
$\theta_y = 113.5°$
$\theta_z = 106.4°$

10-110 $I_{max} = 0.402 \text{ kg} \cdot \text{m}^2$
$\theta_x = 101.3°$
$\theta_y = 76.5°$
$\theta_z = 17.8°$
$I_{int} = 0.378 \text{ kg} \cdot \text{m}^2$
$\theta_x = 52.9°$
$\theta_y = 37.4°$
$\theta_z = 94.1°$
$I_{min} = 0.0546 \text{ kg} \cdot \text{m}^2$
$\theta_x = 39.3°$
$\theta_y = 124.0°$
$\theta_z = 72.7°$

10-113 $I_{max} = 0.0419 \text{ slug} \cdot \text{ft}^2$
$\theta_x = 60.2°$
$\theta_y = 90.0°$
$\theta_z = 29.8°$
$I_{int} = 0.0346 \text{ slug} \cdot \text{ft}^2$
$\theta_x = 90.0°$
$\theta_y = 0°$
$\theta_z = 90.0°$
$I_{min} = 0.0322 \text{ slug} \cdot \text{ft}^2$
$\theta_x = 29.8°$
$\theta_y = 90.0°$
$\theta_z = 119.8°$

10-114 $I_{max} = 2.42 \text{ kg} \cdot \text{m}^2$
$\theta_x = 130.6°$
$\theta_y = 75.8°$
$\theta_z = 44.1°$
$I_{int} = 1.925 \text{ kg} \cdot \text{m}^2$
$\theta_x = 100.6°$
$\theta_y = 29.9°$
$\theta_z = 117.6°$
$I_{min} = 0.892 \text{ kg} \cdot \text{m}^2$
$\theta_x = 42.6°$
$\theta_y = 64.3°$
$\theta_z = 58.7°$

10-117   **a.** $I_x = 51.2 \text{ in.}^4$
b. $I_y = 39.0 \text{ in.}^4$

10-118   **a.** $k_x = 9.43 \text{ mm}$
$k_y = 25.3 \text{ mm}$
b. $k_{xC} = 4.99 \text{ mm}$
$k_{yC} = 10.84 \text{ mm}$

10-121 $I_y = 372ma^2/1120$

10-122 $I_{xy} = mab/20$

Chapter 11

11-1 $U = 31.7(10^6) \text{ ft} \cdot \text{lb}$

11-2 $U = 236 \text{ kJ}$

11-3 $U = 520 \text{ ft} \cdot \text{lb}$

11-4 $U = 1724 \text{ J}$

11-7   **a.** $U = 927 \text{ ft} \cdot \text{lb}$
b. $U = 0$
c. $U = -927 \text{ ft} \cdot \text{lb}$

11-8   **a.** $U = -1314 \text{ J}$
b. $U = 2070 \text{ J}$
c. $U = -759 \text{ J}$

11-11 $U = 28.0 \text{ ft} \cdot \text{lb}$

11-12 $U = -10.00 \text{ J}$

11-15 $\theta = 31.2°$

11-16 $m_A = 100.0 \text{ kg}$

11-17 $\theta = 8.13°$

11-18 $\theta = 27.8°$

11-21 $F = 750 \text{ lb}$

11-22 $F = 375 \text{ N}$

11-25 $\mathbf{B} = 1500 \text{ lb}\uparrow$

11-26 $\mathbf{A} = 6.61 \text{ kN}\uparrow$

11-27 $F = 75.0 \text{ lb}$

11-28 $\mathbf{C}_x = 567 \text{ N}\rightarrow$
$\mathbf{C}_y = 675 \text{ N}\uparrow$

11-31 $\mathbf{C} = 550 \text{ lb}\uparrow$

11-32 $\mathbf{A} = 500 \text{ N}\uparrow$
$\mathbf{M}_A = 750 \text{ N} \cdot \text{m} \downarrow$

11-35 $\mathbf{B}_x = 275 \text{ lb}\leftarrow$
$\mathbf{B}_y = 417 \text{ lb}\uparrow$

11-36 $T_{CD} = 400 \text{ N } T$

11-39 $\mathbf{A}_x = 170 \text{ lb} \leftarrow$
$\mathbf{A}_y = 8.33 \text{ lb}\uparrow$
$\mathbf{M}_A = 5000 \text{ in.} \cdot \text{lb} \downarrow$

11-40 $\mathbf{A}_x = 953 \text{ N}\leftarrow$
$\mathbf{A}_y = 327 \text{ N}\downarrow$

11-43   **a.** $V_s = 213 \text{ in.} \cdot \text{lb}$
b. $V_T = 62.5 \text{ in.} \cdot \text{lb}$

11-44 $V_T = 45.1 \text{ J (at } 30°)$
$V_T = 179.2 \text{ J (at } 60°)$
$V_T = 400 \text{ J (at } 90°)$

11-47 $\theta = 0°$ and $48.2°$

11-48 $\theta = 0°$ and $60.6°$

11-51 $\theta = 31.9°$

11-52 $k = 794 \text{ N/m}$

11-53 $\theta = 44.4°$

11-54 $k = 2550 \text{ N/m}$

11-57 $\theta = 56.7°$

11-58 $\theta = 44.4°$

11-61 $\theta = 32.1°$

11-62 $\theta = 20.6°$

11-65   **a.** $k_{min} = WL/2$
        b. $\theta = 0°$ (unstable)
            $\theta = 36.6°$ (stable)

11-66 $\theta = 0°$ (stable)
        $\theta = 11.19°$ (unstable)

11-69 $\theta = 0°$ (unstable)
        $\theta = 18.45°$ (stable)

11-71 $\mathbf{B} = 1038$ lb↑

11-72   **a.** $\mathbf{E} = 6.50$ kN↑
        b. $T_{CD} = 8.00$ kN C

11-75 $M = 3080$ in. · lb

11-76   **a.** $k = 4370$ N/m
        b. $\theta = 30°$ (stable)

INDEX

PHOTO CREDITS

Chapter 1 Comstock

Chapter 2 Courtesy Columbia Helicopters, Inc.

Chapter 3 Brett Froomer/Image Bank

Chapter 4 Harold Sund/Image Bank

Chapter 5 Richard Negri/Gamma-Liaison

Chapter 6 Richard Pasley/Stock, Boston

Chapter 7 Jim Rudnick/Stock Market

Chapter 8 Steve Solum/Bruce Coleman

Chapter 9 Comstock

Chapter 10 Courtesy NASA

Chapter 11 Kaku Kurita/Gamma-Liaison